Antioxidant and Antimicrobial Activity of Honey Bee Products—2nd Edition

Antioxidant and Antimicrobial Activity of Honey Bee Products—2nd Edition

Guest Editors

Josipa Vlainić

Ivana Tlak Gajger

MDPI

Basel • Beijing • Wuhan • Barcelona • Belgrade • Novi Sad • Cluj • Manchester

Guest Editors
Josipa Vlainić
Laboratory for Oxidative Stress
Institute Ruđer Bošković
Zagreb
Croatia

Ivana Tlak Gajger
Department for Biology and
Pathology of Fish and Bees
University of Zagreb
Zagreb
Croatia

Editorial Office
MDPI AG
Grosspeteranlage 5
4052 Basel, Switzerland

This is a reprint of the Special Issue, published open access by the journal *Antioxidants* (ISSN 2076-3921), freely accessible at: https://www.mdpi.com/journal/antioxidants/special_issues/SP679NDK5I.

For citation purposes, cite each article independently as indicated on the article page online and as indicated below:

Lastname, A.A.; Lastname, B.B. Article Title. *Journal Name* **Year**, *Volume Number*, Page Range.

ISBN 978-3-7258-7148-3 (Hbk)
ISBN 978-3-7258-7149-0 (PDF)
https://doi.org/10.3390/books978-3-7258-7149-0

Contents

About the Editors

Josipa Vlainić

Josipa Vlainić graduated from the Faculty of Veterinary Medicine, University of Zagreb, in 2002. That same year, she was employed at the Laboratory for Neuropharmacology, Rudjer Boskovic Institute, and obtained her PhD degree in 2009 from the Faculty of Veterinary Medicine University of Zagreb. Within RBI, she moved to the Laboratory of Advanced Genomics in 2015 and, recently, to the Laboratory of Oxidative Stress. She gives lectures to students regarding antimicrobial resistance and the effectiveness of antimicrobial substances. Her research includes antimicrobial effectiveness-testing of naturally occurring products as well as their antioxidant activity.

Ivana Tlak Gajger

Ivana Tlak Gajger is a professor at the Faculty of Veterinary Medicine University of Zagreb, and an award-winning scientist and advocate for the health of bees. She graduated in 2005 and received her doctorate in 2010. She is the head of postgraduate specialist studies in Honeybee Health Protection and leads an officially accredited National Reference Laboratory for Bee Diseases, known as APISlab, which is certified under HRN EN ISO/IEC 17025. Additionally, she founded the Educational and Archive Station for Beekeeping, the first skills laboratory of its kind. She works closely with beekeepers and other stakeholders in the beekeeping sector through various science-based activities to identify opportunities that can be used to simultaneously promote beekeeping, healthy food production, and natural biodiversity, as well as improve the implementation of bee health protection measures. In 2022, she received the Croatian Academy of Sciences and Arts Award for her outstanding achievements in medical sciences. In 2023, she became a full member of the Collegium of Veterinary Sciences within the Croatian Academy of Medical Sciences. A Stanford University study lists her among the top 2% of the most influential scientists in the world for 2022, 2023, and 2024.

antioxidants

MDPI

Editorial

Antioxidant and Antimicrobial Activity of Honey Bee Products—2nd Edition

Ivana Tlak Gajger [1,*] and Josipa Vlainić [2]

1 Department for Biology and Pathology of Fish and Bees, NRL for Honeybee Diseases, Faculty of Veterinary Medicine, University of Zagreb, Heinzelova 55, 10000 Zagreb, Croatia
2 Institute Ruđer Bošković, Bijenička Cesta 54, 10000 Zagreb, Croatia; josipa.vlainic@irb.hr
* Correspondence: itlak@vef.unizg.hr; Tel.: +385-2390041

Antioxidants have acquired a central position in modern nutrition and preventive medicine, and honey bee products are increasingly recognized as complex natural matrices that provide a rich and diverse source of these compounds [1]. Building on the success of the previous Special Issue "Antioxidant Activity of Honey Bee Products" (https://www.mdpi.com/journal/antioxidants/special_issues/Honey_Bee_Products, accessed on 31 October 2025), this new edition, "Antioxidant and Antimicrobial Activity of Honey Bee Products—2nd Edition" (https://www.mdpi.com/journal/antioxidants/special_issues/SP679NDK5I, accessed on 1 January 2026), broadens the focus by integrating antioxidant and antimicrobial properties and emphasizing their joint relevance for health promotion and disease prevention in an era dominated by chronic, inflammation- and oxidative stress-driven disorders [2].

Honey and its products, including propolis, honey bee pollen, honey bee bread, royal jelly, beeswax, and honey bee venom, are increasingly regarded not only as traditional remedies, but also as complex functional ingredients capable of modulating oxidative stress, inflammation and immune responses, primarily through a concerted action of their polyphenols, flavonoids, enzymes and bee-derived peptides such as defensin-1 [3–5]. Also, comparative studies of monofloral, multifloral, and honeydew honeys have shown that samples with higher total phenolic content exhibit stronger radical-scavenging capacity together with more pronounced antibacterial effects, in some cases approaching or rivaling the activity of benchmark honeys such as manuka and thus emerge as promising candidates for targeted nutritional and therapeutic applications [6,7].

An attractive feature of this Special Issue was the systematic comparison of antioxidant and antimicrobial properties across multiple honey bee products and their combinations, highlighting that pollen and honey bee bread, rich in phenolic acids such as p-coumaric and rosmarinic acid, can display notable antioxidant potential coupled with inhibition of clinically relevant bacterial strains and opportunistic fungi, supporting their positioning as "superfoods" with potential impact on chronic disease prevention. Equally compelling were the contributions that demonstrated synergistic effects in multi-component formulations. Mixtures of honey with polyphenol-rich plant extracts, as well as combinations of honey, royal jelly, and propolis, exhibited enhanced biochemical, histological, and immunohistochemical markers, protecting against oxidative and inflammatory damage. This illustrates the potential for bee-based interventions in complex pathophysiological conditions. A recurrent and conceptually important theme was the integration of detailed chemical characterization with functional outcomes, enabling the Special Issue content to move

 https://doi.org/10.3390/antiox15010099

beyond simply cataloging antioxidant capacity and minimum inhibitory concentrations. By coupling advanced analytical techniques with a battery of in vitro antioxidant assays and antimicrobial tests against Gram-positive and Gram-negative bacteria, yeasts, and filamentous fungi, it was demonstrated that compositional patterns underlie both enhanced redox activity and antimicrobial performance. This multidimensional, mechanism-oriented perspective, spanning composition, processing effects, bioaccessibility, bioavailability, and biological efficacy across diverse honey bee products, provides this Special Issue a distinctive application-oriented profile as well as a strong scientific foundation for the development of standardized, evidence-based functional products from the hive [see the Contributions list].

This volume is organized into several interconnected sections that guide the reader from fundamental concepts to applied aspects. The first part focuses on the chemical composition, analytical characterization, and mechanistic understanding of antioxidant and antimicrobial activities of individual honey bee products. The second part brings together experimental and clinical studies that explore the impact of bee-derived antioxidants on biomarkers of oxidative stress, inflammation, and selected chronic diseases. The final chapters address technological, processing-related, and regulatory perspectives relevant to the development and standardization of functional products from the hive. This book is intended for a broad scientific and professional audience, including researchers in food and nutritional science, veterinary and human medicine, pharmacy, toxicology, and microbiology, as well as stakeholders from the food, nutraceutical, and pharmaceutical industries and regulatory bodies interested in evidence-based valorization of honey bee products.

This Special Issue features nine original articles and three reviews. The first notable contribution discusses a standardized bee-derived propolis extract, which represents a promising next-generation strategy for treating acne. This extract effectively targets pathogenic biofilms of *Cutibacterium acnes*, reducing its virulence while preserving the diversity of the skin microbiome and supporting antioxidant and anti-inflammatory responses in the epidermis (Contribution 1). In addition, process innovations such as ultrasound-assisted extraction are redefining how propolis bioactive compounds are obtained, with optimized ethanolic conditions yielding higher polyphenol and flavonoid levels, stronger antioxidant performance across multiple assays, and superior antimicrobial effects against foodborne Gram-positive pathogens, while preserving notable bioaccessibility after simulated digestion and thereby expanding the technological space for propolis-enriched health products (Contribution 2). According to an article about an integrated workflow that combines palynological examination with comprehensive physicochemical and phenolic–flavonoid profiling, along with extensive antibacterial and antifungal testing from a single propolis sample, now provides a resource-efficient method to connect the botanical origins of propolis with its bioactive potential. This approach aims to improve the development of standardized, evidence-based propolis preparations (Contribution 3). Advancements in formulation science now allow honey bee propolis to be transformed into stable, consumer-friendly delivery systems, expanding its health applications beyond just topical acne care. By microencapsulating ethanolic propolis extracts with plant and protein carriers and subsequent spray-drying, researchers have achieved improved solubility, thermal stability, and controlled release of phenolic bioactives, while successfully incorporating these microcapsules into functional gummies without loss of sensory quality, thereby opening avenues for complementary, ingestible formats of propolis-based interventions (Contribution 4).

In parallel, emerging work on kombucha beverages fortified with fermented bee-collected pollen highlights a complementary, ingestible route to harness bee-derived phenolics and phenylamides, enhancing antimicrobial functionality and reshaping the underlying

https://doi.org/10.3390/antiox15010099

polyphenol profile towards flavanol- and phenolic-acid-rich matrices that may synergize with propolis-based dermal care (Contribution 5). A long-term study on the storage of bee pollen highlights the importance of carefully controlling drying and freezing processes. This is crucial for preserving phenolic compounds, ascorbic acid, and in vitro antioxidant capacity, which help maintain its value as a nutritionally rich, polyfloral ingredient primarily composed of *Salix* and *Brassica napus* pollens (Contribution 6). Honey bee pollen's remarkable nutritional and antioxidant potential is further underscored by work linking its palynological profile to phenolic richness and radical-scavenging capacity, with highly active monofloral Cistus- and Brassica-type pollens emerging as especially valuable polyphenol sources (Contribution 7). Most recently, cutting-edge chromatographic and effect-directed analyses have revealed N1,N5N10-tricaffeoylspermidine as a shared, highly bioactive marker in bee pollen and tree androecia from *Castanea*, *Salix*, and *Quercus* species, providing a new chemotaxonomic handle and highlighting xanthine oxidase inhibition as an additional mechanistic pathway through which well-characterized bee pollen preparations may contribute to disease-preventive nutraceutical concepts (Contribution 8).

Emerging in vivo data now extend the bee-product paradigm beyond propolis and pollen, showing that beeswax alcohol can outperform coenzyme Q10 in a zebrafish model of high-cholesterol and D-galactose burden by improving survival, dyslipidemia, redox balance, liver integrity, and cognition, thereby positioning structured beeswax lipids as promising candidates for future brain–liver axis-targeted nutraceuticals (Contribution 9).

At an ecological level, these insights highlight the role of pollen-derived polyphenols as crucial nutritional modulators within the honey bee stress response network. They connect diet quality to redox homeostasis, immunity, detoxification capacity, and ultimately, the resilience of honey bee colonies under pressure from pesticides, pathogens, and a changing climate (Contribution 10).

Within the broader apitherapy framework, honey emerges as a complementary antioxidant matrix in which phenolic compounds, enzymes, vitamins, and amino acids cooperate to buffer oxidative stress and inflammation, offering cardiometabolic, neuroprotective, immunomodulatory, and wound-healing benefits that align naturally with the health-promoting potential of propolis, pollen, and beeswax-derived preparations (Contribution 11).

In this context, comparative meta-analytic evidence indicates that the stingless bee honey tends to exhibit higher phenolic loads, stronger ferric-reducing antioxidant power, and distinctively higher moisture and free acidity than *Apis mellifera* honey, reinforcing its position as a particularly potent but still underexplored functional honey type that warrants more systematic characterization and clinical investigation (Contribution 12).

Author Contributions: Writing—original draft preparation, I.T.G.; writing—review and editing, I.T.G. and J.V. All authors have read and agreed to the published version of the manuscript.

Funding: This research received no external funding.

Acknowledgments: Our sincere appreciation goes to all authors whose submissions have enriched this Special Issue dedicated to the antioxidant and antimicrobial activity of honey bee products. The collective effort, diversity of approaches, and high scientific standards of their work have substantially advanced knowledge in this field. We also extend warm thanks to the reviewers for their careful evaluation, insightful comments, and constructive suggestions, which greatly contributed to refining and strengthening the published articles. Finally, special acknowledgment is given to COST Action CA22105—Beekeeping products valorization and biomonitoring for the SAFEty of BEEs and HONEY (BeSafeBeeHoney), whose support in the networking framework has provided an important stimulus and context for research presented in this volume.

Conflicts of Interest: The authors declare no conflicts of interest.

https://doi.org/10.3390/antiox15010099

List of Contributions:

1. Athanasopoulou, S.; Panagiotidou, E.; Spanidi, E.; Gkika, M.; Georgiou, D.; Anagnostopoulos, A.K.; Ganos, C.; Chinou, I.; Beletsiotis, E.; Gardikis, K. Propolis Extract with Activity Against *Cutibacterium acnes* Biofilm Targeting the Expression of Virulence Genes. *Antioxidants* **2025**, *14*, 849. https://doi.org/10.3390/antiox14070849.
2. Mejía, J.; Giovagnoli-Vicuña, C.; Jacob, C.; Montenegro, G.; Moreno-Switt, A.I.; Giordano, A. Antioxidant, Antibacterial, and Bioaccessibility Properties of Ultrasound-Extracted Chilean Propolis. *Antioxidants* **2025**, *14*, 651. https://doi.org/10.3390/antiox14060651.
3. Alcivar-Saldaña, J.J.; Rodriguez-Monroy, M.A.; Aguirre-Gómez, A.; Carrillo-Miranda, L.; Velasco-Bejarano, B.; Canales-Martinez, M.M. Determination of the Botanical Origin and Physicochemical Properties of a Propolis Sample Through an Integrated Methodology. *Antioxidants* **2024**, *13*, 1412. https://doi.org/10.3390/antiox13111412.
4. Ligarda-Samanez, C.A.; Choque-Quispe, D.; Palomino-Rincón, H.; Moscoso-Moscoso, E.; Guzmán Gutiérrez, R.J.; Banda Mozo, I. Microencapsulation of Propolis by Complex Coacervation with Chia Mucilage and Gelatin: Antioxidant Stability and Functional Potential. *Antioxidants* **2025**, *14*, 845. https://doi.org/10.3390/antiox14070845.
5. Kostić, A.Ž.; Sknepnek, A.; Milinčić, D.D.; Gašić, U.; Kilibarda, S.; Pešić, M.B. Influence of Kombucha Fermentation on Antioxidant and Antimicrobial Activity of Monofloral Rapeseed Bee-Collected Pollen. *Antioxidants* **2025**, *14*, 752. https://doi.org/10.3390/antiox14060752.
6. Stebuliauskaitė, R.; Liaudanskas, M.; Žvikas, V.; Čeksterytė, V.; Sutkevičienė, N.; Sorkytė, Š.; Bračiulienė, A.; Trumbeckaitė, S. Changes in Ascorbic Acid, Phenolic Compound Content, and Antioxidant Activity In Vitro in Bee Pollen Depending on Storage Conditions: Impact of Drying and Freezing. *Antioxidants* **2025**, *14*, 462. https://doi.org/10.3390/antiox14040462.
7. Boulfous, N.; Belattar, H.; Ambra, R.; Pastore, G.; Ghorab, A. Botanical Origin, Phytochemical Profile, and Antioxidant Activity of Bee Pollen from the Mila Region, Algeria. *Antioxidants* **2025**, *14*, 291. https://doi.org/10.3390/antiox14030291.
8. Sen, N.B.; Vovk, I.; Kırmızıbekmez, H.; Guzelmeric, E. Phytochemical and Bioactivity Evaluation of Bee Pollen and Androecia of *Castanea*, *Salix*, and *Quercus* Species. *Antioxidants* **2025**, *14*, 40. https://doi.org/10.3390/antiox14010040.
9. Cho, K.-H.; Lee, Y.; Bahuguna, A.; Lee, S.H.; Yang, C.-E.; Kim, J.-E.; Kwon, H.-S. The Consumption of Beeswax Alcohol (BWA, Raydel®) Improved Zebrafish Motion and Swimming Endurance by Protecting the Brain and Liver from Oxidative Stress Induced by 24 Weeks of Supplementation with High-Cholesterol and D-Galactose Diets: A Comparative Analysis Between BWA and Coenzyme Q_{10}. *Antioxidants* **2024**, *13*, 1488. https://doi.org/10.3390/antiox13121488.
10. Tlak Gajger, I.; Cvetkovikj, A. Antioxidant Potential of Pollen Polyphenols in Mitigating Environmental Stress in Honeybees (*Apis mellifera*). *Antioxidants* **2025**, *14*, 1086. https://doi.org/10.3390/antiox14091086.
11. Tlak Gajger, I.; Dar, S.A.; Ahmed, M.M.M.; Aly, M.M.; Vlainić, J. Antioxidant Capacity and Therapeutic Applications of Honey: Health Benefits, Antimicrobial Activity and Food Processing Roles. *Antioxidants* **2025**, *14*, 959. https://doi.org/10.3390/antiox14080959.
12. Zaldivar-Ortega, A.K.; Cenobio-Galindo, A.d.J.; Morfin, N.; Aguirre-Álvarez, G.; Campos-Montiel, R.G.; Esturau-Escofet, N.; Garduño-García, A.; Angeles-Hernandez, J.C. The Physicochemical Parameters, Phenolic Content, and Antioxidant Activity of Honey from Stingless Bees and *Apis mellifera*: A Systematic Review and Meta-Analysis. *Antioxidants* **2024**, *13*, 1539. https://doi.org/10.3390/antiox13121539.

References

1. Martinello, M.; Mutinelli, F. Antioxidant Activity in Bee Products: A Review. *Antioxidants* **2021**, *10*, 71. [CrossRef] [PubMed]
2. Tlak Gajger, I.; Vlainić, J. Antioxidant Activity of Honey Bee Products. *Antioxidants* **2025**, *14*, 64. [CrossRef] [PubMed]
3. El-Seedi, H.R.; Eid, N.; Abd El-Wahed, A.A.; Rateb, M.E.; Afifi, H.S.; Algethami, A.F.; Zhao, C.; Al Naggar, Y.; Alsharif, S.M.; Tahir, H.E.; et al. Honey Bee Products: Preclinical and Clinical Studies of Their Anti-Inflammatory and Immunomodulatory Properties. *Front. Nutr.* **2022**, *8*, 761267. [CrossRef] [PubMed]

 https://doi.org/10.3390/antiox15010099

4. Bucekovà, M.; Sojka, M.; Valachovà, I.; Martinotti, S.; Ranzato, E.; Szep, Z.; Majtan, V.; Klaudiny, J.; Majtan, J. Bee-Derived Antibacterial Peptide, Defensin-1, Promotes Wound Re-Epithelialisation In Vitro and In Vivo. *Sci. Rep.* **2017**, *7*, 7340. [CrossRef] [PubMed]
5. Tlak Gajger, I.; Cvetkovikj, A. Antioxidant Potential of Pollen Polyphenols in Mitigating Environmental Stress in Honeybees (*Apis mellifera*). *Antioxidants* **2025**, *14*, 1086. [CrossRef] [PubMed]
6. Broznić, D.; Ratkaj, I.; Malenica Staver, M.; Kraljević Pavelić, S.; Žurga, P.; Bubalo, D.; Gobin, I. Evaluation of the Antioxidant Capacity, Antimicrobial and Antiproliferative Potential of Fir (*Abies alba* Mill.) Honeydew Honey Collected from Gorski Kotar (Croatia). *Food Technol. Biotechnol.* **2018**, *56*, 533–545. [CrossRef] [PubMed]
7. Attanzio, A.; Tesoriere, L.; Allegra, M.; Livrea, M.A. Monofloral Honeys by Sicilian Black Honeybee (*Apis mellifera* ssp. *sicula*) Have High Reducing Power and Antioxidant Capacity. *Heliyon* **2016**, *2*, e00193. [CrossRef] [PubMed]

https://doi.org/10.3390/antiox15010099

antioxidants

MDPI

Article

Propolis Extract with Activity Against *Cutibacterium acnes* Biofilm Targeting the Expression of Virulence Genes

Sophia Athanasopoulou [1], Eleni Panagiotidou [1], Eleni Spanidi [1], Maria Gkika [1], Danai Georgiou [1], Athanasios K. Anagnostopoulos [2], Christos Ganos [3], Ioanna Chinou [3], Evangelos Beletsiotis [4] and Konstantinos Gardikis [1,5,*]

1 Research and Development Department, APIVITA SA, Industrial Park Markopoulo Mesogaias, 19003 Athens, Greece; sofia.athanasopoulou@puig.com (S.A.); eleni.panagiotidou@puig.com (E.P.); eleni.spanidi@puig.com (E.S.)

2 Biomedical Research Foundation, Academy of Athens, 11527 Athens, Greece

3 Laboratory of Pharmacognosy and Chemistry of Natural Products, Department of Pharmacy, School of Health Sciences, National and Kapodistrian University of Athens, Zografou, 15771 Athens, Greece

4 Molecular Microbiology Department, QACS The Challenge Test Laboratory, Antigonis 1, 14451 Metamorfosi, Greece

5 Department of Pharmacy, Frederick University, Nicosia 1036, Cyprus

* Correspondence: konstantinos.gardikis@puig.com

Abstract

Acne is a highly prevalent skin condition with multifactorial pathophysiology, where *Cutibacterium acnes* (*C. acnes*) overgrowths generate inflammation. *C. acnes* can grow and adhere, through the formation of biofilms, to almost any surface, which enables chronic infections. Acne treatment with antibiotics can induce topical antimicrobial resistance, impair microbiome biodiversity and cause cutaneous dysbiosis. In this study, we assess the effect of a standardized propolis extract (PE) from Greece against *C. acnes*, whilst maintaining skin's microbiome biodiversity, and we investigate its effect against genes related to the attachment and colonization of *C. acnes*, as well as against biofilm formation. The extract has been chemically characterized by GC-MS and was additionally tested for its antioxidant properties by the Folin–Ciocalteu method and the 2,2-Diphenyl-1-Picrylhydrazyl (DPPH) assay and its regulatory activity on the expression of antimicrobial and anti-inflammatory genes in normal human epidermal keratinocytes (NHEKs). The suggested efficacy of PE in targeting pathogenic *C. acnes* biofilm, via downregulation of virulence genes, represents an alternative strategy to modulate the behavior of skin microbiota in acne, paving the way for next-generation acne-targeting products.

Keywords: propolis extract; *Cutibacterium acnes*; biofilm

1. Introduction

Acne vulgaris is a highly prevalent skin condition with a multifactorial pathophysiology. The standard model suggests that follicular plugging occurs due to a sticky epithelium associated with increased sebum, resulting in anaerobic follicular conditions that favor *C. acnes* (formerly known as *Propionibacterium acnes*) bacterial overgrowth, generating inflammation [1]. Moreover, in the last decade many studies have consistently demonstrated that patients with acne experience increased oxidative stress [2].

The acne-causing strains of *C. acnes* inhabit skin-gland cavities and form sessile aggregates known as biofilms, causing inflammation and skin disorders [3]. Biofilms are structured consortia of microorganisms that attach to biotic or abiotic surfaces and are embedded within a self-produced extracellular matrix [4]. Biofilm formation is a multistep process in which microbial cells adhere to the surface (initial reversible adhesion), while the subsequent production of an extracellular matrix results in a more stable attachment. The biofilm of acne-associated *C. acnes* strains consists mainly of 62.6% polysaccharides, 9.6% proteins, 4.0% DNA and 23.8% other compounds (porphyrin precursors and others) [5]. The essential exogenous products secreted by *C. acnes* accumulate within the biofilm's extracellular matrix, where they play a pivotal role in maintaining the structural and functional integrity of the biofilm. Immobilized (biofilm-associated) cells are phenotypically and physiologically different from non-adherent (planktonic) cells, and one of the typical properties of immobilized cells is their increased resistance to antimicrobial agents. It has been verified that *C. acnes* in biofilm is more resistant to antibiotics than bacteria in the planktonic growth phase [4]. *C. acnes* can form a biofilm both in vitro and in vivo. Biofilm formation is considered a primary factor contributing to the failure of antimicrobial therapies, with an estimated 65–80% of all infections involving biofilm-associated microbial communities [6,7].

The prevention of the formation of biofilms and action against the established biofilm is a strategy for the development of new anti-pathogenic treatments. An important role in the above actions is the obstruction of the ability of bacteria to adhere to the host [8].

The initial step in bacterial pathogenesis and biofilm formation involves the colonization of and attachment to host surfaces. Important factors in the fundamental stage of adhesion are bacterial virulence factors (secreted, cytosolic, or membrane-associated). These are bacterial molecules used by pathogens to colonize the host at the cellular level. Secreted bacterial factors play a critical role in evading both the innate and adaptive immune responses of the host. Cytosolic components contribute to metabolic, physiological, and morphological adaptations, whereas membrane-associated virulence factors facilitate bacterial adhesion to host tissues and biomaterial surfaces [9,10].

A number of natural compounds target bacterial virulence factors to disrupt pathogenic potential without affecting bacterial viability. An increasing line of evidence reveals propolis' ability to interfere with virulence factor production and biofilm formation in multiple pathogenic microorganisms [11].

Propolis is a resinous substance produced by honeybees (*Apis melifera*). It exhibits multiple bioactive properties, including antimicrobial, immunostimulating, and antioxidant activities, and is utilized in the formulation of functional foods, cosmetics, and traditional medicinal applications [12–16]. Its composition varies depending on the geographic origin and flora thereof, and on the foraging area of the honeybees and the collection season. Propolis consists of about 50% resins, 30% waxes, 10% aromatic compounds, 5% pollen and 5% other ingredients [14,17,18]. The bioactive components of propolis are mostly phenolic compounds, terpenes, as well as sugars and amino acids [13,19–22]. Among the most abundant polyphenols are flavonoids, phenolic acids, as well as their esters. The structural complexity of propolis, coupled with its compositional variability, influenced by geographic origin and seasonal factors, results in different types of propolis, making it difficult or even impossible to determine the exact quality and formulate it into products for per os or cutaneous use intake (through the mouth) or products applied to the skin [23,24].

The antibacterial properties of propolis have been examined in a great number of Gram-positive and Gram-negative bacteria (>600 bacterial strains—reviewed in [15]) with variations in efficacy, depending on its composition and geographical origin. Propolis extracts either directly affect bacteria or stimulate the immune system of the host [25]. Gram-positive bacteria are more susceptible to propolis antibacterial activity, which affects bacterial mobility, disrupts their membrane potential, increases membrane permeability and decreases adenosine triphosphate (ATP) production [15].

Until today, very few studies have described the activity of propolis against *C. acnes*, while there are no reports of its activity against the biofilm of this bacterium. Recently the antioxidant, anti-inflammatory and anti-acne activities of stingless bee (*Tetragonula biroi*) propolis have been studied thoroughly, and have been reported to stimulate the growth of skin tissue, protect skin from photodamage and free radicals, prevent premature skin ageing, and help extinguish acne, especially for acne-prone skin [26].

The present study assesses the in vitro and ex vivo action of a specific Greek propolis extract on the microflora commonly found on human forehead skin, displays its effect on the formation and growth of biofilms and on genes related to the colonization ability of *C. acnes*, and demonstrates, at the cellular level, its ability to regulate the expression of antimicrobe- and anti-inflammation-associated genes in human keratinocytes.

2. Materials and Methods

2.1. Propolis Extraction

The 1.3 propanediol was supplied by Connect Chemicals (Vimercate (MB), Italy) while 3-[(2-ethylhexyl)oxy]propan-1,2-diol was provided by Ashland (Rotterdam, The Netherlands). Methanol was acquired by PanReac AppliChem (Darmstadt, Germany).

Raw propolis from Greek beekeeping activity from the wider area of the Olympus Mountain (Central Greece) was used. Typical plant families that bees use for propolis collection are *Asteraceae*, *Rosaceae*, *Liliaceae* and *Boraginaceae*, while they collect propolis from nectarless plants as well, such as *Quercus* sp., *Chenopodium* sp., *Scabiosa* sp., etc. [16]. The criterion was a concentration of total polyphenols >900 mg/L gallic acid equivalents (GAE) after dissolving 10% by weight of propolis in ethanol and subsequent measurement in a spectrophotometer using the Folin–Ciocalteu method.

The propolis was shredded into small particles (<1 mm) after cooling it for 24 h at −20 °C. It was then dispersed in deionized water at a temperature of 5–7 °C at a ratio of 1:1. The system was stirred under cooling for 15 min and then the supernatant floating impurities (wax, wood, bee parts) were discarded followed by filtration with filter paper under vacuum. The solid propolis as well as the water-dissolved propolis were then cooled for 24 h at −80 °C and then lyophilized under the following conditions: condenser temperature −50 °C, vacuum 8.2×10^{-2} mbar.

The wax-free propolis was again shredded into small pieces (<1 mm) after cooling for 24 h at −20 °C. It was then dispersed at a rate of 1 kg/min in the solvent system, which was under agitation at 1500 rpm. The concentration of propolis was 20% by weight. The extraction solvent system consisted of 1.3-Propanediol and the glycerylether ethylhexylglycerin (Ethylhexylglycerin) at a ratio of Ethylhexylglycerin:1.3-Propanediol: 3/97.

The resulting suspension was then placed in a pressure-assisted extractor (Timatic Micro, solid–liquid extractor, Tecnolab, Spello, Italy).

The additional extraction took place under a pressure of 6 bar and at a temperature of 25 °C. The process included 16 compression cycles, lasting 6 min and corresponding decompression cycles lasting 6 min. Then the mixture was left in a hermetically closed container at rest at 5–7 °C for 24 h. It was then filtered by an array of cartridge-type filters (Donaldson Filtration Deutschland GmbH, Haan, Germany) with a pore size of 10 μm–5 μm–1 μm–0.45 μm.

2.2. Quantification of Extract Components

Propolis crude material was extracted three times with 70% ethanol (1:10, *w:v*) by maceration at room temperature for 24 h, followed by filtration of the resulting suspension at room temperature using a paper filter and in-vacuum evaporation of the solvent to dryness on a rotary evaporator. For GC/MS analysis approximately 5 mg of each residue were silylated mixed with 40 μL of dry pyridine and 50 μL of BSTFA (bis(trimethylsilyl) trifluoracetamide) 25 mL for gas chromatography (CASno 25561-30-2) 2 CASno 25561-30-2 (Merck) and heated at 80 °C for 20 min [16].

2.3. GC-MS Analysis

The chemical analysis was performed by the technique of gas chromatography coupled with mass spectrometry (gas chromatography–mass spectrometry, GC-MS).

The analysis was performed on an Agilent 7820A gas chromatograph, connected to an Agilent 5977B mass spectrometer system (Agilent Technologies, Santa Clara, CA, USA) based on electron impact (EI) and 70 eV ionization energy. The gas chromatograph was also equipped with a split/splitless injector and a capillary column HP5MS 30 m, internal diameter 0.25 mm and membrane thickness 0.25 μm. The temperature was programmed from 100 to 300 °C at a rate of 5 °C/min. The carrier gas was He at a flow rate of 0.7 mL/min, injection volume of 2 μL, split ratio of 1:10 and injector temperature of 280 °C. The identification was accomplished using Wiley mass spectral databases and a database created by our research team [16].

2.4. Total Phenolic Content (TPC)

The quantification of total phenolic content was performed using the Folin–Ciocalteu reagent, adhering to the methodology outlined by Arnous et al. (2002) [27]. The phenols were determined using a calibration curve for 100 mg/L to 1200 mg/L of gallic acid (GA) and absorbance was 750 nm. The samples (0.02 mL) were diluted in ultra-pure water (1.58 mL) and the followed addition of the Folin-Ciocalteu reagent (0.1 mL) (Merck KGaA, Darmstadt, Germany) and mixed. After 1 min, aqueous Na_2CO_3 solution (20% *w*/*v*) (0.3 mL) was added and mixed by vortex. The mixture was kept for 2 h in darkness and at room temperature for 2 h, and absorbance measurement followed.

2.5. Antioxidant Activity via 2,2-Diphenyl-1-picrylhydrazyl (DPPH)

The antioxidant capacity of the colloidal systems was evaluated using the DPPH free radical assay, as detailed by Brand-Williams et al. (1995) [28]. PE (0.1 mL) was added to 3.9 mL of a 6×10^{-5} mol/L DPPH methanol solution. The decrease in absorbance was determined at 515 nm. The antioxidant activity was estimated from a standard curve of absorbance values derived from standard concentration solutions of Trolox.

$$\text{Trolox Equivalents (mM)} = 0.0177\%\ DA_{515nm} + 0.08\ R^2 = 0.9998 \quad (1)$$

where

$$\%\ DA_{515nm} = Abs_{515nm(t=0)} - Abs_{515nm(t=30)}/Abs_{515nm(t=0)} \times 100 \quad (2)$$

2.6. Stability of PE

PE was subjected to storage for two months under various conditions, including temperatures of 6 °C, 25 °C, 38 °C, and exposure to UV light, to assess its physical and chemical stability. The chemical stability was evaluated by measuring antioxidant activity using the DPPH free radical scavenging method and determining the total phenolic content with the Folin–Ciocalteu assay.

2.7. In Vitro Microbiome Preservation

The PE's effect on the forehead microbiome was assessed with a three-phase in vitro/ex vivo test.

In phase 1, the influence on the growth behavior of microbes that comprise the majority of skin microbiome for this given area was examined on eight ATCC species, namely,

Staphylococcus epidermidis, *Staphylococcus capitis*, *Cutibacterium acnes*, *Corynebacterium tuberculostearicum*, *Micrococcus luteus*, *Malassezia globosa*, *Staphylococcus hominis* and *Streptococcus mitis*. In phase 2, the effect on the growth behavior of specific pathogens was assessed on 12 ATCC species, namely, *Staphylococcus aureus*, *Pseudomonas aeruginosa*, *Klebsiella pneumoniae*, *Escherichia coli*, *Streptococcus mutans*, *Candida albicans*, *Candida glabrata*, *Malassezia furfur*, *Enterobacter cloacae*, *Streptococcus viridans*, *Candida tropicalis* and *Corynebacterium kroppenstedti*. For all microorganisms, an inoculum of 1×10^6 CFU/mL was prepared and exposed for 2 h against the product (10% dilution). Their survival was then determined by serial dilutions in appropriate culture media. Finally, in phase 3, the effect of PE on the whole microbial biodiversity was examined by inoculating swabs taken from the forehead of two healthy volunteers (ex vivo). After a 2 h-exposure against the substance, the survival of the major groups [Total Aerobic Microbial count (Tryptic Soy Agar-TSA, Plate Count Agar-PCA, Brain Heart Infusion-BHI), Total Anaerobic Microbial count and *Staphylococcus* spp.] was determined by appropriate media and conditions.

2.8. Crystal Violet Staining for Quantitative Analysis of Biofilm

Quantification of biofilm was performed using a modified and optimized crystal violet assay method. Both the inhibitory effect on biofilm formation and the PE ability to eliminate established biofilms were assessed.

For the evaluation of the inhibition of biofilm formation, *C. acnes* strain ATCC 6919 was resuscitated and cultured for 72 h in tryptic soy broth (TSB) under anaerobic conditions. The culture was further diluted with TSB to an appropriate concentration (1×10^8 CFU/mL) and 100 µL of this culture was inoculated into the wells of a sterile 96-well flat bottom plastic plate (Nunclon Surface F; Nunc A/S). The following conditions were examined: (i) biofilm positive control where only *C. acnes* was present, (ii) three different concentrations of the propolis substrate −0.1% *v*/*v*, 0.3% *v*/*v* and 0.5% *v*/*v*, (iii) biofilm negative control containing *C. acnes* along with 6 µg/mL of oxytetracycline and (iv) a sterility and background control containing only TSB. All wells had a final volume of 200 µL. Each concentration was tested in six technical replicates and the procedure was repeated twice. The microplate was incubated for 72 h under anaerobic conditions before biofilm staining. After incubation, the culture medium was removed, followed by gentle washing (three times) with 200 µL of phosphate-buffered saline (PBS). The plate was dried for 30 min at 45 °C and then stained with 200 µL of crystal violet (0.1% (*w*/*v*) in water) for 15 min. The stain was removed, and the wells were then washed three times with 200 µL of PBS. The dye was then extracted with 200 µL of acetone–ethanol (20:80 (*v*/*v*)) for 15 min. After the extraction, 100 µL were transferred to new wells and the absorbance was measured at 595 nm (OD_{595}) in an ELISA reader (Biochrom EZ Read 400, Biochrom Ltd., Cambridge, UK). The absorbance of the negative control was subtracted from the absorbance of each well before statistical evaluation.

For the evaluation of the disruption of *C. acnes*-established biofilm, a biofilm of *C. acnes* was allowed to pre-form. Biofilm formation was achieved by spreading 100 µL of a standardized *C. acnes* culture in a 96-well microtiter plate. The microplate was incubated anaerobically at 37 °C for 3 days to allow cell attachment. The following conditions were examined: (i) biofilm positive control where only *C. acnes* was present, (ii) three different concentrations of the propolis substrate −0.1% *v*/*v*, 0.3% *v*/*v* and 0.5% *v*/*v*, (iii) biofilm negative control containing *C. acnes* along with 6 µg/mL of oxytetracycline and (iv) a sterility and background control containing only TSB. All wells had a final volume of 200 µL. The plate was further incubated for 48 h before the crystal violet staining was performed. The staining and cleaning were performed as outlined above.

The percentage of inhibition was determined through the Equation (3).

$$\% \text{ Inhibition} = [(\text{OD negative control} - \text{OD sample})/\text{OD negative control}] \quad (3)$$

2.9. Gene Expression Analysis in *C. acnes*

The strain *C. acnes* ATCC6919 was resuscitated and cultured for 72 h in TSB under anaerobic conditions. The culture was then diluted with TSB under anaerobic conditions to an appropriate concentration (1×10^7 CFU/mL) and incubated for 72 h under anaerobic conditions. Consequently, *C. acnes* cultures were exposed to PE at concentrations of 0.1% *v*/*v* for 24 h. For total RNA isolation as well as cDNA synthesis, Monarch® Total RNA Miniprep Kit (Hitchin, UK) and PrimeScript-RT reagent Kit (Takara Bio, Otsu, Japan) were used, respectively. Equal amounts of cDNA were used for real-time polymerase chain reaction (qRT-PCR) and the expression of biofilm- and virulence-related genes *CAMP1*, *CAMP4*, *hyl*, *roxP* and *luxS* was analyzed. A 16 s rRNA gene was used as a housekeeping gene. The real-time polymerase chain reaction (qRT-PCR) method as well as the gene analysis procedure were performed according to the standard protocols of the Laboratory [29].

2.10. Gene Expression Analysis in Keratinocytes

NHEKs from an adult donor were purchased from Lonza (Basel, Switzerland) and cultured in KGM™ Gold medium, human epidermal growth factor, bovine pituitary extract, hydrocortisone, epinephrine, transferrin, and gentamicin/amphotericin Lonza Bioscience. The cells were left in culture until 70–80% fullness of the flask and then experiments were carried out with their 24-h incubations with the formulation (PE) at concentrations of 0.01% *v*/*v* and 0.03% *v*/*v*.

For total RNA isolation (500 ng) as well as cDNA synthesis, the Nucleospin RNA kit (Macherey-Nagel, Düren, Germany) and the PrimeScript-RT reagent kit (Takara Bio, Otsu, Japan) were used, respectively. Equal amounts of cDNA were used for real-time polymerase chain reaction (qRT-PCR) and the expression of *IL-4*, *DEF1B*, *ITGB2* and *CXCL12* genes was analyzed. The qRT-PCR method as well as the gene analysis procedure were performed according to the standard protocols of the Laboratory [29]. The glyceraldehyde 3-phosphate dehydrogenase gene (GAPDH) was chosen as a reference gene to normalize values. Two conditions are shown: NHEK cells not incubated with the active (N/T), and NHEK cells incubated with the compound (PE) at concentrations of 0.01% *v*/*v* and 0.03% *v*/*v*.

2.11. Statistical Analysis

Results are expressed as the mean ± SEM of two different experiments. For the in vitro microbiome preservation assay, all microbiological data were log-transformed (log10) before further processing. These were then tested for normality using the Shapiro–Wilk test. Variance homogeneity within groups was assessed using Levene's test. When the assumption of homogeneity of variances was violated, Welch's ANOVA was performed, followed by post hoc pairwise comparisons using the Games–Howell test (phase 1 and phase 2). In cases where the assumption of normality was not met, the Kruskal–Wallis test was applied, with Dunn's test used for post hoc pairwise comparisons (phase 3). All data were processed on python v3.8.0 utilizing the NumPy, Pandas, SciPy, Matplotlib, Seaborn, Scikit-learn, Pingouin and Scikit-posthocs libraries. For all the other results, statistical analysis between two individual groups was performed by Student's *t*-test and one-way ANOVA (Dunnett's multiple comparisons test). A $p \leq 0.05$ was considered statistically significant. Statistical analysis and graphs were performed with Sigma Plot Software v.10 and GraphPad Prism 9 (GraphPad software Inc., San Diego, CA, USA).

3. Results

3.1. Propolis Composition

The chemical composition and biological properties of propolis from different geographic regions are extremely valuable to solving the problem of different propolis types, as well as with their standardization (Table 1).

Table 1. Chemical composition and relative concentration (area%) of the studied propolis sample.

No	RT (min)	Compound	Area%	Chemical Type of Compounds
1	5.40	Ethyl phenyl ether	4.90	Ether
2	6.40	Benzoic acid	9.70	Aromatic acid
3	7.16	Butanedioic acid	10.40	Aliphatic acid
4	9.23	Dimethyl-phenanthrene	12.60	Aromatic hydrocarbon
5	9.81	unknown	2.50	-
6	10.28	γ-curcumene	0.64	Sesquiterpene
7	10.34	ar-curcumene	1.27	Sesquiterpene
8	12.79	calarene	1.30	Sesquiterpene
9	13.49	cadinene	0.56	Sesquiterpene
10	13.86	β-eudesmol	0.82	Sesquiterpene
11	14.24	longifolene	0.91	Sesquiterpene
12	17.89–20.51	sugars	6.70	Sugars
13	27.38	ferulic acid ester	0.65	Ester of Aromatic acid
14	28.58	unknown	1.24	-
15	29.18	unknown	1.66	-
16	29.3	pinostrobin chalcone	0.37	Flavonoids/Chalcones
17	29.62	coumarate/caffeate derivative	0.73	Ester of Aromatic acid
18	29.85	pinocembrin chalcone	0.96	Flavonoids/Chalcones
19	29.95	pinocembrin	0.80	Flavonoids/Chalcones
20	30.41	caffeate derivative	0.74	Ester of Aromatic acid
21	30.51	hydroxy methoxy chalcone	5.38	Flavonoids/Chalcones
22	30.9	pinobanksin	3.32	Flavonoids/Chalcones
23	31.6	methoxylflavanone	5.13	Flavonoids/Chalcones
24	32.09	pinobanksin-3-acetate	2.58	Flavonoids/Chalcones
25	32.52	chalcone	3.85	Flavonoids/Chalcones
26	33.18	galangin	2.71	Flavonoids/Chalcones
27	33.68	phenyl ethyl ester of caffeic acid	1.80	Ester of Aromatic acid
28	33.88	dihydroxymethoxyflavone	0.77	Flavonoids/Chalcones
29	34.08	caffeic acid derivative	1.70	Ester of Aromatic acid
30	35.96	cinnamyl ester of isoferulic acid	1.67	Ester of Aromatic acid
31	36.85	cinnamyl ester of caffeic acid	311	Ester of Aromatic acid
32	37.3	alpinon chalcone	1.03	Flavonoids/Chalcones
33	37.46	kaempferol derivative	0.90	Flavonoids/Chalcones
34	40.17	unknown	1.76	-
35	42.32	unknown	2.24	-
		Total	**98.40**	

The different most common types of propolis as they are referred to in the literature [12] are (i) Poplar type (*Populus* spp. originated mainly from Europe, non-tropic regions of Asia, North America and New Zealand), (ii) Birch type (*Betula verrucosa* derived from Russia), (iii) Green type (*Baccharis* spp. mostly characteristic of Brazil), (iv) Red type (*Dalbergia* spp. which is located in Brazil, Mexico and Cuba), (v) Clusia type (from *Clusia* spp. from Venezuela and Cuba), (vi) Pacific type (*Macaranga tanarius* originated from Indonesia, Taiwan and Okinawa), and (vii) Mediterranean type (plants mainly from Cupressaceae family which is located in Greece, Sicily, Malta and other Mediterranean countries).

Of the main chemical classes of the identified compounds (Table 1) in all samples, the most abundant type of metabolites were the flavonoids and chalcones (27.80%), followed by aromatic acid/hydrocarbon (22.30%), aliphatic acid/ether (15.30%) and esters of aromatic acids (11.40%), respectively.

In the current study, the chemical profile of the propolis sample, as captured by the GC/MS method, identified the existence of high content in phenolic compounds (flavonoids and chalcones together with aromatic acids and their esters). Moreover, among the identified flavonoids/chalcones, methoxyflavones, chalcones, galangin, dihydroxy-methoxyflavone, pinocembrin, pinobanksin and its 3-O-acetate, as well as caffeic and ferulic acids and their derivatives are characterized in higher percentages, which are typical constituents representing the European type of propolis [13,16].

The antioxidant activity measured 31.55 ± 1.55 mM Trolox equivalents, and the initial total phenol concentration was 7046 ± 394 μg GA/mL. Throughout a 6-week period, both the antioxidant capacity and total polyphenol levels remained stable, showing no significant variations (Table 2).

Table 2. Evaluation of stability test of DPPH: Antioxidant activity (±SEM) and TPCs (±SEM) of the PE.

Time	Conditions	TPC (μg GA/mL)	DPPH (mM Trolox Equivalent)
T (0)		7046 ± 227	31.55 ± 0.89
2 months	RT	8309 ± 63.5 **	36.84 ± 0.61 *
	6 °C	5782 ± 271.6 **	26.59 ± 1.37 *
	38 °C	6666 ± 176.2	29.01 ± 1.45
	UV	6611 ± 145.5	28.70 ± 0.73

Differences compared to T0 were analyzed using one-way ANOVA ** $p < 0.01$, * $p < 0.05$.

3.2. *Sensitivity of Skin Microbiome to PE In Vitro and Ex Vivo*

To test, in vitro, whether the PE preserves the skin microbiome, we selected a panel of bacteria species that are commonly found in the forehead (normal flora). The panel consisted of *Staphylococcus epidermidis*, *Staphylococcus capitis*, *Cutibacterium acnes*, *Corynebacterium tuberculostearicum*, *Micrococcus luteus*, *Malassezia globosa*, *Staphylococcus hominis* and *Streptococcus mitis* (Table 3). The bacteria were exposed to three different concentrations of PE for 2 h and enumerated on appropriate culture media after the exposure. The results are shown in Table 3. Furthermore, a second panel of 12 pathogenic bacteria was also exposed to PE (Table 4). At a concentration of 0.1% *v*/*v*, PE did not affect the viability of the tested bacteria. At a concentration of 0.3% *v*/*v*, the population of *C. acnes* and *S. mitis*, part of the normal flora, as well as the population of *Candida glabrata* and *Streptococcus viridans*, part of the pathogenic flora, were significantly reduced. Finally, at the concentration of 0.5% *v*/*v*, PE negatively affected the survival of four microorganisms of the normal flora panel and the survival of five of the pathogenic panel. However, all microbiological differences in abundance were considered 'statistically non-significant' in the multiple pairwise comparisons, regardless of the panel or PE concentration (Tables S1 and S2 in Supplementary Materials), indicating that there are no harmful effects in the panel of the tested microorganisms.

Table 3. Microbial count of bacteria species commonly found in the forehead after exposure to various concentrations of PE. Results are shown as log (CFU/mL).

Microorganisms	Control (0 h) (Mean ± SEM)	0.1% *v/v* PE (2 h) (Mean ± SEM)	0.3% *v/v* PE (2 h) (Mean ± SEM)	0.5% *v/v* PE (2 h) (Mean ± SEM)
Staphylococcus epidermidis	7.24 ± 0.05	7.31 ± 0.02	6.97 ± 0.08	6.34 ± 0.09
Staphylococcus capitis	7.79 ± 0.18	7.31 ± 0.04	7.80 ± 0.09	7.77 ± 0.12
Cutibacterium acnes	6.70 ± 0.10	6.74 ± 0.06	5.24 ± 0.04 #	1.70 ± 0.00 #
Corynebacterium tuberculostearicum	5.91 ± 0.04	5.93 ± 0.11	5.32 ± 0.04	3.02 ± 0.22 #
Micrococcus luteus	7.03 ± 0.05	6.92 ± 0.02	6.88 ± 0.07	6.92 ± 0.03
Malassezia globosa	6.08 ± 0.01	6.03 ± 0.02	5.66 ± 0.08	1.70 ± 0.00 #
Staphylococcus hominis	6.56 ± 0.07	6.57 ± 0.09	6.44 ± 0.10	5.89 ± 0.08
Streptococcus mitis	5.65 ± 0.07	4.89 ± 0.01	4.02 ± 0.06 #	1.70 ± 0.00 #

Reduction > 1 log.

Table 4. Microbial count of pathogenic bacteria species after exposure to various concentrations of PE. Results are shown as log (CFU/mL).

Microorganisms	Control (0 h) (Mean ± SEM)	0.1% *v/v* PE (2 h) (Mean ± SEM)	0.3% *v/v* PE (2 h) (Mean ± SEM)	0.5% *v/v* PE (2 h) (Mean ± SEM)
Staphylococcus aureus	7.82 ± 0.12	7.61 ± 0.02	7.34 ± 0.04	6.94 ± 0.06
Pseudomonas aeruginosa	8.08 ± 0.12	8.02 ± 0.08	7.89 ± 0.09	7.74 ± 0.16
Klebsiella pneumoniae	7.95 ± 0.02	8.18 ± 0.01	8.11 ± 0.04	8.07 ± 0.05
Escherichia coli	7.95 ± 0.05	8.16 ± 0.04	8.05 ± 0.11	8.01 ± 0.03
Streptococcus mutans	5.86 ± 0.08	5.78 ± 0.03	5.59 ± 0.02	5.55 ± 0.05
Candida albicans	6.47 ± 0.08	6.51 ± 0.05	6.10 ± 0.00	5.72 ± 0.10
Candida glabrata	6.27 ± 0.13	6.09 ± 0.12	4.96 ± 0.20 #	3.48 ± 0.12 #
Malassezia furfur	7.32 ± 0.03	7.16 ± 0.04	7.01 ± 0.07	5.36 ± 0.07 #
Enterobacter cloacae	8.35 ± 0.04	8.41 ± 0.01	8.33 ± 0.05	8.41 ± 0.06
Streptococcus viridans	6.26 ± 0.01	4.92 ± 0.15 #	3.70 ± 0.18 #	3.60 ± 0.13 #
Candida tropicalis	5.92 ± 0.10	5.82 ± 0.06	5.50 ± 0.22	4.35 ± 0.05 #
Corynebacterium kroppenstedtii	6.32 ± 0.01	6.30 ± 0.01	5.81 ± 0.12	1.70 ± 0.00 #

Reduction > 1 log.

For the ex vivo testing, swabs from the forehead of two healthy volunteers were inoculated and exposed to 0.1% *v*/*v*, 0.3% *v*/*v* and 0.5% *v*/*v* PE for 2 h and the survival of total anaerobic flora, total aerobic flora and *Staphylococcus* spp. was evaluated. According to the results presented in Table 5, the concentration 0.1% *v*/*v* PE had no impact on the microbial biodiversity, at concentration 0.3% *v*/*v* PE slightly affected the total aerobic flora, while at 0.5% *v*/*v*, PE reduced the levels of total aerobic flora. Nonetheless, these effects were found to be 'statistically non-significant' (Table S3 in Supplementary Materials).

Table 5. Microbial count of bacteria isolated from the forehead of two healthy volunteers and exposed to various concentrations of PE. Results are shown as log (CFU/mL).

Microorganisms	Control (0 h) (Mean ± SEM)	0.1% *v/v* PE (2 h) (Mean ± SEM)	0.3% *v/v* PE (2 h) (Mean ± SEM)	0.5% *v/v* PE (2 h) (Mean ± SEM)
Total Aerobic Microbial count (TSA)	4.31 ± 0.74	3.90 ± 0.45	3.36 ± 0.28 #	3.00 ± 0.04 #
Total Aerobic Microbial count (PCA)	4.69 ± 0.44	4.42 ± 0.38	4.14 ± 0.31	3.74 ± 0.09 #
Total Aerobic Microbial count (BHI)	4.58 ± 0.45	4.16 ± 0.40	3.89 ± 0.26	3.65 ± 0.13 #
Total Anaerobic Microbial count (RCA)	6.41 ± 0.11	6.45 ± 0.30	6.34 ± 0.16	6.38 ± 0.26
Growth medium for *Staphylococcus* (MSA)	4.79 ± 0.44	4.43 ± 0.45	4.21 ± 0.24	3.93 ± 0.21

Reduction > 1 log.

3.3. *PE Inhibits Biofilm Growth and Development and Alters the Gene Expression Levels of C. acnes*

C. acnes strains are classified into phylotypes IA1, IA2, IB1, IB2, IB3, IC, II and type III, with types IA1 and IA2 being predominantly found in acne lesions [30]. The *C. acnes* strain ATCC 6919, a strain isolated from facial acne and corresponding to phylotype IA1, was used for the estimation of the antibiofilm activity of PE extract. PE significantly inhibited the biofilm formation (Figure 1a) and decreased the established biofilm (Figure 1b) in a concentration-analogous manner. Interestingly, PE at the concentration of 0.5 v/v% inhibited biofilm formation at a similar percentage with that of tetracycline that was used as positive control (89.07% and 92.12% respectively). Furthermore, qPCR results revealed a decrease in gene expression of the putative virulence factors *CAMP1*, *CAMP4*, *roxP* and *hyl* (Figure 1c).

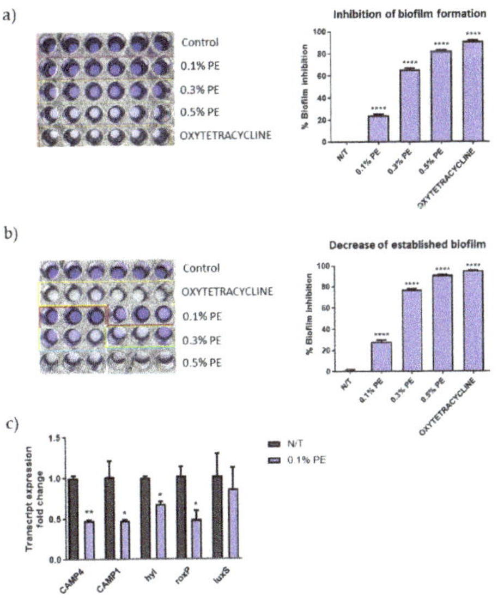

Figure 1. Biofilm disruption and altered gene expression of *C. acnes* by PE. Dose-dependent (**a**) inhibition of biofilm formation and (**b**) decrease of established biofilm as seen by crystal violet staining. The experiment was assessed in two independent replicates. Purple color intensity reflects biofilm biomass. (**c**) Relative gene expression levels for *CAMP4*, *CAMP1*, *hyl*, *roxP* and *luxS* of *C. acnes* cells exposed to 0.1% v/v PE expressed as a fold change ± SEM compared to untreated N/T. The experiment was assessed in two independent replicates. Differences were analyzed using one-way ANOVA with Dunnett's test (**a**,**b**) or the unpaired Student's t test (**c**). **** $p < 0.001$, ** $p < 0.01$, * $p < 0.05$.

3.4. PE Alters the Gene Expression in Keratinocytes

Treatment of NHEKs with PE for 24 h resulted in significant alterations in the expression of genes responsible for a host's antimicrobial response and anti-inflammatory action. Specifically, qPCR analysis demonstrated that *IL-4* gene expression was upregulated up to 55-fold, indicating a robust response to the compound. *DEF1B* expression increased by up to 6-fold.

Conversely, the expression of *ITGB2* was downregulated by up to 5-fold and *CLCX12* by up to 4-fold. The data indicate a dose-response relationship, with significant upregulation observed at higher PE concentrations compared to the control (not-treated cells). Graphs depicting the fold changes in gene expression are shown in Figure 2.

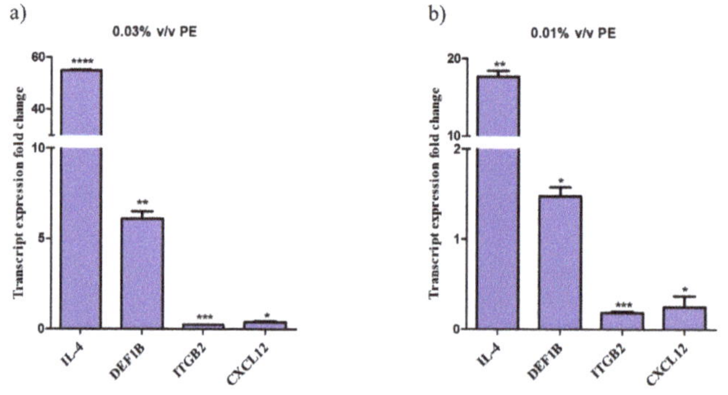

Figure 2. Altered gene expression profile of keratinocytes treated with PE. Relative gene expression levels for *IL-4, DEF1B, ITGB* and *CXCL12* of keratinocytes exposed to (**a**) 0.01% v/v and (**b**) 0.03% v/v PE expressed as a fold change ± SEM compared to untreated N/T. The experiment was assessed in two independent replicates. Differences were analyzed using one-way ANOVA with Dunnett's test. **** $p < 0.0001$, *** $p < 0.001$, ** $p < 0.01$, * $p < 0.05$.

4. Discussion

This study provides an analysis of the bioactivity of a Greek propolis extract. Specifically, we investigated the dose-dependent bioactivity of the extract against pathogenic constituents of the human forehead skin microbiota, while ensuring the preservation of the area's microbial biodiversity. Furthermore, in the case of *C. acnes*, the PE downregulated the expression of key genes involved in attachment, colonization, and biofilm formation. Subsequent analyses confirmed that PE significantly inhibited both biofilm development and the growth of *C. acnes*. Finally, PE was found to modulate the expression of genes associated with antimicrobial defense and anti-inflammatory responses in NHEKs.

In this study we reveal for the first time the inhibitory properties of propolis on *C. acnes* biofilm. Biofilm formation is a well-documented strategy used by many bacterial species to enhance resistance to antibiotics [31]. The formation of biofilm is a multi-stage process where microbial cells first attach to a surface (initial reversible adhesion), followed by the production of an extracellular matrix that strengthens the attachment. The biofilm can negatively affect the entry of antibiotics and is a potent contributing factor to *C. acnes* resistance and failure of treatments, thus biofilm is a promising target for therapeutic strategies that target *C. acnes* [32]. Biofilm inhibition can occur through multiple mechanisms targeting different stages of biofilm development. These include preventing the initial bacterial adhesion to surfaces, interfering with quorum sensing communication pathways essential for biofilm formation, and disrupting the extracellular matrix that stabilizes the biofilm structure. Additionally, biofilm inhibition may involve the direct killing or growth inhibition of sessile cells within the biofilm, as well as impeding biofilm maturation and maintenance processes. Various natural and synthetic compounds have demonstrated efficacy in these roles, often acting through a combination of antimicrobial activity, anti-adhesive effects, and signaling interference, making them promising candidates for controlling biofilm-associated infections [33].

Various propolis extracts have shown antimicrobial activity against pathogenic microorganisms, such as bacteria of the genus *Salmonella* spp., but also against microorganisms found in the microflora of the skin (microbiome), such as bacteria *Staphylococcus epidermidis* [15]. Considering the contribution of the microbiome to the integrity of the skin, products intended for direct application to the skin that aim to inhibit pathogenic or potentially (opportunistically) pathogenic microorganisms, should not affect the qualitative and quantitative composition of the commensal microbiome.

Antibiotics, which have long been a staple of acne therapy, can cause localized antimicrobial resistance, inadvertently affect beneficial microbial communities and induce cutaneous dysbiosis—a disruption in the natural balance of the skin microbiome [3]. Moreover, antibiotics' impact on microbial diversity can weaken the skin's barrier function, making it more susceptible to irritation and infection. Reduced microbial diversity has been linked to other dermatological conditions, such as dermatitis or secondary infections [3,34].

PE was characterized as microbiome-preserving regarding normal microflora. At the tested concentrations of 0.1% v/v and 0.3% v/v demonstrated in vitro and ex vivo, it leaves the natural biodiversity of the forehead area unaffected. Interestingly, when tested at the concentration of 0.5% v/v, PE showcased a "non-friendly" behavior towards commensal microflora, as it significantly altered the forehead's microbiome diversity, and multiple species' growth was affected.

RoxP, uniquely produced by *C. acnes*, plays a crucial role in shaping the lipidome and, consequently, constitutes a significant component of the interaction interfaces between the host and microbes [35,36]. Although RoxP has been suggested to possess antioxidant properties and maintain redox homeostasis in normal skin, there is no data concerning its actual role in inflammatory conditions such as acne. We observed a significant reduction in gene expression levels of *roxP* upon treatment of *C. acnes* with PE. Our data support the idea that *roxP* may be a target for potent therapeutic interventions regarding this specific skin condition, worthy of further investigation. Apart from *roxP*, PE treatment reduced the gene expression levels of *hyl*. Hyl has been reported to be involved in the spread of bacteria [37]. Furthermore, PE significantly reduced the mRNA levels of the virulence factors CAMP1 and CAMP4, that can trigger inflammation in host tissue [38].

The bioactivity of PE was further assessed on human keratinocytes, focusing on the transcriptomic regulation of genes related to antimicrobial response—as propolis is known also for stimulating the host's immune system [15,39]. We observed a significant transcriptional upregulation of interleukin-4, an pleiotropic cytokine that has positive effects on the outcome of infectious diseases, contributing to tissue healing [40,41], and of B-defensin 1 (DEF1B), a factor that exhibits microbicidal activity mainly against Gram-negative bacteria by interfering with the synthesis of their cell wall and causing microbial death in a manner similar to that of antibiotics [42]. Moreover, qPCR results showed a downregulation of β2-integrins (ITGB2), receptors that are exploited by common skin pathogens to establish contact with host cells [43,44], and of C-X-C ligand motif chemokine 12 (CXCL12 or SDF-1), a chemokine with an important role in skin inflammation and inflammatory angiogenesis [45,46].

Collectively, our data highlight that this dual mechanism of PE—targeting biofilm-forming pathogenic strains of C. *acnes*, via direct inhibition of C. *acnes* virulence factors and enhancement of the host's antimicrobial defenses—provides a multifaceted approach to modulate the behavior of skin microbiota in acne, paving the way for next-generation acne-targeting products. Aiming to investigate further the efficacy of PE in vivo, we are in the process of the incorporation of PE into three types of formulas (a cream, a cleansing gel and a lotion), that will be tested on volunteers with acne-prone skin. With respect to skin microbiota, in vivo studies are designed to evaluate the sensitivity of skin microbiota biodiversity using advanced sequencing techniques. These results will provide additional insights into the use of the PE on acne-targeting products.

5. Conclusions

Our findings highlight that this dual approach—targeting biofilm-forming pathogenic strains of *C. acnes* through direct inhibition of its virulence factors and boosting the host's antimicrobial defenses—offers a comprehensive strategy to influence the behavior of skin microbiota in acne.

Supplementary Materials: The following supporting information can be downloaded at: https://www.mdpi.com/article/10.3390/antiox14070849/s1.

Author Contributions: Conceptualization, S.A. and K.G.; methodology, S.A., E.P., E.S., M.G., D.G., C.G., I.C., A.K.A., E.B. and K.G.; formal analysis, S.A., E.S. and E.P.; investigation, S.A., E.P., E.S., M.G., D.G., C.G., I.C., A.K.A., E.B. and K.G.; writing—original draft preparation, S.A., E.S. and E.P.; writing—review and editing, K.G., C.G., I.C., A.K.A. and E.B.; visualization, S.A. and E.P.; supervision, K.G., I.C., A.K.A. and E.B.; funding acquisition, K.G. All authors have read and agreed to the published version of the manuscript.

Funding: This research received no external funding.

Institutional Review Board Statement: The study has been conducted by suitably trained and experienced personnel in accordance with the Efficacy Evaluation Guidelines published by COLIPA—May 2008. Precautions will be taken to avoid the possibility that participants in the study might experience undesirable effects. The test product should comply with the requirements of Regulation 1223/2009. The manufacturer of the test product is responsible for its compliance with the requirements of the current Cosmetics Regulation (EC/1223/2009). Ethical requirements which have been taken into consideration in the planning of the study include: (i) participants are informed volunteers selected after application of inclusion/exclusion criteria. (ii) participants are aware of the purpose and nature of the study and of any foreseeable risks involved in participation in the study and have given written informed consent before the study starts. The informed consent should be in accordance with principles that originated in the Declaration of Helsinki.

Informed Consent Statement: In accordance with the current legislation and the declaration of Helsinki, two volunteers provided informed consent for forehead swabs used in ex vivo biodiversity analysis. The consent forms are managed and archived following QACS S.A.'s internal procedures.

Data Availability Statement: The data presented in this study are available upon request from the corresponding author.

Acknowledgments: We would like to thank all the members of the different research groups that participated in this study.

Conflicts of Interest: Sophia Athanasopoulou, Eleni Panagiotidou, Eleni Spanidi, Maria Gkika, Danai Georgiou were employed by APIVITA SA. The remaining authors declare that the research was conducted in the absence of any commercial or financial relationships that could be construed as a potential conflict of interest.

References

1. Dréno, B. What is new in the pathophysiology of acne, an overview. *J. Eur. Acad. Dermatol. Venereol.* **2017**, *31* (Suppl. 5), 8–12. [CrossRef] [PubMed]
2. Popa, G.L.; Mitran, C.I.; Mitran, M.I.; Tampa, M.; Matei, C.; Popa, M.I.; Georgescu, S.R. Markers of Oxidative Stress in Patients with Acne: A Literature Review. *Life* **2023**, *13*, 1433. [CrossRef]
3. Cavallo, I.; Sivori, F.; Truglio, M.; De Maio, F.; Lucantoni, F.; Cardinali, G.; Pontone, M.; Bernardi, T.; Sanguinetti, M.; Capitanio, B.; et al. Skin dysbiosis and Cutibacterium acnes biofilm in inflammatory acne lesions of adolescents. *Sci. Rep.* **2022**, *12*, 21104. [CrossRef] [PubMed]
4. Coenye, T.; Spittaels, K.-J.; Achermann, Y. The role of biofilm formation in the pathogenesis and antimicrobial susceptibility of Cutibacterium acnes. *Biofilm* **2022**, *4*, 100063. [CrossRef] [PubMed]
5. Gannesen, A.V.; Zdorovenko, E.L.; Botchkova, E.A.; Hardouin, J.; Massier, S.; Kopitsyn, D.S.; Gorbachevskii, M.V.; Kadykova, A.A.; Shashkov, A.S.; Zhurina, M.V.; et al. Composition of the Biofilm Matrix of Cutibacterium acnes Acneic Strain RT5. *Front. Microbiol.* **2019**, *10*, 1284. [CrossRef]
6. Nakamura, K.; O'Neill, A.M.; Williams, M.R.; Cau, L.; Nakatsuji, T.; Horswill, A.R.; Gallo, R.L. Short chain fatty acids produced by Cutibacterium acnes inhibit biofilm formation by Staphylococcus epidermidis. *Sci. Rep.* **2020**, *10*, 21237. [CrossRef]
7. Singh, A.; Amod, A.; Pandey, P.; Bose, P.; Pingali, M.S.; Shivalkar, S.; Varadwaj, P.K.; Sahoo, A.K.; Samanta, S.K. Bacterial biofilm infections, their resistance to antibiotics therapy and current treatment strategies. *Biomed. Mater.* **2022**, *17*, 022003. [CrossRef]
8. Pecoraro, C.; Carbone, D.; Parrino, B.; Cascioferro, S.; Diana, P. Recent Developments in the Inhibition of Bacterial Adhesion as Promising Anti-Virulence Strategy. *Int. J. Mol. Sci.* **2023**, *24*, 4872. [CrossRef]
9. Dufrêne, Y.F.; Persat, A. Mechanomicrobiology: How bacteria sense and respond to forces. *Nat. Rev. Microbiol.* **2020**, *18*, 227–240. [CrossRef]
10. Parrino, B.; Diana, P.; Cirrincione, G.; Cascioferro, S. Bacterial Biofilm Inhibition in the Development of Effective Anti-Virulence Strategy. *Open Med. Chem. J.* **2018**, *12*, 84–87. [CrossRef]
11. Queiroga, M.C.; Laranjo, M.; Andrade, N.; Marques, M.; Costa, A.R.; Antunes, C.M. Antimicrobial, Antibiofilm and Toxicological Assessment of Propolis. *Antibiotics* **2023**, *12*, 347. [CrossRef] [PubMed]
12. Graikou, K.; Popova, M.; Gortzi, O.; Bankova, V.; Chinou, I. Characterization and biological evaluation of selected Mediterranean propolis samples. Is it a new type? *LWT Food Sci. Technol.* **2016**, *65*, 261–267. [CrossRef]
13. Popova, M.; Giannopoulou, E.; Skalicka-Woźniak, K.; Graikou, K.; Widelski, J.; Bankova, V.; Kalofonos, H.; Sivolapenko, G.; Gaweł-Bęben, K.; Antosiewicz, B.; et al. Characterization and biological evaluation of propolis from Poland. *Molecules* **2017**, *22*, 1159. [CrossRef] [PubMed]
14. Kocot, J.; Kiełczykowska, M.; Luchowska-Kocot, D.; Kurzepa, J.; Musik, I. Antioxidant Potential of Propolis, Bee Pollen, and Royal Jelly: Possible Medical Application. *Oxidative Med. Cell. Longev.* **2018**, *2018*, 7074209. [CrossRef]
15. Przybyłek, I.; Karpiński, T.M. Antibacterial Properties of Propolis. *Molecules* **2019**, *24*, 2047. [CrossRef]
16. Pyrgioti, E.; Graikou, K.; Aligiannis, N.; Karabournioti, S.; Chinou, I. Qualitative Analysis Related to Palynological Characterization and Biological Evaluation of Propolis from Prespa National Park (Greece). *Molecules* **2022**, *27*, 7018. [CrossRef]
17. Anjum, S.I.; Ullah, A.; Khan, K.A.; Attaullah, M.; Khan, H.; Ali, H.; Bashir, M.A.; Tahir, M.; Ansari, M.J.; Ghramh, H.A.; et al. Composition and functional properties of propolis (bee glue): A review. *Saudi J. Biol. Sci.* **2019**, *26*, 1695–1703. [CrossRef]
18. El-Guendouz, S.; Lyoussi, B.; Miguel, M.G. Insight on Propolis from Mediterranean Countries: Chemical Composition, Biological Activities and Application Fields. *Chem. Biodivers.* **2019**, *16*, e1900094. [CrossRef]
19. Aminimoghadamfarouj, N.; Nematollahi, A. Propolis diterpenes as a remarkable bio-source for drug discovery development: A review. *Int. J. Mol. Sci.* **2017**, *18*, 1290. [CrossRef]
20. Misir, S.; Aliyazicioglu, Y.; Demir, S.; Turan, I.; Hepokur, C. Effect of Turkish Propolis on miRNA Expression, Cell Cycle, and Apoptosis in Human Breast Cancer (MCF-7) Cells. *Nutr. Cancer* **2020**, *72*, 133–145. [CrossRef]
21. Saavedra, N.; Cuevas, A.; Cavalcante, M.F.; Dörr, F.A.; Saavedra, K.; Zambrano, T.; Abdalla, D.S.P.; Salazar, L.A. Polyphenols from Chilean Propolis and Pinocembrin Reduce MMP-9 Gene Expression and Activity in Activated Macrophages. *BioMed Res. Int.* **2016**, *2016*, 1–8. [CrossRef]
22. Vukovic, N.L.; Obradovic, A.D.; Vukic, M.D.; Jovanovic, D.; Djurdjevic, P.M. Cytotoxic, proapoptotic and antioxidative potential of flavonoids isolated from propolis against colon (HCT-116) and breast (MDA-MB-231) cancer cell lines. *Food Res. Int.* **2018**, *106*, 71–80. [CrossRef] [PubMed]
23. Kapare, H.; Lohidasan, S.; Sinnathambi, A.; Mahadik, K. Standardization, anti-carcinogenic potential and biosafety of Indian propolis. *J. Ayurveda Integr. Med.* **2019**, *10*, 81–87. [CrossRef]
24. Zabaiou, N.; Fouache, A.; Trousson, A.; Baron, S.; Zellagui, A.; Lahouel, M.; Lobaccaro, J.-M.A. Biological properties of propolis extracts: Something new from an ancient product. *Chem. Phys. Lipids* **2017**, *207*, 214–222. [CrossRef] [PubMed]

25. Zulhendri, F.; Lesmana, R.; Tandean, S.; Christoper, A.; Chandrasekaran, K.; Irsyam, I.; Suwantika, A.A.; Abdulah, R.; Wathoni, N. Recent Update on the Anti-Inflammatory Activities of Propolis. *Molecules* **2022**, *27*, 8473. [CrossRef]
26. Zullkiflee, N.; Taha, H.; Abdullah, N.A.; Hashim, F.; Usman, A.; Journal, P. Antibacterial and Antioxidant Activities of Ethanolic and Water Extracts of Stingless Bees Tetrigona binghami, Heterotrigona itama, and Geniotrigona thoracica Propolis Found in Brunei. *Philipp. J. Sci.* **2022**, *151*, 1455–1462. [CrossRef]
27. Arnous, A.; Makris, D.P.; Kefalas, P. Correlation of Pigment and Flavanol Content with Antioxidant Properties in Selected Aged Regional Wines from Greece. *J. Food Compos. Anal.* **2002**, *15*, 655–665. [CrossRef]
28. Brand-Williams, W.; Cuvelier, M.E.; Berset, C. Use of a free radical method to evaluate antioxidant activity. *LWT Food Sci. Technol.* **1995**, *28*, 25–30. [CrossRef]
29. Spanidi, E.; Athanasopoulou, S.; Liakopoulou, A.; Chaidou, A.; Hatziantoniou, S.; Gardikis, K. Royal Jelly Components Encapsulation in a Controlled Release System—Skin Functionality, and Biochemical Activity for Skin Applications. *Pharmaceuticals* **2022**, *15*, 907. [CrossRef]
30. Lomholt, H.B.; Scholz, C.F.P.; Brüggemann, H.; Tettelin, H.; Kilian, M. A comparative study of Cutibacterium (Propionibacterium) acnes clones from acne patients and healthy controls. *Anaerobe* **2017**, *47*, 57–63. [CrossRef]
31. Zhou, J.-W.; Yin, K.-Y.; Luo, W.-Q.; Chen, A.; Liang, Z.-W.; Ji, P.-C.; Wang, Y.-J.; Wang, X.-N. Anti-virulence potential of carvone against Serratia marcescens. *Food Med. Homol.* **2024**, *1*, 9420001. [CrossRef]
32. Ruffier d'Epenoux, L.; Fayoux, E.; Veziers, J.; Dagnelie, M.-A.; Khammari, A.; Dréno, B.; Corvec, S. Biofilm of Cutibacterium acnes: A target of different active substances. *Int. J. Dermatol.* **2024**, *63*, 1541–1550. [CrossRef] [PubMed]
33. Kalia, V.C.; Patel, S.K.S.; Lee, J.-K. Bacterial biofilm inhibitors: An overview. *Ecotoxicol. Environ. Saf.* **2023**, *264*, 115389. [CrossRef]
34. Condrò, G.; Guerini, M.; Castello, M.; Perugini, P. Acne Vulgaris, Atopic Dermatitis and Rosacea: The Role of the Skin Microbiota-A Review. *Biomedicines* **2022**, *10*, 2523. [CrossRef]
35. Holland, C.; Mak, T.N.; Zimny-Arndt, U.; Schmid, M.; Meyer, T.F.; Jungblut, P.R.; Brüggemann, H. Proteomic identification of secreted proteins of Propionibacterium acnes. *BMC Microbiol.* **2010**, *10*, 230. [CrossRef]
36. Allhorn, M.; Arve, S.; Brüggemann, H.; Lood, R. A novel enzyme with antioxidant capacity produced by the ubiquitous skin colonizer Propionibacterium acnes. *Sci. Rep.* **2016**, *6*, 36412. [CrossRef] [PubMed]
37. Marion, C.; Stewart, J.M.; Tazi, M.F.; Burnaugh, A.M.; Linke, C.M.; Woodiga, S.A.; King, S.J. Streptococcus pneumoniae can utilize multiple sources of hyaluronic acid for growth. *Infect. Immun.* **2012**, *80*, 1390–1398. [CrossRef]
38. Spittaels, K.-J.; Coenye, T. Developing an in vitro artificial sebum model to study Propionibacterium acnes biofilms. *Anaerobe* **2018**, *49*, 21–29. [CrossRef]
39. Sforcin, J.M.; Bankova, V. Propolis: Is there a potential for the development of new drugs? *J. Ethnopharmacol.* **2011**, *133*, 253–260. [CrossRef]
40. Gause, W.C.; Wynn, T.A.; Allen, J.E. Type 2 immunity and wound healing: Evolutionary refinement of adaptive immunity by helminths. *Nat. Rev. Immunol.* **2013**, *13*, 607–614. [CrossRef]
41. Allen, J.E. IL-4 and IL-13: Regulators and Effectors of Wound Repair. *Annu. Rev. Immunol.* **2023**, *41*, 229–254. [CrossRef]
42. Raschig, J.; Mailänder-Sánchez, D.; Berscheid, A.; Berger, J.; Strömstedt, A.A.; Courth, L.F.; Malek, N.P.; Brötz-Oesterhelt, H.; Wehkamp, J. Ubiquitously expressed Human Beta Defensin 1 (hBD1) forms bacteria-entrapping nets in a redox dependent mode of action. *PLoS Pathog.* **2017**, *13*, e1006261. [CrossRef] [PubMed]
43. Borisova, M.; Shi, Y.; Buntru, A.; Wörner, S.; Ziegler, W.H.; Hauck, C.R. Integrin-mediated internalization of Staphylococcus aureus does not require vinculin. *BMC Cell Biol.* **2013**, *14*, 2. [CrossRef] [PubMed]
44. Fagerholm, S.C.; Guenther, C.; Llort Asens, M.; Savinko, T.; Uotila, L.M. Beta2-Integrins and Interacting Proteins in Leukocyte Trafficking, Immune Suppression, and Immunodeficiency Disease. *Front. Immunol.* **2019**, *10*, 254. [CrossRef] [PubMed]
45. Zgraggen, S.; Huggenberger, R.; Kerl, K.; Detmar, M. An important role of the SDF-1/CXCR4 axis in chronic skin inflammation. *PLoS ONE* **2014**, *9*, e93665. [CrossRef]
46. Lu, L.; Li, J.; Jiang, X.; Bai, R. CXCR4/CXCL12 axis: "old" pathway as "novel" target for anti-inflammatory drug discovery. *Med. Res. Rev.* **2024**, *44*, 1189–1220. [CrossRef]

antioxidants

MDPI

Article

Antioxidant, Antibacterial, and Bioaccessibility Properties of Ultrasound-Extracted Chilean Propolis

Jessica Mejía [1], Claudia Giovagnoli-Vicuña [2,*], Cristian Jacob [1], Gloria Montenegro [1], Andrea I. Moreno-Switt [3] and Ady Giordano [2,*]

1 Departamento de Ciencias Vegetales, Facultad de Agronomía y Sistemas Naturales, Pontificia Universidad Católica de Chile, Santiago 7820436, Chile; jcmejia@uc.cl (J.M.); cjjacob@uc.cl (C.J.); gmonten@uc.cl (G.M.)
2 Departamento de Química Inorgánica, Facultad de Química y de Farmacia, Pontificia Universidad Católica de Chile, Santiago 7820436, Chile
3 Escuela de Medicina Veterinaria, Facultad de Agronomía y Sistemas Naturales, Facultad de Ciencias Biológicas, Facultad de Medicina, Pontificia Universidad Católica de Chile, Santiago 7820436, Chile; andrea.moreno@uc.cl
* Correspondence: ccgiovagnoli@uc.cl (C.G.-V.); agiordano@uc.cl (A.G.)

Abstract: Propolis is a honeybee product with significant biological properties, strongly influenced by geographic and floral origin as well as extraction method. Ultrasound-assisted extraction (UAE) is an emerging alternative for propolis preparation. This study aimed to determine the optimal UAE conditions for total polyphenol content (TPC) and total flavonoid content (TFC) in Chilean propolis and compare its antioxidant and antimicrobial capacities with a conventional method. UAE was performed with varying ethanol concentrations (50–90%), temperatures (20–60 °C), and extraction times (10–50 min), keeping the solid/liquid ratio constant at 1:10 (w/v). The optimal ethanolic UAE extract (OE) was compared to the conventional ethanolic extract (CE) for antimicrobial activity against *Salmonella enterica* and *Listeria monocytogenes* and antioxidant capacity (DPPH, ABTS, FRAP assays). Optimal UAE conditions were 80% ethanol, 30 °C, and 30 min. The OE showed significantly higher ($p < 0.05$) TPC (22.4 ± 0.3 mg GAE/mL), TFC (15.7 ± 0.7 mg QE/mL), antioxidant capacity (ABTS: 35.7 ± 0.9; DPPH: 62.9 ± 0.3; FRAP: 49.7 ± 2.1 µmol TE/mL), and antimicrobial activity in the evaluated Gram-positive bacteria (>15 mm inhibition halo, MBC = 3.1 mg/mL) compared to the CE. The bioaccessibility analysis revealed that the OE maintained 20.1% of its polyphenol content and 69.5%, 60.5%, and 61.9% of DPPH, ABTS, and FRAP, respectively, after simulated gastrointestinal digestion. The established optimal UAE conditions generated extracts with increased polyphenol content, antioxidant capacity, antimicrobial activity, and bioaccessibility, indicating its potential as an extraction method for propolis with enhanced bioactivity.

Keywords: polyphenols; ultrasound-assisted extraction; bioactive properties; digestion

1. Introduction

Propolis is produced by honeybees (*Apis mellifera*) from the exudates of the bark and various tissues of plants to control the growth of microorganisms [1]. This natural substance is characterized by a complex chemical composition [2], generally consisting of 50% resin, 30% wax, 10% essential oils, 5% pollen, and 5% other substances [3]. The chemical composition of propolis, and, consequently, its biological properties, varies substantially depending on multiple phytogeographic factors, such as vegetation, season, and environmental conditions [2,4].

Ultrasound-assisted extraction (UAE) is an effective extraction method based on acoustic cavitation, where ultrasonic waves generate microbubbles that collapse violently, producing high localized pressures and temperatures. This phenomenon induces shear forces, microstreaming, and microjetting, which promote matrix disruption and enhance solvent penetration and mass transfer [5,6]. By contrast, conventional extraction methods have lower selectivity for compounds of interest, low extraction yields, and long extraction periods (i.e., maceration for 28 days) [7]. Therefore, switching from a conventional extraction to an optimized UAE method would allow for obtaining extracts with a high proportion of their free bioactive compounds and greater biological properties such as antioxidant capacity and antimicrobial activity.

Chilean propolis is characterized by its botanical origin in native plant species such as *Lithraea caustica*, *Quillaja saponaria*, and *Baccharis linearis* [8,9]. These species are part of the evergreen matorral vegetation of central Chile, located in a Mediterranean semiarid region. These unique floral sources contribute to a distinctive chemical profile rich in phenolic compounds, flavonoids, and diterpenes, which varies significantly from that of propolis in other geographical origins [9,10]. Despite its promising biological potential, Chilean propolis remains underexplored in the global scientific literature, particularly in terms of its phytochemical diversity and functional applications.

Propolis has been widely recognized for its strong biological properties, largely attributed to its high content of phenolic compounds, which exhibit both antioxidant and antimicrobial activities [11,12]. These bioactive compounds play a key role in free radical-scavenging agents, thus reducing oxidative stress [13]. In addition, propolis can effectively inhibit the growth of microorganisms depending on the propolis's geographical origin and the resistance of these microorganisms [1]. This makes propolis an important natural agent in food preservation and health-related applications [14]. Therefore, enhancing the extraction process can significantly increase the concentration of these bioactive compounds, thereby improving the extract's overall biological properties, such as antioxidant capacity and antimicrobial effectiveness, and, consequently, its bioaccessibility.

Bioaccessibility refers to the fraction of the bioactive compound that is released from the food matrix to the gastrointestinal tract for luminal absorption [15]. To date, there have been in vitro models to evaluate how digestion acts on these compounds and, thus, predict their release from different food matrices. For this, simulations are carried out in different parts of the digestive system such as the oral cavity, stomach, small intestine, and large intestine [16].

The objective of this study was to optimize the extraction of propolis using UAE and compare the optimal ethanolic UAE extract (OE) with a conventional ethanolic extract (CE). Under optimal parameters, UAE exhibited greater antioxidant capacity, antibacterial activity, and bioaccessibility of bioactive compounds as compared to extracts obtained with a conventional method. These findings reveal the potential of UAE as an efficient technique for extracting Chilean propolis with enhanced bioactivity.

2. Materials and Methods

2.1. Raw Material and Sample Preparation

Propolis from *Apis mellifera* was acquired in 2023 from the central zone of Chile. The sample was stored at −20 °C until the extraction processes. The propolis sample had a moisture content of 2.0 ± 0.0 g/100 g, crude protein of 4.6 ± 0.3 g/100 g, total lipids of 56.8 ± 1.4 g/100 g, and crude ash of 1.3 ± 0.2 g/100 g. Analyses were performed according to AOAC methods [17].

2.2. Botanical Analysis of Propolis

For the palynological and plant structure analyses, 10 g of propolis samples was weighed and mixed with 20 mL of distilled water and stirred until a homogeneous dispersion was obtained. The mixture was then centrifuged at 1006× g for 10 min, and the supernatant was discarded. The sediment was resuspended in 20 mL of distilled water and vortexed to ensure uniformity. A 100 µL aliquot was taken and placed on a glass slide, and this process was repeated four times for consistency. A drop of Calberla's fuchsin stain was added to facilitate the visualization of pollen grains under a light microscope. The identification and quantification of pollen grains and plant structures (trichomes, vessels, and epidermis) were performed by comparing them with reference materials, including specialized literature and the botanical bee pollen catalog at the Laboratory of Botany (Department of Plant Sciences, Faculty of Agronomy and Natural Systems, Pontificia Universidad Católica de Chile, Santiago, Chile). Additional comparisons were made using photographs and permanent preparations available at the same laboratory. The proportion of each structure in relation to the total counted structures was then estimated.

2.3. Ethanolic Extraction of Propolis

Conventional and ultrasound methods were used to obtain ethanolic extracts from the propolis sample. The conventional method for preparation of ethanolic propolis extract consisted of procedures routinely conducted by local beekeepers. Briefly, a mixture of 1 g of ground propolis and 10 mL of 70% (v/v) ethanol was placed in an amber Falcon tube and stored at room temperature for 28 days. During the initial 14 days, the mixture was manually shaken twice daily, followed by a 14-day undisturbed maceration period. Finally, the extract was centrifuged for 10 min at 1807× g (Bioprocen 22 R model, ORTO ALRESA, Madrid, Spain), and the resulting supernatant was recovered and stored at 4 °C for subsequent analysis.

To determine the OE, single-factor tests were conducted in an ultrasonic bath (Ultrasonic Cleaners, BIOBASE, Jinan, China; 80 W of power and a 40 kHz operating frequency). Three variables were evaluated: ethanol concentration (50–90%, v/v), temperature (20–60 °C), and extraction time (10–50 min). For each test, the remaining parameters were kept constant at 80% (v/v) ethanol, 25 °C, and 30 min. All experiments were conducted using a 1:10 (w/v) sample-to-solvent ratio, as described for the conventional extraction protocol. The ultrasonic power density was calculated by dividing the nominal power by the extract volume, yielding a value of 8 W/mL.

2.4. Phytochemical Screening

Ethanolic extracts of propolis were subjected to standard screening tests to determine the presence of phytochemicals. Secondary metabolites, such as tannins, flavonoids, coumarins, alkaloids, phenols, proteins, carbohydrates (sugars), and saponins, were quantified through procedures previously described [18] (Table 1). The relative concentrations of the compounds were determined using visual color intensity as an indicator compared to the positive control. +++ indicated high intensity and high concentration, ++ mild intensity and mild concentration, + low intensity and low concentration, and − indicated no change, with the compound not detected. The positive controls were tannic acid (1000 ppm), catechin (1000 ppm), coumarin (500 ppm), oxymatrine (800 ppm), gallic acid (1000 ppm), casein (10% w/v), sucrose (1% w/v), and oats (1% w/v). The negative control was distilled water.

Table 1. Phytochemical screening tests conducted to characterize the ethanolic propolis extracts.

Phytochemical	Test	Positive Result
Tannins [a]	500 µL extract + 500 µL H_2O + 1–2 drops of $FeCl_3$ (10%)	Blue-Green color precipitate
Flavonoids [b]	1 mL extract + 2 mL of 2% NaOH solution+ drops diluted HCl	Colorless
Coumarins [c]	1 mL extract + 10% NaOH + Chloroform	Yellow color
Alkaloids [d]	2 mL extract + drops of Dragendorff's reagents	Reddish-brown precipitate
Phenols [e]	Extract + drops of $K_2Cr_2O_7$	Dark color
Proteins [f]	2 mL extract + 1 drop of $CuSO_4$ (2%) + 1 mL of Ethanol (95%) + 1 pellet KOH	Pink color
Sugars [g]	1 mL extract + 1 mL Fehling's solution A+ 1 mL Fehling's solution B + boiled in water bath	Red precipitate
Saponins [h]	1 mL extract + 2 mL water (vigorously shaken)	Foam for ten minutes

The positive controls were [a] tannic acid, [b] catechin, [c] coumarin, [d] oxymatrine, [e] gallic acid, [f] casein, [g] sucrose, [h] oats. The negative control was distilled water.

2.5. Determination of TPC and TFC

The total polyphenol content (TPC) in the propolis extracts was determined using the Folin–Ciocalteu (FC) method described by Singleton et al. [19]. Propolis extracts (20 µL) were mixed with diluted FC reagent (100 µL at a 1:10 (*v*/*v*) ratio) and 7.5% sodium carbonate solution (80 µL). The spectrophotometric readings were taken at 765 nm after 40 min of dark incubation in a microplate reader (FlexA-200, ALLSHENG Instrument Co., Ltd., Hangzhou, China). Gallic acid was utilized to create a standard curve (50–400 µg/mL; y = 0.0026x + 0.0859; R^2 = 0.9914), and the TPC was quantified as mg gallic acid equivalents (GAEs) per milliliter of the ethanolic propolis extract.

For the total flavonoid content (TFC) in the propolis samples, the aluminum chloride colorimetric assay was employed [20]. Propolis samples (100 µL) were mixed with 2% $AlCl_3$ (100 µL), and, after homogenization and dark incubation (30 min), the absorbance at 420 nm was quantified. The calibration curve was established using a quercetin standard (10–60 µg/mL; y = 0.0093x − 0.0239; R^2 = 0.9955), and the TFC was expressed as mg quercetin equivalents (QEs) per milliliter of the ethanolic propolis extract. All measurements were performed in triplicate (n = 3).

2.6. Antioxidant Capacity Assay

The antioxidant capacity was assessed using 1,1-Diphenyl-2-picryl-hydrazyl (DPPH) [21], 2,2-azino-bis-3-ethylbenzothiazoline-6-sulfonic acid (ABTS) [22], and ferric reducing antioxidant power (FRAP) [23]. For each measurement, 50 µL of the propolis extract was mixed with 150 µL of the appropriate solution (DPPH, ABTS, or FRAP). The absorbance was measured after 30 min of incubation. The calibration curves for the tests were established using a Trolox standard at 517 nm for the DPPH test, 732 nm for the ABTS test, and 593 nm for the FRAP test. The calibration equations were as follows: for DPPH, 30–70 µg/mL (y = −0.0108x + 0.8178; R^2 = 0.9990); for ABTS, 8–60 ug/mL (y = −0.0103x + 0.8091; R^2 = 0.9933); and for FRAP, 6–20 ug/mL (y = 0.0274x + 0.031; R^2 = 0.9940). The antioxidant capacity was expressed as mg Trolox equivalents (TEs) per milliliter of the ethanolic propolis extract. Each measurement was repeated three times.

2.7. HPLC–MS/MS

The phenolic profile, including both identification and quantification, was determined following the methodology previously described [20]. The CE and OE, both before and after digestion, were analyzed using an ultra-high-pressure liquid chromatograph (Ekspert Ultra LC 100-XL system, Eksigent Technologies, Dublin, CA, USA) coupled to a triple quadrupole mass spectrometer operating in positive electrospray ionization (ESI) mode (AB Sciex Triple Quad 4500, Framingham, MA, USA). Chromatographic separation was performed on a LiChrospher 100 RP-18 endcapped column (125 mm × 4 mm i.d., 5 µm particle size) (Merck, Darmstadt, Germany) maintained at 30 °C, using a mobile phase composed of 0.1% formic acid and methanol in gradient mode, with a flow rate of 0.5 mL min^{-1}. The gradient program was as follows: 0–1 min, 15% B; 1–17 min, 15–100% B; 17–21 min, 100% B; 21–22 min, 100–15% B; and 22–25 min, 15% B. The LC–MS/MS system was controlled using Analyst 1.6.2 software, and data processing was performed with Multiquant 3.0. For the quantitative analysis of phenolic compounds, commercial standards obtained from Sigma-Aldrich (Sigma-Aldrich, St. Louis, MO, USA) were employed. The analysis utilized a primary transition for quantification and a secondary transition for identification. Results were expressed as mean values ± standard deviation (SD) based on triplicate measurements.

2.8. Antibacterial Assay

The bacterial strains used in the antimicrobial assays were *Listeria monocytogenes* BM-01-02 1 (serotype 1/2a), BM-14 (serotype 1/2b), and BM-01-02 4 (serotype 4b) and *Salmonella enterica* strains FA0519 (ser. Typhimurium), FA0208 (ser. Infantis), and FA0032 (ser. Enteritidis). *Staphylococcus aureus* strain ATCC 25923 was also included in the assays.

Antibacterial activity was evaluated by measuring the inhibition zone diameter on Mueller–Hinton agar plates using the Clinical and Laboratory Standards Institute (CLSI) method [24]. Bacterial suspensions were standardized to a concentration of 0.5 McFarland standard (1.5 × 10^8 CFU/mL, Becton Dickinson Company, Franklin Lakes, NJ, USA). An aliquot from each suspension was used for mass seeding on Petri dishes, and wells with a 6 mm diameter were prepared. In each well, 100 µL of propolis extract was added with a concentration of 200 mg/mL. The plates were incubated for 24 h at 37 °C. The assays were conducted in triplicate.

For the minimum bactericidal concentration (MBC) assay, bacterial suspensions were prepared at a concentration of 0.5 McFarland standard. A volume of 150 µL of the extract was added to 150 µL of Mueller–Hinton broth in a 96-well microplate for 1:2 serial dilutions, then 50 µL of the bacterial suspension was added to each well. Microplates were incubated for 24 h at 37 °C. From each well, 4.0 µL was plated on Mueller–Hinton agar plates and incubated for 24 h at 37 °C. The MBC was considered to be the lowest concentration that inhibited growth. The controls for both tests were Streptomycin (1 mg/mL) and ethanol 80% (v/v).

2.9. In Vitro Digestion and Relative Bioaccessibility

To simulate the oral, gastric, and intestinal phases of digestion, we conducted assays based on the protocol previously described by INFOGEST [25]. During the oral phase simulation, 1.5 mL of propolis extracts was mixed with simulated digestion fluids and the corresponding enzyme for 2 min at 37 °C. The gastric and intestinal phase simulations used the sample from the abovementioned oral phase, incubating for 2 h at 37 °C with continuous mixing. Each phase included a control (simulated digestion fluids without sample). All measurements were carried out in triplicate.

To estimate bioaccessibility, the total bioactive compounds and functional properties were measured in samples before and after in vitro digestion. The relative bioaccessibility (RB) percentages were calculated using the following formula previously described [26]:

$$RB(\%) = \frac{CD}{CI} \times 100$$

where CD represents the results after digestion (bioaccessible fraction) and CI represents the results before digestion.

2.10. Statistical Analysis

Data analysis was performed using Statgraphics Plus® 5.1 software. Analysis of variance (ANOVA) was employed to assess the effect of treatments on dependent variables (i.e., extraction time on TPC). Significant differences between multiple means were determined with Tukey's test, while paired mean comparisons were carried out using Student's *t*-test.

3. Results and Discussion

3.1. Botanical Origin of Propolis

The analysis of propolis pollen grains obtained from the central zone of Chile indicates a multifloral origin (Table 2) with predominant species such as *Lithraea caustica* (28.2%), *Escallonia pulverulenta* (18.4%), and *Populus nigra* (13.6%). It is expected that propolis presents antioxidant and anti-inflammatory properties due to the high flavonoid and phenolic content of *L. caustica* [27]. Additionally, the *Populus* genus, to which *P. nigra* belongs, is one of the most documented botanical sources of propolis, particularly in temperate regions [28]. Their high content of phenolic acids and flavonoids enhances the antimicrobial efficacy of propolis [29,30]. Other important bioactive properties are expected due to other contributors, which include *Peumus boldus* (8.7%), *Cryptocarya alba* (7.8%), and *Eucalyptus* sp. (7.8%). *P. boldus* is widely recognized for its alkaloid and flavonoid composition, which could enhance the bioactivity of the propolis [31]. Similarly, *Eucalyptus* sp. is associated with antibacterial and expectorant properties, attributed to its high content of eucalyptol (1, 8-cineole) and terpenoids, which reinforces the potential use of this propolis in respiratory health formulations [32,33].

Table 2. Botanical origin of propolis from the central zone of Chile.

Botanical Attributes		Plant Source	Contribution (%)
Pollen		*Cryptocarya alba*	7.8
		Peumus boldus	8.7
		Escallonia pulverulenta	18.4
		Melilotus indicus	4.9
		Quillaja saponaria	2.9
		Populus nigra	13.6
		Schinus latifolius	7.8
		Eucalyptus sp.	7.8
		Lithraea caustica	28.2
Plant structures	Trichomes	*Lithraea caustica*	28.0
		Retanilla trinervia	12.0
	Vessels	*Populus nigra*	16.0
		Eucalyptus camaldulensis	18.0
		Nothofagus obliqua	20.0
	Epidermis	*Carduus picnocephalus*	6.0

Analysis of other plant structures indicated that the high contribution of *L. caustica* trichomes (28.0%) supports the hypothesis that this species is a key contributor to the chemical profile of the evaluated propolis. The contribution of *Eucalyptus camaldulensis* vessels (18.0%) is also particularly relevant, as it has been widely documented for its antiseptic, anti-inflammatory, and wound-healing properties due to its phenolic-rich extracts [34]. These findings suggest that the botanical diversity of this propolis sample could enhance its pharmacological versatility, making it a potential candidate for therapeutic and nutraceutical applications.

3.2. Optimized Parameters for UAE Extraction Method

Beekeepers widely use ethanol to extract propolis due to its food-grade quality and effectiveness in extracting polyphenols from the matrix [35,36]. Therefore, we determined the TPC of the extracts obtained using various ethanol concentrations (50, 60, 70, 80, and 90%), temperatures (20, 30, 40, 50, and 60 °C), and times (10, 20, 30, 40, and 50 min) (Figure 1). All assays were conducted at 80 W, with a propolis/solvent ratio of 1:10 (*w*/*v*) g/mL. The maximum TPC ($p < 0.05$) was achieved with an ethanol concentration of 80% (*v*/*v*) (Figure 1A). The TPC increased progressively from 50 to 80% (*v*/*v*) ethanol concentration, whereas it significantly decreased from 80 to 90%. This behavior is consistent with findings previously reported [37,38]. Interestingly, the highest TPC at an 80% ethanol concentration has also been determined in propolis extracted by ultrasound at 100 W and 40 °C for 5 h [39].

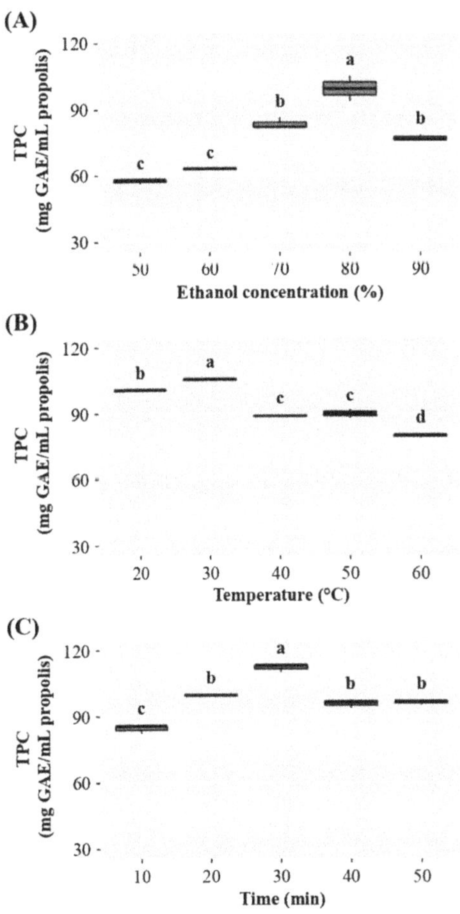

Figure 1. Extraction parameters impact on total polyphenol content (TPC) of ethanolic propolis extract. (**A**) Solvent concentration (% ethanol *v*/*v*), (**B**) Temperature (°C), and (**C**) Time (min). The effect of each extraction variable on TPC was assessed using ANOVA. Differences between means were determined with Tukey's test, and means sharing the same letters indicate no significant differences ($p < 0.05$).

The highest values of TPC were exhibited at an extraction temperature of 30 °C, whereas increasing temperatures (40, 50, and 60 °C) significantly decreased the TPC content (Figure 1B). By contrast, higher temperatures (59 °C) have been reported as optimal for Malaysian propolis [37]. In the case of propolis collected in Turkey, the optimal extraction temperature was 65 °C, using a 1:10 ratio, an ethanol concentration of 80%, and a 25-min ultrasound-assisted method [40].

An increase in TPC was observed with longer extraction times in the interval from 10 to 30 min, whereas the process conducted for 40 and 50 min yielded similar TPC as that conducted for 20 min (Figure 1C). The optimal extraction time of 30 min agrees with that reported in Italian propolis when a 1:10 (w/v) ratio of propolis/ethanol (70% v/v) was used [41].

The extraction of propolis using ethanol demonstrates optimal concentrations at 80% and an extraction time of 30 min, consistent with previous findings. However, the optimal temperature for this extraction was identified as 30 °C, which contrasts with reported parameters for propolis from other investigations that suggest higher temperatures.

3.3. *Phytochemical Screening*

According to previous reports, the chemical composition of propolis varies depending on the local geographical location and flora [2,42]. The central zone has a warm and subhumid Mediterranean-type climate, which allows the existence of native vegetation. Specifically, in the Mediterranean agroecological landscape, the predominant native vegetation that results is the renewal type of sclerophyll forest, characterized mainly by *Cryptocarya alba* (Chilean peumo), *Quillaja saponaria* Mol. (Quillay), and *Lithrea caustica* Mol. (Liters) species [43]. Consequently, evaluating a qualitative phytochemical screening provides an initial insight into the compounds present in Chilean propolis.

In the propolis obtained with both extraction methods, before and after in vitro digestion, we observed the presence of all the phytochemicals evaluated except for saponins (Table 3). The most abundant compounds corresponded to flavonoids and phenols, followed by coumarins and sugars. There were no differences in the intensity of color between the CE and OE before digestion, which exhibited an overall similar detected abundance of phytochemicals. Degradation of most determined compounds was observed after digestion, except for alkaloids and proteins, which exhibited higher stability during in vitro digestion (Table 3).

Table 3. Qualitative phytochemical screening of propolis samples before and after in vitro digestion.

Phytochemical	Negative Control	Positive Control	Before Digestion		After Digestion	
			Conventional Extract	Optimized Extract	Conventional Extract	Optimized Extract
Tannins	-	+	+	+	-	-
Flavonoids	-	+	+++	+++	++	++
Phenols	-	+	+++	+++	++	++
Coumarins	-	+	++	++	+	+
Alkaloids	-	+	+	+	+	+
Proteins	-	+	+	+	+	+
Sugars	-	+	++	++	+	+
Saponins	-	+	-	-	-	-

+++ highly abundant, ++ moderately present, + present, - absent.

Similar results were found in roasted coffee beans, where tannins were not detected after the intestinal digestion stage [44]. This behavior can be explained by the formation of tannin–protein precipitates through hydrogen bonds and hydrophobic interactions during the oral phase [45]. It has also been reported that no intact proteins were visually detected in the soluble or insoluble fractions once the intestinal phase was completed, either in isolated proteins or in their food matrix [46]. This phenomenon might be the result of the action of enzymes such as pepsin and pancreatin, which can convert proteins into simpler peptide structures [47]. In the case of alkaloids, these compounds have also shown high stability in their bioaccessibility in hydroethanolic extracts of herbs, exhibiting no significant concentration decrease after in vitro intestinal digestion [48]. These findings emphasize the significance of extraction methods and digestive stability in assessing the bioactive potential of propolis.

3.4. Total Polyphenols, Flavonoid Content, and HPLC–MS/MS

Over 500 different metabolites have been identified in propolis, with polyphenols being the predominant compounds [49]. The TPC of the OE was 24.4 ± 0.4 mg GAE/mL, which is comparable to other Chilean propolis with values greater than 20.0 mg GAE/mL [9]. The optimization process for obtaining the ethanolic propolis extract resulted in a significant increase in TPC (23.4%) and TFC (45.2%) (Table 4). Likewise, ultrasound-assisted extraction in Romanian propolis with 70% ethanol, a 1:50 (*w*/*v*) ratio, and 15 min increased TPC by 7.2%, as compared to maceration [38]. In Thai propolis, ultrasound extraction also increased the TFC by 6.0% at 25 °C with 70% ethanol and a 1:10 (*w*/*v*) ratio [50]. These results indicate that ultrasound extraction improves both the yield of bioactive compounds and the efficiency of the extraction process (i.e., lower extraction temperature and time). These characteristics contribute to the environmental sustainability of the overall process [51]. Ultrasound waves produce a cavitation effect that facilitates the breakdown of cell walls, thereby promoting the release of phenolics and flavonoids into solvent [52]. As a result, ultrasound-assisted propolis extraction proves to be a better method than conventional extraction techniques, such as maceration.

Table 4. Phenolic content and antioxidant capacity of ethanolic propolis extract obtained with conventional and optimized method, before and after simulated digestion.

Analysis		Before Digestion		After Digestion		Bioaccessibility (%)	
		CE	OE	CE	OE	CE	OE
Phenolic compounds	TPC (mg GAE/mL)	18.7 ± 0.1 ^{b}	24.4 ± 0.4 ^{a}	3.8 ± 0.1 ^{d}	4.9 ± 0.0 ^{c}	20.3	20.1
	TFC (mg QE/mL)	8.6 ± 0.4 ^{b}	15.7 ± 0.7 ^{a}	0.6 ± 0.0 ^{d}	1.2 ± 0.1 ^{c}	7.0	7.6
Antioxidant Capacity (mg TE/mL)	DPPH	60.1 ± 0.3 ^{b}	62.9 ± 0.3 ^{a}	34.0 ± 0.7 ^{d}	43.7 ± 0.4 ^{c}	56.6	69.5
	ABTS	21.6 ± 0.6 ^{b}	35.7 ± 1.0 ^{a}	13.5 ± 0.3 ^{c}	21.6 ± 1.4 ^{b}	62.5	60.5
	FRAP	36.2 ± 1.2 ^{b}	51.2 ± 0.1 ^{a}	23.4 ± 3.3 ^{c}	31.7 ± 0.5 ^{b}	64.6	61.9
HPLC–MS/MS (ppb)	Cinnamic acid	1900 ± 100 ^{b}	3500 ± 100 ^{a}	nd	500 ± 10 ^{c}	-	14.3
	Ferulic acid	4835 ± 400 ^{b}	8350 ± 600 ^{a}	695 ± 40 ^{c}	790 ± 60 ^{c}	14.4	9.5
	Chlorogenic acid	78.0 ± 3.5 ^{b}	168.5 ± 0.5 ^{a}	-	-	-	-
	Caffeic acid	73.0 ± 2.0 ^{b}	88.0± 1.5 ^{a}	-	-	-	-
	Coumaric acid	51.0 ± 0.5 ^{d}	77.5 ± 1.5 ^{c}	94.5 ± 1.5 ^{b}	123.0 ± 1.5 ^{a}	185.3	158.7
	Rutin	86.5 ± 2.5 ^{b}	212.0 ± 4.5 ^{a}	3.0 ± 0.5 ^{c}	4.5 ± 1.1 ^{c}	3.5	2.1

Results are shown as mean ± standard deviation (n = 3). The effect of the extraction method and digestion process on the extract properties was assessed using ANOVA. Differences between means were determined with Tukey's test, and means sharing the same letters within each row indicate no significant differences ($p < 0.05$). OE: optimal ethanolic UAE extract; CE: conventional ethanolic extract; TPC: total polyphenol content; TFC: total flavonoid content; DPPH: 1,1-Diphenyl-2-picryl-hydrazyl; ABTS: 2,2-azino-bis-3-ethylbenzothiazoline-6-sulfonic acid; FRAP: ferric reducing antioxidant power; -: not detected.

The TPC relative bioaccessibility (RB) was 20.3% for the CE and 20.1% for the OE, whereas the RB of the TFC in the CE and OE was 7.0% and 7.6%, respectively (Table 4). During in vitro digestion, TPC experiences molecular modifications and degradation by digestive enzymes under acidic and alkaline conditions [53]. A similar behavior can be observed when analyzing the concentrations of certain model compounds as determined by chromatography. For both polyphenolic acids (e.g., chlorogenic acid) and flavonoids (e.g., rutin), a decrease in the available amount after digestion can be observed, regardless of the extraction method. Research has suggested that the intestinal phase may break down phenolic compounds like flavan-3-ol through processes such as autooxidation, polymerization, or complex formation in alkaline conditions [54]. Furthermore, bonds can break during digestion, leaving less stable free phenols that can form complex phenolic derivatives that spectrophotometric methods are unable to detect [55].

In conclusion, although the optimized extraction process significantly increased TPC and TFC in propolis, the relative bioaccessibility during in vitro digestion remained similar across methods. This indicates that molecular modifications and degradation in phenolic compounds occur uniformly, affecting bioaccessibility regardless of the extraction technique.

3.5. Antioxidant Capacity

The antioxidant capacity was assessed in the CE and OE, before and after simulated digestion, using DPPH, ABTS, and FRAP. The antioxidant capacity was significantly lower in the CE than in the OE ($p < 0.05$) before digestion (Table 4). The UAE resulted in increases of 4.5%, 39.5%, and 29.3% in DPPH, ABTS, and FRAP values, respectively. A similar trend was observed in Chinese propolis, where optimal UAE conditions led to a 95.5% increase in DPPH values and a 65.5% increase in ABTS values compared to conventional extraction methods [56]. Additionally, Thai propolis also exhibited an 18.5% increase in DPPH values under UAE conditions [51].

After digestion, the RB of the antioxidant capacity in terms of DPPH, ABTS, and FRAP was 56.6%, 62.5%, and 64.6% for the CE and 69.5%, 60.5%, and 61.9% for the OE, respectively (Table 4). By contrast, studies on Turkish propolis reported lower bioaccessibility values for ABTS (10.9%) and FRAP (5.6%) [57]. Additionally, research on Mexican propolis indicated substantial reductions of 98.1% in DPPH and 77.8% in ABTS during in vitro digestion, alongside an 80% loss of total polyphenol content (TPC) and a 90% loss of total flavonoid content (TFC) [58]. Our findings align with those of Wanyo et al. [59], who reported reductions of 43.0% and 24.0% in DPPH and FRAP values, respectively, in crude extracts of Purple Rice Bran subjected to the intestinal phase. Collectively, in vitro digestion studies have revealed a relevant variability in antioxidant stability and bioaccessibility across different propolis sources and extraction methods.

In addition to the observed differences in antioxidant capacity and its retention following digestion, correlation analysis provided further insight into the relationship between bioactive compound content and functional performance (Figure 2). Statistically significant associations between TPC, TFC, and antioxidant capacity were consistently observed both before and after digestion. The analysis was performed using all replicates from both extraction methods (CE and OE), offering a broader perspective on compound–activity relationships. These findings indicate that, despite the partial degradation of phenolic and flavonoid compounds, their contribution to antioxidant activity is largely retained in the bioaccessible fraction. This underscores the relevance of UAE not only for enhancing the initial concentration of bioactives but also for preserving their functional efficacy under simulated gastrointestinal conditions.

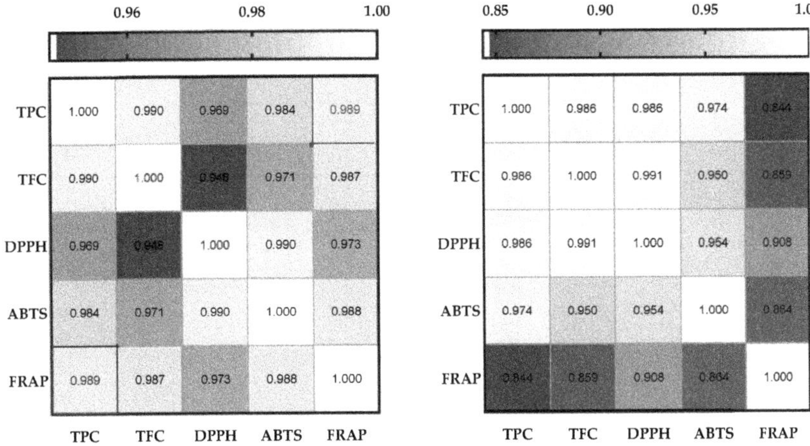

Figure 2. Pearson correlation heatmaps between total phenolic content (TPC), total flavonoid content (TFC), and antioxidant activity (DPPH, ABTS, FRAP) in ethanolic propolis extracts before (**left**) and after (**right**) simulated digestion (n = 6). All correlations shown are statistically significant ($p < 0.05$). The color scale represents the strength of the correlation coefficient (r).

3.6. Antibacterial Activity

Ethanolic extracts of propolis are promising sources of bioactive compounds with potential antibacterial properties, especially against Gram-positive bacteria [60]. The OE and CE showed, in general, higher antimicrobial activity against the *L. monocytogenes* strains as compared to the *S. enterica* strains (Table 5). These differences have also been reported in propolis extract obtained with an ethanol concentration of 80% (v/v) from Eastern Algeria, which exhibited greater inhibition of Gram-positive bacteria (*Staphylococcus aureus* and *Streptococcus agalactiae*) compared to Gram-negative bacteria (*Klebsiella pneumoniae*, *Escherichia coli*, and *Pseudomonas aeruginosa*) [61]. This variation can be explained by the fact that Gram-negative bacteria possess a complex cell wall that includes an outer membrane and other components [62]. Additionally, it has been observed that Gram-negative bacteria have efflux pumps that prevent the accumulation of propolis compounds, thereby contributing to their relatively higher tolerance [63].

The inhibition diameter results indicated no significant differences between the OE and CE for all evaluated bacteria (Table 5). However, the MBC of the OE for the *L. monocytogenes* strains and for *S. aureus* was 3.1 mg/mL, which was significantly lower than that found for the CE (6.3 mg/mL) (Table 5). Overall, the ultrasound extraction method for 30 min enhanced bioactivity against both *S. aureus* and *L. monocytogenes* compared to the CE, which could be related to the increase in phenolic acids and rutin in the OE.

Although the tested ethanol concentrations showed a low inhibition zone diameter, ethanol at 20% (v/v) was the MBC against all evaluated strains (Table 5). The ethanolic propolis extract concentration of 50 mg/mL corresponded to an ethanol concentration of 20% (v/v); therefore, the observed antimicrobial activity against Gram-negative bacteria could not be attributed uniquely to the effect of propolis. By contrast, the ethanol concentrations in the propolis dilutions controlling the Gram-positive bacteria were 1.3 and 2.5% (v/v) for the MBC of 3.1 and 6.3 mg/mL, respectively.

After the simulated digestion, both the OE and CE completely lost their antimicrobial activity against the tested bacteria, according to the inhibition zone and the MBC assays (Table 5). This loss may be attributed to a reduction in phytochemical diversity and concentration, as well as to the structural degradation or transformation of active compounds during gastrointestinal processing. Polyphenols, in particular, can be metabolized into derivatives with reduced or no bioactivity or may be rapidly degraded and eliminated. Furthermore, interindividual differences in gut microbiota composition can influence the metabolic fate of phenolic compounds, potentially altering their biological activity in vivo [64]. As this study was conducted using an in vitro digestion model, such variability could not be accounted for, which represents a limitation when extrapolating these results to physiological conditions.

Table 5. Antibacterial capacity of the ethanolic propolis extract quantified through inhibition zone diameter (mm) and minimum bactericidal concentration (MBC) (mg/mL) before and after simulated digestion.

Assay	Bacteria	Before Digestion		After Digestion		Streptomycin	Ethanol
		Conventional Extract	Optimized Extract	Conventional Extract	Optimized Extract		
Inhibition diameter (mm)	*S. enterica* ser. Infantis	7.7 ± 0.5	7.0 ± 0.0 ns	-	-	23.0 ± 0.5	7.0 ± 0.0
	S. enterica ser. Enteritidis	8.0 ± 0.8	9.0 ± 0.0 ns	-	-	24.0 ± 0.5	7.7 ± 0.5
	S. enterica ser. Typhimurium	8.3 ± 0.5	9.0 ± 0.0 ns	-	-	27.0 ± 0.9	7.0 ± 0.0
	L. monocytogenes ser. 1/2a	15.3 ± 0.5	15.3 ± 0.5 ns	-	-	27.0 ± 0.8	8.7 ± 0.5
	L. monocytogenes ser. 1/2 b	14.0 ± 0.0	16.5 ± 0.4 *	-	-	29.0 ± 0.0	8.7 ± 0.5
	L. monocytogenes ser. 4b	14.0 ± 1.6	16.0 ± 0.0 ns	-	-	28.0 ± 0.5	9.0 ± 0.0
	S. aureus	24.7 ± 1.9	23.0 ± 2.2 ns	-	-	28.0 ± 0.0	7.0 ± 0.0
MBC (mg/mL)	*S. enterica* ser. Infantis	50.0 ± 0.0	50.0 ± 0.0 ns	-	-	1.0 ± 0.0	20.0 ± 0.0 [a]
	S. enterica ser. Enteritidis	50.0 ± 0.0	50.0 ± 0.0 ns	-	-	0.3 ± 0.0	20.0 ± 0.0 [a]
	S. enterica ser. Typhimurium	50.0 ± 0.0	50.0 ± 0.0 ns	-	-	0.5 ± 0.0	20.0 ± 0.0 [a]
	L. monocytogenes ser. 1/2a	6.3 ± 0.0	3.1 ± 0.0 ***	-	-	0.03 ± 0.0	20.0 ± 0.0 [a]
	L. monocytogenes ser. 1/2 b	6.3 ± 0.0	3.1 ± 0.0 ***	-	-	0.02 ± 0.0	20.0 ± 0.0 [a]
	L. monocytogenes ser. 4b	6.3 ± 0.0	3.1 ± 0.0 ***	-	-	0.03 ± 0.0	20.0 ± 0.0 [a]
	S. aureus	6.3 ± 0.0	3.1 ± 0.0 ***	-	-	0.03 ± 0.0	20.0 ± 0.0 [a]

Assays results are shown as mean ± standard deviation ($n = 3$). The activity of conventional and optimized extracts, for each bacterium and parameter, was compared using Student's t-test (ns = not significant $p \geq 0.05$, * = $p < 0.05$, *** = $p < 0.001$). -: no activity. [a] MBC in ethanol% (v/v).

4. Conclusions

Under the established optimal conditions, ultrasound-assisted extraction was shown to be an efficient technique for obtaining ethanolic extracts from propolis. This method yielded extracts with enhanced bioactive compound concentrations and biological activity compared to conventional maceration-based extraction. Although the simulated digestion process resulted in a decrease in total polyphenol and flavonoid content and in antioxidant capacity, these properties remained higher in the extracts with the optimized ultrasound system. Overall, this study highlights the importance of developing optimized protocols for novel extraction techniques to generate propolis extracts with superior bioactive attributes. Future research may also benefit from integrating complementary analyses, such as solid content quantification, to strengthen comparisons across different extraction formats.

Author Contributions: Conceptualization, J.M., C.G.-V., C.J., G.M. and A.G.; Data curation, J.M., C.G.-V. and C.J.; Formal analysis, J.M., C.G.-V. and C.J.; Investigation, J.M. and C.G.-V.; Methodology, C.G.-V.; Resources, C.G.-V., C.J., G.M., A.I.M.-S. and A.G.; Supervision, C.G.-V., C.J., G.M. and A.G.;

Writing—original draft, J.M.; Writing—review and editing, C.G.-V., C.J. and A.G. All authors have read and agreed to the published version of the manuscript.

Funding: The APC was funded by the FIC-Coquimbo 2021 Project BIP Code 40041020-0 and ANID-Fondecyt Postdoctorado Project No. 3220661.

Institutional Review Board Statement: Not applicable.

Informed Consent Statement: Not applicable.

Data Availability Statement: Data will be made available on request.

Acknowledgments: The authors thank the FIC-Coquimbo 2021 Project BIP Code 40041020-0, ANID-Fondecyt Postdoctorado Project No. 3220661, ANID-Fondecyt Regular Project No. 1231835, and Karol Rojas of the Universidad de La Serena for their technical support.

Conflicts of Interest: The authors declare no conflicts of interest.

References

1. Rebaza, R.; Amaya, L.; Gutiérrez, A.; Haro, R.; Tumbajulca, M.; Valera, F.; Vargas, Y.; Barraza-Jauregui, G.; León Vargas, J.M.; Sánchez-González, J.A. Aplicación del propóleo en envasado activo. *Agroindustrial Sci.* **2016**, *6*, 239–252. [CrossRef]
2. Zulhendri, F.; Chandrasekaran, K.; Kowacz, M.; Ravalia, M.; Kripal, K.; Fearnley, J.; Perera, C.O. Antiviral, antibacterial, antifungal, and antiparasitic properties of propolis: A review. *Foods* **2021**, *10*, 1360. [CrossRef]
3. Al-Waili, N.; Al-Ghamdi, A.; Ansari, M.J.; Al-Attal, Y.; Salom, K. Synergistic effects of honey and propolis toward drug multi-resistant *Staphylococcus aureus, Escherichia coli* and *Candida albicans* isolates in single and polymicrobial cultures. *Int. J. Med. Sci.* **2012**, *9*, 793–800. [CrossRef] [PubMed]
4. Silva-Carvalho, R.; Baltazar, F.; Almeida-Aguiar, C. Propolis: A complex natural product with a plethora of biological activities that can be explored for drug development. *Evid.-Based Complement. Altern. Med.* **2015**, *2015*, 206439. [CrossRef]
5. Sethi, S.; Rathod, V.K. Recent advancements in ultrasound-assisted biomolecule extraction from prokaryotic and eukaryotic cells: A review. *Prep. Biochem. Biotechnol.* 2024; 1–27, *online ahead of print*. [CrossRef]
6. Contieri, L.S.; Mesquita, L.M.d.S.; Sanches, V.L.; Chaves, J.; Pizani, R.S.; da Silva, L.C.; Viganó, J.; Ventura, S.P.; Rostagno, M.A. Recent progress on the recovery of bioactive compounds obtained from propolis as a natural resource: Processes, and applications. *Sep. Purif. Technol.* **2022**, *298*, 121640. [CrossRef]
7. Reis, J.H.d.O.; Barreto, G.d.A.; Cerqueira, J.C.; dos Anjos, J.P.; Andrade, L.N.; Padilha, F.F.; Druzian, J.I.; Machado, B.A.S. Evaluation of the antioxidant profile and cytotoxic activity of red propolis extracts from different regions of northeastern Brazil obtained by conventional and ultrasound-assisted extraction. *PLoS ONE* **2019**, *14*, e0219063. [CrossRef]
8. Barrientos, L.; Herrera, C.L.; Montenegro, G.; Ortega, X.; Veloz, J.; Alvear, M.; Cuevas, A.; Saavedra, N.; Salazar, L.A. Chemical and botanical characterization of Chilean propolis and biological activity on cariogenic bacteria Streptococcus mutans and Streptococcus sobrinus. *Braz. J. Microbiol.* **2013**, *44*, 577–585. [CrossRef]
9. Alvear, M.; Santos, E.; Cabezas, F.; Pérez-Sanmartín, A.; Lespinasse, M.; Veloz, J. Geographic area of collection determines the chemical composition and antimicrobial potential of three extracts of chilean propolis. *Plants* **2021**, *10*, 1543. [CrossRef] [PubMed]
10. Veloz, J.; Cabezas, F.; Fernández, C.; Lespinasse, M. The structural diversity of flavonoid present in chilean propolis determines their biological activity. *Ann. Chem. Sci. Res.* **2023**, *4*, 1–2. [CrossRef]
11. Wagh, V.D. Propolis: A wonder bees product and its pharmacological potentials. *Adv. Pharmacol. Sci.* **2013**, *2013*, 308249. [CrossRef]
12. Herrera, C.L.; Alvear, M.; Barrientos, L.; Montenegro, G.; Salazar, L.A. The antifungal effect of six commercial extracts of Chilean propolis on *Candida* spp. *Cienc. Investig. Agrar.* **2010**, *37*, 75–84. [CrossRef]
13. Zullkiflee, N.; Taha, H.; Usman, A. Propolis: Its role and efficacy in human health and diseases. *Molecules* **2022**, *27*, 6120. [CrossRef]
14. El-Sakhawy, M.; Salama, A.; Mohamed, S.A.A. Propolis applications in food industries and packaging. *Biomass Convers. Biorefinery* **2024**, *14*, 13731–13746. [CrossRef]
15. Flores, F.P.; Kong, F. In vitro release kinetics of microencapsulated materials and the effect of the food matrix. *Annu. Rev. Food Sci. Technol.* **2017**, *8*, 237–259. [CrossRef] [PubMed]
16. Wojtunik-Kulesza, K.; Oniszczuk, A.; Oniszczuk, T.; Combrzyński, M.; Nowakowska, D.; Matwijczuk, A. Influence of in vitro digestion on composition, bioaccessibility and antioxidant activity of food polyphenols—A non-systematic review. *Nutrients* **2020**, *12*, 1401. [CrossRef]
17. Association of Official Analytical Chemists—AOAC. *Official Method of Analysis*, 15th ed.; Association of Official Analytical Chemists: Washington, DC, USA, 1990.

18. Shaikh, J.R.; Patil, M. Qualitative tests for preliminary phytochemical screening: An overview. *Int. J. Chem. Stud.* **2020**, *8*, 603–608. [CrossRef]
19. Singleton, V.L.; Orthofer, R.; Lamuela-Raventós, R.M. Analysis of total phenols and other oxidation substrates and antioxidants by means of Folin-Ciocalteu reagent. *Methods Enzymol.* **1999**, *299*, 152–178. [CrossRef]
20. Fuentes, Y.; Giovagnoli-Vicuña, C.; Faúndez, M.; Giordano, A. Microencapsulation of chilean papaya waste extract and its impact on physicochemical and bioactive properties. *Antioxidants* **2023**, *12*, 1900. [CrossRef]
21. Brand-Williams, W.; Cuvelier, M.E.; Berset, C. Use of a free radical method to evaluate antioxidant activity. *LWT Food Sci. Technol.* **1995**, *28*, 25–30. [CrossRef]
22. Van den Berg, R.; Haenen, G.R.M.M.; van den Berg, H.; Bast, A. Applicability of an improved Trolox equivalent antioxidant capacity (TEAC) assay for evaluation of antioxidant capacity measurements of mixtures. *Food Chem.* **1999**, *66*, 511–517. [CrossRef]
23. Benzie, I.F.F.; Strain, J.J. The ferric reducing ability of plasma (FRAP) as a measure of "antioxidant power": The FRAP assay. *Anal. Biochem.* **1996**, *239*, 70–76. [CrossRef] [PubMed]
24. *M02-A11*; Performance Standards for Antimicrobial Disk Susceptibility Tests. Approved Standard—Eleventh Edition. Clinical and Laboratory Standards Institute: Wayne, PA, USA, 2012. Available online: www.clsi.org (accessed on 1 October 2024).
25. Brodkorb, A.; Egger, L.; Alminger, M.; Alvito, P.; Assunção, R.; Ballance, S.; Bohn, T.; Bourlieu-Lacanal, C.; Boutrou, R.; Carrière, F.; et al. INFOGEST static in vitro simulation of gastrointestinal food digestion. *Nat. Protoc.* **2019**, *14*, 991–1014. [CrossRef]
26. Giovagnoli-Vicuña, C.; Briones-Labarca, V.; Romero, M.S.; Giordano, A.; Pizarro, S. Effect of extraction methods and in vitro bio-accessibility of microencapsulated lemon extract. *Molecules* **2022**, *27*, 4166. [CrossRef] [PubMed]
27. Muñoz-Ramírez, A.; Mascayano-Collado, C.; Barriga, A.; Echeverría, J.; Urzúa, A. Inhibition of soybean 15-lipoxygenase and human 5-lipoxygenase by extracts of leaves, stem bark, phenols and catechols isolated from *Lithraea caustica* (Anacardiaceae). *Front. Pharmacol.* **2020**, *11*, 594257. [CrossRef]
28. Dezmirean, D.S.; Paşca, C.; Moise, A.R.; Bobiş, O. Plant sources responsible for the chemical composition and main bioactive properties of poplar-type propolis. *Plants* **2020**, *10*, 22. [CrossRef]
29. Hossain, R.; Quispe, C.; Khan, R.A.; Saikat, A.S.M.; Ray, P.; Ongalbek, D.; Yeskaliyeva, B.; Jain, D.; Smeriglio, A.; Trombetta, D.; et al. Propolis: An update on its chemistry and pharmacological applications. *Chin. Med.* **2022**, *17*, 100. [CrossRef] [PubMed]
30. Avula, B.; Sagi, S.; Masoodi, M.H.; Bae, J.-Y.; Wali, A.F.; Khan, I.A. Quantification and characterization of phenolic compounds from northern indian propolis extracts and dietary supplements. *J. AOAC Int.* **2020**, *103*, 1378–1393. [CrossRef]
31. Torres-Vega, J.; Gómez-Alonso, S.; Pastene-Navarrete, E.; Pérez-Navarro, J. Green extraction of alkaloids and polyphenols from *Peumus boldus* leaves with natural deep eutectic solvents and profiling by HPLC-PDA-IT-MS/MS and HPLC-QTOF-MS/MS. *Plants* **2020**, *9*, 242. [CrossRef]
32. Shao, J.; Yin, Z.; Wang, Y.; Yang, Y.; Tang, Q.; Zhang, M.; Jiao, J.; Liu, C.; Yang, M.; Zhen, L.; et al. Effects of different doses of *Eucalyptus* oil from *Eucalyptus globulus* labill on respiratory tract immunity and immune function in healthy rats. *Front. Pharmacol.* **2020**, *11*, 1287. [CrossRef]
33. El Shiekh, R.A.; Atwa, A.M.; Elgindy, A.M.; Mustafa, A.M.; Senna, M.M.; Alkabbani, M.A.; Ibrahim, K.M. Therapeutic Applications of Eucalyptus Essential Oils. *Inflammopharmacology* **2024**, *33*, 163–182. [CrossRef]
34. Singh, S.; Syukri, D.M.; Ushir, Y.V.; Mishra, A.; Ontong, J.C.; Nwabor, O.F.; Darekar, S.M.; Samee, W.; Chidrawar, V.R.; Chittasupho, C. Post-operative wound healing efficacy of eucalyptus camaldulensis phenolic-rich extracts incorporated hydrogel with enhanced antioxidant, antibacterial, and anti-inflammatory activities. *J. Polym. Environ.* **2024**, *33*, 814–839. [CrossRef]
35. Liu, Y.; Liu, H.-Y.; Xia, Y.; Guo, H.; He, X.-Q.; Li, H.; Wu, D.-T.; Geng, F.; Lin, F.-J.; Li, H.-B.; et al. Screening and process optimization of ultrasound-assisted extraction of main antioxidants from sweet tea (*Lithocarpus litseifolius* [Hance] Chun). *Food Biosci.* **2021**, *43*, 101277. [CrossRef]
36. Canale, A.; Cosci, F.; Canovai, R.; Giannotti, P.; Benelli, G. Foreign matter contaminating ethanolic extract of propolis: A filth-test survey comparing products from small beekeeping farms and industrial producers. *Food Addit. Contam. Part A* **2014**, *31*, 2022–2025. [CrossRef] [PubMed]
37. Yusof, N.; Munaim, M.S.A.; Kutty, R.V. The effects of different ethanol concentration on total phenolic and total flavonoid content in Malaysian propolis. *IOP Conf. Ser. Mater. Sci. Eng.* **2020**, *991*, 012033. [CrossRef]
38. Oroian, M.; Ursachi, F.; Dranca, F. Influence of ultrasonic amplitude, temperature, time and solvent concentration on bioactive compounds extraction from propolis. *Ultrason. Sonochem.* **2020**, *64*, 105021. [CrossRef]
39. Sun, C.; Wu, Z.; Wang, Z.; Zhang, H. Effect of ethanol/water solvents on phenolic profiles and antioxidant properties of Beijing propolis extracts. *Evid.-Based Complement. Altern. Med.* **2015**, *2015*, 595393. [CrossRef]
40. Bakkaloglu, Z.; Arici, M.; Karasu, S. Optimization of ultrasound-assisted extraction of turkish propolis and characterization of phenolic profile, antioxidant and antimicrobial activity. *Food Sci. Technol.* **2021**, *41*, 687–695. [CrossRef]

41. Trusheva, B.; Trunkova, D.; Bankova, V. Different extraction methods of biologically active components from propolis: A preliminary study. *Chem. Cent. J.* **2007**, *1*, 13. [CrossRef]
42. Chavda, V.P.; Vuppu, S.; Balar, P.C.; Mishra, T.; Bezbaruah, R.; Teli, D.; Sharma, N.; Alom, S. Propolis in the management of cardiovascular disease. *Int. J. Biol. Macromol.* **2024**, *266*, 131219. [CrossRef]
43. Salazar-Rojas, A.; Castro-Huerta, R.; Altieri, M. The main agroecological structure, a methodology for the collective analysis of the Mediterranean agroecological landscape of San Clemente, Region del Maule, Chile. *Front. Sustain. Food Syst.* **2023**, *7*, 1241648. [CrossRef]
44. Wu, H.; Liu, Z.; Lu, P.; Barrow, C.; Dunshea, F.R.; Suleria, H.A.R. Bioaccessibility and bioactivities of phenolic compounds from roasted coffee beans during in vitro digestion and colonic fermentation. *Food Chem.* **2022**, *386*, 132794. [CrossRef] [PubMed]
45. Laib, I.; Laib, D.E.; Semouma, D.; Cheriet, N.; Aouzal, B.; Barkat, M.; Zaidi, N.; Benredouane, Z.M.; Bennaga, I.; Bendjama, K.; et al. Optimization of extraction and study of the in vitro static simulation of INFOGEST gastrointestinal digestion and in vitro colonic fermentation on the phenolic compounds of dandelion and their antioxidant activities. *J. Food Meas. Charact.* **2023**, *17*, 5660–5682. [CrossRef]
46. Sousa, R.; Portmann, R.; Dubois, S.; Recio, I.; Egger, L. Protein digestion of different protein sources using the INFOGEST static digestion model. *Food Res. Int.* **2020**, *130*, 108996. [CrossRef]
47. Kautzmann, C.; Castanha, E.; Aloísio Johann Dammann, C.; Andersen Pereira de Jesus, B.; Felippe da Silva, G.; de Lourdes Borba Magalhães, M.; Turnes Pasini Deolindo, C.; Pinto Kempka, A. Roasted yerba mate (Ilex paraguariensis) infusions in bovine milk model before and after in vitro digestion: Bioaccessibility of phenolic compounds, antioxidant activity, protein–polyphenol interactions and bioactive peptides. *Food Res. Int.* **2024**, *183*, 114206. [CrossRef]
48. Thumann, T.A.; Pferschy-Wenzig, E.-M.; Aziz-Kalbhenn, H.; Ammar, R.M.; Rabini, S.; Moissl-Eichinger, C.; Bauer, R. Application of an in vitro digestion model to study the metabolic profile changes of an herbal extract combination by UHPLC–HRMS. *Phytomedicine* **2020**, *71*, 153221. [CrossRef] [PubMed]
49. Šturm, L.; Ulrih, N.P. Advances in the propolis chemical composition between 2013 and 2018: A review. *eFood* **2020**, *1*, 24–37. [CrossRef]
50. Khacha-Ananda, S.; Tragoolpua, K.; Chantawannakul, P.; Tragoolpua, Y. Antioxidant and anti-cancer cell proliferation activity of propolis extracts from two extraction methods. *Asian Pac. J. Cancer Prev.* **2013**, *14*, 6991–6995. [CrossRef]
51. Yusof, N.; Munaim, M.S.A.; Kutty, R.V. Optimization of total phenolic compounds extracted from propolis by ultrasound- assisted extraction. *Chem. Eng. Commun.* **2020**, *208*, 564–572. [CrossRef]
52. Rodríguez, Ó.; Bona, S.; Stäbler, A.; Rodríguez-Turienzo, L. Ultrasound-assisted extraction of polyphenols from olive pomace: Scale up from laboratory to pilot scenario. *Processes* **2022**, *10*, 2481. [CrossRef]
53. Pinto, D.; López-Yerena, A.; Lamuela-Raventós, R.; Vallverdú-Queralt, A.; Delerue-Matos, C.; Rodrigues, F. Predicting the effects of in-vitro digestion in the bioactivity and bioaccessibility of antioxidant compounds extracted from chestnut shells by supercritical fluid extraction—A metabolomic approach. *Food Chem.* **2024**, *435*, 137581. [CrossRef]
54. Jakobek, L.; Ištuk, J.; Barron, A.R.; Matić, P. Bioactive phenolic compounds from apples during simulated in vitro gastrointestinal digestion: Kinetics of their release. *Appl. Sci.* **2023**, *13*, 8434. [CrossRef]
55. Cavia, M.M.; Arlanzón, N.; Busto, N.; Carrillo, C.; Alonso-Torre, S.R. The impact of in vitro digestion on the polyphenol content and antioxidant activity of spanish ciders. *Foods* **2023**, *12*, 1861. [CrossRef] [PubMed]
56. Peng, S.; Zhu, M.; Li, S.; Ma, X.; Hu, F. Ultrasound-assisted extraction of polyphenols from Chinese propolis. *Front. Sustain. Food Syst.* **2023**, *7*, 1131959. [CrossRef]
57. Mergen Duymaz, G.; Duz, G.; Ozkan, K.; Karadag, A.; Yilmaz, O.; Karakus, A.; Cengiz, O.; Akyildiz, I.E.; Basdogan, G.; Damarlı, E.; et al. The evaluation of L-arginine solution as a solvent for propolis extraction: The phenolic profile, antioxidant, antibacterial activity, and in vitro bioaccessibility. *Food Sci. Nutr.* **2024**, *12*, 2724–2735. [CrossRef]
58. González-Montiel, L.; Figueira, A.C.; Medina-Pérez, G.; Fernández-Luqueño, F.; Aguirre-Álvarez, G.; Pérez-Soto, E.; Pérez-Ríos, S.; Campos-Montiel, R.G. Bioactive compounds, antioxidant and antimicrobial activity of propolis extracts during in vitro digestion. *Appl. Sci.* **2022**, *12*, 7892. [CrossRef]
59. Wanyo, P.; Chamsai, T.; Toontom, N.; Nghiep, L.K.; Tudpor, K. Differential effects of in vitro simulated digestion on antioxidant activity and bioaccessibility of phenolic compounds in purple rice bran extracts. *Molecules* **2024**, *29*, 2994. [CrossRef]
60. Bouchelaghem, S. Propolis characterization and antimicrobial activities against *Staphylococcus aureus* and *Candida albicans*: A review. *Saudi J. Biol. Sci.* **2022**, *29*, 1936–1946. [CrossRef]
61. Bouzahouane, H.; Ayari, A.; Guehria, I.; Riah, O. Propolis: Antimicrobial Activity and Chemical Composition Analysis. *J. Microbiol. Biotechnol. Food Sci.* **2021**, *10*, e3211. [CrossRef]
62. Ramata-Stunda, A.; Petriņa, Z.; Valkovska, V.; Borodušķis, M.; Gibnere, L.; Gurkovska, E.; Nikolajeva, V. Synergistic Effect of Polyphenol-Rich Complex of Plant and Green Propolis Extracts with Antibiotics against Respiratory Infections Causing Bacteria. *Antibiotics* **2022**, *11*, 160. [CrossRef]

63. Petruzzi, L.; Corbo, M.R.; Campaniello, D.; Speranza, B.; Sinigaglia, M.; Bevilacqua, A. Antifungal and antibacterial effects of propolis: A comparative study on food-borne *Pseudomonas*, *Enterobacteriaceae*, and fungi. *Foods* **2020**, *9*, 559. [CrossRef]
64. Makarewicz, M.; Drożdż, I.; Tarko, T.; Duda-Chodak, A. The interactions between polyphenols and microorganisms, especially gut microbiota. *Antioxidants* **2021**, *10*, 188. [CrossRef] [PubMed]

antioxidants

MDPI

Article

Determination of the Botanical Origin and Physicochemical Properties of a Propolis Sample Through an Integrated Methodology

Jose Juan Alcivar-Saldaña [1,2], Marco Aurelio Rodriguez-Monroy [3], Arturo Aguirre-Gómez [4], Liborio Carrillo-Miranda [5], Benjamin Velasco-Bejarano [6] and Maria Margarita Canales-Martinez [1,*]

1 Laboratorio de Farmacognosia, Unidad de Biotecnología y Prototipos (UBIPRO), Facultad de Estudios Superiores Iztacala, Universidad Nacional Autónoma de México (UNAM), Avenida de los Barrios Número 1, Colonia Los Reyes Iztacala, Tlalnepantla de Baz 54090, CP, Mexico; bio.alcivar@gmail.com

2 Posgrado en Ciencias Biológicas, Universidad Nacional Autónoma de México, Unidad de Posgrado, Edificio D, 1° piso, Circuito de Posgrado, Ciudad Universitaria, Coyoacán 04510, CP, Mexico

3 Laboratorio de Investigación Biomédica en Productos Naturales, Carrera de Medicina, Facultad de Estudios Superiores Iztacala, Universidad Nacional Autónoma de México (UNAM), Avenida de los Barrios Número 1, Colonia Los Reyes Iztacala, Tlalnepantla de Baz 54090, CP, Mexico; dr.marcorodriguezmonroy@gmail.com

4 Laboratorio de Análisis e Investigación en Química Agrícola y Ambiental, L-15 UIM, Departamento de Química Campus 4 Facultad de Estudios Superiore Cuautitlán, Universidad Nacional Autónoma de México, Carretera Cuautitlán-Teoloyucan km. 2.5, Col. San Sebastián Xhala, Cuautitlán Izcalli 54714, CP, Mexico; aag@unam.mx

5 Módulo de Apicultura, Centro de Enseñanza Agropecuaria (CEA), Facultad de Estudios Superiores Cuautitlan, Universidad Nacional Autónoma de México, Carretera Cuautitlán-Teoloyucan km. 2.5, Col. San Sebastián Xhala, Cuautitlán Izcalli 54714, CP, Mexico; liboriocami@gmail.com

6 Laboratorio de Química Verde Medicinal, Departamento de Ciencias Químicas, Facultad de Estudios Superiores Cuautitlan, Universidad Nacional Autónoma de México, Av. Primero de Mayo s/n, Sta. María Guadalupe las Torres, Cuautitlán Izcalli 54740, CP, Mexico; qfbbenjamin.velascob@cuautitlan.unam.mx

* Correspondence: dra.margaritacanales@gmail.com; Tel.: +52-55-27-69-21-73

Abstract: The growing interest in products of natural origin has led to the implementation of products such as propolis because they possess biological properties that are useful in the treatment of various ailments, so it is relevant to know the botanical origin of the physicochemical compounds that provide propolis with its biological properties. To identify the floral sources that provide resources to bees for the manufacture of propolis, several methodologies have been implemented, such as palynological analysis, which, through pollen content, has made it possible to identify the plant species that provide resources to bees. On the other hand, analysis of the physicochemical components of propolis has revealed that phenols and flavonoids are mainly responsible for the biological activity of propolis. Evaluation of the antibacterial and antifungal potential of propolis revealed the inhibitory potential of both Gram (+) and Gram (−) bacteria, as well as *Candida albicans*. However, all these investigations have been carried out individually and not always with the same sample. Therefore, the objective of this research was to design a methodology that allows the use of a single sample of propolis and uses sufficient resources for different research areas to evaluate most of the physical and chemical properties of propolis, as well as its botanical origin. With the implemented methodology, it was possible to obtain sufficient resources that provided results for each of the research areas, taking advantage of the propolis sample.

Keywords: *Apis mellifera*; plant sources; pollen; phenols; flavonoids

1. Introduction

The growing interest in products of natural origin for food, therapeutic or medicinal purposes has led to the use of those generated by honeybees and other native bee species (*Melipona quadrifasciata, Melipona compressipes, Tetragonisca angustula* and *Nannotrigona* sp.) [1–5]. Among these products, propolis stands out; it is a dark-colored resin manufactured by bees from resins and balsams produced by various plant species and processed by *Apis mellifera* L., which functions as a thermal insulator and antiseptic for this species [3,6]. After the process of collection, transport and storage, in the hive, the bees add enzymes and glandular secretions from the hypopharynx, as well as waxes from the wax glands present in the sternites of the abdomen, forming a final product that differs in its chemical composition from one hive to another and from one species to another [3,4].

Propolis and geopropoleum are recognized for their antifungal, antibacterial, anticancer, antiprotozoal, anesthetic, tissue regeneration, antiviral and natural antioxidant properties [3,4]. However, both the physical and chemical properties of propolis are closely related to the vegetation that provides honeybee resources, as well as the climatic conditions in which this product is manufactured [2,7]. This vegetation provides phytochemical compounds such as phenolic compounds, esters, phenolic acids and flavonoids such as apigenin, pinocembrin, chrysin, hesperidin, quercetin, kaempferol and galangin, which, together with the enzymatic contributions of honeybees added during the manufacturing process, enrich propolis [7,8].

Several studies have aimed to identify the vegetation and source of resources for the manufacture of this product, such as observing the foraging activity of honeybees and performing palynological analysis and genetic determination of the pollen content of propolis [2,9–12]. Palynological analysis is the most widely used method suitable for the melizopalynological analysis of honey, which has a relatively liquid consistency and is easy to dissolve in water or other solvents such as ethanol, facilitating the extraction of pollen, but in the case of propolis, which has a gummy and resinous consistency, the implementation of this method is somewhat complicated for the extraction of pollen. Therefore, various methodologies have been designed with the aim of extracting the greatest amount of pollen from propolis [13].

One of the methodologies implemented for pollen extraction is sonication, which, by means of ultravibrations, disintegrates the components of propolis [14]. Another regularly implemented method is total cleaning by Soxhlet, in which the propolis sample is leached with ether and then with 96° ethanol [14]. At the end of both techniques, the residues are treated via the acetolysis technique [15], which eliminates any organic residue from the sample, leaving only pollen structures, which are resistant because they are composed of sporopolenin, a protein that resists acid treatment [16]. However, these methods have several drawbacks, such as damage to the pollen structure caused by sonication, making the taxonomic identification of pollen difficult. In turn, in studies that implement the acetolysis technique [15], which is correct for performing more accurate determinations of pollen types, its drawback is the acidic nature of the process, which eliminates other biological structures of interest when it comes to gaining knowledge about the constitution and quality of propolis, such as spores, fungi, bacteria and other structures that provide information of interest [17]. For the Soxleth method, the drawback lies in the amount of sample to be analyzed, since with small samples (1 g), the pollen content obtained is low, with a minimum representation of the total diversity of pollen types present in the sample [14].

Therefore, it is necessary to implement extraction methods that focus on the physicochemical characteristics of propolis, which manage to disintegrate the conglomerate of compounds of which propolis is constituted and allow not only the extraction of pollen but also the acquisition of useful resources in the analysis of other areas of study. Some methods that could be of interest have been designed and implemented in biochemistry to extract compounds from samples of different herbaceous plants, roots or bark, in which the polarity of the phytochemical components is considered to implement solvents that extract these compounds [18]. Therefore, considering the constitution of propolis is composed mainly of phenols, flavonoids and waxes, several investigations have implemented the use of solvents such as water, ethanol, methanol or hexane to obtain a soft extract of propolis, which is implemented in analyses such as phytochemical characterization by high-performance liquid chromatography (HPLC), in antibacterial and antifungal evaluation or as an active compound in the elaboration of products for therapeutic purposes [4,9,17,18]. However, although each area provides results of interest for gaining knowledge of the constitution and activity of propolis, there are few investigations that work on a propolis sample in an integral way. Therefore, the present research proposes a methodology based on the physicochemical characteristics of propolis, which allows for the analysis of resources for phytochemical evaluation and characterization, antibacterial and antifungal evaluation, and sufficient material for palynological analysis, which allows for an integral analysis of the relevant sample, resulting in the physicochemical characterization and botanical origin of the studied propolis.

2. Materials and Methods

2.1. Study Area

The propolis sample was provided by the Universidad Nacional Autónoma de México (UNAM) at the Facultad de Estudios Superiores Cuautitlan (FES-C), located in the Estado of Mexico, municipality of Cuautitlan Izcalli, Mexico (19°41′; north, 99°11′ west, 2260 m a.s.l.), within the highland beekeeping region [19].

2.2. Sample Collection Method

The sample was obtained by means of a propolis collector (PC) (Brand propolistrap, MI, USA) (Figure 1); such a collector is a high-density polystyrene grid of black color that is placed over the brood chamber or supers according to the case, which, after a certain time and at the beekeeper's consideration or the filling of the PC, is removed. Once the PC is out of the apiary, it is frozen at −20 °C; for its extraction, the propolis frozen in the PC is removed via cryofracture, resulting in what is called propolis in shreds (PS) (Figure 1). Due to the implementation of the PC, the collected propolis presented less than 10% impurities; thus, 100 g of this material was taken for the extraction process according to the proposed work scheme (Figure 2) [20].

Figure 1. Propolis collection method at FES-C. (**A**) PC; (**B**) location of PC; (**C**) removal of PC; (**D**) sample obtained without PC with too many impurities; (**E**) sample obtained with PC with few impurities.

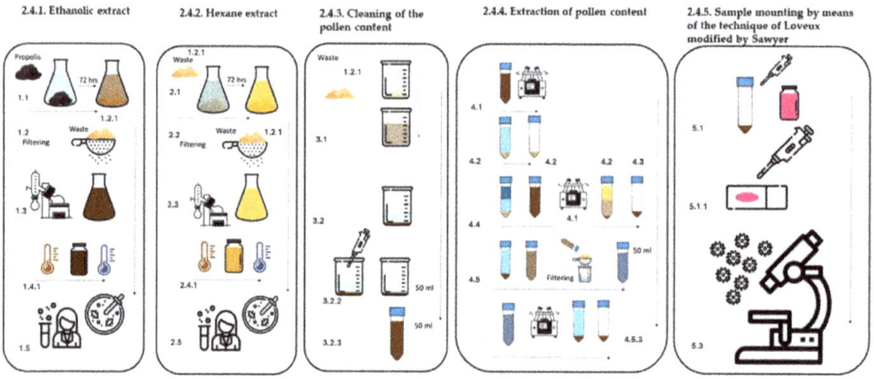

Figure 2. Methodology for obtaining EEP and EHP, as well as pollen content.

2.3. Pollen Extraction via the Soxhlet Method

The propolis sample was processed via Soxhlet extraction, which removes both resins and waxes from a gram of propolis by leaching it with 96% ethanol to remove resins and petroleum ether to remove waxes and fats [21,22].

2.4. Extraction Methods with Solvents of Different Polarities

2.4.1. Ethanolic Extract of Propolis (EEP), (Figure 2) [18]

1.1 One hundred grams of PS was placed in a flask with 70% ethanol at a 1:5 ratio and covered at room temperature in the dark for 72 h.

1.2. After 72 h, the contents of the flask were filtered through Whatman filter paper to obtain an ethanolic extract.

  1.2.1. The residue obtained after the filtration process was preserved.

1.3. The ethanolic extract obtained was distilled via a rotary evaporator at 40 °C to ensure that the extract was dry.

1.4. The extract obtained from the distillation mixture was brought to total dryness at room temperature, after which it was spread in glass containers and named ethanolic extract of propolis (EEP).

  1.4.1. Once the totality of the solvent was evaporated, it was stored in closed glass jars and kept frozen at −20 °C.

1.5. The extract obtained was used in individual portions for physical and chemical analysis, according to the quantities indicated in each technique.

2.4.2. Hexane Extract of Propolis (EHP) (Figure 2) [18]

2.1. The residues obtained from the filtration process in Step 1.2.1 were placed in a flask with hexane at a 5:1 ratio and covered at room temperature in the dark for 72 h.

2.2. After 72 h and inside an extraction hood, the contents of the flask were filtered through Whatman filter paper to obtain a hexanic extract.

  2.2.1. The residue obtained after the filtration process was preserved.

2.3. The obtained hexanic extract was distilled Via a rotary evaporator at 26 °C to ensure that the extract was dry and named hexane extract of propolis (EHP).

2.4. The hexanic extract obtained from the distillation was brought to total dryness in an extraction hood at room temperature, and the extract was extended in glass containers.

  2.4.1. Once the totality of the solvent was evaporated, it was stored in closed glass jars and kept frozen at −20 °C.

2.5 The obtained hexanic extract was used in individual portions for physical and chemical analysis according to the quantities indicated in each technique.

2.4.3. Cleaning of the Pollen Content (Figure 2)

3.1 The residues obtained from the filtering process described in Step 2.2.1 were disintegrated in 70% ethanol and shaken vigorously until the mixture was homogenized.

3.2. The obtained mixture was filtered through sterile gauze, which retained persistent impurities, as well as resins not dissolved in the medium.

  3.2.1. The previous process was carried out many times until a fine residue that was left to precipitate was obtained.

  3.2.2. Once the precipitate formed, the solvent was withdrawn until a volume of 50 mL was obtained.

  3.2.3. The obtained volume was deposited in a 50 mL centrifuge tube.

2.4.4. Extraction of Pollen Content (Figure 2)

4.1. The mixture obtained at point 3.2.3 was centrifuged at 2500 rpm for 10 min.

4.2. At the end of the centrifugation process, the supernatant was removed, and the pellet, which was a wax matrix with pollen, was retained (Figure 2).

4.3. Recovery of the pollen content of the wax–pollen matrix.

4.4. The pellet (wax-pollen matrix) inside the 50 mL centrifuge tube was disintegrated in 25 mL of 96% ethanol. Once the pellet was dissolved, 25 mL of hexane was added to the tube, and it was shaken gently until both solvents were incorporated.

  4.4.1. Once the sample was homogenized, it was centrifuged at 2500 rpm for 10 min.

  4.4.2. After this time, a double phase is formed where the elements are separated in the following order from bottom to top: pollen tablet and some impurities, ethanol, hexane with dissolved waxes, and persistent resins (Figure 2).

  4.4.3. To finalize the process of obtaining pollen, the liquid phase was removed, and the pellet was preserved.

4.5. The obtained pellet disintegrated in 70% ethanol, with the purpose of removing the impurities that the pollen grains could have accumulated on the ornamentation of the exine of each species.

  4.5.1. The disintegrated pellet was filtered with gauze, the filtrate obtained was deposited in a centrifuge tube, and if necessary, it was gauged to 50 mL with 70% ethanol.

  4.5.2. Later, the sample was centrifuged at 2500 rpm for 10 min.

  4.5.3. After centrifugation, the pellet was recovered, and the supernatant was removed.

2.4.5. Sample Mounting via the Technique of Loveux [21] Modified by Sawyer [22]

5.1 With the aid of a micropipette, 100 µL of glycerogelatin with basic fuchsin was taken and added to the pellet.

  5.1.1. With the aid of a micropipette, the gelatin is retracted with pieces of the pellet many times until the mixture is homogenized.

5.2. Place the homogenized mixture on an object holder with the help of an object cover to distend the mixture.

  5.2.1. Once the sample is mounted, it is sealed with transparent varnish.

5.3. The obtained slides were reserved for palynological analysis.

The slides were examined under a Nikon Labophot-2 optical microscope at 40× and 100× [21,22]. To determine the pollen grains, dichotomous keys and specialized studies in the literature were used. With the determination of pollen content, the percentage of pollen types was determined via the method proposed by Loveux [21] and modified by Sayer [22]; four equidistant transects in the width of the preparations were marked, which served as guides where 500 pollen grains were counted per transect, in which the occurrence of each pollen type was recorded, to establish the pollen percentage.

2.5. Physicochemical Analysis

2.5.1. Waxes

The concentration of wax was determined by the difference in weight between the weight of the propolis in the shag and the weight of the EHP extract, with the percentage obtained not being greater than 25% [23,24]. The yield was calculated according to the following equation [24]:

$$R(\%) = \frac{P}{m}(100)$$

R = yield in percent
P = weight of the extract in (g)
m = Initial weight of the sample (g)

2.5.2. Total Phenols

The concentration of total phenols was determined through the modified Singleton method [25] by spectrophotometry based on an oxidation–reduction colorimetric reaction using Folin–Ciocalteu's reagent as the oxidizing agent and measuring absorbance at 760 nm [26,27].

Calibration Curve

A 0.2 mg/mL gallic acid standard solution was used for the calibration curve. From this solution, serial aliquots of 0.00625, 0.0125, 0.05, 0.1, and 0.2 mg/mL of gallic acid were made, with roughly 1 mL with distilled water [26,27].

Preparation of EEP

An amount of 250 mg of EEP dissolved in 2 mL of distilled water was prepared by taking an aliquot of 250 μL and adding 750 μL of distilled water to obtain a concentration of 1 mL [26,27].

Analysis

To each concentration of gallic acid and the propolis (EEP), 7 mL of distilled water and 500 mL of the Folin–Ciocalteu reagent were added. After 5 min, 1.5 mL of Na_2CO_3 solution (200 g/L) was added, allowing it to stand for 2 h at room temperature and then taking the absorbance reading at 760 nm [26,27].

2.5.3. Total Flavonoids

Dowd's method determined the total flavonoid content of propolis [27,28].

Calibration Curve

An amount of 3 mg of quercetin dissolved in 3 mL of HPLC grade methanol were prepared, then aliquots were taken to form a standard curve with concentrations ranging from 1 to 100 ppm.

Preparation of EEP

A total of 100 mg of EEP was weighed and dissolved in 3 mL of methanol to form the stock. Subsequently, 1 mL of the stock was taken and 1 mL of 2% $AlCl_3$ was added. The blank sample consisted of 1 mL of the stock without $AlCl_3$ [27].

Analysis

The standard curve and the sample were allowed to stand at room temperature for 10 min, and then the absorbance was read at 450 nm [27].

2.5.4. Antioxidant Capacity

To evaluate this parameter and calculate the antioxidant capacity of propolis at 50% (IC_{50}), the reduction of the 2,2-diphenyl-1-1picrihydrazyl radical (DPPH) was carried out by means of the technique proposed by [29–31]. For the preparation of the sample, 100 mg of EEP dissolved in 5 mL of methanol (HPLC) was used, while the blank consisted of 200 μL of methanol. The absorbance of each sample was subsequently measured at 540 nm on an ELISA plate reader (Multiskan-Ex Thermo Scientific, Waltham, MA, USA). The results are reported as the percent reduction via the following equation:

$$\%\ \text{Reduction} = \frac{\text{C} - \text{E}}{\text{C}}(100)$$

where

C = absorbance of DPPH;

E = Absorbance of the experiment (DPPH + propolis mixture).

2.6. *Chromatography Analysis of the Propolis*

2.6.1. Gas Chromatography–Mass Spectrometry Ethanolic Extract

The ethanolic extract that was obtained from the maceration of the propolis in Section 2.4.1 was analyzed by gas chromatography–mass spectrometry using a Model 6850 chromatograph (Agilent Technologies, Santa Clara, CA, USA) coupled to a Model 5975C mass spectrometer (Agilent Technologies) with an HP-5MS column (30 m × 0.25 mm, 0.25 μm Agilent Technologies). An amount of 1 μL of the sample was analyzed from a 3 mg/mL solution injected in split mode at an initial temperature of 70 °C for 2 min followed by a heating ramp of 15 °C min^{-1} at 290 °C for 6 min, using helium as a carrier gas for a total time of 31 min. The mass range detected was 35–600 m/z, ionizing by electron impact at 70 eV at an ionization temperature of 230 °C. The compounds were identified by comparison with the NIST library database version 8.0 (National Institute of Standards and Technology Gaithersburg, MD, USA).

2.6.2. High-Performance Liquid Chromatography–Diode Array (HPLC-DAD) and HPLC-ESI-TOF-MS

The ethanolic extract of the studied propolis was also analyzed by means of the HPLC-DAD system (Hewlett Packard, Agilent Technologies 1100 Wilmington, DE, USA) equipped with a diode array detector (DAD) 1100 operated by Chen Station A0903. A total of 20 μL of the propolis sample was analyzed from a 3 mg/500 μL injected solution. The mobile phase consisted of methanol–acetonitrile–H_3PO_4–H_2O (25:25:0.1:50) under isocratic conditions for 35 min, a procedure that provided better resolution when injecting the flavonoid standards from the database, using an Allsphere ODS-1 column (250 mm × 4.6 mm, with a particle size of 5 μm) at a pressure of 269 bar and a temperature between 22 and 23 °C. In comparison, the flow rate was 1 mL/min. A 280 nm wavelength diode array detector (DAD) with a full 200–400 nm scan was used. Detected compounds were identified by comparing the retention time and their absorption maxima (λmax) under ultraviolet light of the standards. The standards that made up the database were luteolin, genistein, caffeine, apigenin, myricetin, chrysin, acacetin, kaempferol, catechin, pinocembrin, baicalein, naringenin, naringin, catechol, and quercetin of the Sigma-Aldrich brand (St. Louis, MO, USA). HPLC-ESI-TOF-MS was performed with an Agilent 1200 Infinity coupled to an Agilent 6230 TOF mass spectrometer with an Agilent dual ESI source (ESI SG1 4289023) and the software Mass Hunter Workstation, version B.05.01, Build 5.01 5125.3, in negative ionization mode. The capillary voltage was 4000 V; the dry gas temperature was 250 °C; nitrogen was used at a flow rate of 6 l min^{-1}; the nebulizer pressure was 60 psi; the MS range was 50–1300 m/z; and the Ms acquisition rate was one spectra/s. Chromatographic separation was performed using an HPLC system (Infinity Series 1200, Agilent Technologies, Waldbronn, Germany) equipped with a Kinetx 2.6 U, C18 100 Å (150 × 2.1 mm) column (Phenomenex, Torrance, CA, USA).

2.7. *Evaluation of Antimicrobial Activity*

The antimicrobial activity of EEP against *Gram* (+) and *Gram* (−) bacterial species as well as against yeast fungi was evaluated. This effect was evaluated via the Kirby–Baüer method of the CLSI [32,33], in which agar plates were inoculated with known concentrations of the microorganisms *Staphylococcus aureus* ATCC 25923, *Escherichia coli* ATCC 123 and *Candida albicans* ATCC 10231.

The suspension with microorganisms was prepared as follows: a roast of the colony to be evaluated was taken and immersed in Müeller–Hinton (MH) broth for bacteria and Sabouraud broth for yeasts. The tubes containing the inoculated broth were incubated at 37 °C for 24 hr. The surface of each agar plate was subsequently inoculated (bacteria 1–1.5 × 10^8 CFU/mL; yeast (1–1.5 × 10^6 CFU/mL)), seeded via the streaking method in three directions, and the plate was rotated at 60° angles at the end of each streak. Once the inoculum was dry, 6 mm diameter sensidiscs impregnated with 4 mg (400 mg/mL) of EEP per sensidisc were placed; as a positive control, sensidiscs with 25 µg of chloramphenicol for bacteria and with 25 µg of Nystatin for yeast fungi were placed; as a negative control, sensidiscs with 10 µL of 80% ethanol were placed.

3. Results

3.1. *Methodology*

3.1.1. Pollen Extraction via Soxhlet

The Soxhlet method resulted in the efficient extraction of pollen grains under good conditions for observation and the absence of waxes and resin, and no extracts were obtained for another analysis (Figure 3).

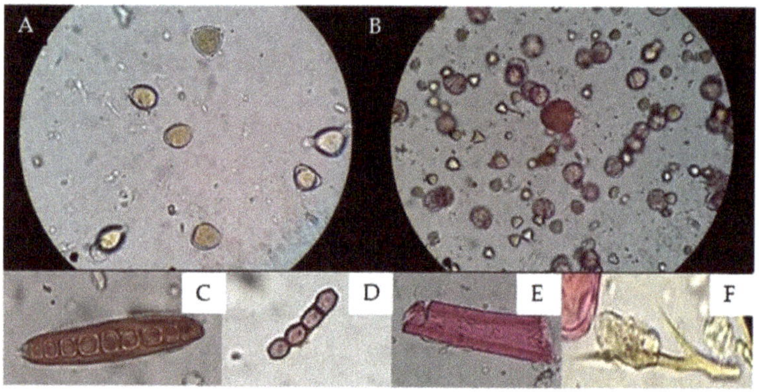

Figure 3. Comparisons of the results obtained via microscopic observations between the (**A**) Soxhlet method [21,22] and (**B**) proposed methodology. (**A**) Field 400X magnification without observing coloration changes due to the action of fuchsin on pollen grains. (**B**) Field 400X magnification coloration changes due to the action of fuchsin, (**C**) fungal spores, (**D**) spores, (**E**) plant structure, and (**F**) animal str.

3.1.2. Extraction Methods with Solvents of Different Polarities

The methodology proposed in Figure 2 allowed the pollen contained in the propolis to be obtained via extraction of its constituents, such as phenols, flavonoids, and waxes, via solvents according to the polarity of these components. With this methodology, two ethanol (EEP) and hexane (EHP) extracts were also obtained, which from 100 g of propolis sample in shaggy propolis yielded 25.20% EEP, with which physicochemical and biological analyses were carried out. For the EHP yield, 11.11% was obtained as the wax concentration in the 100 g propolis sample in the shaggy propolis. The extracts to be implemented in other research protocols, such as physicochemical evaluation and antibacterial and antifungal activity, were preserved and frozen at −20 °C.

The implementation of gauze resulted in the extraction of pollen content, since the adsorbent capacity of the gauze retained both resins and waxes and even persistent impurities that prevented adequate observation, as shown in Figure 3; in turn, the wide light of the gauze gave way to other different particles, such as pollen, fungi, and bacteria.

The double solvent phase process, 96° ethanol and hexane, resulted in the disintegration of the phenolic and serum components of the propolis residues, separating the pollen content from the wax–pollen matrix, which, together with washing with 70% ethanol, decreased the concentration and cleaning of the pollen content, which was reflected in the higher optical resolution of the pollen grains observed in Figure 4.

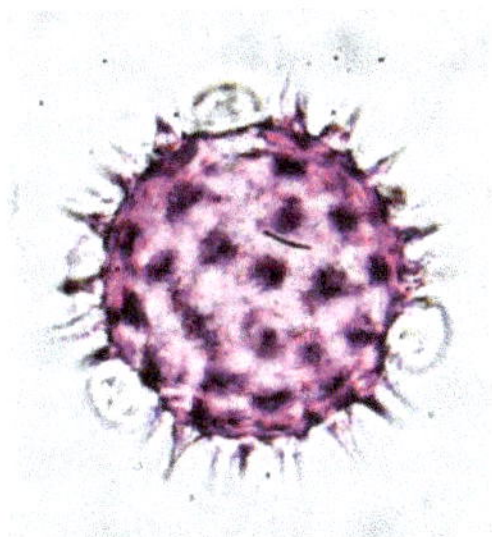

Figure 4. Pollen grain of the family Asteraceae in the genus *Bidens* obtained via the proposed extraction method 1000X magnification.

Staining with basic fuchsin incorporated into glycerogelatin (mounting medium) allows palynological analysis to differentiate between pollen grains and other structures, such as fungi and bacteria, because their composition is not stained by basic fuchsin (Figure 3).

3.2. *Palynological Analysis*

3.2.1. Soxhlet Method

Mounting of the residues with pollen was carried out by means of glycerogelatin with basic fuchsin to obtain semipermanent samples, which were used for optical analysis, resulting in a total of 13 families, 17 genera, and 13 species (Table 1), not finding a main pollen type according to the classification proposed by Loveux [21] and modified by Sayer [22], with the pollen of *Brassica rapa* L. (IZTA-3814) being the most represented species, with 40%, followed by *Melaleuca citrina* (Curtis) Dum. (IZTA-3814), with a value of 10.56% (Table 2).

Table 1. Palynological spectrum obtained from the analysis of the sample processed by Soxhlet extraction.

Family	Genus	Species
Amaranthaceae	*Amaranthus*	*Amaranthus hybridus* L. (IZTA 3801)
Anacardiaceae	*Schinus*	*Schinus molle* L. (IZTA 3802)
Asteraceae	*Bidens*	*Bidens odorata* Cav. (IZTA 3806)
	Taraxacum	*Taraxacum campylodes* G.E. Haglund (IZTA 3811)
	Titonia	*Titonia tuubeaformis* (Jacq.) Cass. (IZTA 3812)
Brassicaceae	*Brassica*	*Brassica rapa* L. (IZTA 3814)
Convulvulaceae	*Ipomoea*	
Euphorbiaceae	*Ricinus*	*Ricinus communis* L. (FESC-6294)
Fabaceae	*Medicago*	*Medicago sativa* L. (IZTA 3817)
Moraceae	*Ficus*	

Table 1. *Cont.*

Family	Genus	Species
Myrtaceae	*Callistemon*	*Melaleuca citrina* (Curtis) Dum. Cours (IZTA 3820)
	Eucalyptus	*Eucalyptus globulus* Labill. (FESC-11044)
Olaceae	*Fraxinus*	*Fraxinus uhdei* (Wenz.) Lingelsh (IZTA 42812)
	Ligustrum	*Ligustrum japonicum* Thumb (IZTA 42812)
Pinaceae	*Pinus*	
Poaceae	*Zea*	*Zae mays* L.
Rosaceae	*Pronus*	

Table 2. The percentage of pollen types determined was obtained from the analysis of the sample processed by Soxhlet.

	Pollen Type	%
	Brassica rapa L.	40.00
	Melaleuca citrina (Curtis) Dum. Cours	10.56
	Bidens odorata Cav.	8.44
	Eucalyptus	6.22
	Medicago sativa L.	5.44
	Schinus molle L.	4.44
	Titonia tuubeaformis (Jacq.) Cass.	4.22
	Ricinus communis L.	4.11
Complementary	*Pronus*	3.33
	Fraxinus uhdei (Wenz.) Lingelsh	2.56
	Amaranthus hybridus L.	2.00
	Ligustrum japonicum Thumb	2.00
	Ipomoea	1.89
	Ficus	1.67
	Taraxacum campylodes G.E. Haglund	1.44
	Zae mays L.	1.22
	Pinus	0.44

3.2.2. Extraction Methods with Solvents of Different Polarities

By means of this method, a total of 22 families, 32 genera, and 21 species were determined (Table 3), with the *Brassica* genus being the predominant pollen type with 63.8%, followed by the Mirtaceae family with 5%. The percentages of the other taxonomic groups were less than 5% (Table 4).

Table 3. Palynological spectrum by means of the component disaggregation method.

Family	Genus	Species
Altingiaceae	*Liquidambar*	
Amaranthaceae	*Amaranthus*	*Amaranthus hybridus* L. (IZTA 3801)
Anacardiaceae	*Schinus*	*Schinus molle* L. (IZTA 3802)
Asparagaceae	*Yucca*	
Asteraceae	*Bidens*	*Bidens aurea* (Aiton) Sherff (IZTA 3805)
		Bidens odorata Cav. (IZTA 3806)
		Bidens pilosa L. (IZTA 3807)
	Helminthotheca	*Helminthotheca echioides* (L) Holub (IZTA 3809)
	Helianthus	
	Taraxacum	*Taraxacum campylodes* G.E. Haglund (IZTA 3811)
Betulacaeae	*Alnus*	
Brassicaceae	*Brassica*	*Brassica rapa* L. (IZTA 3814)
	Raphanus	*Raphanus raphanistrum* L. (IZTA 3815)
	Sisymbrium	*Sisymbrium irio* L. (IZTA 3816)

Table 3. *Cont.*

Family	Genus	Species
Burseraceae	*Bursera*	
Casuarinaceae	*Casuarina*	*Casuarina equisetifolia* L.
Convulvulaceae	*Ipomoea*	
Cucurbitaceae	*Sicyus*	
Cupressaceae	*Cupressus*	
Euphorbiaceae	*Ricinus*	*Ricinus communis* L.(FESC-6294)
Fabaceae	*Acacia*	
	Medicago	*Medicago sativa* L. (IZTA 3817)
	Trifolium	*Trifolium amabile* Kunth (IZTA 3818)
Moraceae	*Ficus*	
Myrtaceae	*Callistemun*	*Melaleuca citrina* (Curtis) Dum. Cours (IZTA 3820)
	Eucalyptus	*Eucalyptus globulus* Labill (FESC-11044)
Olaceae	*Fraxinus*	*Fraxinus uhdei* (Wenz.) Lingelsh (IZTA 42812)
	Ligustrum	*Ligustrum japonicum* Thumb (IZTA 42812)
Pinaceae	*Pinus*	
Poaceae	*Zea*	*Zea mays* L.
Rosaceae	*Pronus*	
Salicaceae	*Populus*	
Solanaceae	*Nicotiana*	*Nicotiana glauca* Graham (IZTA 3831)
	Solanum	*Solanum rostratum* Dunal (IZTA 3833)

Table 4. Percentage of pollen types obtained through the method of component disaggregation.

Pollen Type		%
Brassica rapa L.		63.08
Melaleuca citrina (Curtis) Dum. Cours		5.00
Eucalyptus globulus Labill		4.83
Fraxinus uhdei (Wenz.) Lingelsh		4.00
Trifolium amabile Kunth		3.75
Cupresus		3.17
Ricinus communis L.		2.83
Acacia		2.00
Alnus		1.75
C. equisetifolia L.		1.75
Schinus molle L.		1.08
Complementary	*Ficus*	0.83
	Pinus	0.83
	Helianthus	0.75
	Ipomoea	0.75
	Nicotiana	0.67
	Taraxacum	0.67
	Liquidambar	0.33
	Bidens	0.33
	Amarantaceae	0.33
	Pronus	0.25
	Yucca	0.25
	Sicyos	0.25
	Populus	0.17
	Zea maiz	0.17
	Ligustrum	0.08
	Bursera	0.08
	Total complementary	6.75

3.3. Physicochemical Analysis

The results for total phenols showed a concentration of 62.79 mg GAE-g^{-1}, which is expressed as a percentage equivalent to 6.27%, whereas the result for total flavonoids the concentration was 6.72 µg Qe-g^{-1}, which is 0.61% of the concentration of this type of compound. For antioxidant activity, an IC_{50} of 486.43 µg/mL was determined.

3.4. Chromatographyl Analysis of the Propolis

3.4.1. High-Performance Liquid Chromatography–Diode Array (HPLC-DAD) and HPLC-ESI-TOF-MS)

The HPLC-DAD analysis showed 17 compounds, 5 of which coincided with the database, compounds that are in the group of phenols and flavonoids. In the HPLC-MS analysis, 17 compounds were also determined, with 4 matches with the database, as shown in Table 5 and Figures 5 and 6. Mass spectra of the compounds are available in Figure S1 of the Supplementary Material.

Table 5. HPLC-DAD and HPLC-MS analysis of ethanolic extract of propolis from Cuautitlan.

Compound	Retention Time		λmax (nm)	Parention (m/z) [M-H]−	Relative Error (ppm)
	HPLC-DAD	HPLC-MS			
Naringin	2.494		220, 284		
Cantechol	2.494		216, 270		
Naringenin	4.096	20.49	224, 282	271.03	3.1
Kaempferol	7.406		220, 366		
Chrysin	12.284	28.08	268, 314, 348	253.03	−1.58
Luteolin		23.25		285.02	−10.74
Pinocembrin		26.55		255.04	3.95

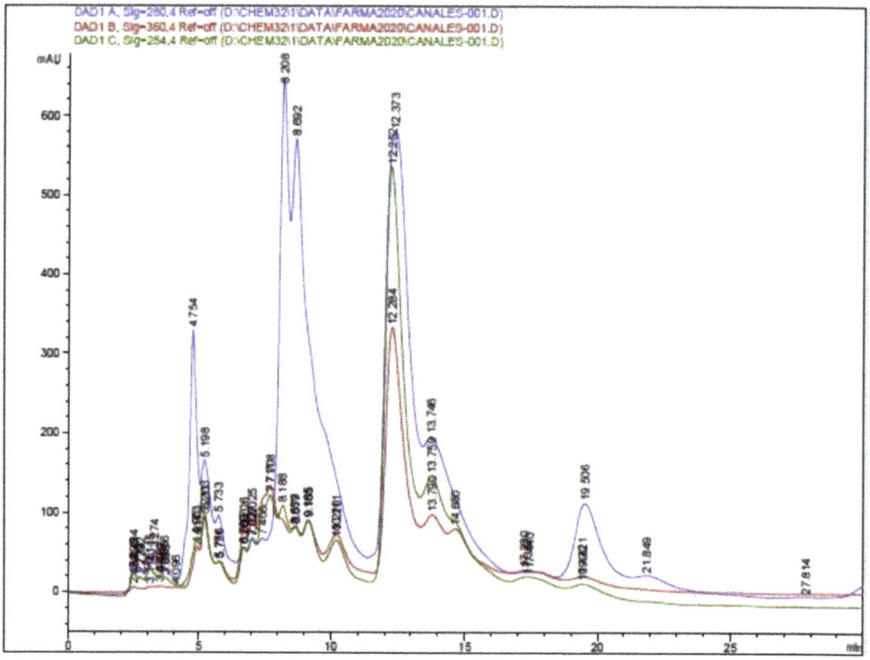

Figure 5. Chromatogram of ethanolic extract of propolis HPLC-DAD.

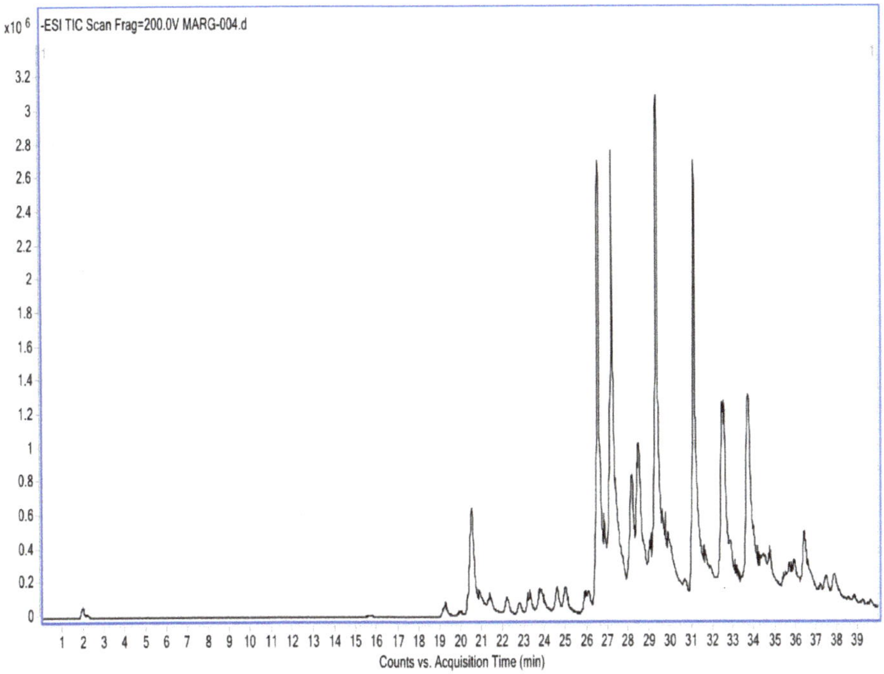

Figure 6. Chromatogram of the ethanolic extract of propolis HPLC-MS.

3.4.2. Gas Chromatography–Mass Spectrometry (GC-MS)

In the gas chromatography–mass spectrometry analysis of the ethanolic propolis extract sample, 16 compounds were identified, including pinocembrin, as shown in Table 6 and Figure 7.

Table 6. Compounds present in Cuautitlan propolis identified by gas chromatography–mass spectrometry.

Compound	RT (min)	%
Benzyl methyl ketone	7.92	1.12
m-Eugenol	11.93	3.97
Alpha bisabolol	16.91	1.06
Palmitic acid, methyl ester	19.97	7.75
Palmitic acid, ethyl ester	20.80	0.91
Linoleic acid, methyl ester	22.02	1.17
Oleic acid, methyl ester	22.09	9.33
Methyl stearate	22.36	1.94
Oleic acid, ethyl ester	22.84	0.94
Palmitamide	23.02	1.14
Methyl 18-methylnonadecanoate	24.56	2.29
Stearamide	25.24	0.97
Pinocembrin	26.51	30.29
Chrysin	28.39	15.77
6-O-Methylemodin, physcion	28.70	1.94
Galangin	28.83	5.67

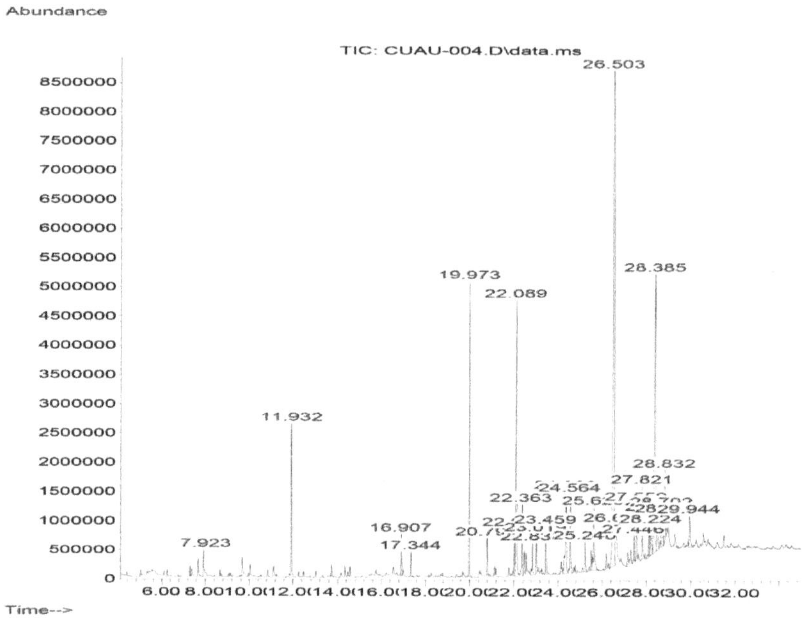

Figure 7. Chromatogram of the ethanolic extract of propolis GC-MS. Pinocembrin (RT = 26.5 min) and chrysin (RT = 28.39 min)are the most abundant flavonoids.

3.4.3. Biological Analysis

The antibacterial analyses revealed an inhibitory effect on *S. aureus*, with an inhibition halo of 8.71 mm; however, no inhibitory activity was observed for *E. coli*. In the evaluation of antifungal activity, no inhibitory activity was observed against *C. albicans*.

4. Discussion

The methodology suggested in this research allowed the pollen content with which qualitative and quantitative palynological analyses were carried out to be obtained from the propolis samples because the pollen grains were obtained under optimal physical conditions for determination. In turn, the differential extraction of the propolis components allowed the extraction of extracts with which biological and physicochemical tests were performed [18,22,26,34–36]. Therefore, this tool is proposed as a practical method for studies aimed at the analysis and palynological characterization of the physicochemical and biological properties of propolis [18,36].

The pollen sample obtained was abundant, representative, and in good condition for palynological analysis, which allowed more options for comparison and the precise taxonomic determination of each of the pollen species present in the sample [37,38]. With taxonomic knowledge of the pollen content, it was possible to infer the botanical origin of propolis, which is useful in the study, classification, and commercialization of propolis, as well as a better knowledge of its biomedical properties [37].

The washing process with solvents of different polarities described in point 2.4 of the proposed methodology allowed a greater number of pollen grains to be released from the conglomerate of waxes, resins, and balsams that constitute the propolis. In turn, the pollen grains were observed to be more clean and to have fewer impurities than with other methods, allowing high-resolution images in which the pollen structures useful for determination with dichotomous keys are detailed to be obtained, so that the results obtained both in concentration and resolution differ from the results obtained by the

Soxhlet method proposed by Loveux [21] and modified by Sayer [22], while presenting the advantage of obtaining extracts useful for physicochemical and biological analysis [18,22].

In the design of the processes of this methodology, the acetolysis technique was not included, which, although it has the purpose of concentrating and specifying the determinations of pollen content by means of an acid train, also eliminates the totality of the biologicals present in the sample, eliminating the possibility of identifying organisms and structures whose determination and quantification would be a parameter that would promote the quality of the propolis. Thus, with the mounting process described in this method, it is possible to observe fungal spores, yeasts, and plant and animal structures under an optical microscope [15,33]. However, in future studies and to improve the determination of pollen types at specific taxonomic levels, such as genus and species, the pollen material obtained at point 4.5.3 of the methodology should be reserved and subjected to acetolysis [15].

The EEP and EHP obtained by using solvents concentrate the main constituents of propolis, such as phenols, flavonoids and waxes [6,18]. Therefore, such extracts are useful in various research protocols, especially in the determination of the physical and chemical composition of propolis. In turn, the same extracts used in biological activity protocols generate results for antibacterial action [39].

The analysis of EHP is a quality parameter of propolis, since the concentration of wax in propolis does not present components that can promote any biological property; thus, it can be taken as a residue, included by the bees in the manufacturing of this product or included by the beekeeper in processes not suitable for obtaining propolis, because its inclusion reduces the volume of resins and balsams that are sources of phenolic compounds, which promote the biological properties of propolis. Therefore, the propolis sample studied with a wax concentration of 11.11% can be considered good-quality propolis since it is less than 25%, which is the maximum limit for considering the sample to be of good quality [26].

The pollen content allowed for the palynological analysis of the sample, revealing that the main pollen type that constitutes the propolis of the FES-C corresponds to the pollen of the *Brassica* genus, which is the main pollen type, followed by *M. citrina* and *E. globulus*, suggesting that the constitution of the propolis sample of the FES-C presents not only resins belonging to these two species but also a high content of balsams from species of the *Brassica* genus, such as *B. rapa*, which would explain the green color of this propolis, as suggested by Londoño et al. [40], who specify that the coloration of propolis depends largely on the propolis flora so that propolis that includes balsams in their constitution usually presents traces or green tones instead of a brown color, which is more common in propolis. These results contrast with the findings of Rodriguez et al. [30], who suggested that *E. globulus*, *R. communis*, and *M. citrina* species are the main sources of resources for the manufacture of propolis.

As mentioned above, with the implementation of this methodology, it was possible to obtain other analysis resources, such as ethanolic and hexanic extracts, which, in the case of EEP, were implemented in different research protocols, such as the analysis of antimicrobial activity, which resulted in inhibitory activity against *S. aureus* but not against *E. coli*, which is in agreement with several works [25] reporting that the inhibitory action of propolis is affected by the formation of the usual cell wall in this type of bacteria, as well as the absence of certain components that could be responsible for the antibacterial activity. Thus, the absence or presence of the components is largely defined by the biotic and abiotic resources in the area or region where the apiary is located [2,36].

The results of both the palynological and physicochemical analyses of the same sample allowed for greater certainty in the comparison of results, in contrast with those obtained by Rodriguez et al. [30], which, unlike this research, showed that propolis from the same location presents inhibitory activity against Gram (+), Gram (−) bacteria as well as against yeasts such as *C. albicans*, whereas in our results, such activity was observed only with *S. aureus*. This is largely due to the vegetation source of resources for the manufacture of this product, as well as the climatic conditions, season, and time of the year of foraging and collection of propolis [2,41]. Moreover, this finding explains the absence of antifungal activity against *C. albicans* in the studied propolis [25,30,42].

In terms of quality, the phenolic load is a relevant factor since the antioxidant activity of these compounds contributes to the maintenance and recovery of human health [36–38]. The results obtained for phenols such as flavonoids, as well as IC_{50} in comparison with those obtained by Rodriguez et al. [30], are lower; however, they comply with the values requested in the Mexican standard [43], as well as in the Argentine and Brazilian statutes [26], which indicates that it is good-quality propolis. However, optical analysis revealed the presence of both fungal spores and yeasts, and their identification and quantification could reduce the quality of this propolis. In turn, the quality of the ethanolic extracts obtained allowed for HPLC-DAD, HPLC-MS, and CG-MS analysis, showing similar compounds between analyses that coincide with various investigations [44,45], coinciding in components such as pinocembrin and chrysin regular flavones in propolis temperate zones.

Other protocols that benefit from the implementation of this method include those of molecular genetics, since for implementing the necessary techniques for genetic determination, the integrity of the biological material, especially the internal content of the pollen grain, is necessary. Therefore, the method becomes a practical tool since it safeguards the biological integrity of the pollen, which allows the use of DNA extraction methodologies, which can be implemented in metagenomic studies [11,12,46].

5. Conclusions

In conclusion, this method represents a practical tool for achieving accurate and complete characterization of propolis. The application of our method allows us to obtain clean pollen under optimal conditions for pollen analysis, which can be used to classify propolis and determine its botanical origin. In addition, the use of this method can be useful for the analysis of propolis for other research purposes, such as the biomedical and physicochemical properties of propolis, because this technique provides materials such as extracts that are useful in other disciplines, such as pharmacognosy, phytochemistry, chemistry, and cosmetics.

Supplementary Materials: The following supporting information can be downloaded at: https://www.mdpi.com/article/10.3390/antiox13111412/s1

Author Contributions: Conceptualization, J.J.A.-S. and M.M.C.-M.; methodology, J.J.A.-S., L.C.-M., B.V.-B. and A.A.-G.; software, J.J.A.-S. and M.A.R.-M.; investigation, J.J.A.-S., M.A.R.-M. and M.M.C.-M.; resources, M.A.R.-M. and M.M.C.-M.; writing—original draft preparation, J.J.A.-S.; writing—review and editing, M.A.R.-M. and M.M.C.-M.; project administration, M.A.R.-M. and M.M.C.-M.; funding acquisition, M.A.R.-M. and M.M.C.-M. All authors have read and agreed to the published version of the manuscript.

Funding: This research was funded by the UNAM PAPIIT IN205020 project and IN204821 project.

Institutional Review Board Statement: Not applicable.

Informed Consent Statement: Not applicable.

Data Availability Statement: Data are contained within the article.

Acknowledgments: Jose Juan Alcivar-Saldaña is a doctoral student in the Programa de Doctorado en Ciencias Biológicas, Universidad Nacional Autónoma deMéxico (UNAM), and received a fellowship from CONACYT (scholarship holder number 631329).

Conflicts of Interest: The authors declare no conflicts of interest.

References

1. Chazaro, M. *Melliferous or Apicultural Flora of the Coffee Basin of Coatepec, Veracruz*; Instituto Mexicano del Café: Tuxtla, Mexico, 1980.
2. Alcivar-Saldaña, J.J.; Rodriguez-Monroy, M.A.; Canales-Martinez, M.M. Current Knowledge of the Melliferous Florae in Mexico Using Methodologies to Understand Bee–Plant–Human Interactions. *Appl. Sci.* **2023**, *13*, 8591. [CrossRef]

3. Salamanca Grosso, G. *Origin, Nature, Physicochemical Properties and Therapeutic Value of Propolis*; Sello Editorial Universidad del Tolima: Ibagué, Colombia, 2017.
4. Gil, M.; Perelli, A.; Alvarado, R.; Arias, Y.; Blumenthal, E. Bacteriostatic and bactericidal activity of propolis tincture on enteropathogenic bacteria. *Salus* **2012**, *16*, 021–025.
5. Manrique, A.J.; Santana, W.C. Flavonoids, antibacterial and antioxidant activities of stingless bee propolis, *Melipona quadrifasciata*, *Melipona compressipes*, *Tetragonisca angustula* and *Nannotrigona* sp. from Brazil and Venezuela. *Zootec. Trop.* **2008**, *26*, 157–166.
6. Peña, R. Standardization in propolis: Chemical and biological background. *Cienc. Inv. Agr.* **2008**, *35*, 17–26.
7. Menezes, H. Propolis: A review of recent studies into its pharmacological properties. *Arq. Inst. Biológico* **2005**, *72*, 405–411. [CrossRef]
8. Marcucci, M.; Ferreres, F.; García-Viguera, C.; Bankova, V.; De Castro, S.; Dantas, A. Phenolic compounds from Brazilian propolis with pharmacological activities. *J. Ethnopharmacol.* **2001**, *74*, 105–112. [CrossRef]
9. Salatino, A.; Teixeira, E.W.; Negri, G.; Message, D. Origin and Chemical Variation of Brazilian Propolis. *Evid. Based Complement. Altern. Med.* **2005**, *2*, 33–38. [CrossRef] [PubMed] [PubMed Central]
10. Ramírez, E.; Navarro, L.; Díaz, E. Botanical characterization of Mexican honeys from a subtropical region (Oaxaca) based on pollen analysis. *Grana* **2011**, *50*, 50–54. [CrossRef]
11. Cornman, R.S.; Otto, C.R.V.; Iwanowicz, D.; Pettis, J.S. Taxonomic Characterization of Honey Bee (*Apis mellifera* L.) Pollen Foraging Based on Non-Overlapping Paired-End Sequencing of Nuclear Ribosomal Loci. *PLoS ONE* **2015**, *10*, e0145365. [CrossRef]
12. Quirasco, M.; Estrada, C.A.; Gálvez, A. Detection of genetically modified plant pollen in honey. In *Phase I. Facultadde Química. Informe Final SNIB-CONABIO, Proyecto KE007*; Universidad Nacional Autónoma de México: Mexico City, Mexico, 2013; Volume 430, pp. 1–36.
13. Barth, O.M.; Freitas, S.A. Palynology as a tool to distinguish between propolis and geopropolis: Southern Brazilian samples. *Open Access Libr. J.* **2015**, *2*, 2217. [CrossRef]
14. Córdova-Córdova, C.I.; Ramírez-Arriaga, E.; Martínez-Hernández, E.; Zaldívar-Cruz, J.M. Botanical characterization of bee honey (*Apis mellifera* L.) from four regions of the state of Tabasco, Mexico, by means of melisopalynological techniques. *Univ. Y Cienc.* **2013**, *29*, 163–178.
15. Erdtman, G. An introduction to the study of pollen grains and spores. In *Editorial Hafner*; Scandinavian University Books: Oslo, Norway, 1969.
16. Salamanca, G.; Osorio, P. Pollen analysis of red propolis from the insular zone of San Andres, Columbia. *Rev. Académica Colomb. Cienc.* **2019**, *43*, 689–698. [CrossRef]
17. Barth, O.M. Pollen analysis of Brazilian propolis. *Grana* **1998**, *37*, 97–101. [CrossRef]
18. Domínguez, A. *Methods of Phytochemical Research Mexico*; Limusa: Durban, South Africa, 1979; pp. 45–47.
19. García, L.E.; Meza, E. Opportunities and Obstacles for the Development of Beekeeping in Nayarit. Master's Thesis, Universidad Autónoma de Nayarit, Tepic, Mexico, 2012.
20. Morales, J.; Landa, M.; Guadalupe, S.; Lavoignet, M. Intelligent harvesting techniques for bee products applied in beehives in the Misantla region. *Rev. Ingeniantes* **2019**, *6*, 3.
21. Loveux, A.; Vorwhol, G.; Maurizio, A. Methods of melissopalinology. *Int. Comm. Bee Bot. IUBS* **1970**, *51*, 125–138.
22. Sawyer, R. *Honey Identification*; Cardiff Academic Press: Cardiff, UK, 1988; pp. 55–73.
23. Puerto, N.; Prieto, G.; Castro, S. Proximal and physical-chemistry analysis of propolis from by boyacenses apiaries. *Rev. Fac. Cienc. Básicas* **2016**, *14*, 126–140.
24. Alcivar-Saldaña, J.J.; Rodriguez-Monroy, M.A.; Carrillo-Miranda, L.; Canales-Martinez, M.M. Botanical Origin and Biological Properties of Honey and Propolis from Cuautitlan, State of Mexico, Mexico. *Antioxidant* **2024**, *13*, 874. [CrossRef]
25. Singleton, V.L.; Orthofer, R.; Lamuela-Raventos, R.M. Analysis of total phenols and other oxidation substrates and antioxidants by Folin-Ciocalteu reagent. *Métodos Enzimol.* **1999**, *299*, 152–178. [CrossRef]
26. Oropeza, G. *Total Alkaloids and Antioxidant Activity of Metabolic Extracts of Ipomea murocoides (Peanut) Leaves*; Tesis Universidad Tecnológica de la Mixteca: Oaxaca, Mexico, 2012.
27. Portela-Márquez, M.A.; Ruiz Cruz, S.; Morán-Palacio, E.F.; Chaidez-Quiroz, C.; Silva-Beltrán, N.P. Phenolic composition, antihemolytic, anti-inflammatory and antibacterial activity of propolis from southern Sonora. *Biotecnia* **2022**, *24*, 77–86. [CrossRef]
28. Ramamoorthy, P.; Bono, A. Antioxidante activity, total phenolic and Flavonoid conted of Morinda citriflora fruit extracts from various extraction processes. *J. Eng. Sci. Technol.* **2007**, *1*, 70–80.
29. Krishnaiah, D.; Bono, A.; Sarbatly, R.; Anisuzzaman, S.M. Antioxidant activity and total phenolic content of an isolated *Morinda citrifolia* L. methanolic extract from Poly-ethersulphone (PES) membrane separator. *J. King Saud Univ.—Eng. Sci.* **2015**, *27*, 63–67. [CrossRef]
30. Rodríguez, B.; Canales, M.; Penieres, J.; Cruz, T. Chemical composition, antioxidant properties and antimicrobial activity of Mexican propolis. *Acta Univ.* **2020**, *30*, e2435. [CrossRef]
31. Okusa, P.N.; Penge, O.; Devleeschouwer, M.; Duez, P. Direct and indirect antimicrobial effects and antioxidant activity of *Cordia gilletii* De Wild (Boraginaceae). *J. Ethnopharmacol.* **2007**, *112*, 476–481. [CrossRef] [PubMed]
32. Vanden, D.; Vlietnck, A. Screening methods for antibacterial agents from higher plants. *Methods Plant Biochem.* **1991**, *6*, 47–71.
33. Bucio-Villalobos, C.M.; Martínez-Jaime, O.A. Antibacterial activity of an aqueous extract of propolis from the municipality of Irapuato, Guanajuato, Mexico. *Mesoamerican Agron.* **2017**, *28*, 223–227. [CrossRef]

34. Barth, O.M.; Limeria, V. Pollen analysis of some propolis samples from southern Brazil. *Cienc. Rural Santa María* **1999**, *29*, 663–667. [CrossRef]
35. Vargas-Sánchez, R.D.; Peñalba-Garmendia, M.C.; Sánchez-Escalante, J.J.; Torrescano-Urrutia, G.R.; Sánchez-Escalante, A. Pollen profile of propolis produced in the eastern edge of the Sonoran Desert in central Sonora, Mexico. *Acta Botanica Mexicana* **2016**, *114*, 69–86. [CrossRef]
36. Muñoz-Muñoz, O.; Copaja, S.; Speisky-Cosoy, H.; Peña, R.C.; Montenegro, G. Content of flavonoids and phenolic compounds in chilean honeys. *Orac Index Química Nova* **2007**, *30*, 848–851. [CrossRef]
37. Medina, S.; Portillo, M. *Bioeconomía Aplicad a la Apicultura*; Universidad Autonoma de Chapingo: Texcoco, Mexico, 2017.
38. Barth, O.M.; da Pinto Luz, C.F. Palynological analysis of Brazilian red propolis samples. *J. Apic. Res.* **2009**, *48*, 181–188. [CrossRef]
39. Oliveira, A.C.; Shinobu, C.S.; Longhini, R.; Franco, S.L.; Svidzinski, T.I. Antifungal activity of propolis extract against yeasts isolated from onychomycosis lesions. *Mem. Inst. Oswaldo Cruz* **2006**, *101*, 493–497, Erratum in *Mem. Inst. Oswaldo Cruz* **2006**, *101*, 696. [CrossRef] [PubMed]
40. Londoño-Orozco, A.; Ávila-Acevedo, J.G.; Canales-Martínez, M.M.; Hernández-Delgado, C.T.; Serrano, P.R.; Flores-Ortíz, C.M.; Durán-Diaz, A.; Penieres-Carrillo, J.G.; García Tovar, C.G.; Cruz-Sánchez, T.A. Antibacterial comparative study extracts of Mexican propolis and of three plants which use *Apis mellifera* for its production. *J. Anim. Vet. Adv.* **2010**, *9*, 1250–1254. [CrossRef]
41. Muñoz, A.M.; Alvarado-Ortíz, C.; Blanco, T.; Castañeda, B.; Ruiz, J.; Alvarado, A. Determination of phenolic compounds, total flavonoids and antioxidant capacity in Peruvian honey from different floral sources. *Rev. Soc. Química Perú* **2014**, *80*, 4.
42. Flores-Rodríguez, I.S.; Moreno-Monteagudo, M.; Londoño-Orozco, A.; Cruz Sánchez, T.A. Use of Mexican Propolis for the Topical Treatment of Dermatomycosis in Horses. *Open J. Vet. Med.* **2016**, *6*, 1–8. [CrossRef]
43. *Norma Oficial Mexicana NOM-003-SAG/GAN-2017*; Propóleos, Producción y Especificaciones Para su Procesamiento. Diario Oficial de la Federación: Ciudad de México, Mexico, 2017.
44. Efraín, L.S.R.; Karla, M.R.; Ramírez, M.; Georgina, D.L.R.L.R.; Antonio, N.R.J. Determinación del Origen Botánico de los Propóleos Sonorenses (CABORCA) Mediante HPLC. *Investigación* **2009**, *4*, 24–30.
45. Rodríguez-Canales, M.; Medina-Romero, Y.M.; Rodríguez-Monroy, M.A.; Nava-Solis, U.; Bolaños-Cruz, S.I.; Mendoza-Romero, M.J.; Canales-Martínez, M.M. Actividad del propóleo de México sobre los factores de proliferación y virulencia de Candida albicans. *BMC Microbiol.* **2023**, *23*, 325. [CrossRef]
46. Peña, C.; Quirasco, M.; Estrada, C.; Gálvez, A. *Detección de Polen de Plantas Genéticamente Modificadas En Miel. Fase I*; CONABIO: Mexico City, Mexico, 2013.

antioxidants

MDPI

Article

Microencapsulation of Propolis by Complex Coacervation with Chia Mucilage and Gelatin: Antioxidant Stability and Functional Potential

Carlos A. Ligarda-Samanez [1,2,3,4,*], David Choque-Quispe [1,2,4,5], Henry Palomino-Rincón [1,2], Elibet Moscoso-Moscoso [1,2,3,4], Rodrigo J. Guzmán Gutiérrez [1,2,3] and Ismael Banda Mozo [1,2]

1 Nutraceuticals and Biomaterials Research Group, Universidad Nacional José María Arguedas, Andahuaylas 03701, Peru; dchoque@unajma.edu.pe (D.C.-Q.); hpalomino@unajma.edu.pe (H.P.-R.); emoscoso@unajma.edu.pe (E.M.-M.); 1008820181@unajma.edu.pe (R.J.G.G.); 1003320212@unajma.edu.pe (I.B.M.)
2 Agroindustrial Engineering School, Universidad Nacional José María Arguedas, Andahuaylas 03701, Peru
3 Food Nanotechnology Research Laboratory, Universidad Nacional José María Arguedas, Andahuaylas 03701, Peru
4 Research Group in the Development of Advanced Materials for Water and Food Treatment, Universidad Nacional José María Arguedas, Andahuaylas 03701, Peru
5 Water and Food Treatment Materials Research Laboratory, Universidad Nacional José María Arguedas, Andahuaylas 03701, Peru
* Correspondence: caligarda@unajma.edu.pe; Tel.: +51-985822085

Abstract

Propolis is a bee-derived resin rich in phenolic compounds known for their antioxidant, anti-inflammatory, and antimicrobial properties; however, its limited solubility and stability hinder its incorporation into food matrices. This study aimed to optimize the microencapsulation of ethanolic propolis extract through complex coacervation using chia mucilage and gelatin as wall materials, followed by spray drying. A 3^2 factorial design was applied to evaluate the effects of coacervate concentration and inlet temperature on various microcapsule properties. The optimal formulation (3.13% coacervate and 120 °C) exhibited high phenolic retention (15.36 mg GAE/g), notable antioxidant capacity (60.10 µmol TE/g), good solubility, thermal stability, and sustained in vitro release. Phenolic compounds were identified and quantified by UPLC-PDA-QDa, including gallic acid, catechin, epicatechin, epigallocatechin gallate, rutin, myricetin, resveratrol, quercetin, and kaempferol. Incorporating the microcapsules into functional gummy candies significantly enhanced their antioxidant activity without compromising sensory attributes. These findings support the use of complex coacervation as an effective strategy for stabilizing propolis bioactives, with promising applications in the development of functional foods that offer potential health benefits.

Keywords: spray drying; controlled release; phenolic compounds; functional gummies; thermal stability

1. Introduction

Beekeeping has undergone significant advancements in recent years, enabling the production of a range of products, including honey, pollen, royal jelly, and propolis. The latter, in particular, is of great interest due to its bioactive compounds, which have recognized health benefits [1–3]. Propolis is a resinous substance collected by honeybees (*Apis mellifera*) from the surrounding flora. It is noted for its remarkable antioxidant, anti-inflammatory, and antimicrobial properties, which are highly valued in the food, pharmaceutical, and cosmetic industries [4,5]. However, the direct application of propolis faces substantial limitations due to its poor solubility and low bioavailability, which restrict its effectiveness in practical applications [2,4,6].

Microencapsulation has been utilized as an effective method to preserve the bioactive properties of propolis, thereby expanding its applicability in various industries. Techniques such as spray drying, emulsification, and freeze-drying have been employed to protect its phenolic compounds; however, they present limitations, including low encapsulation efficiency, uncontrolled release, and poor colloidal stability [5,7,8]. Given these challenges, complex coacervation has garnered growing attention as a promising alternative, as it enables the formation of microcapsules through electrostatic interactions between wall biopolymers, thereby enhancing antioxidant retention and facilitating sustained release under controlled conditions [9,10]. However, a knowledge gap remains regarding the use of combined natural biopolymers, such as chia mucilage and gelatin, which represent an innovative strategy for stabilizing propolis bioactives in functional products.

Chia mucilage (*Salvia hispanica* L.) is an anionic polysaccharide with a complex structure, gelling and emulsifying properties, and film-forming abilities due to its high water retention capacity and the presence of uronic functional groups. These features make it a promising plant-based wall material for controlled release systems. Gelatin, on the other hand, is a positively charged animal-derived protein that forms stable complexes with polysaccharides in acidic media, making it widely used in complex coacervation processes. It provides structural cohesion, good solubility, and compatibility with various industrial applications. The synergistic combination of both biopolymers allows the formation of spherical microcapsules with high encapsulation efficiency (above 90%) and suitable colloidal stability, supporting their use as a functional encapsulation system in diverse food matrices [11–13].

In this context, this study aimed to evaluate the microencapsulation of phenolic extracts from propolis through complex coacervation and spray drying, using a mixture of chia mucilage and gelatin as wall materials. The hypothesis was that this combination would improve the physicochemical stability, controlled release, and antioxidant functionality of propolis. To achieve this, a factorial design was applied to optimize process conditions, and the resulting microcapsules were characterized through physicochemical, instrumental, and functional analyses. Additionally, their application in a functional gummy-type food matrix was validated by evaluating both sensory acceptance and functional performance. Ultimately, this work contributes to the advancement of clean and sustainable technologies for the valorization of beekeeping products.

2. Materials and Methods

2.1. Materials

Propolis samples were collected in the Chumbao River Valley, in the locality of Cuncataca (Andahuaylas, Apurímac, Peru), and were kindly provided by the Chumbao River Beekeepers Association. Chia seeds and red prickly pears were purchased from the municipal market of Andahuaylas, and the gelatin used was of commercial origin (passionfoods, Lima, Peru; manufactured in China). The following reagents and auxiliary materials were used in the study, organized according to their function:

Reagents for phenolic compound and antioxidant analysis: Folin–Ciocalteu reagent (Himedia, Mumbai, India), gallic acid (Sigma-Aldrich, Taufkirchen, Germany), quercetin

(Sigma-Aldrich, Taufkirchen, Germany), aluminum chloride hexahydrate (Sigma-Aldrich, Taufkirchen, Germany), Trolox [6-hydroxy-2,5,7,8-tetramethylchroman-2-carboxylic acid] (Sigma-Aldrich, Taufkirchen, Germany), DPPH [2,2-diphenyl-1-picrylhydrazyl] (Himedia, Mumbai, India), and ABTS [2,2′-azino-bis(3-ethylbenzothiazoline-6-sulfonic acid)] (Sigma-Aldrich, Taufkirchen, Germany).

Salts and buffers: sodium carbonate (Na_2CO_3), potassium persulfate ($K_2S_2O_8$), and sodium chloride (NaCl), all from commercial sources (Spectrum, New Brunswick, NJ, Canada).

Solvents: 96% ethanol (Coderal, Lima, Peru) and 99% isopropyl alcohol (Martell, Lima, Peru).

Analytical standards for chromatography: gallic acid, catechin, epicatechin, epigallocatechin gallate, rutin, myricetin, resveratrol, quercetin, and kaempferol (Sigma-Aldrich, Taufkirchen, Germany).

Ingredients for the formulation of functional gummy candies: glucose (Jofsac, Lima, Peru), isomalt (passionfoods, Lima, Peru; manufactured in Germany), potassium benzoate, glycerin (Alkofarma, Lima, Peru), and commercial sucrose.

2.2. Extraction of Chia Mucilage

The chia seeds were first sorted to remove any visible debris and impurities. They were then soaked in distilled water at a ratio of 1:25 (*w*/*v*), stirring constantly for 4 to 5 h at 40 °C in a water bath (WTB50, Memmert, Germany, Schwabach). During this time, the seeds absorb water, causing the outer mucilaginous layer to expand. Once optimal hydration was achieved, the mucilage was separated from the liquid by filtration using a fine cloth. The extract obtained was dried in a forced convection oven (FED 115, Binder, Germany, Tuttlingen) at 50 °C for 24 h. It was then ground using a cyclone impact mill (Retsch Cyclone, Retsch GmbH, Mettmann, Germany), with a 2 mm mesh used to standardize the particle size. The dry powder was stored in airtight jars inside a desiccator until it was used in the subsequent experimental stages [11,12].

2.3. Obtaining the Ethanolic Extract of Propolis

The ethanolic extract of propolis was obtained by dissolving 15 g of raw propolis in 100 mL of 80% ethanol (*v*/*v*) under constant stirring at 1100 rpm for 24 h. The mixture was filtered using a vacuum pump (Model WP6122050, Millipore, MA, USA) equipped with a 0.42 mm filter. The filtrate was centrifuged in a refrigerated centrifuge (TDL-5M, Bioridge, Shanghai, China), adding a few drops of distilled water to facilitate the removal of waxy residues. The clarified extract was concentrated in a rotary evaporator (HS-2005V-N, HAHNVAPOR, Shanghai, China) at 40 °C until its volume was reduced to 20% of the initial volume, promoting the evaporation of ethanol. Finally, the concentrate was stored at 4 °C in amber bottles until it was used. After ethanol evaporation, the extract presented a propolis solids content of 30% *w*/*w*, which served as the basis for formulating the microcapsules [4,6,14].

2.4. Microencapsulation by Complex Coacervation

The ethanolic wall material was formulated using a 1:1 (*w*/*w*) ratio of chia mucilage and gelatin, and it was used to prepare coacervate solutions at concentrations of 3%, 4%, and 5% (*w*/*v*). For each treatment, the corresponding amount of biopolymers was weighed and reconstituted in 70 mL of distilled water, with constant stirring at 350 rpm and 40 °C until completely dissolved. Homogenization was performed using an HG-15D mixer (Daihan Scientific, Gangwon, Republic of Korea) at 5000 rpm for 20 s [11,12].

Next, 10 mL of concentrated ethanolic propolis extract was added, and homogenization was repeated at 20 s intervals at 5000 rpm to ensure uniform mixing. The final volume was adjusted to 100 mL with distilled water, maintaining the coacervate concentration defined for each treatment.

Microencapsulation was carried out using a Büchi B-290 spray dryer (Büchi Labortechnik AG, Flawil, Switzerland) equipped with a 0.7 mm double-fluid nozzle. The solutions were processed under drying inlet temperatures of 120 °C, 130 °C, and 140 °C, following a 3^2 factorial design that evaluated coacervate concentrations of 3%, 4%, and 5%. The following parameters were kept constant: suction at 95%, peristaltic pump at 15%, and stirring at 350 rpm. Drying was completed when the entire volume of each solution had been processed.

Statistical analysis was based on a second-order quadratic model, which describes the effects and interactions of the factors. The equation used was

$$Y = \beta_0 + \beta_A X_A + \beta_B X_B + \beta_{A,B} X_A X_B + \beta_{A,A} X_A^2 + \beta_{B,B} X_B^2 \quad (1)$$

where Y represents the response variable; X_A and X_B are the coded variables corresponding to the coacervate concentration and inlet temperature, respectively; β_0 is the constant; β_A and β_B are the linear coefficients; $\beta_{A,B}$ is the interaction coefficient; $\beta_{A,A}$ and $\beta_{B,B}$ are the quadratic coefficients of the factors.

2.5. Preparation of Ethanolic and Methanolic Extracts for Bioactive Compound Analysis

For raw propolis, 3 g of finely ground sample was mixed with 10 mL of 80% ethanol (v/v) and subjected to maceration on a magnetic stirrer for 24 h at room temperature. Afterward, the extract was centrifuged and stored for subsequent analysis [15].

For microcapsules and gummy candies, a total of 0.2 g of the sample was weighed into Falcon tubes covered with aluminum foil. Then, 10 mL of 80% methanol (v/v) was added, and the mixture was homogenized using a vortex at 3000 rpm for 60 s. The samples were kept in the dark at room temperature for 24 h. Subsequently, they were centrifuged in a refrigerated centrifuge (TDL-5M, Bioridge, Shanghai, China) at 3500 rpm for 3 min to separate the solids. The clarified methanolic extract was stored at 4 °C until analysis [4].

2.6. Determination of Total Phenolic Compounds

The total phenolic content was determined using the Folin–Ciocalteu method. A standard calibration curve (R^2 = 0.99) was constructed using gallic acid at concentrations ranging from 1 to 30 mg/L, with the following calibration points: 1, 2, 5, 10, 15, 20, 25, and 30 mg/L. For the analysis, a 0.25 N Folin–Ciocalteu reagent solution and a 20% (w/v) solution of anhydrous sodium carbonate (Na_2CO_3) were prepared. In test tubes covered with aluminum foil, 0.90 mL of methanolic extract was added, followed by 2.4 mL of ultrapure water (dilution factor: 3.67). Then, 0.15 mL of 20% Na_2CO_3 and 0.3 mL of Folin–Ciocalteu reagent were added. The mixture was allowed to react for 10 min at room temperature. Finally, absorbance was measured at 755 nm using a UV-Vis spectrophotometer (Genesys 150, Thermo Fisher Scientific, Madison, WI, USA) [16].

The absorbance values were interpolated using the gallic acid calibration curve (y = 0.0379x − 0.0046), where y represents the absorbance and x represents the concentration of phenolic compounds in mg/L. These values were adjusted considering the dilution factor and the sample mass, and the final results were expressed as milligrams of gallic acid equivalents per gram of sample (mg GAE/g); these values were expressed on a dry weight basis, using the moisture content of each sample to perform the conversion.

The encapsulation efficiency (%EE) was calculated under optimal treatment conditions by comparing the total phenolic content in the ethanolic propolis extract with that in the resulting microcapsules. This evaluation allowed for the estimation of the proportion of compounds retained after the complex coacervation encapsulation process [6].

2.7. Determination of Flavonoids

The flavonoid content was determined using a standard calibration curve (R^2 = 0.99) constructed using quercetin as the reference compound at concentrations ranging from 0.2 to 1.2 mg/mL, with the following calibration points: 0.20, 0.40, 0.60, 0.80, 1.00, and 1.20 mg/mL. For the assay, 90.0 µL of methanolic extract, 1.91 mL of 80% ethanol, 100 µL of 5% (*w*/*v*) aluminum chloride ($AlCl_3$), and an additional 2.90 mL of 80% ethanol were added in that order. The mixture was then incubated at room temperature in the dark for 30 min. Measurements were taken at a wavelength of 425 nm using a UV-Vis spectrophotometer (Genesys 150, Thermo Fisher Scientific, Madison, WI, USA) [17].

The absorbance values were interpolated using the quercetin calibration curve (y = 1.54499x − 0.0061) to determine the flavonoid concentration in mg/mL. These values were corrected based on the dilution factor and sample mass, and the results were expressed as milligrams of quercetin equivalents per gram of sample (mg QE/g). These values were expressed on a dry weight basis, using the moisture content of each sample to perform the conversion.

2.8. Determination of Antioxidant Capacity by the DPPH Method

Trolox was used as the reference compound, and a standard calibration curve was obtained in the range of 20 to 700 µmol/L using the following calibration points: 20, 30, 40, 50, 100, 200, 300, 400, 500, 600, and 700 µmol/L with a determination coefficient of R^2 = 0.99. The diluted DPPH solution was adjusted to an initial absorbance of 1.10 at 515 nm. For the assay, 150.0 µL of the methanolic extract was mixed with 2.85 mL of the DPPH solution and allowed to react in the dark for 15 min. As a control, a mixture of 150.0 µL of 80% ethanol and 2.85 mL of DPPH solution was used. Absorbance measurements were taken at 515 nm using a UV-Vis spectrophotometer (Genesys 150, Thermo Fisher Scientific, Madison, WI, USA) [4]. The antioxidant capacity of DPPH was expressed as micromoles of Trolox equivalent per gram of sample (µmol TE/g), calculated on a dry weight basis by using the moisture content of each sample to perform the conversion.

2.9. Determination of Antioxidant Capacity by the ABTS Method

The $ABTS^+$ radical was generated by mixing 250 µL of a 2.45 mM $K_2S_2O_8$ solution with 25 mL of a 7.00 mM ABTS solution and allowing the mixture to stand in the dark for 8 h. For the assay, 300 µL of the methanolic extract was combined with 2.7 mL of the $ABTS^+$ solution. Absorbance was measured at 734 nm using a UV-Vis spectrophotometer (Genesys 150, Thermo Fisher Scientific, Madison, WI, USA). Trolox was used as the reference compound for the standard curve, which was constructed using concentrations of 10, 50, 100, 200, and 300 µmol/L. The resulting calibration curve was described by the equation y = 0.0017x + 0.1306, with a determination coefficient of R^2 = 0.98 [4]. The antioxidant capacity of ABTS was expressed as µmol TE/g on a dry weight basis, using the moisture content of each sample to perform the conversion.

2.10. Color Determination

Color analysis was performed using a CR-5 benchtop colorimeter (Konica Minolta, Tokyo, Japan), employing a 40.5 × 60.0 mm cell and the reflectance accessory. The results were expressed in the CIELAB system (L*, a*, and b*) [18].

2.11. Moisture Determination

The moisture content was determined using the official AOAC method 950.10. A total of 0.2 g of sample was weighed and dried in a forced convection oven (FED 115, BINDER GmbH, Tuttlingen, Germany) at 105 °C until a constant weight was achieved [18].

2.12. Hygroscopicity Determination

A total of 0.2 g of sample was weighed and placed in a sealed container with a saturated NaCl solution, maintaining controlled conditions for 7 days. After this period, the final mass was recorded, and hygroscopicity was calculated using the following formula:

$$I = \left(\frac{M_3 - M_2}{M_2 - M_1} \right) \times 100 \tag{2}$$

where I is the percentage increase in mass; M_1 is the mass of the empty Petri dish; M_2 is the mass of the Petri dish with the sample at the beginning; and M_3 is the mass of the Petri dish with the sample after seven days [17].

2.13. Bulk Density Determination

A known amount of dry, unpacked sample was placed in a 10.0 mL graduated cylinder, and the volume occupied was recorded. The mass was then measured using an analytical balance. Bulk density was calculated as the ratio of the sample mass to the displaced volume and expressed in g/mL [4].

2.14. Solubility Determination

Solubility was quantified by dissolving 0.5 g of microcapsules in 20 mL of distilled water. The mixture was vortexed at 3000 rpm for 2 min and then incubated in a water bath at 80 °C for 30 min. After this period, the sample was allowed to cool to room temperature and then centrifuged at 3500 rpm for 10 min. The supernatant was transferred to a Petri dish and dried in an oven at 90 °C for 24 h. Solubility was calculated using the following formula:

$$S = \left(\frac{M_2}{M_1} \right) \times 100 \tag{3}$$

where S is the solubility percentage; M_1 is the initial mass of the microcapsules (g); and M_2 is the mass of the dry residue obtained from the supernatant (g) [4].

2.15. Particle Size Determination

Particle size was determined by laser diffraction using a Mastersizer 3000 instrument (Malvern Instruments, Worcestershire, UK). A total of 0.02 g of microcapsules was dissolved in 50 mL of isopropyl alcohol, and the mixture was stirred at 3000 rpm for 2 min. It was then subjected to ultrasound in a DU-220S ultrasonic bath (ARGO LAB, Milan, Italy) for 10 min at 40 kHz. Measurements were taken at a wavelength of 600 nm. The polydispersity index (PDI) was calculated using the following equation:

$$PDI = \frac{D_{(0.9)} - D_{(0.1)}}{D_{(0.5)}} \tag{4}$$

where $D_{(0.1)}$, $D_{(0.5)}$, and $D_{(0.9)}$ are the diameters corresponding to 10%, 50%, and 90% of the cumulative particle size distribution, respectively [19].

2.16. *ζ Potential Determination*

ζ potential was determined by dynamic light scattering (DLS) using a Zetasizer ZU3100 (Malvern Instruments, Worcestershire, UK). A total of 0.02 g of sample was dissolved in the appropriate medium and subjected to ultrasound for 10 min at 40 kHz. Measurements were performed in a DTS1070 cell under the instrument's standard conditions [19].

2.17. *In Vitro Release Evaluation of Phenolic Compounds*

The release was evaluated using the Folin–Ciocalteu method. A microcapsule dispersion at a concentration of 10 mg/mL was prepared and incubated at 25 °C under static conditions for 6, 12, 18, 24, 30, 36, 42, and 48 h. At the end of each interval, the phenolic compound content was determined by spectrophotometric reading at 755 nm using a UV-Vis spectrophotometer (Genesys 150, Thermo Fisher Scientific, Madison, WI, USA) [16].

2.18. *Scanning Electron Microscopy (SEM) Analysis*

The morphology of the microcapsules was evaluated by scanning electron microscopy (SEM) using a Prisma E microscope (Thermo Fisher Scientific, Brno, Czech Republic). The samples were mounted on aluminum stubs using 12 mm carbon adhesive tape. Observations were carried out under low-vacuum conditions using ABS and LVD detectors at a pressure of 0.07 Torr [6].

2.19. *Fourier Transform Infrared (FTIR) Spectroscopy Analysis*

Functional group analysis was performed by FTIR spectroscopy in transmission mode using a Nicolet iS50 spectrophotometer (Thermo Fisher Scientific, Waltham, MA, USA). Pellets were prepared by pressing a mixture of 200 mg of KBr and 2 mg of sample under 10 tons of pressure for 40 s. Spectra were acquired in the mid-infrared range using a KBr beam splitter, with a resolution of 8 cm^{-1} and 32 scans [6].

2.20. *Thermogravimetric Analysis (TGA)*

Functional thermogravimetric analysis was performed using a TGA 550 instrument (TA Instruments, New Castle, DE, USA). A total of 10 mg of microcapsules was heated from room temperature to 800 °C at a heating rate of 10 °C/min under a nitrogen atmosphere [16].

2.21. *Differential Scanning Calorimetry (DSC)*

Differential scanning calorimetry analysis was performed using a DSC 2500 instrument (TA Instruments, New Castle, DE, USA). A total of 2 mg of microcapsules was used, applying a heating program from 20 °C to 300 °C at a rate of 3 °C/min [16].

2.22. *Ultra-High-Performance Liquid Chromatography Analysis (UPLC-PDA-QDa)*

The microcapsules from the optimal treatment were obtained from ethanolic propolis extract previously concentrated by rotary evaporation, followed by a complex coacervation process using chia mucilage and gelatin, and finally spray-dried. The identification and quantification of phenolic compounds were carried out using external calibration curves specific to each analytical standard. Analyses were performed in triplicate, and the results were expressed as mean ± standard deviation in mg/g of dry sample.

An ultra-high performance liquid chromatography (UPLC) system coupled to a photodiode array detector (PDA) and a single quadrupole mass detector (QDa) (ACQUITY UPLC-PDA-QDa, Waters Corporation, Milford, MA, USA) was used. The separation was performed on a C18 column (2.1 × 100 mm, 1.8 μm) maintained at 40 °C, using a mobile phase consisting of water (94.8%) with 80% methanol (5%) and formic acid (0.2%) (phase A), and acetonitrile with 0.2% formic acid (phase B). The elution gradient was programmed from 5% to 85% of phase B over 11.2 min, with a flow rate of 0.5 mL/min and an injection volume of 5.0 μL.

Detection was performed at 254 nm with the PDA and in both positive (ESI^+) and negative (ESI^-) electrospray ionization modes with the QDa within an m/z range of 100 to 600. Compound identification was based on the comparison of retention times, UV spectra, and exact m/z values with those of commercial analytical standards. Limits of detection (LODs) and quantification (LOQs) were determined for each compound to ensure the reliability and sensitivity of the quantification results. All identified compounds matched the reference data of the corresponding standards used [20–22].

2.23. Incorporation into Gummy Candies

Microcapsules obtained by complex coacervation were incorporated into three formulations of functional gummy candies using different concentrations of microencapsulated material and red cactus pear juice, as shown in Table 1. The mixture was prepared at 65 °C with constant stirring until a homogeneous mass was obtained. It was then poured into silicone molds and allowed to set at room temperature until solidified, followed by storage at 18 °C [23].

Table 1. Formulation of functional gummy candies.

Formulation	Microcapsules	Red Cactus Pear Juice	Water	Neutral Gelatin (200° Bloom)	Sucrose	Glucose	Isomaltose	Potassium Benzoate	Glycerin
F1	3.00%	35.00%	16.50%	11.00%	11.00%	17.00%	5.00%	1.00%	0.50%
F2	4.00%	30.00%	20.50%	11.00%	11.00%	17.00%	5.00%	1.00%	0.50%
F3	5.00%	40.00%	9.50%	11.00%	11.00%	17.00%	5.00%	1.00%	0.50%

All ingredients are expressed as percentage by weight (% *w*/*w*). F1, F2, and F3 correspond to different formulations.

2.24. Sensory Evaluation

Sensory evaluation was conducted two weeks after storage of the gummy candies enriched with propolis microcapsules, which were obtained by complex coacervation. Portions of 20 g were offered to 70 untrained panelists, aged between 16 and 22 years, selected by convenience sampling, ensuring a balanced gender distribution (50% female and 50% male), and with no prior experience with functional products.

Each sample was assigned a random three-digit code to ensure unbiased evaluation. The tests were conducted in a well-lit environment, free from interfering odors and with adequate ventilation.

The study included two consumer-type tests [23]:

A preference test, in which each participant selected the formulation they liked the most.

An acceptability test using a five-point structured hedonic scale, where 1 = dislike very much, 2 = dislike, 3 = neither like nor dislike, 4 = like, and 5 = like very much. The attributes evaluated were color, aroma, flavor, and texture.

The panel size was determined using Cochran's formula for proportions, considering a 95% confidence level and a 10% margin of error, which validated that those 70 participants constituted a representative sample size. All participants provided informed consent, and the Ethics Committee of the National University José María Arguedas approved the test.

2.25. Texture Profile Analysis (TPA)

Texture profile analysis of the gummy candies was performed using a Brookfield AMETEK Texture Analyzer equipped with a 5000 g load cell. Samples were subjected to a compression test using a cylindrical TA3/1000 probe with 40% deformation at a speed of 1 mm/s. A TPA test with two compression cycles was used to evaluate hardness, cohesiveness, adhesiveness, springiness, and resilience. For each formulation (F1, F2, and F3), three replicates were analyzed, and average values with standard deviations were obtained. Data analysis was performed using Texture Pro software version 1.0.19.

2.26. Statistical Analysis

Physicochemical and instrumental properties were analyzed using ANOVA and Tukey's multiple comparison test, with a significance level of 5% and three replicates per treatment. For the sensory data, the Kolmogorov–Smirnov normality test was applied first, followed by the non-parametric Friedman and Wilcoxon tests. In the microencapsulation study, a 3^2 factorial design was employed, evaluated using a quadratic model, and optimized through the desirability function. Statistical analysis was performed using Design Expert v9.0 software (Trial version, Stat-Ease Inc., Minneapolis, MN, USA) and Origin Pro 2025 software (OriginLab Corporation, Northampton, MA, USA).

3. Results and Discussions

3.1. Characterization of Raw Propolis and Ethanolic Propolis Extract

Table 2 presents the physicochemical properties of raw propolis and its ethanolic extract, showing significant differences between them. The results demonstrated that the extraction process increased the phenolic compounds (17.66 mg gallic acid equivalents (GAE)/g), flavonoids (13.72 mg quercetin equivalents (QE)/g), and antioxidant capacity by DPPH (102.88 µmol Trolox equivalents (TE)/g). Additionally, the extract exhibited higher lightness (L^*) and greater values of chroma (a^* and b^*). A higher L^* indicates a lighter color, while increases in a^* and b^* reflect greater intensity in red and yellow tones. These results demonstrate a more vivid and saturated color in the ethanolic extract compared to the raw propolis. These changes are attributed to the ethanolic extraction process, which removes waxes and other non-functional fractions present in raw propolis; this technique facilitates the concentration of bioactive metabolites responsible for the higher antioxidant capacity of the extract [4,24].

Table 2. Physicochemical properties of raw propolis and ethanolic propolis extract.

Property	Raw Propolis			Ethanolic Extract of Propolis		
	$\bar{x}$	±	SD	$\bar{x}$	±	SD
Phenolic compounds (mg GAE/g)	6.48 ^{a}	±	0.15	17.66 ^{b}	±	0.05
Flavonoids (mg QE/g)	6.07 ^{a}	±	0.18	13.72 ^{b}	±	0.63
Antioxidant capacity by DPPH (µmol TE/g)	45.48 ^{a}	±	0.08	102.88 ^{b}	±	0.46
L^*	33.73 ^{a}	±	0.01	41.98 ^{b}	±	0.25
a^*	3.97 ^{a}	±	0.04	16.92 ^{b}	±	0.05
b^*	9.21 ^{a}	±	0.07	66.27 ^{b}	±	0.37

Values are expressed as mean ($\bar{x}$) ± standard deviation (SD), n = 3. Different letters in the same row indicate statistically significant differences ($p \leq 0.05$) according to Tukey's test.

3.2. Characterization of Wall Materials

Figure 1 shows the physicochemical and structural properties of the studied wall materials. First, chia mucilage (Figure 1a) exhibited smaller particle size, lower lightness, and higher polydispersity, indicating a more heterogeneous distribution. Second, gelatin (Figure 1b) exhibited a larger particle size, higher lightness, and lower polydispersity, indicating a more homogeneous distribution. Finally, the 1:1 mixture of chia mucilage and gelatin (Figure 1c) presented intermediate values of particle size, lightness, and polydispersity, resulting in an encapsulating matrix with a favorable structural balance for conducting the complex coacervation process.

Moreover, ζ potential values indicated varying colloidal stability, which was higher in the case of chia mucilage. Additionally, SEM-EDS analysis revealed differences in particle morphology and elemental composition, characteristic of naturally derived polysaccharides. These findings are consistent with previous studies that highlight the complementary behavior of chia mucilage and gelatin in complex coacervation systems, where their structural and colloidal properties promote the formation of stable and functional capsules [11–13].

Figure 1. Characterization of wall materials: chia mucilage (**a**), gelatin (**b**), and 1:1 mixture of chia mucilage and gelatin (**c**).

3.3. Physicochemical Properties of Microcapsules Obtained by Complex Coacervation and Spray Drying

Table 3 presents the results of the 3^2 factorial design, which evaluated the effect of the coacervate concentration (3%, 4%, and 5%) and inlet temperature (120 °C, 130 °C, and

140 °C) on the physicochemical properties of the obtained microcapsules. The results showed wide ranges for the evaluated dependent variables, including phenolic compounds (9.16–15.36 mg GAE/g), flavonoids (2.67–7.77 mg QE/g), antioxidant capacity measured by DPPH (36.78–60.10 μmol TE/g) and ABTS (22.19–50.10 μmol TE/g), solubility (41.49–79.72%), hygroscopicity (12.58–20.50%), moisture content (2.70–9.57%), bulk density (177.75–284.25 g/cm^3), particle size (5.39–7.80 μm), and ζ potential (−37.29 to −13.20 mV). These results demonstrate that the operating conditions during spray drying significantly affect the structural and functional stability of the microcapsules, providing a valuable framework for their optimization [13,17].

Table 3. Factorial design treatments and results of the dependent variables.

Run	A	B	Phenolic Compounds	Flavonoids	DPPH	ABTS	Solubility	Hygroscopicity	Moisture	Bulk Density	Particle Size	ζ Potential
	%	°C	mg GAE/g	mg QE/g	µmol TE/g	µmol TE/g	%	%	%	g/mL	µm	mV
			$\bar{x} \pm$ SD	$\bar{x} \pm$ SD	$\bar{x} \pm$ SD	$\bar{x} \pm$ SD	$\bar{x} \pm$ SD	$\bar{x} \pm$ SD	$\bar{x} \pm$ SD	$\bar{x} \pm$ SD	$\bar{x} \pm$ SD	$\bar{x} \pm$ SD
T1	3	120	15.36 ± 0.57	7.77 ± 0.93	56.25 ± 1.98	46.33 ± 0.12	49.64 ± 0.00	15.83 ± 0.68	9.57 ± 0.00	206.71 ± 0.00	6.41 ± 0.03	−34.23 ± 1.55
T2	4	120	10.41 ± 0.06	3.83 ± 0.07	40.79 ± 1.96	25.65 ± 0.71	79.72 ± 0.00	13.08 ± 0.79	5.10 ± 0.10	284.25 ± 0.00	5.39 ± 0.09	−30.30 ± 2.32
T3	5	120	9.84 ± 0.18	3.62 ± 0.21	36.78 ± 1.47	23.09 ± 0.19	66.80 ± 0.00	12.58 ± 1.11	2.70 ± 0.20	274.50 ± 0.00	6.66 ± 0.06	−31.27 ± 0.94
T4	3	130	13.12 ± 0.23	5.18 ± 0.20	60.10 ± 1.99	50.10 ± 0.07	41.49 ± 0.00	15.23 ± 0.07	8.52 ± 0.00	197.07 ± 0.00	6.90 ± 0.09	−37.29 ± 1.89
T5	4	130	11.71 ± 0.12	3.79 ± 0.61	47.34 ± 1.35	46.75 ± 0.07	44.30 ± 0.00	14.65 ± 1.15	7.67 ± 0.00	238.07 ± 0.00	7.80 ± 0.03	−29.25 ± 2.42
T6	5	130	10.77 ± 0.38	2.67 ± 0.07	44.53 ± 1.36	47.98 ± 0.58	47.39 ± 0.00	18.54 ± 0.70	8.62 ± 0.00	239.56 ± 0.00	7.23 ± 0.07	−28.05 ± 0.95
T7	3	140	11.92 ± 0.47	4.31 ± 0.14	59.38 ± 0.35	45.45 ± 0.32	53.90 ± 0.00	20.50 ± 1.67	8.99 ± 0.00	177.75 ± 0.00	6.83 ± 0.04	−34.78 ± 1.57
T8	4	140	10.83 ± 0.28	4.72 ± 0.18	47.04 ± 2.43	25.79 ± 0.21	67.53 ± 0.00	12.72 ± 0.92	5.34 ± 0.46	237.75 ± 0.00	5.64 ± 0.02	−13.20 ± 0.75
T9	5	140	9.16 ± 0.95	2.80 ± 0.48	37.89 ± 2.83	22.19 ± 0.09	56.67 ± 0.00	13.65 ± 0.62	4.29 ± 0.45	234.55 ± 0.00	6.59 ± 0.07	−34.63 ± 0.96

Values are expressed as mean ($\bar{x}$) ± standard deviation (SD), $n = 3$. A represents the percentage of coacervate and B the inlet temperature.

Table 4 presents the results of the analysis of variance (ANOVA) for the response variables of the factorial design. The experimental data showed a significant fit to the predicted mathematical models, with coefficients of determination (R^2) ranging from 0.51 to 0.97. These results allowed the experimental values to be fitted to the proposed mathematical model, considering the R^2 values and statistical significance ($p \leq 0.05$) to optimize the conditions of the spray-drying microencapsulation process [17,25].

Table 4. ANOVA of the factorial design for each dependent variable.

Parameter	*p*-Value Model	R^2	Intercept	A	B	AB	A^2	B^2
Phenolic compounds	<0.0001	0.85	11.39	−1.77 *	−0.62 *	0.69 *	0.71 *	−0.61
p-values				<0.0001	0.004	0.01	0.04	0.08
Flavonoids	<0.0001	0.79	4.30	−1.36 *	−0.56 *	0.66 *	0.28	0.63
p-values				<0.0001	0.008	0.01	0.39	0.06
DPPH	<0.0001	0.95	47.93	−9.42 *	1.75 *	−0.50	4.10 *	−4.30 *
p-values				<0.0001	0.004	0.45	0.0002	0.0001
ABTS	<0.0001	0.86	43.97	−8.10 *	−0.27	−0.003	6.46 *	−16.86 *
p-values				<0.0001	0.82	0.99	0.004	<0.0001
Solubility	<0.0001	0.87	51.86	4.31 *	−3.01 *	−3.60 *	−11.20 *	17.98 *
p-values				0.001	0.02	0.02	<0.0001	<0.0001
Hygroscopicity	0.008	0.51	14.43	−1.13 *	0.90	−0.90	2.57 *	−1.41
p-values				0.04	0.10	0.16	0.008	0.12
Moisture	<0.0001	0.75	7.55	−1.91 *	0.21	0.54	1.08	−2.27 *
p-values				<0.0001	0.51	0.17	0.06	0.0003
Bulk density	<0.0001	0.97	246.01	27.85 *	−19.24 *	−2.75	−31.67 *	11.02 *
p-values				<0.0001	<0.0001	0.18	<0.0001	0.0007
Particle size	0.0003	0.65	6.98	0.06	0.10	−0.12	0.50 *	−1.06 *
p-values				0.62	0.38	0.39	0.02	<0.0001
ζ-Potential	0.0032	0.55	−25.45	2.06	2.20	−0.70	−9.12 *	1.80
p-values				0.12	0.09	0.64	0.0003	0.41

A and A^2 represent the linear and quadratic terms of the coacervate percentage; B and B^2 the linear and quadratic terms of the inlet temperature; and AB the interaction between both variables. R^2 indicates the coefficient of determination. An asterisk (*) indicates that the corresponding coefficient in the regression model is statistically significant at $p \leq 0.05$.

The response surface plots shown in Figure 2 illustrate the influence of coacervate concentration (A) and inlet temperature (B) on the properties of the microcapsules. A significant fit was observed between the mathematical models and the experimental data, indicating good predictive capacity for most of the evaluated variables.

Phenolic compounds (Figure 2a) showed high statistical significance (*p*-value $\leq$ 0.0001) and a satisfactory model fit (R^2 = 0.85). Their retention was mainly influenced by the linear effects of A and B, the quadratic effect of A, and the interaction of AB. In the case of flavonoids (Figure 2b), the model also revealed a significant relationship (*p*-value $\leq$ 0.0001, R^2 = 0.79), with the linear effects of A and B, and the interaction AB, being significant. Regarding antioxidant capacity measured by DPPH (Figure 2c), a strong correlation was found (*p*-value $\leq$ 0.0001, R^2 = 0.95), with significant linear effects of A and B, and significant quadratic effects of A and B. In the case of the antioxidant capacity determined by ABTS (Figure 2d), a strong correlation was also observed (*p*-value $\leq$ 0.0001, R^2 = 0.86), with a significant linear effect of A and significant quadratic effects of both A and B. Previous studies have shown that moderate inlet temperatures and adequate encapsulant ratios enhance the stability of bioactive compounds by reducing thermal degradation [4,6]. These findings emphasize the importance of avoiding high temperatures during encapsulation, as the wall material serves as a protective barrier against oxidation and heat, thereby preserving the antioxidant activity [17,25–28].

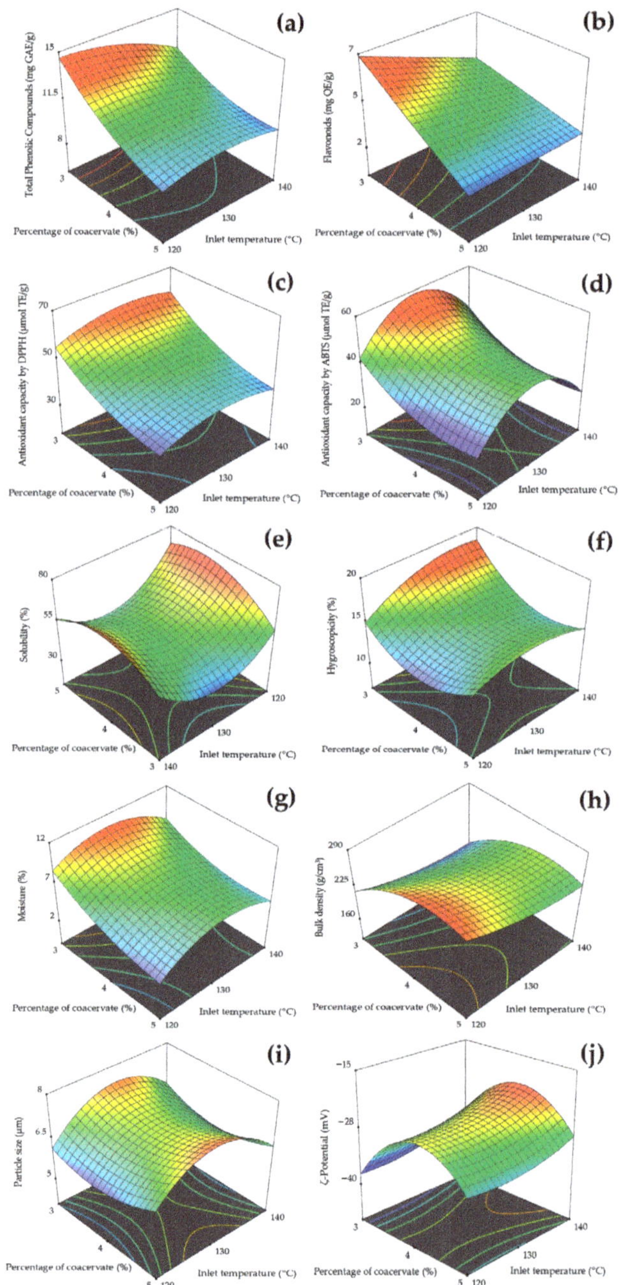

Figure 2. Response surface plots showing the effect of coacervate concentration and inlet temperature on: (**a**) total phenolic compounds, (**b**) flavonoids, (**c**) antioxidant capacity by DPPH, (**d**) antioxidant capacity by ABTS, (**e**) solubility, (**f**) hygroscopicity, (**g**) moisture, (**h**) bulk density, (**i**) particle size, and (**j**) ζ-potential.

The physical properties were also significantly affected by the process parameters. Solubility (Figure 2e) showed a significant relationship with the model (p-value ≤ 0.0001, $R^2 = 0.87$), with significant effects from the linear terms, interaction, and quadratic effects of A and B. For hygroscopicity (Figure 2f), a moderately significant correlation was observed (p-value ≤ 0.008, $R^2 = 0.51$), with both the linear and quadratic effects of A being influential. Moisture content (Figure 2g) showed a strong, significant correlation (p-value ≤ 0.0001, $R^2 = 0.75$), influenced by the linear effect of A and the quadratic effect of B. On the other hand, bulk density (Figure 2h) showed a highly significant correlation (p-value ≤ 0.0001, $R^2 = 0.97$), with significant contributions from the linear and quadratic effects of both A and B. For particle size (Figure 2i), a moderate correlation was found (p-value ≤ 0.0003, $R^2 = 0.65$), where the quadratic effects of A and B were relevant. Finally, ζ potential (Figure 2j) showed a moderately significant correlation (p-value ≤ 0.0032, $R^2 = 0.55$), with a significant effect from the quadratic term of A.

The aforementioned results demonstrate an overall improvement in powder quality under controlled spray-drying conditions, with positive effects also observed on the stability and functionality of the microcapsules. Controlling temperature and encapsulant concentration is thus key to optimizing the final product [4,25,29–31]. Unlike conventional encapsulating systems such as maltodextrin or gum Arabic, the combination of chia mucilage and gelatin offers additional functional advantages. This mixture provides greater gelling capacity, contributes soluble fiber, and may exert prebiotic effects. Furthermore, the mucilage acts as a natural stabilizer, reducing the thermal degradation of phenolic compounds during drying. This structural synergy could explain the high phenolic retention and sustained antioxidant activity observed in the propolis microcapsules [11,12].

To determine the optimal experimental conditions, a multiple-response analysis based on the desirability function was applied, with values ranging from 0 to 1 [32]. Table 5 shows that the optimal treatment was achieved with a 3.13% coacervate concentration and an inlet temperature of 120 °C, resulting in an overall desirability of 0.59. This result was obtained by maximizing the values of phenolic compounds, flavonoids, antioxidant capacity, and solubility while minimizing the values of hygroscopicity, moisture content, bulk density, particle size, and zeta potential.

The experimental results for the optimal treatment fell within the 95% confidence interval. They achieved prediction accuracy values above 90%, indicating the robustness of the factorial design used and the importance of adjusting process conditions to enhance the functionality of propolis microcapsules for potential applications in the food, pharmaceutical, and cosmetic industries.

The encapsulation efficiency (%EE) of phenolic compounds under optimal treatment was 80.75 ± 0.01%. This result demonstrates the effectiveness of the complex coacervation process, using chia mucilage and gelatin, in retaining bioactive compounds. This value is attributed to the moderate drying conditions and the appropriate proportion of biopolymers used, which helped minimize thermal degradation [6,12,21].

Table 5. Optimal parameters of the response variables.

Variable	Experimental Range			Optimal Value			Desirability
Independent variables	Low	High					
A: Percentage of coacervate (%)	3	5		3.13			0.59
B: Inlet temperature (°C)	120	140		120			
Dependent variables	Low	High	95% PI low	Experimental value	95% PI high	Predicted value	Assertiveness
Phenolic compounds (mg GAE/g)	9.16	15.36	13.00	14.26 ± 0.15	15.00	14.00	98%
Flavonoids (mg QE/g)	2.67	7.77	5.38	6.09 ± 0.56	7.85	6.61	92%
DPPH (μmol TE/g)	36.78	60.10	49.21	54.10 ± 0.23	56.15	52.68	97%
ABTS (μmol TE/g)	22.19	50.10	31.72	38.15 ± 0.68	46.79	39.26	97%
Solubility (%)	41.49	79.72	50.15	54.67 ± 0.13	65.02	57.59	95%
Hygroscopicity (%)	12.58	20.50	10.93	13.12 ± 0.81	17.56	14.25	92%
Moisture (%)	2.70	9.57	6.00	8.12 ± 0.25	10.02	8.01	99%
Bulk density (g/cm^3)	177.75	284.25	215.53	229.28 ± 0.86	236.41	225.97	99%
Particle size (μm)	5.39	7.80	5.31	5.68 ± 0.13	6.77	6.04	94%
ζ-Potential (mV)	−37.29	−13.20	−43.10	−33.51 ± 0.77	−27.09	−35.10	95%

PI: Prediction interval (95% confidence). Assertiveness refers to the percentage of experimental values that fall within the prediction interval. Experimental values are expressed as mean ± standard deviation (n = 3).

3.4. Identification and Quantification of Phenolic Compounds in the Optimal Treatment

Table 6 presents the phenolic compounds identified in the microcapsules obtained under optimal treatment conditions. Identification was performed by comparing retention times, molecular masses, and spectra with analytical standards. Nine phenolic compounds were identified: gallic acid, catechin, epicatechin, epigallocatechin gallate, rutin, myricetin, resveratrol, quercetin, and kaempferol. Retention times ranged from 0.76 to 9.12 min, with molecular weights from 170.12 to 610.52 g/mol. Catechin and epicatechin showed the highest concentrations, at 2.303 and 2.275 mg/g, respectively. Meanwhile, quercetin was detected at the lowest concentration (0.005 mg/g), possibly due to its high susceptibility to thermal degradation [21]. The presence of catechin, epicatechin, and resveratrol in these samples reflects the botanical variability of propolis from this Andean region (Andahuaylas), influenced by local species such as *Eucalyptus globulus* (eucalyptus), *Escallonia myrtilloides* (tasta), *Baccharis lanceolata* (chilca), *Prunus serotina* (capulí), *Alnus jorullensis* (alder), *Cupressus macrocarpa* (cypress), *Brassica napus* (wild mustard), and *Pinus radiata* (pine) [4]. Moreover, the use of ethanol as an extraction solvent favors the solubilization of flavanols and stilbenes, as previously reported in studies involving ethanolic extracts of propolis [33–35]. Likewise, the encapsulation system using complex coacervation with chia mucilage and gelatin favored the stabilization and recovery of the bioactive compounds during spray drying [10].

The corresponding chromatogram is shown in Figure 3, where the numbered peaks correlate with the phenolic compounds listed in Table 6. The detected compounds belong to the classes of phenolic acids, flavanols, flavonols, and stilbenes, all of which are recognized for their antioxidant, anti-inflammatory, and cardioprotective properties [20,21]. Rutin stands out due to its high molecular weight (610.52 g/mol) and glycosylated structure, making it particularly susceptible to thermal and oxidative degradation. Its clear detection after spray drying confirms the protective effect of the microencapsulation system [21].

Table 6. Phenolic compounds identified in the microcapsules.

Phenolic Compound	Class	Molecular Formula	Molecular Weight (g/mol)	*m/z*	Retention Time (min)	Quantity (mg/g)
Gallic acid	Phenolic acid	$C_7H_6O_5$	170.12	169.10	0.76	0.743 ± 0.01
Catechin	Flavanol	$C_{15}H_{14}O_6$	290.27	291.09	2.18	2.303 ± 0.04
Epicatechin	Flavanol	$C_{15}H_{14}O_6$	290.27	289.02	3.40	2.275 ± 0.03
Epigallocatechin gallate	Flavanol	$C_{22}H_{18}O_{11}$	458.37	457.08	3.65	0.043 ± 0.01
Rutin	Flavonol (glycoside)	$C_{27}H_{30}O_{16}$	610.52	609.15	4.97	0.028 ± 0.01
Myricetin	Flavonol	$C_{15}H_{10}O_7$	318.24	317.04	6.76	0.007 ± 0.00
Resveratrol	Stilbene	$C_{14}H_{12}O_3$	228.24	229.08	7.39	0.181 ± 0.02
Quercetin	Flavonol	$C_{15}H_{10}O_7$	302.24	301.04	8.36	0.005 ± 0.00
Kaempferol	Flavonol	$C_{15}H_{10}O_6$	286.23	285.05	9.12	0.181 ± 0.02

Phenolic compounds were identified based on retention times (RTs), UV absorption spectra, molecular weights, and mass-to-charge (*m/z*) ratios, using certified external standards. Quantification was performed by UPLC-PDA-QDa in triplicate, based on compound-specific calibration curves, and the results are expressed as mean ± standard deviation in mg/g of dry microcapsules.

The identified compounds (gallic acid, epicatechin, rutin, resveratrol, quercetin, and kaempferol) are consistent with those reported by Duca et al. (2019) in multi-floral propolis from Western Romania, where resveratrol was detected in all analyzed samples, reaching up to 19.77% of total polyphenols [33]. Similarly, Volpi (2004) identified resveratrol, catechin, quercetin, rutin, and kaempferol in commercial propolis extracts from Italy, demonstrating that these compounds are also present in propolis from different regions and extraction matrices [34]. A global review on the chemical composition of propolis conducted by Huang et al. (2014) reported the detection of prenylated stilbenes, such as 5-farnesyl-3′-hydroxyresveratrol, 4-prenyldihydroresveratrol, and 3-prenylresveratrol, in propolis samples collected from Australia, Brazil, Greece, Indonesia, and Kenya, attributing their botanical origin to local plant species used by *Apis mellifera* [36].

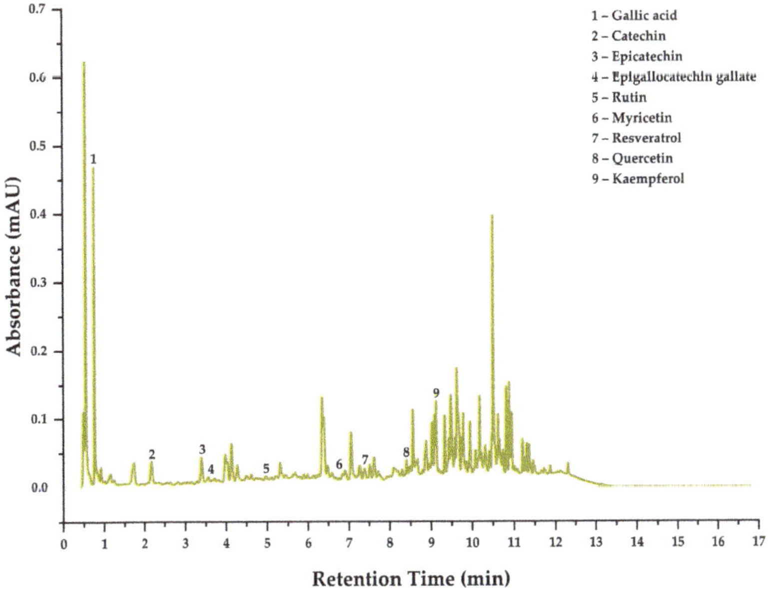

Figure 3. Chromatogram of the phenolic compounds identified in the microcapsules.

In addition, Rocha et al. (2024) [35] identified gallic acid, epicatechin, rutin, resveratrol, quercetin, and kaempferol in stingless bee propolis from Brazil, confirming the recurrence of these key metabolites across different regions and bee species. These findings demonstrate that the phenolic profile is strongly influenced by the botanical environment and bee biology, granting unique characteristics to the encapsulated propolis [4]. The detection of resveratrol is particularly relevant, as it suggests the influence of the local flora, especially *Eucalyptus globulus*, as a probable botanical source. This compound is found in a wide range of plants, including species of the *Myrtaceae* family, of which eucalyptus is a prominent representative and a floral source frequently visited by bees in various ecosystems [4,33,35,37].

Altogether, the simultaneous detection of these compounds after the encapsulation and spray-drying process confirms the effectiveness of this technique in preserving sensitive functional metabolites, particularly those with high reactivity and bioactive value, such as resveratrol and rutin [22].

3.5. Morphological, Structural, and Thermal Evaluation of the Optimal Treatment

Figure 4 shows the morphological, structural, and thermal characterization of the microcapsules obtained under optimal conditions. These analyses enabled the identification of key properties that support the effectiveness of the microencapsulation process.

Figure 4a displays the particle size distribution, with an average value of 5.68 μm and a polydispersity index of 2.17, indicating a relatively uniform distribution suitable for food, pharmaceutical, and cosmetic applications [6,38]. The ζ potential analysis yielded a value of −33.51 mV, indicating good colloidal stability, which is essential for preventing agglomeration during storage and enhancing the solubility of the final product [4,27]. In addition, color analysis revealed that the chromatic coordinates resulted in a light and homogeneous color associated with the stability of the bioactive compounds encapsulated under optimal conditions [4,6]. The SEM micrograph showed a spherical morphology with continuous and well-defined surfaces. Complementary EDS analysis confirmed that the predominant elemental composition was carbon (50.8%) and oxygen (49.2%), consistent with the organic nature of propolis and the biopolymers used as wall materials [4,6].

In Figure 4b, the FTIR spectrum confirms the presence of physical interactions between propolis and the wall materials, with no evidence of chemical reactions. Characteristic peaks were observed at 3342 cm^{-1} and 2932 cm^{-1}, associated with hydroxyl groups, amino acids, and carboxylic acids; the peak at 1652 cm^{-1} corresponds to phenolic compounds and flavonoids, while the one at 1045 cm^{-1} indicates the presence of ether and alcohol groups. Other minor peaks are related to aromatic rings and possible modifications induced by spray drying [4,6]. These results confirm that microencapsulation is a physical process that preserves the stability of bioactive compounds, validating their structural integrity for industrial applications [39,40].

Figure 4c, corresponding to the TGA, shows an initial weight loss of 4.54% at 105 °C, attributed to the removal of moisture. Continuous thermal degradation begins at 594 °C, indicating a significant improvement in the thermal stability of encapsulated propolis compared to its free form [4,6].

Finally, Figure 4d, corresponding to the DSC analysis, shows an endothermic transition at 146.97 °C, associated with microcapsule dehydration, and a main transition at 204.85 °C, indicating the thermal stability of the encapsulating matrix under controlled conditions [16,19].

Taken together, these results demonstrate the effectiveness of the microencapsulation process in enhancing the thermal, colloidal, and structural stability of propolis, consolidating this technique as a promising strategy for its application in the food, pharmaceutical, and cosmetic industries.

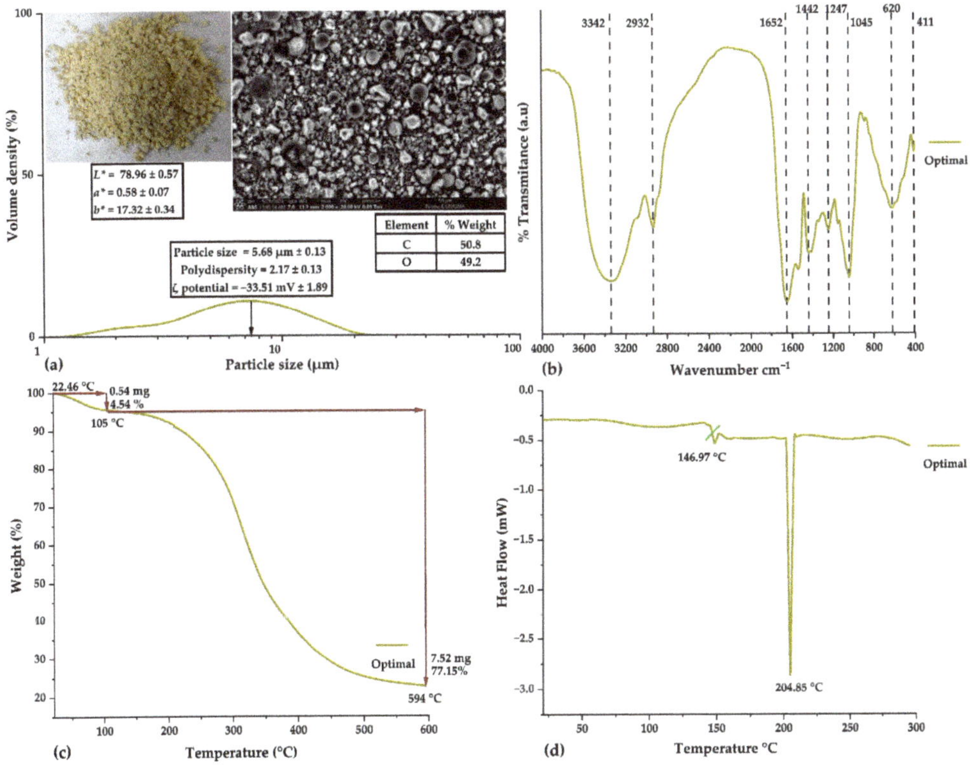

Figure 4. Physical, chemical, and structural properties of the optimal treatment: (**a**) particle size distribution, color, ζ potential, and micrograph (SEM and EDS); (**b**) FTIR spectrum; (**c**) thermogravimetric analysis (TGA); (**d**) differential scanning calorimetry (DSC).

3.6. In Vitro Release of Phenolic Compounds

Figure 5 shows the release profile of phenolic compounds in the optimal treatment, with a maximum concentration of 28.52 mg GAE/g reached at 24 h. Afterward, a gradual decline in the released concentration was observed, reaching 21.77 mg GAE/g at 48 h. This behavior demonstrates the encapsulating matrix's ability to control the release of bioactive compounds and protect them from degradation, thereby ensuring sustained delivery. These findings highlight the potential of the encapsulation system for use in functional products with controlled release in the pharmaceutical, cosmetic, and food industries [16,17,41].

The observed controlled release may be attributed to the semi-structured network formed through complex coacervation between chia mucilage and gelatin, whose electrostatic interaction created a diffusive barrier that modulated the release of phenolic compounds [11,12].

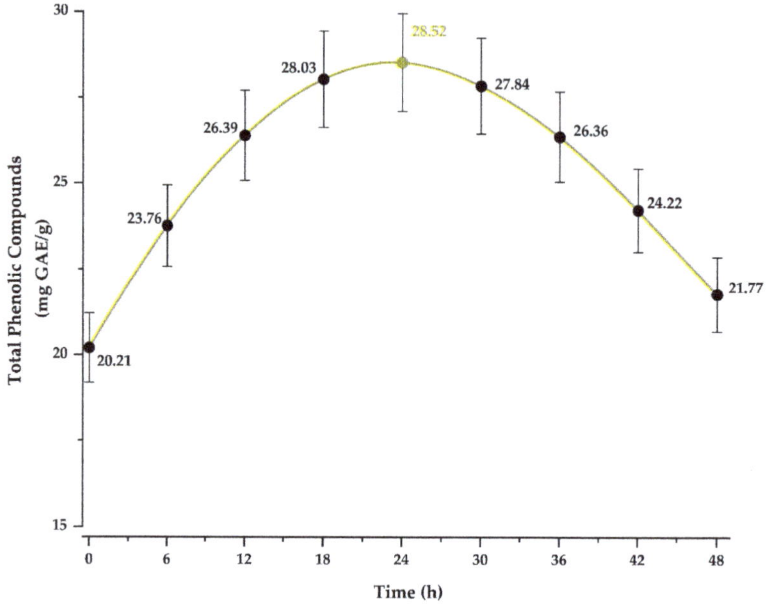

Figure 5. Release of total phenolic compounds in the optimal treatment.

3.7. Incorporation of Microcapsules into Functional Gummy Candies

Figure 6 shows the effect of adding propolis microcapsules obtained from the optimal treatment on the functional properties of gummy candies. It was observed that formulation F3, with the highest content of microcapsules and red cactus pear juice, reached the highest levels of phenolic compounds (2.64 mg GAE/g, Figure 6a), flavonoids (0.62 mg QE/g, Figure 6b), and antioxidant capacity by DPPH (10.73 μmol TE/g, Figure 6c). Conversely, a higher percentage of water significantly reduced these values, indicating a dilution effect on the bioactive compounds. The Pareto charts and response surface plots confirm that the optimal combination of microcapsules and red cactus pear juice significantly ($p \leq 0.05$) enhances the functional properties of the gummy candies. In contrast, a higher proportion of water negatively affects their functionality.

The phenolic compound levels exceeded those reported by Cedeño et al. [42], who obtained values ranging from 0.165 to 0.271 mg GAE/g in gelatin candies enriched with green propolis. Similarly, the antioxidant capacity was higher than that reported by the same authors (0.55 to 1.35 μmol TE/g). Additionally, the phenolic compound levels were also higher than those reported by Kaewpetch et al. [38], who developed honey-based gummies with propolis extract (*Apis cerana*), reaching a maximum of 2.13 mg GAE/g.

Figure 7 shows the sensory evaluation and instrumental color characterization of the functional gummy candies containing propolis microcapsules.

Figure 7a presents the results of the preference test, where formulation F3 (50%) was the most preferred by the panelists. Figure 7b shows the attribute-based sensory analysis (color, aroma, texture, and flavor), where F3 also received the highest scores in most categories, except for aroma. The panelists' preference for F3 in both suggests a more favorable overall sensory perception. This trend may be attributed to the encapsulation process, which could have helped mask undesirable flavors and improve texture, thereby enhancing product acceptance [43,44].

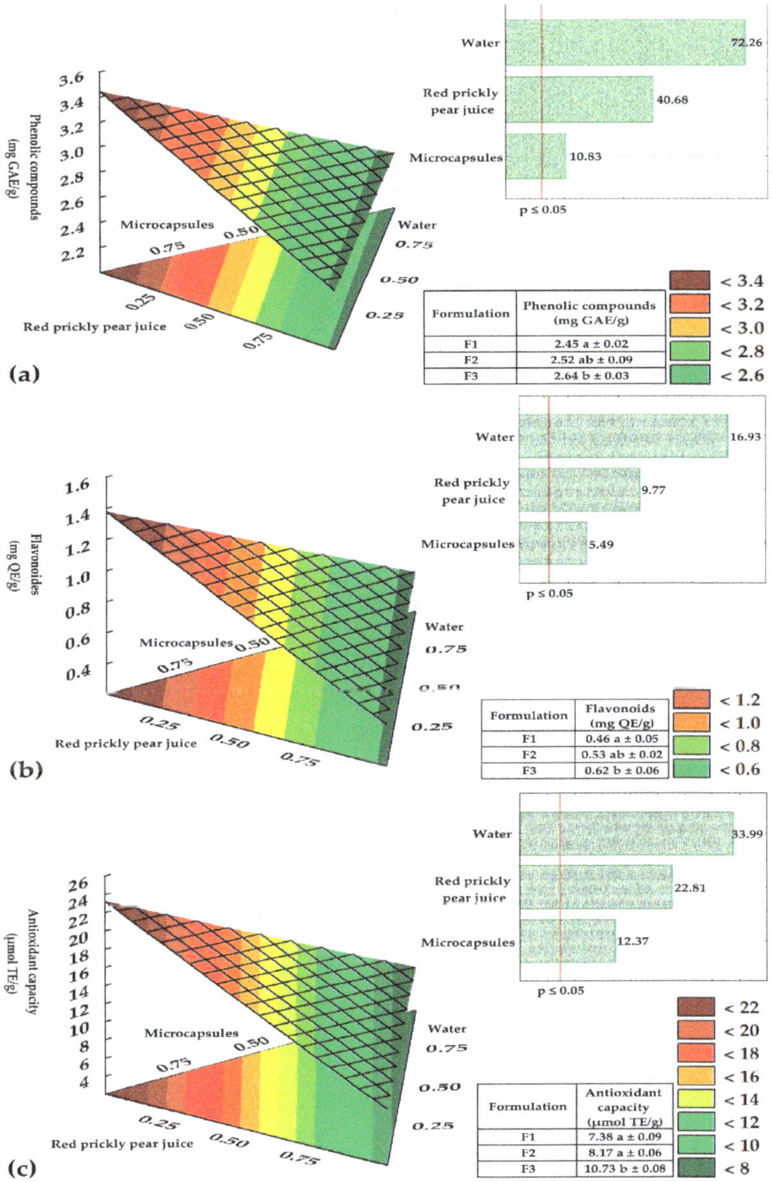

Figure 6. Functional properties of gummy candies: (**a**) phenolic compounds, (**b**) flavonoids, and (**c**) antioxidant capacity. Different letters (a, b) within the same column in the tables indicate statistically significant differences ($p \leq 0.05$) according to Tukey's test.

Figure 7c illustrates a balanced gender distribution among panelists, ensuring a well-balanced sensory analysis.

Figure 7d presents the instrumental color values, where a progressive decrease in lightness (L^*) is observed with increasing microcapsule content, indicating a darkening of the product. The a^* value increased in F3, reflecting a more reddish hue, while the b^*

value was higher in F1, corresponding to greater yellow intensity. These changes reflect the influence of propolis compounds on the final color.

Similar results were reported by Cedeño et al. [42] and Kaewpetch et al. [38], who also found that higher propolis content negatively affected the aroma attribute but improved color perception. Moreover, both studies reported decreases in L^* and variations in a^* and b^*, which is consistent with our findings and confirms the visual impact of propolis in gel-based matrices.

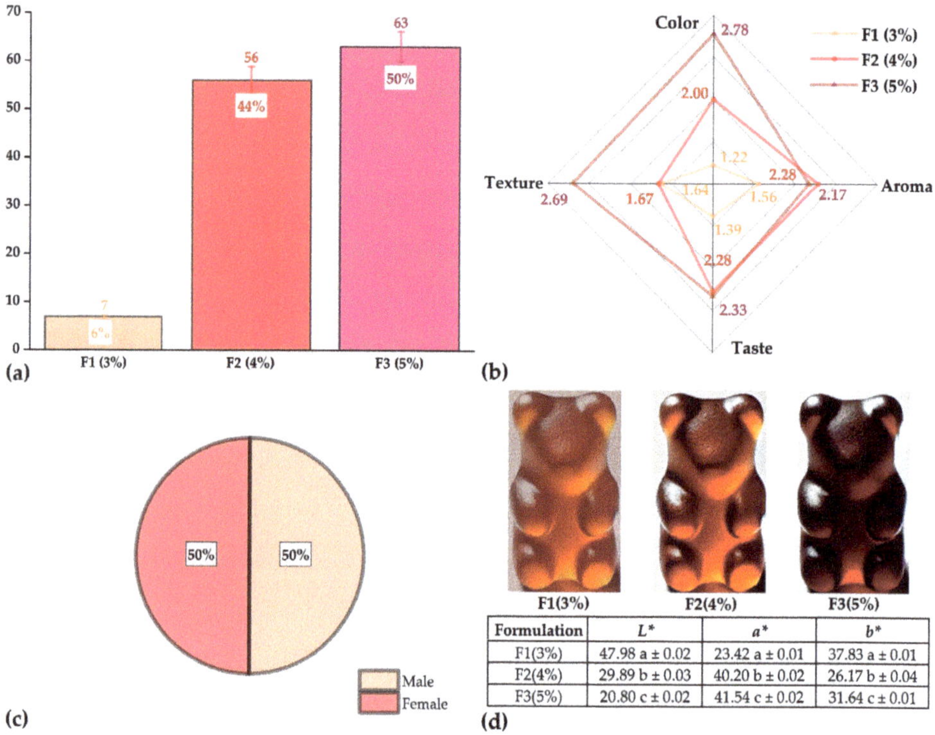

Formulation	L^*	a^*	b^*
F1(3%)	47.98 a ± 0.02	23.42 a ± 0.01	37.83 a ± 0.01
F2(4%)	29.89 b ± 0.03	40.20 b ± 0.02	26.17 b ± 0.04
F3(5%)	20.80 c ± 0.02	41.54 c ± 0.02	31.64 c ± 0.01

Figure 7. Characterization of gummy candies: (**a**) preference test, (**b**) acceptability test, (**c**) gender distribution, and (**d**) instrumental color of the gummies. L^*, a^*, and b^* represent the CIELAB color space parameters. Different letters (a, b, c) within the same column indicate statistically significant differences ($p \leq 0.05$) according to Tukey's test.

Table 7 presents the texture profile results of the functional gummy candies. Formulation F3 showed the highest hardness (2.58 N), attributed to its lower water content and higher proportion of microcapsules, whereas F2 recorded the lowest hardness (1.41 N), reflecting a less compact matrix. Regarding cohesiveness, adhesiveness, and springiness, F2 reached the highest values, likely due to the plasticizing effect of water. Similarly, resilience was greater in F2, indicating a better ability of the gel to recover its shape after deformation. In contrast, F3 showed the lowest values in all these parameters, which may be associated with the influence of microencapsulated propolis, which limits the gel's flexibility and recovery.

These results are consistent with those reported by Cedeño et al. [42] and Kaewpetch et al. [38], who found that incorporating propolis into gelatin-based products and gummy candies can significantly alter the texture, either reducing hardness or increasing stickiness, depending on the concentration and matrix. Moreover, the results showed lower hardness compared to the gummy candies developed by González et al. [45], who reported values ranging from 25.79 to 39.30 N. This behavior is possibly due to the higher water content and the interaction of bioactive compounds with the gelling matrix.

Table 7. Texture profile analysis in the gummy candies.

Formulation	Hardness (N)			Cohesiveness			Adhesiveness (g)			Springiness (mm)			Resilience		
	$\bar{x}$	±	SD	$\bar{x}$	±	SD	$\bar{x}$	±	SD	$\bar{x}$	±	SD	$\bar{x}$	±	SD
F1	1.49 [ab]	±	0.17	0.92 [ab]	±	0.03	3.40 [a]	±	0.80	3.05 [a]	±	0.03	0.51 [a]	±	0.04
F2	1.41 [a]	±	0.33	0.97 [a]	±	0.08	4.07 [a]	±	0.33	3.12 [a]	±	0.03	0.56 [a]	±	0.01
F3	2.58 [b]	±	0.89	0.87 [b]	±	0.02	1.53 [b]	±	0.31	3.02 [a]	±	0.12	0.50 [a]	±	0.03

Values are expressed as arithmetic mean ($\bar{x}$) ± standard deviation (SD), $n = 3$. Different letters in the same column indicate statistically significant differences ($p \leq 0.05$) according to Tukey's test.

Although the results obtained are promising, this study focused on the in vitro evaluation of antioxidant functionality. It did not include simulated gastrointestinal release assays, which would provide deeper insights into the bioavailability of the encapsulated compounds. Additionally, the formulation was applied only in a gummy candy matrix; therefore, future research could evaluate its behavior in other functional food systems, considering key factors such as storage stability and effective consumer dosage.

4. Conclusions

The microencapsulation of propolis by complex coacervation using chia mucilage and gelatin proved to be an effective strategy for protecting and stabilizing phenolic compounds in a gummy candy matrix. The optimal formulation exhibited high levels of total phenolics, flavonoids, and antioxidant capacity, along with favorable sensory acceptance. Liquid chromatography analysis confirmed the presence of bioactive compounds, including gallic acid, catechin, epicatechin, epigallocatechin gallate, rutin, myricetin, resveratrol, quercetin, and kaempferol, which remained detectable even after spray drying, demonstrating the efficiency of the encapsulation system. Overall, the results support the application of this technology in developing functional foods with enhanced stability and potential health benefits. It is worth noting that the use of Generally Recognized As Safe (GRAS) solvents, in combination with emerging extraction techniques, is increasingly promoted as a more sustainable approach to obtaining bioactive compounds. This avenue will be further explored in future studies. Additionally, future research should broaden the panel of analytical standards to achieve a more comprehensive characterization of the phenolic profile.

Author Contributions: Conceptualization, C.A.L.-S. and E.M.-M.; methodology, D.C.-Q. and H.P.-R.; software, C.A.L.-S. and R.J.G.G.; validation, E.M.-M. and D.C.-Q.; formal analysis, C.A.L.-S. and D.C.-Q.; investigation, C.A.L.-S., E.M.-M., I.B.M., and R.J.G.G.; data curation, I.B.M. and H.P.-R.; writing—original draft preparation, C.A.L.-S. and R.J.G.G.; writing—review and editing, C.A.L.-S. and R.J.G.G.; supervision, C.A.L.-S.; project administration, H.P.-R. All authors have read and agreed to the published version of the manuscript.

Funding: The research group on nutraceuticals and biomaterials of the UNAJMA supported the project.

Institutional Review Board Statement: Not applicable.

Informed Consent Statement: Informed consent was obtained from all subjects involved in the study.

Data Availability Statement: They are available in the same article.

Acknowledgments: The authors acknowledge the Food Nanotechnology Research Laboratory of UNAJMA.

Conflicts of Interest: The authors declare no conflicts of interest.

References

1. Braakhuis, A. Evidence on the Health Benefits of Supplemental Propolis. *Nutrients* **2019**, *11*, 2705. [CrossRef]
2. Chavda, V.P.; Vuppu, S.; Balar, P.C.; Mishra, T.; Bezbaruah, R.; Teli, D.; Sharma, N.; Alom, S. Propolis in the management of cardiovascular disease. *Int. J. Biol. Macromol.* **2024**, *266*, 131219. [CrossRef]
3. Pasupuleti, V.R.; Sammugam, L.; Ramesh, N.; Gan, S.H. Honey, Propolis, and Royal Jelly: A Comprehensive Review of Their Biological Actions and Health Benefits. *Oxidative Med. Cell. Longev.* **2017**, *2017*, 1259510. [CrossRef]
4. Ligarda-Samanez, C.A.; Choque-Quispe, D.; Moscoso-Moscoso, E.; Huamán-Carrión, M.L.; Ramos-Pacheco, B.S.; Peralta-Guevara, D.E.; Cruz, G.D.; Martínez-Huamán, E.L.; Arévalo-Quijano, J.C.; Muñoz-Saenz, J.C.; et al. Obtaining and Characterizing Andean Multi-Floral Propolis Nanoencapsulates in Polymeric Matrices. *Foods* **2022**, *11*, 3153. [CrossRef]
5. Jansen-Alves, C.; Martins Fonseca, L.; Doring Krumreich, F.; Zavareze, E.D.R. Applications of propolis encapsulation in food products. *J. Microencapsul.* **2023**, *40*, 567–586. [CrossRef]
6. Ligarda-Samanez, C.A.; Choque-Quispe, D.; Moscoso-Moscoso, E.; Huamán-Carrión, M.L.; Ramos-Pacheco, B.S.; De la Cruz, G.; Arévalo-Quijano, J.C.; Muñoz-Saenz, J.C.; Muñoz-Melgarejo, M.; Quispe-Quezada, U.R.; et al. Microencapsulation of Propolis and Honey Using Mixtures of Maltodextrin/Tara Gum and Modified Native Potato Starch/Tara Gum. *Foods* **2023**, *12*, 1873. [CrossRef]
7. Tavares, L.; Smaoui, S.; Lima, P.S.; de Oliveira, M.M.; Santos, L. Propolis: Encapsulation and application in the food and pharmaceutical industries. *Trends Food Sci. Technol.* **2022**, *127*, 169–180. [CrossRef]
8. Maroof, K.; Lee, R.S.F.; Siow, S.F.; Gan, S.H. Microencapsulation of propolis by spray drying: A review. *Dry. Technol.* **2022**, *40*, 1083–1102. [CrossRef]
9. Muhoza, B.; Yuyang, H.; Uriho, A.; Harindintwali, J.D.; Liu, Q.; Li, Y. Spray-and freeze-drying of microcapsules prepared by complex coacervation method: A review. *Food Hydrocoll.* **2023**, *140*, 108650. [CrossRef]
10. Sukri, N.; Marin, P.T.T.; Mahani; Nurhadi, B. Characteristics of propolis encapsulated with gelatin and sodium alginate by complex coacervation method. *Int. J. Food Prop.* **2023**, *26*, 696–707. [CrossRef]
11. Hernández-Nava, R.; López-Malo, A.; Palou, E.; Ramírez-Corona, N.; Jiménez-Munguía, M.T. Encapsulation of oregano essential oil (Origanum vulgare) by complex coacervation between gelatin and chia mucilage and its properties after spray drying. *Food Hydrocoll.* **2020**, *109*, 106077. [CrossRef]
12. Hernández-Nava, R.; López-Malo, A.; Palou, E.; Ramírez-Corona, N.; Jiménez-Munguía, M.T. Complex Coacervation Between Gelatin and Chia Mucilage as an Alternative of Encapsulating Agents. *J. Food Sci.* **2019**, *84*, 1281–1287. [CrossRef]
13. Bustamante, M.; Laurie-Martínez, L.; Vergara, D.; Campos-Vega, R.; Rubilar, M.; Shene, C. Effect of Three Polysaccharides (Inulin, and Mucilage from Chia and Flax Seeds) on the Survival of Probiotic Bacteria Encapsulated by Spray Drying. *Appl. Sci.* **2020**, *10*, 4623. [CrossRef]
14. Busch, V.M.; Pereyra-Gonzalez, A.; Šegatin, N.; Santagapita, P.R.; Poklar Ulrih, N.; Buera, M.P. Propolis encapsulation by spray drying: Characterization and stability. *LWT* **2017**, *75*, 227–235. [CrossRef]
15. Kurek-Górecka, A.; Kara, Y.; Pokajewicz, K.; Ben Hammouda, I.; Wieczorek, P.P.; Marciniak, D.; Balwierz, R.; Kłósek, M.; Czuba, Z.P.; Kolaylı, S. Phenolic content, volatile compounds and antioxidant activity in pooled propolis samples from Turkey and Poland. *Eur. Food Res. Technol.* **2025**. [CrossRef]
16. Ligarda-Samanez, C.A.; Choque-Quispe, D.; Moscoso-Moscoso, E.; Pozo, L.M.F.; Ramos-Pacheco, B.S.; Palomino-Rincón, H.; Gutiérrez, R.J.G.; Peralta-Guevara, D.E. Effect of Inlet Air Temperature and Quinoa Starch/Gum Arabic Ratio on Nanoencapsulation of Bioactive Compounds from Andean Potato Cultivars by Spray-Drying. *Molecules* **2023**, *28*, 7875. [CrossRef] [PubMed]
17. Ligarda-Samanez, C.A.; Choque-Quispe, D.; Moscoso-Moscoso, E.; Palomino-Rincón, H.; Taipe-Pardo, F.; Aguirre Landa, J.P.; Arévalo-Quijano, J.C.; Muñoz-Saenz, J.C.; Quispe-Quezada, U.R.; Huamán-Carrión, M.L.; et al. Nanoencapsulation of Phenolic Extracts from Native Potato Clones (*Solanum tuberosum* spp. *andigena*) by Spray Drying. *Molecules* **2023**, *28*, 4961. [CrossRef]
18. Ligarda-Samanez, C.A.; Palomino-Rincón, H.; Choque-Quispe, D.; Moscoso-Moscoso, E.; Arévalo-Quijano, J.C.; Huamán-Carrión, M.L.; Quispe-Quezada, U.R.; Muñoz-Saenz, J.C.; Gutiérrez-Gómez, E.; Cabel-Moscoso, D.J.; et al. Bioactive Compounds and Sensory Quality in Chips of Native Potato Clones (*Solanum tuberosum* spp. *andigena*) Grown in the High Andean Region of PERU. *Foods* **2023**, *12*, 2511. [CrossRef]
19. Moscoso-Moscoso, E.; Ligarda-Samanez, C.A.; Choque-Quispe, D.; Huamán-Carrión, M.L.; Arévalo-Quijano, J.C.; De la Cruz, G.; Luciano-Alipio, R.; Calsina Ponce, W.C.; Sucari-León, R.; Quispe-Quezada, U.R.; et al. Preliminary Assessment of Tara Gum as a

Wall Material: Physicochemical, Structural, Thermal, and Rheological Analyses of Different Drying Methods. *Polymers* **2024**, *16*, 838. [CrossRef]
20. Berretta, A.A.; Zamarrenho, L.G.; Correa, J.A.; De Lima, J.A.; Borini, G.B.; Ambrósio, S.R.; Barud, H.D.; Bastos, J.K.; De Jong, D. Development and Characterization of New Green Propolis Extract Formulations as Promising Candidates to Substitute for Green Propolis Hydroalcoholic Extract. *Molecules* **2023**, *28*, 3510. [CrossRef]
21. Andrade, J.K.S.; Denadai, M.; Andrade, G.R.S.; da Cunha Nascimento, C.; Barbosa, P.F.; Jesus, M.S.; Narain, N. Development and characterization of microencapsules containing spray dried powder obtained from Brazilian brown, green and red propolis. *Food Res. Int.* **2018**, *109*, 278–287. [CrossRef]
22. Andrade, J.K.S.; Denadai, M.; de Oliveira, C.S.; Nunes, M.L.; Narain, N. Evaluation of bioactive compounds potential and antioxidant activity of brown, green and red propolis from Brazilian northeast region. *Food Res. Int.* **2017**, *101*, 129–138. [CrossRef] [PubMed]
23. Ligarda-Samanez, C.A.; Villano-Limache, E.; Pichihua-Oscco, W.; Choque-Quispe, D.; Sucari-León, R.; Calderón Huamaní, D.F.; Cruz, G.D.; Luciano-Alipio, R.; Calsina Ponce, W.C.; Aroquipa-Durán, Y.; et al. Physicochemical and Sensory Evaluation of Gummy Candies Fortified with Microcapsules of Guinea Pig (*Cavia porcellus*) Blood Erythrocytes and Tumbo (*Passiflora tarminiana*) Juice. *Appl. Sci.* **2025**, *15*, 917. [CrossRef]
24. Tumbarski, Y.; Ivanov, I.; Todorova, M.; Apostolova, S.; Tzoneva, R.; Nikolova, K. Phenolic Content, Antioxidant Activity and In Vitro Anti-Inflammatory and Antitumor Potential of Selected Bulgarian Propolis Samples. *Biomedicines* **2025**, *13*, 334. [CrossRef]
25. Rosales-Chimal, S.; Navarro-Cortez, R.O.; Bello-Perez, L.A.; Vargas-Torres, A.; Palma-Rodríguez, H.M. Optimal conditions for anthocyanin extract microencapsulation in taro starch: Physicochemical characterization and bioaccessibility in gastrointestinal conditions. *Int. J. Biol. Macromol.* **2023**, *227*, 83–92. [CrossRef] [PubMed]
26. Das, A.B.; Goud, V.V.; Das, C. Microencapsulation of anthocyanin extract from purple rice bran using modified rice starch and its effect on rice dough rheology. *Int. J. Biol. Macromol.* **2019**, *124*, 573–581. [CrossRef] [PubMed]
27. Pedrozo, R.C.; Antônio, E.; Khalil, N.M.; Mainardes, R.M. Bovine serum albumin-based nanoparticles containing the flavonoid rutin produced by nano spray drying. *Braz. J. Pharm. Sci.* **2020**, *56*, e17692. [CrossRef]
28. Mousavi Kalajahi, S.E.; Ghandiha, S. Optimization of spray drying parameters for encapsulation of Nettle (*Urtica dioica* L.) extract. *LWT* **2022**, *158*, 113149. [CrossRef]
29. Kyriakoudi, A.; Tsimidou, M.Z. Properties of encapsulated saffron extracts in maltodextrin using the Büchi B-90 nano spray-dryer. *Food Chem.* **2018**, *266*, 458–465. [CrossRef]
30. Nayak, C.A.; Rastogi, N.K. Effect of Selected Additives on Microencapsulation of Anthocyanin by Spray Drying. *Dry. Technol.* **2010**, *28*, 1396–1404. [CrossRef]
31. Vergara, C.; Pino, M.T.; Zamora, O.; Parada, J.; Pérez, R.; Uribe, M.; Kalazich, J. Microencapsulation of Anthocyanin Extracted from Purple Flesh Cultivated Potatoes by Spray Drying and Its Effects on In Vitro Gastrointestinal Digestion. *Molecules* **2020**, *25*, 722. [CrossRef]
32. Bezerra, M.A.; Santelli, R.E.; Oliveira, E.P.; Villar, L.S.; Escaleira, L.A. Response surface methodology (RSM) as a tool for optimization in analytical chemistry. *Talanta* **2008**, *76*, 965–977. [CrossRef] [PubMed]
33. Duca, A.; Sturza, A.; Moacă, E.-A.; Negrea, M.; Lalescu, V.-D.; Lungeanu, D.; Dehelean, C.-A.; Muntean, D.-M.; Alexa, E. Identification of Resveratrol as Bioactive Compound of Propolis from Western Romania and Characterization of Phenolic Profile and Antioxidant Activity of Ethanolic Extracts. *Molecules* **2019**, *24*, 3368. [CrossRef] [PubMed]
34. Volpi, N. Separation of flavonoids and phenolic acids from propolis by capillary zone electrophoresis. *Electrophoresis* **2004**, *25*, 1872–1878. [CrossRef]
35. Rocha, V.M.; Portela, R.W.; Lacerda, L.E.; Sokolonski, A.R.; de Souza, C.O.; dos Anjos, J.P.; Nascimento, R.Q.; Umsza-Guez, M.A. Propolis from different Brazilian stingless bee species: Phenolic composition and antimicrobial activity. *Food Prod. Process. Nutr.* **2023**, *6*, 12. [CrossRef]
36. Huang, S.; Zhang, C.-P.; Wang, K.; Li, G.Q.; Hu, F.-L. Recent Advances in the Chemical Composition of Propolis. *Molecules* **2014**, *19*, 19610–19632. [CrossRef]
37. Freitas, M.O.; Ponte, F.A.; Lima, M.A.S.; Silveira, E.R. Flavonoids and triterpenes from the nest of the stingless bee Trigona spinipes. *J. Braz. Chem. Soc.* **2008**, *19*, 532–535. [CrossRef]
38. Heng, D.; Lee, S.H.; Ng, W.K.; Tan, R.B.H. The nano spray dryer B-90. *Expert Opin. Drug Deliv.* **2011**, *8*, 965–972. [CrossRef]
39. Delia, S.-C.; Chávez, G.M.; León-Martínez Frank, M.; Araceli, S.-G.P.; Irais, A.-L.; Franco, A.-A. Spray drying microencapsulation of betalain rich extracts from Escontria chiotilla and Stenocereus queretaroensis fruits using cactus mucilage. *Food Chem.* **2019**, *272*, 715–722. [CrossRef]

40. Pashazadeh, H.; Zannou, O.; Ghellam, M.; Koca, I.; Galanakis, C.M.; Aldawoud, T.M.S. Optimization and Encapsulation of Phenolic Compounds Extracted from Maize Waste by Freeze-Drying, Spray-Drying, and Microwave-Drying Using Maltodextrin. *Foods* **2021**, *10*, 1396. [CrossRef]
41. Fredes, C.; Becerra, C.; Parada, J.; Robert, P. The Microencapsulation of Maqui (*Aristotelia chilensis* (Mol.) Stuntz) Juice by Spray-Drying and Freeze-Drying Produces Powders with Similar Anthocyanin Stability and Bioaccessibility. *Molecules* **2018**, *23*, 1227. [CrossRef]
42. Cedeño-Pinos, C.; Marcucci, M.C.; Bañón, S. Contribution of Green Propolis to the Antioxidant, Physical, and Sensory Properties of Fruity Jelly Candies Made with Sugars or Fructans. *Foods* **2021**, *10*, 2586. [CrossRef] [PubMed]
43. Calderón-Oliver, M.; Ponce-Alquicira, E. The Role of Microencapsulation in Food Application. *Molecules* **2022**, *27*, 1499. [CrossRef] [PubMed]
44. Otálora, M.C.; Wilches-Torres, A.; Gómez Castaño, J.A. Microencapsulation of Betaxanthin Pigments from Pitahaya (*Hylocereus megalanthus*) By-Products: Characterization, Food Application, Stability, and In Vitro Gastrointestinal Digestion. *Foods* **2023**, *12*, 2700. [CrossRef]
45. González-Montiel, L.; Miranda-Altamirano, D.; Bautista-Marcial, A.S.; Güemes-Vera, N.; Soto-Simental, S.; Franco-Fernández, M.J.; Sánchez-Hernández, C.; Campos-Pastelín, J.M. Análisis de perfil de textura y color en gomitas elaboradas a partir de una deccoción de plantas medicinales. *Investig. Desarro. Cienc. Tecnol. Aliment.* **2019**, *1*, 756–760.

antioxidants

MDPI

Article

Influence of Kombucha Fermentation on Antioxidant and Antimicrobial Activity of Monofloral Rapeseed Bee-Collected Pollen

Aleksandar Ž. Kostić [1,*,†], Aleksandra Sknepnek [2,†], Danijel D. Milinčić [1], Uroš Gašić [3], Sofija Kilibarda [4] and Mirjana B. Pešić [1]

[1] Chair of Chemistry and Biochemistry, Faculty of Agriculture, University of Belgrade, Nemanjina 6, 11080 Belgrade, Serbia; danijel.milincic@agrif.bg.ac.rs (D.D.M.); mpesic@ag.rif.bg.ac.rs (M.B.P.)
[2] Chair of Industrial Microbiology, Faculty of Agriculture, University of Belgrade, Nemanjina 6, 11080 Belgrade, Serbia; alekavram@gmail.com
[3] Institute for Biological Research "Siniša Stanković", National Institute of Serbia, University of Belgrade, 11000 Belgrade, Serbia; uros.gasic@ibiss.bc.ac.rs
[4] Department for Field and Vegetable Crops, Faculty of Agriculture, University of Belgrade, Nemanjina 6, 11080 Belgrade, Serbia; sofija.kilibarda@agrif.bg.ac.rs
* Correspondence: akostic@agrif.bg.ac.rs
† These authors contributed equally to this work.

Abstract: Bee-collected pollen (BCP) can serve as an excellent enhancer of functional food bioactivity, particularly when it is fermented. The aim of this study was to prepare a novel kombucha-based beverage (KPE) enriched with fermented monofloral rapeseed (*Brassica napus* L.) BCP. To characterize the obtained samples, a proximate phytochemical composition analysis (including total phenolic and flavonoid content) was performed, as well as a detailed untargeted UHPLC-Q-ToF-MS profiling of phenolics and phenylamides. To biologically characterize KPE, antioxidant and antimicrobial activities were monitored. The total phenolic and flavonoid content, enhanced by the addition of BCP to the kombucha green tea beverage, was dose-dependent. The control sample showed a strong predominance of flavan-3-ols, distinguishing it from the KPE samples, where flavonol predominance and an increased content of phenolic acids were observed. Notably, the most significant markers of BCP were phenylamides, which were completely absent in the control. Although antioxidant activity was proximately highest in the control sample, KPE samples exhibited significantly improved antimicrobial activity.

Keywords: *Brassica napus* L. pollen; kombucha; phenolics; phenylamides; UHPLC Q-ToF-MS; biological activity

1. Introduction

Different foodstuffs can serve as sources of nutrients and bioactive compounds in the human diet. This is particularly true for plant-based food [1–3], and it is difficult to find a food that provides a better source of these nutrients than pollen. Pollen contains a balanced amount of sugars, proteins, and lipids, as well as an entire palette of various bioactive compounds, including phenolics, carotenoids, vitamins, sterols, alkaloids, etc. [4–6]. These properties make it an excellent functional food supplement and an "almost perfect food" [7,8]. Additionally, bee-collected pollen (BCP) has been recognized as an excellent source of pharmaceuticals [9] with notable potential for preventing various inflammatory processes, particularly neuroinflammation, as well as managing metabolic

 https://doi.org/10.3390/antiox14060752

disorders such as obesity, diabetes, and certain cardiovascular disorders [6,9]. However, due to the presence of two resilient membranes (intended to protect sensitive pollen grains from unfavorable environmental factors), along with a non-polar layer called sporopollenin, the bioaccessibility and bioavailability of pollen's nutrients and phytochemicals are often limited [10]. Recent research has confirmed that grinding BCP grains (a physical method) increases the bioaccessibility of various nutrients in camellia and lotus BCP, as measured after in vitro digestion of the samples [11].

The fermentation process is widely recognized as an enhancer of food quality. Through this process, new probiotics can be identified and incorporated into the diet, providing positive effects on human health [12]. Studies have shown that the nutritional quality of several vegetables improves after fermentation [13]. Among nutrients, proteins are particularly susceptible to fermentation, transforming into more desirable bioactive peptides with enhanced biological activity [14]. Interestingly, fermentation also reduces the content of antinutrients, including tannins, phytates, and certain proteins [13,14]. Given the various ways to improve the nutritional quality and bioactivity of BCP through different physical, chemical, or biological methods, fermentation is the most logical choice, as it naturally occurs in the hive after bees collect pollen. Through fermentation, bees transform BCP into a product known as bee bread (BB), or perga. Due to their shared botanical origin, BCP and BB have similar chemical compositions. However, fermentation is the primary factor driving the observed differences between these two bee products, especially regarding their free amino acids, free fatty acids, minerals, and phenolic content [15,16]. Studies have documented that BB possesses improved biological activity (particularly antioxidant activity) [17]. The most recent research also confirms BB's antimicrobial properties, with potential to prevent microbial adhesion to inert surfaces [18].

Kombucha is a well-known beverage produced through the fermentation of black or green tea (*Camelia sinensis* L.), although other types of tea can also serve as substrates [19]. Kombucha is slightly acidic and carbonated, which contributes to its popularity among consumers. It can be made as a low-alcohol beverage (similar to sparkling wine) or as a non-alcoholic drink (with less than 0.5% (v/v)) [19]. Kombucha offers several benefits for consumers, including the presence of different probiotic microbes (such as acetic acid bacteria), amino acids, phenolics from tea, sugars, organic acids, ethanol, water soluble vitamins, etc. [19]. There have been efforts to enhance the quality of BCP through fermentation processes using lactic acid bacteria (LAB) strains on an industrial pilot-scale level [20] including the application of kombucha [21,22]. However, data on kombucha fermentation of monofloral BCP are still lacking, including a detailed profile of bioactive compounds and the biological activity of the resulting kombucha pollen-enriched (KPE) beverage. The importance of using monofloral BCP lies in ensuring a relatively consistent chemical composition. Therefore, the aim of this research was to prepare a KPE beverage by fermenting monofloral rapeseed (*Brassica napus* L.) BCP. The selected samples (control and samples with the highest BCP addition) were subjected to detailed untargeted UHPLC Q-ToF-MS analysis to obtain a precise profile of phenolics and phenylamides, the most important bioactive constituents of BCP. In addition, the antioxidant properties and potential antimicrobial activity of the KPE beverage were evaluated. The findings of this study could be useful to future researchers in developing a novel functional beverage with enhanced bioactivity, which could be popular among consumers who strive toward a modern but healthier lifestyle.

2. Materials and Methods

2.1. List of Reagents

For kombucha fermentation and microbial analysis, the sucrose used was provided by Sunoko d.o.o. (Kovačica, Serbia) and the green tea (*Camellia sinensis* L.) substrate was obtained from Welton, Jadar-pak d.o.o., Osečina (Serbia). For HPLC analysis, methanol and acetonitrile were purchased from Carlo Erba (Val-de-Reuil, France). The sodium hydroxide for all of the experiments was purchased from Zorka Šabac (Sabac, Serbia) as well as sodium carbonate. Folin–Cicolteu reagent was obtained from Sigma Aldrich (Steinheim, Germany). Sodium nitrite was purchased from Alkaloid, Skoplje (North Macedonia) while aluminum chloride hexahydrate was purchased from Sigma Aldrich (Steinheim, Germany). For antioxidant assays, the following chemicals were used: sulfuric acid (Zorka Šabac, Serbia), sodium phosphate dodecahydrate (Alkaloid, Skoplje, North Macedonia), sodium molybdate (Uni-chem, Belgrade, Serbia), ferric chloride hexahydrate (Sigma Aldrich, Steinheim, Germany), sodium dihydrogenphosphate and sodium hydrogenphosphate (Merck, Darmstadt, Germany), trichloroacetic acid (Merck, Darmstadt, Germany), cupric chloride dehydrate (Merck, Darmstadt, Germany), neocuproine (Sigma Aldrich, Steinheim, Germany), ammonium acetate (Merck, Darmstadt, Germany), DPPH (2,2-Diphenyl-1-picrylhydrazyl) radical (Sigma Aldrich, Steinheim, Germany), ABTS (2,2′-azino-bis(3-ethylbenzothiazoline-6-sulfonic acid)) radical cation (Sigma Aldrich, Steinheim, Germany), and methanol (Zorka Šabac, Serbia). For all experiments, mili-Q water was obtained with Smart-DUV apparatus (Amtast USA Inc., Lakeland, FL, USA). All reagents were of analytical grade except for the chemicals used for HPLC analysis, which were of HPLC purity grade.

2.2. Bee-Collected Pollen Sample Collection

The BCP sample was obtained from a rapeseed plantation near Belgrade, courtesy of Professor N. Nedić. BCP was collected using pollen traps placed on the hive entrance. Pollen was collected over several days when the most intensive rapeseed (*Brassica napus* L.) was blossoming. Each day, after they were gathered from the traps, the grains were preliminary separated by shape and color and stored for preservation. At the end of the collection process, one representative sample was prepared and used for further palynological and chemical analyses. Its uniformity was confirmed (Figure 1) through color analysis, which showed a consistent light-yellow color. This was followed by palynological analysis, which verified that more than 99% of the pollen grains were from *B. napus* L., confirming the sample's monofloral nature, with over 80% of pollen grains originating from rapeseed plants [23]. The sample was then transferred to the chemical laboratory at the Faculty of Agriculture, where it was carefully dried at a mild temperature (~40 °C) to achieve a desirable humidity level (~10%) to prevent spoilage. The sample was then vacuum-sealed and stored at 4 °C until the beverage preparation and further analyses.

2.3. Pollen-Enriched Kombucha Sample Preparation

To prepare enriched and non-enriched kombucha samples, 70 g/L of sucrose was added to water and heated to boiling. Then, 3 g/L of green tea (*Camellia sinensis* L.) was added and allowed to infuse for 15 min. The prepared tea (180 mL) was transferred to sterile 0.5 L glass bottles, which were sealed with cotton plugs, and cooled to room temperature. Enrichment with BCP was performed by mixing in various concentrations of pollen—5 g/L, 10 g/L, and 20 g/L. The enriched and non-enriched teas were inoculated with 20 mL of actively fermenting kombucha beverage. At the same time, tea samples enriched with BCP but not inoculated with fermenting kombucha were prepared for spontaneous fermentation. All samples were prepared in three replicates. The fermentation of all the samples was carried out at 25 ± 2 °C for 7 days. Spontaneously fermented enriched samples containing 5 g/L, 10 g/L, and 20 g/L BCP were labelled 1, 2, and 3, respectively, while their kombucha-fermented counterparts were labelled K1, K2, and K3. Kombucha-fermented, non-enriched green tea was designated to be sample 0. The titratable acidity of the samples was determined using 0.1 M NaOH with phenolphthalein as an indicator and expressed as grams of acetic acid per liter. The cellulose pellicles that formed were removed from the surface, and the samples were filtered and stored in the freezer until further analysis. A summary of KPE preparation is presented in Figure 1.

Figure 1. Appearance of rapeseed monofloral bee-collected pollen (BCP) and illustration of kombucha-fermented BCP preparation.

2.4. Proximate Phytochemical Characterization and UHPLC Q-ToF-MS Analysis

To broadly describe the phytochemical composition of the samples, the total phenolic content (TPC) and total flavonoid content (TFC) were determined in accordance with the literature [24]. The results were expressed in mg/100 mL. Following this, three representative samples were selected and subjected to solid phase extraction (SPE) pretreatment for UHPLC Q-ToF-MS analysis. These samples included the control sample (in order to determine the influence of green tea substrate on the phytochemical profile) and two samples with the highest BCP content (20 g/L), with and without kombucha fermentation, to try to identify and quantify all possible phytochemicals originating from BCP. The control and KBE samples (10 mL) were passed through preconditioned SPE cartridges (CLEAN-UPR, C18 Extraction columns, Unendcapped-PKG50, UCT, Bristol, UK) to concentrate bioactive compounds and remove residual sugars and other impurities. The SPE cartridges were conditioned with 5 mL of acidified methanol and 5 mL of milliQ water, while adsorbed phenolics and phenylamides were eluted with acidified methanol (methanol containing 0.1% HCl). The samples were then filtered through 0.22 μm syringe filters and analyzed using UHPLC Q-ToF-MS. The kombucha samples were analyzed with an Agilent 1290 Infinity ultra-high-performance liquid chromatography (UHPLC) system coupled with a quadrupole time-of-flight mass spectrometer (6530C Q-ToF-MS, Santa Clara, CA, USA). All parameters and conditions for UHPLC Q-ToF-MS profiling were as previously reported [25]. Briefly, a Zorbax C18 column (2.1 × 50 mm, 1.8 μm) was used for chromatographic separation. Mobile phases comprised the following: (A) ultrapure water+0.1% HCOOH and (B) 98% acetonitrile+0.1% HCOOH) (MS grade). The flow rate was set at 0.3 mL/min, and the injection volume was 5 μL. The ESI operation parameters were the same as previously reported by Kostić et al. [25]. These included a nebulizer pressure of 45 psi, a drying gas temperature of 225 °C, a flow rate of 8 L/min, a sheath gas temperature of 300 °C, a sheath gas flow rate of 10 L/min, a capillary voltage of 2500 V, a fragmentor energy value of 175 V, a skimmer voltage of 65 V, and an octopole RF peak at 750 V. Phenolic compounds were detected in negative ionization mode, while the analysis of phenylamides was performed in positive ionization mode, using auto MS/MS acquisition (m/z = 50–1700, scan rate 1 spectra/s). All bioactive compounds were identified based on their monoisotopic mass, MS fragmentation, and literature data [26–32].

Phenolic compounds were quantified, and the results were expressed as μg/100 mL of the obtained beverages. Due to the lack of specific standards, the content of some phenolic derivatives was expressed in equivalents of the structurally closest (available) standard (gallic acid for phenolic acids and derivatives; epicatechin for flavan-3-ols and derivatives; quercetin for flavanol aglycones and glycosides; and naringenin for flavanones). Quantification parameters (equation parameters, coefficient of determination, and limit of quantification) for the applied standards are given in Table S1. The relative content of individual phenylamides (%) was calculated as the ratio of the area of each individual phenylamide to the total identified phenylamides.

2.5. Determination of Antioxidant Properties

To fully understand the bioactivity of the obtained samples, six different standard antioxidant assays were employed: in vitro phosphomolybdenum total antioxidant capacity (TAC), ferric reducing power (FRP), cupric reducing antioxidant capacity (CUPRAC), 2,2-diphenyl-1-picrylhydrazyl ($DPPH^{\bullet}$) radical quenching ability, (2,2-azino-bis-3-ethylbenzothiazoline-6-sulphonic acid) radical cation ($ABTS^{\bullet+}$) quenching ability, and Fe^{2+} chelating ability. Detailed descriptions of the experimental procedures for TAC, FRP, CUPRAC, $DPPH^{\bullet}$, and $ABTS^{\bullet+}$ assays are available in prior research [24].

2.6. Determination of Antimicrobial Activity

To assess the antimicrobial properties, pollen-enriched kombucha samples were prepared and labelled as described in Section 2.2. To remove the antimicrobial influence of the synthesized acids, neutralized samples were also prepared by adding 0.1 M NaOH to achieve a pH of 7.0, and the samples were labelled as 0N, 1N, 2N, 3N, K1N, K2N, and K3N. Prior to analysis, all samples were filtered through 0.22 μm filters (Sartorius, Goettingen, Germany). For antimicrobial testing, four Gram-positive bacteria (*Bacillus spizizenii* ATCC 6633, *Staphylococcus aureus* ATCC 25923, *Staphylococcus aureus* clinical isolate, *Listeria monocytogenes* ATCC 19111), five Gram-negative bacteria (*Pseudomonas aeruginosa* ATCC 35302, *Escherichia coli* ATCC 25922, *Salmonella enterica* subsp. *enterica* serovar Enteritidis ATCC 13073, *Klebsiella aerogenes* ATCC 13048, *Acinetobacter baumannii* ATCC 19606), and one pathogenic yeast (*Candida albicans* ATCC 10231) were used. The antimicrobial activity of the samples was determined using a microdilution assay as previously described [33]. The results were expressed as minimal inhibitory concentration (MIC), minimal bactericidal concentration (MBC), or minimal fungicidal concentration (MFC) in % of the sample content, which ranged from 50% to 0.39%. Antibiotic penicillin and antimycotic nystatin were used to analyse the susceptibility of tested strains in concentrations that ranged from 20 to 0.05 mg/mL.

2.7. Statistical Analysis

Statistical analysis was conducted using R Studio 4.3.1 software, employing analysis of variance (ANOVA) and Tukey's post hoc test to determine the statistically significant differences ($p < 0.05$).

3. Results and Discussion

3.1. Proximate Phytochemical Composition of Prepared Samples

Previous studies confirm that extended kombucha fermentation (over 10 days) may diminish the beverage's refreshing and pleasant sensory properties, resulting in a vinegar-like taste [34]. Therefore, a 7-day fermentation period was selected for this study. The results for proximate phytochemical composition, obtained through total phenolic (TPC) and flavonoid (TFC) content determination, are presented in Table 1.

Table 1. Proximate phytochemical composition and antioxidant characterization of kombucha enriched with BCP, expressed as mg equivalents **/100 mL of sample. All results are expressed as mean value ± st. dev.

Sample	TPC * (GAE **)	TFC (QE)	TAC (AAE)	FRP (AAE)	CUPRAC (AAE)	DPPH• (TE)	ABTS•+ (TE)
0	16.98 ± 0.11 e	31.97 ± 0.14 a	156.46 ± 0.59 c	46.70 ± 0.13 a	424.20 ± 16.85 a	31.34 ± 0.89 a	34.87 ± 0.51 a
1	21.94 ± 0.17 d	12.84 ± 0.93 c	212.66 ± 0.37 a	13.96 ± 0.16 d	213.76 ± 4.06 c	10.14 ± 0.10 b,c	18.08 ± 0.71 d
K1	23.45 ± 0.33 c	11.99 ± 0.21 c	188.42 ± 1.58 b	15.85 ± 0.36 b	272.35 ± 14.47 b	10.69 ± 0.07 b	22.16 ± 0.54 b
2	25.31 ± 0.35 b	12.70 ± 0.36 c	180.70 ± 4.42 b	11.10 ± 0.19 f	196.52 ± 27.28 c	9.81 ± 0.06 b,c	16.66 ± 0.57 e
K2	24.66 ± 0.21 b	11.77 ± 1.29 c	159.81 ± 3.78 c	12.67 ± 0.36 e	204.05 ± 12.82 c	10.33 ± 0.01 b,c	18.68 ± 0.37 c,d
3	27.03 ± 0.59 a	12.56 ± 0.07 c	155.93 ± 1.34 c	13.66 ± 0.37 d	212.44 ± 5.03 c	9.53 ± 0.01 c	18.21 ± 0.03 d
K3	26.46 ± 0.32 a	14.77 ± 0.43 b	114.43 ± 3.88 d	14.88 ± 0.14 c	212.44 ± 5.03 c	9.99 ± 0.03 b,c	19.56 ± 0.03 c

0—control sample (contains only tea substrate); 1, 2, 3—samples fermented without kombucha enriched with 5, 10, and 20 g/L pollen, respectively; K1, K2, K3—samples fermented with kombucha enriched with 5, 10, and 20 g/L pollen, respectively; * TPC—total phenolic content; TFC—total flavonoid content; TAC—in vitro phosphomolybdenum total antioxidant capacity; FRP—ferric reducing power; CUPRAC—cupric reducing antioxidant capacity; DPPH•—2,2-Diphenyl-1-picrylhydrazyl radical quenching ability; ABTS•+—(2,2-azino-bis-3-ethylbenzothiazoline-6-sulphonic acid) radical cation quenching ability; ** GAE—gallic acid equivalents; QE—quercetin equivalents; AAE—ascorbic acid equivalents; TE—Trolox equivalents. Different lowercase letters (a–f) in the same column indicate significant differences among samples, as determined using analysis of variance (ANOVA) and Tukey's post hoc test, with a significance level of $p < 0.05$.

Pollen addition clearly enriched the prepared samples with additional bioactive compounds belonging to both groups. The highest TPC values were observed in the samples with 20 g of added pollen, regardless of kombucha fermentation, the results of which were not statistically different. However, for TFC, kombucha fermentation with the highest pollen content (20 g) resulted in the highest measured TFC. This outcome is expected, as flavonoids and their derivatives are known to be the most important class of phenolic compounds in BCP [35–37]. Similarly, authors from Romania observed increased TPC content in kombucha after pollen addition [21].

3.2. The Obtained UHPLC Q-ToF-MS Profile of Selected Samples

A detailed comparative UHPLC-Q-ToF-MS analysis was conducted on the control sample (kombucha-fermented green tea) alongside BCP samples, both without and with kombucha fermentation (samples 3 and K3, respectively) to determine specific biomarkers. To the best of the authors' knowledge, this is the first report with detailed untargeted phenolics/phenylamides analysis of KPE samples available in the literature. Based on the results presented in Table 2, the prepared beverages displayed distinct phytochemical profiles.

Table 2. Identification and quantification (μg/100 mL of sample) of phenolic compounds and their derivatives in kombucha beverages with/without bee-collected pollen, using UHPLC-QToF-MS.

RT	Compound Name	Formula	Calculated Mass	*m/z* Exact Mass	mDa	Fragments (MS²)	Refs. *	Samples (μg/100 mL)		
								0	3	K3
					Phenolic acids and derivatives					
5.73	**Hydroxybenzoic acid isomer I** [b]	$C_7H_5O_3^-$	137.02390	137.02480	−0.90	/	[25,38]	<LOQ	<LOQ	25.73
8.83	**Hydroxybenzoic acid isomer II** [b]	$C_7H_5O_3^-$	137.02390	137.02509	−1.19	/	[25,38]	<LOQ	<LOQ	<LOQ
7.75	**Cinnamic acid** [b]	$C_9H_7O_2^-$	147.04460	147.04543	−0.83	101.0412(58), 103.05682(7), **117.03663(100)**, 118.03995(10), 119.05082(7)	[35,39]	-	46.58	74.58
3.44	**Dihydroxybenzoic acid** [b]	$C_7H_5O_4^-$	153.01880	153.01970	−0.90	**108.02333(100)**, 109.03079(77)	[25,37]	-	45.29	<LOQ
1.68	**Gallic acid** [a]	$C_7H_5O_5^-$	169.01370	169.01470	−1.00	107.01514(10), 108.02348(9), 123.00827(12), **124.01806(84)**, **125.02585(100)**	Standard [27,35]	<LOQ	51.81	26.48
5.69	**Methyl gallate** [b]	$C_8H_7O_5^-$	183.02930	183.02985	−0.55	123.01056(35), **124.01831(100)**, 125.02362(6), **168.00384(2)**	[29]	<LOQ	<LOQ	<LOQ
6.33	**Sinapic acid** [b]	$C_{11}H_{11}O_5^-$	223.06060	223.06260	−2.00	109.03047(46), 121.032(21), 123.04656(11), 125.02461(9), 135.04836(13), **137.02591(100)**, 138.03402(99), 151.04375(10), 161.02543(10), 179.03436(8)	[40]	<LOQ	<LOQ	<LOQ
2.39	**Gallic acid hexoside** [b]	$C_{13}H_{15}O_{10}^-$	331.06650	331.06970	−3.20	124.0185(18), **125.02615(96)**, **168.00922(100)**, **169.01397(17)**, 331.07253(7)	[27]	<LOQ	27.37	34.50
6.47	**Coumaroylquinic acid isomer I** [b]	$C_{16}H_{17}O_8^-$	337.09230	337.09549	−3.19	**119.05185(100)**, 120.05465(10), **163.04224(45)**, 164.0453(5), 173.04689(3), **191.05832(15)**, 192.06292(2)	[27,28]	<LOQ	-	<LOQ
7.35	**Coumaroylquinic acid isomer II** [b]	$C_{16}H_{17}O_8^-$	337.09230	337.09515	−2.85	111.0467(23), **119.05188(52)**, 137.0275(11), **163.04257(28)**, **173.04812(100)**, **191.05846(7)**	[27,28]	<LOQ	-	<LOQ
2.09	**Galloylquinic acid isomer I** [b]	$C_{14}H_{15}O_{10}^-$	343.06650	343.06970	−3.20	**125.02662(1)**, 127.04233(3), 137.02634(6), 173.04847(2), **191.05879(100)**	[27,28]	<LOQ	27.74	26.54
3.30	**Galloylquinic acid isomer II** [b]	$C_{14}H_{15}O_{10}^-$	343.06650	343.06970	−3.20	107.01476(10), 111.04708(8), 124.01873(3), **125.02625(89)**, 137.02579(5), **169.01627(100)**, 173.0471(27), **191.05852(45)**	[27,28]	<LOQ	<LOQ	<LOQ
5.93	**Caffeoylquinic acid** [b]	$C_{16}H_{17}O_9^-$	353.08730	353.09044	−3.14	134.04159(10), **135.047(97)**, 136.05015(12), 161.02425(6), 173.04671(5), **179.03619(43)**, **191.05764(100)**, 192.06487(10)	[27,41]	<LOQ	<LOQ	-

Table 2. *Cont.*

RT	Compound Name	Formula	Calculated Mass	*m/z* Exact Mass	mDa	Fragments (MS^2)	Refs. *	Samples (µg/100 mL)		
								0	3	K3
					Phenolic acids and derivatives					
7.55	**Feruloylquinic acid** [b]	$C_{17}H_{19}O_9^-$	367.10290	367.10760	−4.70	111.04851(12), 129.05678(7), 130.08874(19), **134.04015(41)**, 173.04998(24), 174.04921(9), 175.04191(7), **191.05723(100)**, 192.05995(10), **193.05302(23)**	[41]	-	<LOQ	<LOQ
6.88	**Galloyl-HHDP-hexose** [b]	$C_{27}H_{21}O_{18}^-$	633.07280	633.07770	−4.90	169.01712(3), 249.04386(3), 275.02233(9), **301.00272(100)**, 302.00625(18), 331.07316(2), 463.05704(8), 633.0797(33)	[28,29]	26.09	<LOQ	<LOQ
					Σ			**26.09**	**198.78**	**187.83**
					Flavan-3-ols and derivatives					
6.74	**Catechin** [a]	$C_{15}H_{13}O_6^-$	289.07120	289.07420	−3.00	109.03155(95), 121.03162(28), 122.03861(16), **123.04722(100)**, 125.02644(43), 137.02677(28), 151.04223(32), 161.06179(13), 203.07402(20), 221.08493(13)	Standard [27]	7628.17	711.36	1283.74
7.35	**Epicatechin** [a]	$C_{15}H_{13}O_6^-$	289.07120	289.07420	−3.00	109.03111(93), 121.03105(28), **123.04687(100)**, 125.02618(45), 137.02672(25), 149.02669(15), 151.04176(29), 161.06232(13), 203.07438(19), 221.08592(11)	Standard [27]	1504.70	447.58	510.25
3.91	**Gallocatechin** [c]	$C_{15}H_{13}O_7^-$	305.06610	305.06920	−3.10	109.03095(15), 111.04691(22), 121.03207(5), 123.01092(12), 124.018(9), **125.02624(100)**, 137.02557(33), 139.04209(29), 165.02075(14), 167.03728(21), 219.06784(7)	[27]	709.41	783.37	581.65
4.92	**Epigallocatechin** [c]	$C_{15}H_{13}O_7^-$	305.06610	305.06920	−3.10	109.03141(15), 111.04717(20), 123.01039(11), 124.01854(9), **125.02647(100)**, 137.02679(36), 139.04187(33), 165.02131(19), 167.03747(21), 219.06795(7)	[27]	1109.42	856.09	1233.48
8.83	**Epiafzelechin 3-*O*-gallate** [c]	$C_{22}H_{17}O_9^-$	425.08730	425.09098	−3.68	124.01793(45), **125.02621(57)**, 137.02573(16), 149.02736(10), 168.00816(10), **169.01729(74)**, 187.08013(18), 189.05883(21), 205.08878(23), 211.07995(15), 229.08791(19), 255.07135(37), **273.07992(100)**, 274.08329(13)	[27]	107.99	-	-
8.29	**Epicatechin gallate** [c]	$C_{22}H_{17}O_{10}^-$	441.08220	441.08710	−4.90	109.03134(5), 124.01855(11), **125.02618(45)**, **169.0168(100)**, 170.01997(8), 179.03774(4), 203.07401(6), 205.05352(5), 245.085(13), **289.07557(20)**, 290.07856(4)	[27]	2203.25	234.44	282.18
7.48	**Epigallocatechin gallate** [c]	$C_{22}H_{17}O_{11}^-$	457.07710	457.08234	−5.24	124.01823(2), 125.02634(48), **169.01665(100)**, 170.01985(9), 305.07011(3)	[27]	829.36	138.32	158.26

Table 2. *Cont.*

RT	Compound Name	Formula	Calculated Mass	*m/z* Exact Mass	mDa	Fragments (MS[2])	Refs. *	Samples (µg/100 mL)		
								0	3	K3
						Flavan-3-ols and derivatives				
4.18	**Gallocatechin hexoside** [c]	$C_{21}H_{23}O_{12}^-$	467.11900	467.12140	−2.40	**125.02606(91)**, 137.02621(47), 165.0216(37), 167.03661(54), 179.03724(28), 219.06938(27), 261.08123(35), **305.07077(100)**, 306.07254(21), 467.12336(10)	[27]	71.56	40.61	25.07
6.40	**Epigallocatechin hexoside** [c]	$C_{21}H_{23}O_{12}^-$	467.11900	467.12390	−4.90	**125.02594(76)**, 137.02617(51), 161.02611(20), 165.02483(39), 167.0384(86), 169.02013(19), 179.0352(51), 261.07905(32), **305.07145(100)**	[27]	73.80	-	-
7.01	**Procyanidin dimer B type** [c]	$C_{30}H_{25}O_{12}^-$	577.13460	577.14030	−5.70	125.02592(91), 137.0266(16), 151.04409(11), 161.02718(26), 205.05107(12), 245.08291(24), 273.04275(10), **289.07448(100)**, 407.0819(73), 408.08588(28)	[27]	404.62	26.77	54.98
7.35	**Chalcan-flavan 3-ol dimer** [c]	$C_{30}H_{27}O_{12}^-$	579.15080	579.15600	−5.20	125.02547(15), 137.02661(10), 167.03743(10), 205.05567(14), 245.08484(38), **289.07504(100)**, 290.07943(22), 369.06652(9), 459.10279(6), 579.14044(6)	[28]	-	60.44	-
6.26	**Epigallocatechin-epigallocatechin 3-*O*-gallate** [c]	$C_{37}H_{29}O_{18}^-$	761.13540	761.14290	−7.50	125.02502(12), **169.01804(11)**, 297.04335(15), 299.05776(11), 327.056(34), 339.05854(16), 435.07646(11), 453.08687(47), 465.08847(25), 471.09785(13), 573.10382(11), **591.11999(100)**, 592.12505(36), 609.13553(70), 610.13481(32)	[27]	199.77	-	-
					Σ			**14,842.1**	**3298.98**	**4129.60**
						Flavonol aglycones and glycosides				
10.74	**Kaempferol** [d]	$C_{15}H_9O_6^-$	285.03990	285.04190	−2.00	143.05241(8), 151.00639(6), 159.04741(9), 171.04757(7), 185.06338(13), 187.04283(11), 211.04312(8), 229.05402(11), 239.03849(9), **285.04486(100)**, 286.04783(19)	[38]	<LOQ	1635.77	1617.70
9.91	**Quercetin** [a]	$C_{15}H_9O_7^-$	301.03480	301.03600	−1.20	107.01558(46), 121.03135(43), 149.02665(9), **151.0061(100)**, 152.00988(10), 179.00107(13), 187.04291(3), 245.04782(4), 301.04008(4)	Standard [39]	88.55	461.76	449.41
10.85	**Isorhamnetin** [a]	$C_{16}H_{11}O_7^-$	315.05050	315.05420	−3.70	107.01561(19), 151.00586(49), 152.01066(7), 227.03772(6), 243.03308(6), 255.03296(7), 271.02837(11), 283.02806(8), **300.03166(100)**, 301.03448(20)	Standard [42]	<LOQ	99.61	105.72

Table 2. *Cont.*

RT	Compound Name	Formula	Calculated Mass	*m/z* Exact Mass	mDa	Fragments (MS^2)	Refs. *	Samples (µg/100 mL)		
								0	3	K3
					Flavonol aglycones and glycosides					
9.06	**Kaempferol-3-*O*-rhamnoside** [d]	$C_{21}H_{19}O_{10}^-$	431.09780	431.10553	−7.73	227.04101(24), 255.03527(41), 256.03883(21), 257.04198(7), 271.08023(5), **284.03663(100)**, 285.04091(75), 431.10621(6)	[38]	<LOQ	-	-
8.56	**Kaempferol-3-*O*-hexoside** [d]	$C_{21}H_{19}O_{11}^-$	447.09270	447.09790	−5.20	227.03763(21), 228.04433(3), 255.03369(32), 257.04587(3), **284.03646(100)**, 285.04159(41), 447.10039(13)	[37]	29.65	120.87	163.17
8.22	**Quercetin 3-*O*-hexoside** [d]	$C_{21}H_{19}O_{12}^-$	463.08770	463.09320	−5.50	151.00521(4), 179.00053(3), 243.03319(2), 255.03379(6), 256.03636(1), 271.02862(11), **300.03139(100)**, 301.03764(48)	[27,38]	63.82	<LOQ	<LOQ
7.75	**Myricetin 3-*O*-hexoside** [d]	$C_{21}H_{19}O_{13}^-$	479.08260	479.08780	−5.20	151.0069(1), 179.00116(3), 271.02806(8), 287.02337(4), **316.02649(100)**, 317.03185(29), 318.0332(5), 479.08636(4)	[27]	30.94	-	-
8.89	**Kaempferol 3-*O*-(6″-acetyl) hexoside** [d]	$C_{23}H_{21}O_{12}^-$	489.10330	489.10871	−5.41	227.04146(5), 229.05163(5), 255.03556(11), 256.03807(5), 257.04548(7), **284.03729(100)**, 285.04252(96), 489.28126(3)	[29]	<LOQ	<LOQ	<LOQ
8.89	**Kaempferol 3-*O*-(6″-*O*-malonyl) hexoside** [d]	$C_{24}H_{21}O_{14}^-$	533.09310	533.10034	−7.24	227.03607(2), 229.05413(2), 255.03368(7), 257.04871(2), **284.03563(94)**, **285.04255(100)**, 286.04646(18), 287.04729(2), 489.10903(5), 490.11036(2)	[26]	<LOQ	-	<LOQ
8.14	**Kaempferol 3-*O*-(2″-*O*-pentosyl) hexoside** [d]	$C_{26}H_{27}O_{15}^-$	579.13500	579.13861	−3.61	227.03675(3), 255.03401(10), 257.04532(4), **284.03524(100)**, 285.04112(35), 429.08863(4), 579.14137(23), 580.14395(9)	[26]	-	<LOQ	<LOQ
8.36	**Kaempferol 3-*O*-(6″-*O*-rhamnosyl) hexoside** [d]	$C_{27}H_{29}O_{15}^-$	593.15060	593.15660	−6.00	227.03779(2), 255.03327(5), 256.03779(2), 257.04991(2), **284.03638(50)**, **285.04383(100)**, 286.04755(17), 593.1586(44), 594.16158(16)	[28,38]	30.94	32.55	60.58
7.75	**Isorhamnetin 3-*O*-(6″-*O*-hexosyl)hexoside** [d]	$C_{28}H_{31}O_{17}^-$	639.15610	639.16235	−6.25	271.02798(6), 299.02295(44), 300.02956(19), 301.03072(3), **314.04657(100)**, 315.0526(42), 459.09849(5), 639.16315(58), 640.16622(23)	[26]	<LOQ	-	<LOQ
8.35	**Kaempferol 3-*O*-(2″-pentosyl)-malonyl-pentoside** [d]	$C_{29}H_{31}O_{17}^-$	651.15610	651.15920	−3.10	135.04763(40), 145.03209(11), 255.0313(9), **284.03597(100)**, 285.04397(45), 399.20994(52), 509.28402(13), 519.26712(52), 535.2655(8), 651.16299(40)	/	-	<LOQ	<LOQ

Table 2. *Cont.*

RT	Compound Name	Formula	Calculated Mass	*m/z* Exact Mass	mDa	Fragments (MS²)	Refs. *	Samples (µg/100 mL)		
								0	3	K3
					Flavonol aglycones and glycosides					
8.02	**Kaempferol 3-*O*-(2″-rhamnosyl-6″-hexosyl) hexoside** [d]	$C_{33}H_{39}O_{20}^{-}$	755.20350	755.20606	−2.56	229.05162(2), 255.03363(2), 284.03595(12), **285.04393(100)**, 286.04858(17), 449.05875(4), 467.06532(5), 605.10631(2), 755.21071(84), 756.21563(38)	[26,43]	-	142.53	142.68
7.82	**Quercetin 3-*O*-(6″-rhamnosyl-hexosyl)hexoside** [d]	$C_{33}H_{39}O_{21}^{-}$	771.19840	771.20540	−7.00	151.00583(1), 179.00097(1), 271.02833(1), 300.03146(16), **301.03879(24)**, **771.20752(100)**, 772.21109(44), 773.21283(13)	[27,43]	212.29	30.26	81.36
6.57	**Kaempferol 3-*O*-(6″-hexosyl)hexoside-7-*O*-hexoside** [d]	$C_{33}H_{39}O_{21}^{-}$	771.19840	771.20839	−9.99	255.03458(2), 283.0282(6), **284.03554(13)**, 285.04319(10), 446.09042(9), 447.09619(4), **609.15384(100)**, 610.15678(36), 651.16692(1), 771.20778(1)	[38]	-	<LOQ	47.14
6.40	**Quercetin 3-*O*-(2″-hexosyl)hexoside-7-*O*-hexoside** [d]	$C_{33}H_{39}O_{22}^{-}$	787.19330	787.20056	−7.26	178.99502(3), 299.02313(26), 300.02955(9), **301.04217(11)**, 445.08287(3), 462.08742(58), **463.09198(49)**, **625.14705(88)**, 626.15142(31), **787.20256(100)**, 788.20381(38)	[38]	-	<LOQ	<LOQ
8.02	**Quercetin 3-*O*-(6″-rhamnosyl) hexoside** [d]	$C_{27}H_{29}O_{16}^{-}$	609.14569	609.15110	−5.41	151.00589(3), 179.00092(3), 255.03315(3), 271.02923(6), **300.03140(100)**, 301.03816(73)	[27,37]	164.79	155.44	193.02
7.88	**Kaempferol 3-*O*-(2″-hexosyl)hexoside** [d]	$C_{27}H_{29}O_{16}^{-}$	609.14569	609.15299	−7.30	227.03677(3), 255.03261(8), **284.03562(100)**, 285.04182(49), 429.08801(5)	[38]	-	167.83	181.06
7.54	**Quercetin 3-*O*-(2″-hexosyl)hexoside** [d]	$C_{27}H_{29}O_{17}^{-}$	625.14058	625.14793	−7.35	179.00092(4), 271.02771(5), **300.03166(100)**, **301.03668(30)**, 445.08254(2), 463.09269(2)	[38]	-	225.47	293.15
9.30	**Quercetin 3-*O*-(2″-rhamnosyl-3″-pentosyl-6″-coumaroyl)hexoside** [d]	$C_{41}H_{43}O_{22}^{-}$	887.22465	887.23370	−9.05	151.00495(3), 179.00133(3), 299.02251(12), **300.03151(29)**, **301.03918(83)**, 723.18393(3), 741.1968(20), **887.23568(100)**	[28]	39.06	-	-
9.23	**Quercetin 3-*O*-(2″-rhamnosyl-3″-hexosyl-6″-coumaroyl) hexoside** [d]	$C_{42}H_{45}O_{23}^{-}$	917.23527	917.24240	−7.13	169.01477(3), 178.99951(3), 299.01998(9), **300.03042(21)**, **301.03909(49)**, 609.14184(3), 753.19589(4), 771.20409(8), **917.24439(100)**	[28]	<LOQ	<LOQ	<LOQ
					Σ			**660.03**	**3072.07**	**3334.97**

Table 2. *Cont.*

RT	Compound Name	Formula	Calculated Mass	*m/z* Exact Mass	mDa	Fragments (MS^2)	Refs. *	Samples (µg/100 mL)		
								0	3	K3
						Flavone aglycone				
10.58	**Apigenin [a]**	$C_{15}H_9O_5^-$	269.04500	269.04763	−2.63	107.01539(26), **119.05147(20)**, 133.03093(53), 135.04631(16), **151.0058(12)**, 159.04666(22), 181.06752(15), 183.04727(15), 201.05826(11), 224.05064(12), 225.05784(9), **269.04913(100)**, 270.05167(21)	Standard	<LOQ	-	<LOQ
9.87	**Luteolin [a]**	$C_{15}H_9O_6^-$	285.03990	285.04310	−3.20	107.01603(18), 121.03196(5), **133.03111(100)**, 134.03446(11), 151.00573(31), 175.04358(15), 199.04331(12), 217.05309(7), 285.04402(45), 286.04861(8)	Standard	364.29	-	-
					Σ			**364.29**	**0**	**0**
						Flavanone aglycone				
10.58	**Naringenin [a]**	$C_{15}H_{11}O_5^-$	271.06060	271.06123	−0.63	107.01562(30), **119.05205(100)**, 120.05542(11), **151.00569(31)**, 152.00874(3), 161.06313(3), 165.02226(2), 177.02197(3), 187.04301(5)	Standard	<LOQ	477.61	395.35
9.06	**Eridictyol [e]**	$C_{15}H_{11}O_6^-$	287.05550	287.05723	−1.63	107.01722(11), **125.02643(100)**, 126.02753(11), 131.05415(11), 151.00542(17), 152.01339(16), 173.06236(8), 177.06071(27), 201.05888(6), 241.04914(6), 259.06193(10)	[37]	-	<LOQ	<LOQ
					Σ			**0**	**477.61**	**395.35**
					ΣΣ			**15,892.5**	**7047.44**	**8047.76**

Abbreviations: Bolded fragments are the main ones for every compounds; 0—control sample (kombucha-fermented (7 days) green tea sample,); 3—sample with pollen (20 g) fermented (7 days) without kombucha; K3—beverage with pollen (20 g) fermented (7 days) with kombucha; "LOQ"—less of limit quantification; "-"—nonidentified compounds; compound quantities expressed using available standards [a]; compounds expressed as gallic acid equivalents [b]; compounds expressed as epicatechin equivalents [c]; compounds expressed as quercetin equivalents [d]; compounds expressed as naringenin equivalents [e]. * Previously reported compounds in rapeseed/other bee pollens and/or *Camellia sinensis* tea.

The control sample (0), consisting of green tea fermented with kombucha, had a significantly different profile compared to the samples with 20 g of pollen addition (3 and K3). In total, sample 0 contained 15 flavonols and derivatives, 12 phenolic acids and derivatives, eleven flavan-3-ols and derivatives, and two flavones. However, a strong quantitative predominance of flavan-3-ols/derivatives was observed (14,842 μg/100 mL) compared to flavonols/derivatives (660 μg/100 mL), flavones (364 μg/100 mL), and phenolic acids/derivatives (26.1 μg/100 mL). The profile observed for sample 0 was expected due to the presence of green tea, in which flavan-3-ols/derivatives are known to comprise the main phenolic subclass [44]. Luteolin (flavone, 364 μg/100 mL) was quantified only in sample 0, suggesting its relation to the tea substrate. Among other, more specific compounds identified and/or quantified only in sample 0, galloyl-HHDP-hexose (26 μg/100 mL) was the only phenolic acid derivative. Additionally, three flavan-3-ols were determined as specific to sample 0: epiafzelechin-3-*O*-gallate (108 μg/100 mL), epigallocatechin-hexoside (74 μg/100 mL), and epigallocatechin-epigllocatechin 3-*O*-gallate (199 μg/100 mL). Interestingly, epiafzelechin-3-*O*-gallate is a highly specific metabolite of the *Camelia sinensis* plant, i.e., green tea [45]. Among flavonol derivatives, only two tea-related compounds were detected: myricetin-3-*O*-hexoside (31 μg/100 mL) and quercetin 3-*O*-(2″-rhamnosyl-3″-pentosyl-6″-coumaroyl)-hexoside (39 μg/100 mL). Kaempferol-3-*O*-rhamnoside was also identified, but it was below the LOQ. In contrast, samples 3 and K3 were characterized by the presence of naringenin, a flavanone not found in sample 0, while eriodyctiol was present but at a level below the LOQ. Sample 3 showed a qualitative phenolic profile with 17 flavonols/derivatives, 13 phenolic acids/derivatives, and nine flavan-3-ols/derivatives. In sample K3, there were additional increases in flavonols/derivatives (18 compounds) and phenolic acids/derivatives (14 compounds), while flavan-3-ol diversity decreased to eight compounds. However, compared to the control, the quantitative profile was completely altered. While the total amount of flavan-3-ols was significantly reduced (~3300–4130 μg/100 mL), flavonols were quantified in samples 3 and K3 at 3070 and 3330 μg/100 mL, respectively. This confirmed that kombucha fermentation positively influenced the enriched profile and bioavailability of flavonols in the obtained beverages. The total amount of phenolic acids/derivatives was 199 μg/100 mL in sample 3 and 187 μg/100 mL in sample K3. Previous studies indicate that the amount of phenolic acid in fermented BCP fluctuates depending on specific compounds. For example, in one study, caffeic, ferulic acid, and ellagic acid contents increased, while chlorogenic acid content decreased after fermentation [46]. Regarding specific biomarkers, a higher diversity was observed originating from pollen compared to the control. Three phenolic acids were identified/quantified: cinnamic acid (45.3 μg/100 mL), dihydroxybenzoic acid (26.5–58 μg/100 mL), and feruloylquinic acid (below LOQ). Among flavonols/derivatives, as many as seven compounds were identified/quantified, including, among others, quercetin 3-*O*-(2″-hexosyl)-hexoside (225–293 μg/100 mL), kaempferol 3-*O*-(2″-hexosyl)-hexoside (168–181 μg/100 mL), kaempferol 3-*O*-(2″-rhamnosyl-6″-hexosyl)-hexoside (142 μg/100 mL in both samples), kaempferol 3-*O*-(2″-*O*-pentosyl)-hexoside (below LOQ), and quercetin 3-*O*-(2″-hexosyl)hexoside-7-*O*-hexoside (below LOQ). Notably, kaempferol 3-*O*-(6″-hexosyl)-hexoside-7-*O*-hexoside (47 μg/100 mL) was quantified only in the kombucha-fermented sample, possibly due to kombucha activity.

However, more informative and conclusive data were obtained through UHPLC monitoring of the phenylamide profiles (Table 3). This group of bioactive compounds was completely absent in the control sample, confirming its exclusive connection to BCP. Phenylamides were overlooked in BCP for decades, likely due to the absence of appropriate standards and highly sophisticated analytical techniques. However, in recent years, several reports [25,47,48] have documented their presence in BCP, with profiles strongly influenced by the botanical origin of the pollen. Even more interestingly, previous authors have reported that phenylamides are the most abundant group of bioactive compounds in BCP, compared to flavonoids/derivatives [47]. To the best of our knowledge, this study is the first to identify and characterize phenylamides derived from rapeseed BCP. In the current study, 13 derivatives from this group of compounds were identified. Based on their polyamine core, these phenylamides can be classified into three groups: (1) spermine derivatives (eight compounds), (2) spermidine derivatives (four compounds), and (3) a putrescine derivate (one compound). The structures and names of all identified phenylamides, along with the most likely positions of phenolic acid moieties on the spermidine/spermine core, were proposed based on typical fragments (Table 3). Individual or combined coumaroyl, caffeoyl, feruloyl, and hydroxyferuloyl derivatives of spermidine and spermine were confirmed primarily in KPE beverages, with the coumaroyl moiety distinctly dominant. All identified spermidine derivatives contained one, two, or three coumaroyl moieties (compounds **2–4**, Table 3). Key fragments for the identification of these compounds were observed at *m/z* 147 (coumaroyl moiety), *m/z* 204 (coumaroyl propenamide), *m/z* 292 (coumaroyl spermidine)→275 (-NH_3), and *m/z* 438 (di-coumaroyl spermidine)→421 (-NH_3). These coumaroyl spermidine derivatives were previously found in various monofloral and polyfloral BCP samples [25,32,49,50]. Fragments observed at *m/z* 177, *m/z* 234, and *m/z* 322 indicate the presence of a feruloyl moiety in the spermidine structure (compound **4**, Table 3). In contrast to previous studies that mostly identified spermidine derivatives in pollen, this study detected eight specific spermine derivatives which could serve as potential chemotaxonomic markers for rapeseed BCP. All identified derivatives contained coumaroyl, caffeoyl, feruloyl, or hydroxyferuloyl moieties in various positions on the spermine core. The presence of tricoumaroyl spermine in pollen has been confirmed in several studies [25,32], while dicoumaroyl feruloyl spermine and dicoumaroyl caffeoyl spermine were identified and reported by [32] in analyses of eight monofloral BCP samples from China. Of particular interest are spermine derivatives containing a hydroxyferuloyl residue, indicated by fragments at *m/z* 193 (hydroxyferuloyl moiety), *m/z* 250 (hydroxyferuloyl propenamide), and *m/z* 321. Compound **11** (Table 3) was identified as N^1-coumaroyl-N^5-hydroxyferuloyl-N^{10}-caffeoyl spermine, with its proposed fragmentation pathway shown in Figure 2.

The structures of the other spermine derivatives were proposed based on typical fragments, as detailed in Table 3. Moreover, tentatively identified compounds and their fragments were additionally confirmed and supported by available literature, related to previously publish mass spectrometry data [38,51–55]. However, the proposed phenylamide structures should be confirmed in future studies using NMR or other analytical techniques.

Table 3. Identification, proposed structure/name and relative content (%) of phenylamides in kombucha samples with/without bee-collected pollen, using UHPLC-QToF-MS.

No	RT	Compounds (Proposed Phenylamide Name)	Formula	Calculated Mass	*m/z* Exact Mass	mDa	Fragments (MS^2)	Refs. *	Samples (%)			Ratio 3/K3
									0	3	K3	
1	7.41	**Benzoiyl putrescin (*N*[1]-benzoyl putrescin)**	$C_{11}H_{17}N_2O^+$	193.13410	193.13470	−0.60	105.07078(9), 108.08172(12), 120.08237(11), 134.09717(8), 136.11229(27), 148.11206(13), 150.0926(43), **164.10833(100)**, 165.11111(19)	/	-	0.46	2.72	0.17
2	7.92	**Coumaroyl spermidine (*N*[1]-coumaroyl spermidine)**	$C_{16}H_{26}N_3O_2{}^+$	292.20250	292.20449	−1.99	**119.05032(15)**, 146.16608(1), **147.04553(100)**, 148.0485(14), **204.10387(7)**	[38]	-	2.55	2.62	0.96
3	8.01	**Dicoumaroyl spermidine (*N*[1],*N*[5]-dicoumaroyl spermidine)**	$C_{25}H_{32}N_3O_4{}^+$	438.23930	438.24160	−2.30	**119.05045(8)**, 129.14018(4), **147.04577(95)**, **204.10406(100)**, 205.10683(22), 218.11927(16), 275.17809(15), **292.20439(39)**, 293.20749(10)	[26,38]	-	56.04	55.41	1.00
4	8.02	**Feruloyl coumaroyl spermidine (*N*[1]-feruloyl-*N*[5]-coumaroyl spermidine)**	$C_{26}H_{34}N_3O_5{}^+$	468.24980	468.25352	−3.72	145.02926(19), **147.04528(51)**, **177.0557(100)**, 178.05894(13), **204.10298(71)**, 205.10605(12), **234.11379(63)**, 235.11715(12), 275.17737(8), 292.20353(32), 293.20609(7), 322.21375(22), 468.25352(15)	[38]	-	1.65	1.59	1.02
5	9.90	**Tricoumaroyl spermidine (*N*[1],*N*[5],*N*[10]-tri-coumaroyl spermidine)**	$C_{34}H_{38}N_3O_6{}^+$	584.27610	584.28269	−6.59	119.05086(3), **147.0457(41)**, **204.10425(50)**, 275.1783(15), 292.20513(32), 293.20743(8), 420.23277(15), 421.22839(9), **438.24348(100)**, 439.2466(40)	[26,38]	-	0.02	0.01	1.63
6	8.96	**Tricoumaroyl spermine (*N*[1],*N*[5],*N*[10]-tri-coumaroyl spermine)**	$C_{37}H_{45}N_4O_6{}^+$	641.33390	641.33760	−3.70	129.13967(7), **147.04515(36)**, 203.1194(14), **204.10365(84)**, 205.10674(13), 275.17735(91), 477.28941(17), **495.29994(54)**, 496.30279(23), **641.33788(100)**, 642.34076(59)	[38]	-	2.04	1.96	1.03
7	8.62	**Dicoumaroyl caffeoyl spermine (*N*[1],*N*[5]-di-coumaroyl-*N*[10]-caffeoyl spermine)**	$C_{37}H_{45}N_4O_7{}^+$	657.32880	657.33260	−3.80	**147.04542(22)**, **204.10404(45)**, 205.1068(7), **220.09925(17)**, 275.1777(45), 276.18055(11), 291.17267(27), 292.17816(6), 349.2620(2), 365.2564(1), **511.29547(35)**, 512.29859(15), 640.3124(2), **657.334(100)**, 658.33662(57)	[26,38]	-	4.82	4.64	1.03
8	8.96	**Dicoumaroyl feruloyl spermine (*N*[1],*N*[5]-di-coumaroyl-*N*[10]-feruloyl spermine)**	$C_{38}H_{47}N_4O_7{}^+$	671.34450	671.34820	−3.70	**147.04525(19)**, **177.056(11)**, 204.10379(41), 234.11506(16), 275.17771(43), 305.18806(23), 451.2254(3), 495.2997(10), 525.31078(32), 526.31495(15), **671.34824(100)**, 672.35241(61)	[26,38]	-	3.14	3.08	1.01

Table 3. *Cont.*

No	RT	Compounds (Proposed Phenylamide Name)	Formula	Calculated Mass	*m/z* Exact Mass	mDa	Fragments (MS^2)	Refs. *	Samples (%)			Ratio 3/K3
									0	3	K3	
9	8.62	**Dicoumaroyl hydroxyferuloyl spermine (*N*^{1},*N*^{10}-di-coumaroyl-*N*^{5}-hydroxyferuloyl spermine)**	$C_{38}H_{47}N_4O_8^+$	687.33940	687.34340	−4.00	**147.04555(16)**, **193.05083(5)**, 204.10339(39), 205.10675(6), 250.11019(14), 275.17709(39), 321.18326(23), 495.30019(11), 541.30598(32), **687.34338(100)**, 688.34674(63)	[26,38]	-	16.96	16.39	1.02
10	8.96	**Diferuloyl coumaroyl spermine (*N*^{1},*N*^{10}-di-feruloyl-*N*^{5}-coumaroyl spermine)**	$C_{39}H_{49}N_4O_8^+$	701.35500	701.35750	−2.50	**147.0452(11)**, **177.05649(12)**, 204.10372(30), 234.1151(9), 275.17743(34), 276.18067(7), 305.18736(14), 335.20036(9), 525.30953(13), 555.3214(18), **701.35917(100)**, 702.36096(65), 703.35572(30)	[38]	-	1.37	1.21	1.12
11	8.36	**Coumaroyl hydroxyferuloyl caffeoyl spermine (*N*^{1}-coumaroyl-*N*^{5}-hydroxyferuloyl-*N*^{10}-caffeoyl spermine)**	$C_{38}H_{47}N_4O_9^+$	703.33430	703.33870	−4.40	**147.0451(7)**, 204.1034(3), **220.0986(24)**, **250.1098(11)**, 275.1770(4), 291.1721(26), 321.1830(17), 467.2221(2), 511.2947(8), 541.3054(11), 557.3008(17), 686.3193(1), **703.3387 (100)**, 704.3422(64)	[26,38]	-	4.38	4.32	1.00
12	8.69	**Coumaroyl hydroxyferuloyl feruloyl spermine (*N*^{1}-coumaroyl-*N*^{5}-hydroxyferuloyl-*N*^{10}-feruloyl spermine)**	$C_{39}H_{49}N_4O_9^+$	717.35000	717.35390	−3.90	**147.04485(5)**, **177.05619(10)**, 204.10388(10), 234.11428(15), **250.11079(9)**, 275.17762(13), 305.18827(15), 321.1835(15), 525.31035(7), 541.30497(11), 553.30564(5), 571.31718(12), **717.35426(100)**, 718.35733(62)	[26,38]	-	6.10	5.64	1.07
13	9.57	**Tricaffeoyl hydroxyferuloyl spermine**	$C_{47}H_{53}N_4O_{13}^+$	881.36090	881.36498	−4.08	**220.09787(4)**, **250.10682(3)**, 291.17122(6), 321.18306(4), 557.29922(7), 689.32273(4), 701.32262(5), **719.33302(100)**, 720.33609(64), 881.36498(9)	[26,38]	-	0.47	0.41	1.14
							∑Total area 3/∑Total area K3					0.99

Abbreviations: Bolded fragments are the main ones for every compounds; 0—control sample (kombucha-fermented (7 days) green tea sample); 3—sample with pollen (20 g) fermented (7 days) without kombucha; K3—beverage with pollen (20 g) fermented (7 days) with kombucha; "-"—nonidentified compounds; The relative content (%) of individual phenylamide in each sample is given. Target compounds, expected retention time (RT), molecular formula, calculated mass, exact mass, and MS^2 fragments are presented. * Previously reported compounds in rapeseed or other bee pollens.

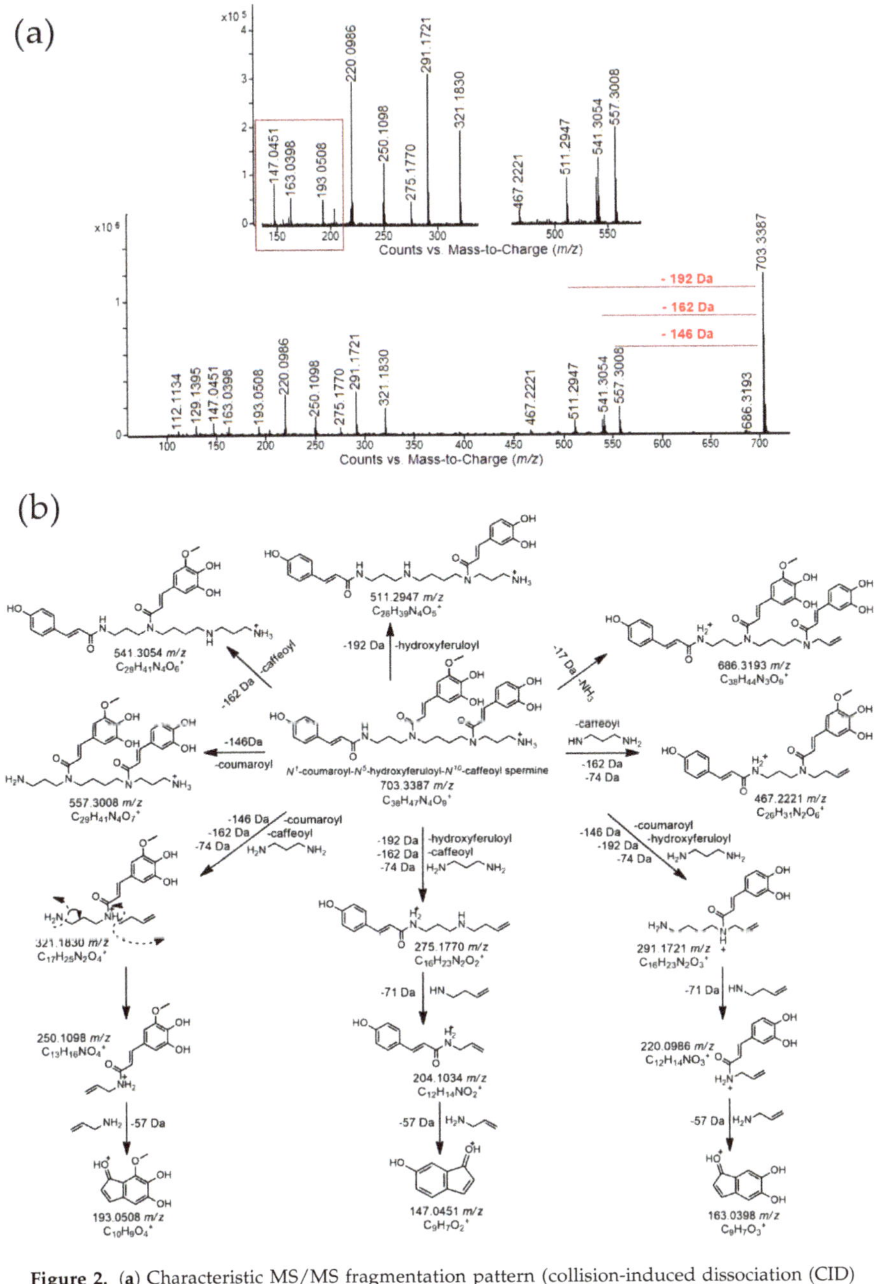

Figure 2. (**a**) Characteristic MS/MS fragmentation pattern (collision-induced dissociation (CID) mass spectra) and (**b**) proposed fragmentation pathway corresponding to N^1-coumaroyl-N^5-hydroxyferuloyl-N^{10}-caffeoyl spermine (703 m/z) (Agilent, Q-ToF, ESI(+), CE = 30 eV).

Due to the lack of appropriate standards, the relative content of individual phenylamides was calculated to indicate their presence/dominance in the samples. In both samples containing pollen, the phenylamide identified as N^1,N^5-dicoumaroyl spermidine was the predominant compound, accounting for 55–56% of the total phenylamides detected. This was followed by compounds recognized as N^1,N^5-di-coumaroyl-N^{10}-hydroxyferuloyl spermine (16–17%) and N^1-coumaroyl-N^5-hydroxyferuloyl-N^{10}-feruloyl spermine (5–5.6%). These results indicate that coumaroyl derivatives were the most predominant, originating from the initial BCP obtained from rapeseed. Since there are no data available on pollen-enriched beverages for comparison, the phenylamides' profile will be matched to previously published monofloral BCP samples. For instance, coconut BCP was characterized by the presence of tricoumaroyl spermidine [36] unlike rapeseed BCP, where this compound was only present in trace amounts (0.01–0.02%). Additionally, *Castanea* sp. BCP from Spain contained tricoumaroyl spermidine and a qualitative predominance of caffeoyl spermidine derivatives [56], though quantification was not performed. Corn poppy monofloral BCP from Serbia showed significantly higher diversity in these compounds, with 27 different derivatives and a strong predominance of coumaroyl spermine derivatives [25]. Research from Portugal also reported a tetracoumaroyl spermine isomer (3.3 mg/g) as the most predominant bioactive compound in BCP, primarily from *Plantago* sp. and *Crepis capillaris* plants [47]. The provided data and observed differences highlight the importance of phenylamides as carriers of BCP bioactivity and underscore their potential as a strong tool for BCP chemotaxonomy.

3.3. *Antioxidant Properties of Obtained Samples*

Analyzing the antioxidant properties of a foodstuff is an important means of gaining insights into its biological activity. Consequently, the KPE sample, the pollen-enriched sample without kombucha, and the control sample were analyzed using five standard antioxidant assays, and the results are presented in Table 1. It can be observed that for all applied assays, except the in vitro phosphomolybdenum total antioxidant capacity (TAC), the control sample exhibited the highest activities. In all pollen-enriched samples, there were significant decreases in the obtained values, which is likely linked to the reduction in flavan-3-ols content in all pollen-enriched samples, discussed in detail in the previous section. Indeed, flavan-3-ols are exceptional antioxidants [57]. Although pollen-enriched samples contained a significantly higher content of flavonols and phenolic acids, it has been reported that the antioxidant activity of different phenolic subclasses decreases in the following order: procyanidin dimers (derivatives of catechins) < flavanols < flavonols < hydroxycinnamic acid derivatives < simple phenolic acids [58]. It can be speculated that the decrease in flavan-3-ol content may be due to increased microbial activity in pollen-enriched samples, resulting from the high nutrient availability in pollen. Flavan-3-ols from tea, for instance, are known to be metabolized by microbes in the human colon, transforming into several metabolites, including methyl, glucuronide, and sulphate derivatives [59]. For TAC values, the highest antioxidant capacity was observed in the sample without kombucha fermentation (31) containing 5 g/L of pollen with 212.7 mg/100 mL AAE. It is possible that the high sugar content in BCP (significantly higher than in green tea) contributed to the higher or comparable antioxidant values in all pollen-enriched samples relative to the control. Rapeseed BCP, in particular, contains sugars as the predominant nutrients, comprising more than 62% d.w. [60]. In pollen-enriched samples, kombucha fermentation was associated with increased antioxidant activity in all KPE samples across assays (FRP; CUPRAC, DPPH$^{\bullet}$, ABTS$^{\bullet+}$) compared to the samples without kombucha, except in TAC, as explained earlier. Interestingly, in all assays except TAC, the KPE beverage with 5 g/L of pollen exhibited the highest antioxidant activity, suggesting that this pollen concentration may be optimal for enhancing antioxidant capacity. With higher pollen content, potential interferences may arise due to the complexity of the food matrix, which includes various nutrients from pollen, phytochemicals from multiple classes, and microbes from the kombucha consortium.

3.4. Antimicrobial Activity of Obtained Samples

Minimal inhibitory, minimal bactericidal, and minimal fungicidal concentrations were determined for the obtained samples of green tea kombucha, pollen-enriched green tea kombucha, and spontaneously fermented green tea with BCP. The results are presented in Table 4. Additionally, these values were measured for neutralized samples to exclude the influence of product acidity. The acidity levels in the fermented samples were found to be 0.45 g/L, 1.1 g/L, 1.6 g/L, 1.9 g/L, 2.2 g/L, 2.3 g/L, and 2.8 g/L for samples 0, 1, 2, 3, K1, K2, and K3, respectively.

The spontaneously fermented sample 1, with the lowest concentration of added pollen (5 g/L), exhibited antimicrobial and antifungal activity against all tested microorganisms, with inhibition values lower than those of microbicidal activity. The neutralized sample 1 (1N) retained inhibitory activity against all Gram-negative bacterial strains and against *L. monocytogenes* among the Gram-positive bacteria, though it did not show any microbicidal properties. Following kombucha fermentation of the pollen-enriched green tea (K1), the antimicrobial activity was partially improved compared to sample 1, especially regarding the MBC values. This improvement was mainly due to the higher acidity, as the neutralized K1N sample demonstrated better activity with a lower MIC value only against *P. aeruginosa*, but it showed weaker activity against *K. aerogenes* and *E. coli* compared to sample 1N. The addition of a medium concentration of pollen (10 mg/L) did not improve the antibacterial activity of the spontaneously fermented sample. The improved activity of sample 2 compared to sample 1 is attributed to its higher acidity, as shown by the equal activity of neutralized samples 1N and 2N, except in the case of *S. aureus* ATCC, where samples 1 and 1N exhibited better activity. The fermented kombucha beverage with 10 mg/L of added pollen (sample K2) expressed improved inhibitory activity against *A. baumannii*, better bactericidal activity against *S. aureus* ATCC 25293, and better fungicidal activity compared to sample 2. However, sample 2 showed better inhibitory activity against the clinical isolate of *S. aureus* and better bactericidal activity against *S.* Enteritidis. After neutralization, sample K2N showed better inhibitory activity against *A. baumannii* and both *S. aureus* species compared to all neutralized samples at 5 mg/L and 10 mg/L (1N, K1N, and 2N) and a lower MIC for *K. aerogenes* than K1N. The inhibitory effect of sample K2N on other species remained the same as for the K1N sample.

Sample 3, with the highest pollen concentration (20 mg/L) and spontaneous fermentation, showed lower MIC values against *K. aerogenes*, *A. baumannii*, and *S. aureus* ATCC 25293 compared to sample 2. Additionally, MBC values for *P. aeruginosa* and both *S. aureus* species, as well as MFC values for *C. albicans*, were lower.

However, sample 3 demonstrated weaker bactericidal activity against *S.* Enteritidis compared to sample 2. After neutralization, sample 3N showed better inhibitory activity than sample 2N against *P. aeruginosa*, *A. baumannii*, and both *S. aureus* strains, although bactericidal and fungicidal activities were absent. The kombucha beverage with 20 mg/L pollen (K3) showed improved inhibitory activity against *K. aerogenes*, both *S. aureus* species, and *B. subtilis* and better bactericidal activity against *S. aureus* and *L. monocytogenes* compared to sample K2.

Table 4. Antimicrobial activity of kombucha samples enriched with BCP expressed as MIC, MBC, and MFC.

Sample		*P. aeruginosa*	S. Enteritidis	*K. aerogenes*	*A. baumannii*	*E. coli*	*S. aureus* ATCC	*S. aureus* Clinical Isolate	*L. monocytogenes*	*B. spizizenii*	*C. albicans*
Antibiotic/ Antimycotic * (mg/mL)	MIC	2.5 ± 0.0	<0.05	1.7 ± 0.7	1.6 ± 0.0	<0.05	<0.05	<0.05	<0.05	<0.05	0.3 ± 0.0
	MBC	13.3 ± 5.7	0.05 ± 0.0	6.7 ± 2.9	2.5 ± 0.0	0.05 ± 0.0	<0.05	<0.05	<0.05	<0.05	0.3 ± 0.0
0	MIC	50.0 ± 0.0% [a]	-	6.3 ± 0.0% [d]	-	-	-	50.0 ± 0.0% [a]	-	-	-
	MBC/ MFC	-	-	-	-	-	-	-	-	-	-
0N	MIC	-	-	-	-	-	-	50.0 ± 0.0% [a]	-	-	-
	MBC/ MFC	-	-	-	-	-	-	-	-	-	-
1	MIC	6.3 ± 0.0% [d]	1.6 ± 0.0% [e]	6.3 ± 0.0% [d]	6.3 ± 0.0% [d]	3.1 ± 0.0% [c]	3.1 ± 0.0% [e]	3.1 ± 0.0% [e]	12.5 ± 0.0% [c]	25.0 ± 0.0% [b]	50.0 ± 0.0% [a]
	MBC/ MFC	12.5 ± 0.0% [c]	25.0 ± 0.0% [a]	25.0 ± 0.0% [b]	12.5 ± 0.0% [c]	12.5 ± 0.0% [a]	25.0 ± 0.0% [b]	25.0 ± 0.0% [b]	50.0 ± 0.0% [a]	25.0 ± 0.0% [b]	50.0 ± 0.0% [a]
1N	MIC	50.0 ± 0.0% [a]	3.1 ± 0.0% [d]	12.5 ± 0.0% [c]	50.0 ± 0.0% [a]	6.3 ± 0.0% [b]	-	-	25.0 ± 0.0% [b]	-	-
	MBC/ MFC	-	-	-	-	-	-	-	-	-	-
K1	MIC	3.1 ± 0.0% [e]	1.6 ± 0.0% [e]	6.3 ± 0.0% [d]	6.3 ± 0.0% [d]	3.1 ± 0.0% [c]	12.5 ± 0.0% [c]	12.5 ± 0.0% [c]	6.3 ± 0.0% [d]	12.5 ± 0.0% [c]	50.0 ± 0.0% [a]
	MBC/ MFC	6.3 ± 0.0% [d]	6.3 ± 0.0% [c]	25.0 ± 0.0% [b]	12.5 ± 0.0% [c]	12.5 ± 0.0% [a]	50.0 ± 0.0% [a]	12.5 ± 0.0% [c]	25.0 ± 0.0% [b]	12.5 ± 0.0% [c]	-
K1N	MIC	50.0 ± 0.0% [a]	3.1 ± 0.0% [d]	-	50.0 ± 0.0% [a]	12.5 ± 0.0% [a]	-	-	25.0 ± 0.0% [b]	-	-
	MBC/ MFC	-	-	-	-	-	-	-	-	-	-
2	MIC	3.1 ± 0.0% [e]	1.6 ± 0.0% [e]	6.3 ± 0.0% [d]	6.3 ± 0.0% [d]	3.1 ± 0.0% [c]	12.5 ± 0.0% [c]	1.6 ± 0.0% [f]	6.3 ± 0.0% [d]	12.5 ± 0.0% [c]	25.0 ± 0.0% [b]
	MBC/ MFC	6.3 ± 0.0% [d]	6.3 ± 0.0% [c]	12.5 ± 0.0% [c]	6.3 ± 0.0% [d]	6.3 ± 0.0% [b]	50.0 ± 0.0% [a]	25.0 ± 0.0% [b]	25.0 ± 0.0% [b]	12.5 ± 0.0% [c]	50.0 ± 0.0% [a]
2N	MIC	50.0 ± 0.0% [a]	3.1 ± 0.0% [d]	12.5 ± 0.0% [c]	50.0 ± 0.0% [a]	6.3 ± 0.0% [b]	-	-	25.0 ± 0.0% [b]	-	-
	MBC/ MFC	-	-	-	-	-	-	-	-	-	-

Table 4. *Cont.*

Sample		*P. aeruginosa*	S. Enteritidis	*K. aerogenes*	*A. baumannii*	*E. coli*	*S. aureus* ATCC	*S. aureus* Clinical Isolate	*L. monocytogenes*	*B. spizizenii*	*C. albicans*
K2	MIC	3.1 ± 0.0% ^{e}	1.6 ± 0.0% ^{e}	6.3 ± 0.0% ^{d}	3.1 ± 0.0% ^{e}	3.1 ± 0.0% ^{c}	12.5 ± 0.0% ^{c}	6.3 ± 0.0% ^{d}	6.3 ± 0.0% ^{d}	12.5 ± 0.0% ^{c}	25.0 ± 0.0% ^{b}
	MBC/MFC	6.3 ± 0.0% ^{d}	12.5 ± 0.0% ^{b}	12.5 ± 0.0% ^{c}	6.3 ± 0.0% ^{d}	6.3 ± 0.0% ^{b}	50.0 ± 0.0% ^{a}	6.3 ± 0.0% ^{d}	25.0 ± 0.0% ^{b}	12.5 ± 0.0% ^{c}	25.0 ± 0.0% ^{b}
K2N	MIC	50.0 ± 0.0% ^{a}	3.1 ± 0.0% ^{d}	12.5 ± 0.0% ^{c}	25.0% ^{b}	12.5 ± 0.0% ^{a}	50.0 ± 0.0% ^{a}	50.0 ± 0.0% ^{a}	25.0 ± 0.0% ^{b}	-	-
	MBC/MFC	-	-	-	-	-	-	-	-	-	-
3	MIC	3.1 ± 0.0% ^{d}	1.6 ± 0.0% ^{e}	3.1 ± 0.0% ^{e}	3.1% ^{e}	3.1 ± 0.0% ^{c}	6.3 ± 0.0% ^{d}	1.6 ± 0.0% ^{f}	6.3 ± 0.0% ^{d}	12.5 ± 0.0% ^{c}	25.0 ± 0.0% ^{b}
	MBC/MFC	3.1 ± 0.0% ^{e}	12.5 ± 0.0% ^{b}	12.5 ± 0.0% ^{c}	6.3% ^{d}	6.3 ± 0.0% ^{b}	25.0 ± 0.0% ^{b}	12.5 ± 0.0% ^{c}	25.0 ± 0.0% ^{b}	12.5 ± 0.0% ^{c}	25.0 ± 0.0% ^{b}
3N	MIC	12.5 ± 0.0% ^{c}	3.1 ± 0.0% ^{d}	12.5 ± 0.0% ^{c}	25.0% ^{b}	6.3 ± 0.0% ^{b}	25.0 ± 0.0% ^{b}	50.0 ± 0.0% ^{a}	25.0 ± 0.0% ^{b}	-	-
	MBC/MFC	-	-.	-	-	-	-	-	-	-	-
K3	MIC	3.1 ± 0.0% ^{e}	1.6 ± 0.0% ^{e}	3.1 ± 0.0% ^{e}	3.1% ^{e}	3.1 ± 0.0% ^{c}	6.3 ± 0.0% ^{d}	1.6 ± 0.0% ^{f}	6.3 ± 0.0% ^{d}	6.3 ± 0.0% ^{d}	25.0 ± 0.0% ^{b}
	MBC/MFC	6.3 ± 0.0% ^{d}	12.5 ± 0.0% ^{b}	12.5 ± 0.0% ^{c}	6.3% ^{d}	6.3 ± 0.0% ^{b}	50.0 ± 0.0% ^{a}	6.3 ± 0.0% ^{d}	12.5 ± 0.0% ^{c}	12.5 ± 0.0% ^{c}	25.0 ± 0.0% ^{b}
K3N	MIC	12.5 ± 0.0% ^{c}	3.1 ± 0.0% ^{d}	12.5 ± 0.0% ^{c}	25% ^{b}	12.5 ± 0.0% ^{a}	50.0 ± 0.0% ^{a}	50.0 ± 0.0% ^{a}	25.0 ± 0.0% ^{b}	50.0 ± 0.0% ^{a}	-
	MBC/MFC	25.0 ± 0.0% ^{b}	-	50.0 ± 0.0% ^{a}	-	-	-	-	-	-	-

Different superscript letters in the same column indicate statistically significant differences ($p < 0.05$) among samples; * Penicillin was used to test the susceptibility of bacteria to antibiotics. Nystatin was used to test the susceptibility of yeast.

None of the neutralized samples, whether spontaneously fermented or fermented with kombucha, exhibited bactericidal activity until the pollen concentration was raised to 20 mg/L and fermented by a kombucha consortium. Sample K3N showed bactericidal activity against *P. aeruginosa* at a concentration of 25% and against *K. aerogenes* at a concentration of 50%. In addition, K3N showed inhibitory activity against all bacteria, with the strongest effect observed against *B. subtilis*. Chemical analysis of sample K3 indicated the highest content of total phenolic compounds (Table 1). UHPLC-QToF-MS analysis (Table 2) further revealed that sample K3 contained higher amounts of cinnamic acid, gallic acid hexoside, and hydroxybenzoic acid isomer I compared to sample 3. Moreover, K3 had an increased total content of flavonol aglycones and glycosides, which likely contributed to its enhanced antimicrobial activity. Bioactive compounds in pollen, such as phenolic and flavonoid compounds, are known to be antimicrobial agents. Additionally, pollen-related bacterial strains such as *Streptomyces*, known as antibiotic-producing strains, are associated with the suppression of pathogenic microorganisms.

It is important to emphasize that all pollen-enriched samples showed better antimicrobial activity compared to green tea kombucha samples (0 and 0N). The only exception was with the clinical isolate of *S. aureus*, where only samples K2N, 3N, and K3N had the same effect as the neutralized green tea kombucha sample. *S.* Enteritidis was the most sensitive to both acidic and neutralized samples, with MIC values of 1.6% and 3.1%, respectively. Sample K3 also showed effectiveness against the clinical isolate of *S. aureus*, with an MIC of 1.6%. Generally, Gram-negative bacteria were more sensitive than Gram-positive, whereas antibiotic penicillin was primarily effective against Gram-positive bacterial species. This may indicate that the kombucha samples possessed bioactive compounds that act through mechanisms different from those of β-lactam antibiotics, with the potential to overcome the outer membrane of Gram-negative bacteria. As the neutralized samples did not affect *C. albicans*, it can be concluded that the antifungal activity was due only to the synthesized acids. These results show that the production of acids during the kombucha fermentation process is a suitable way to inhibit the growth of microorganisms and increase the shelf life and safety of a product by preventing spoilage and inhibiting pathogenic microorganisms [33].

4. Conclusions

Beverages with improved health functions are gaining popularity. In this study, the addition of fermented rapeseed bee-collected pollen (BCP) to a kombucha green tea-based beverage aimed to create a novel, healthier beverage. BCP was found to enhance the total phenolic content and increase the diversity of bioactive compounds, as identified and quantified using UHPLC-Q-ToF-MS analysis. Although the control sample (without BCP) exhibited the highest antioxidant activity in almost all assays (mostly due to the presence of flavan-3-ols and derivatives), KPE samples demonstrated superior antimicrobial activity compared to the control sample. Gram-positive bacteria were generally more susceptible to KPE's antibacterial activity compared to Gram-negative strains. At the highest dose of BCP in the kombucha beverage, improved activity was observed against *K. aerogenes*, *S. aureus*, and *B. subtilis*, as well as enhanced bactericidal activity against *L. monocytogenes*. Further research should focus on the sensory analysis of KPE, as well as further investigation into potential anticancer and anti-inflammatory properties of the BCP-enriched kombucha beverage to support its potential health benefits.

Supplementary Materials: The following supporting information can be downloaded at: https://www.mdpi.com/article/10.3390/antiox14060752/s1, Table S1: Equation parameters of phenolic standards used for quantification.

Author Contributions: Conceptualization, A.Ž.K. and A.S.; methodology, A.Ž.K., A.S. and D.D.M.; software, S.K., D.D.M. and U.G.; validation, D.D.M. and U.G.; formal analysis, A.Ž.K., A.S., D.D.M.; investigation, A.Ž.K., A.S. and D.D.M.; resources, A.Ž.K. and M.B.P.; data curation, A.Ž.K., A.S., D.D.M. and S.K.; validation, U.G. and D.D.M.; writing—original draft preparation, A.Ž.K., A.S. and D.D.M.; writing—review and editing, A.Ž.K.; visualization, A.Ž.K. and A.S.; supervision, A.Ž.K.; project administration, A.Ž.K. and M.B.P.; funding acquisition, M.B.P. All authors have read and agreed to the published version of the manuscript.

Funding: This research received external funding from the Science Fund of the Republic of Serbia, #GRANT No. 7744714, FUNPRO.

Institutional Review Board Statement: Not applicable.

Informed Consent Statement: Not applicable.

Data Availability Statement: Data will be made available on request.

Acknowledgments: This work was supported by the Ministry of Education, Science and Technological Development of the Republic of Serbia (contract no. 451-03-65/2025-03/200116). The authors are grateful to N. Nedić (University of Belgrade, Faculty of Agriculture) for providing bee-collected pollen samples.

Conflicts of Interest: The authors declare no conflicts of interest.

References

1. Tuso, P.J.; Ismail, M.; Ha, B.P.; Bartolotto, C. Nutritional Update for Physicians: Plant-Based Diets. *Perm. J.* **2013**, *17*, 61–66. [CrossRef] [PubMed]
2. Hever, J.; Cronise, R.J. Plant-Based Nutrition for Healthcare Professionals: Implementing Diet as a Primary Modality in the Prevention and Treatment of Chronic Disease. *J. Geriatr. Cardiol.* **2017**, *14*, 355–368. [CrossRef] [PubMed]
3. Nájera Espinosa, S.; Hadida, G.; Jelmar Sietsma, A.; Alae-Carew, C.; Turner, G.; Green, R.; Pastorino, S.; Picetti, R.; Scheelbeek, P. Mapping the Evidence of Novel Plant-Based Foods: A Systematic Review of Nutritional, Health, and Environmental Impacts in High-Income Countries. *Nutr. Rev.* **2025**, *83*, e1626–e1646. [CrossRef]
4. Thakur, M.; Nanda, V. Composition and Functionality of Bee Pollen: A Review. *Trends Food Sci. Technol.* **2020**, *98*, 82–106. [CrossRef]
5. Ecem Bayram, N.; Kostic, A.Ž.; Can Gercek, Y. (Eds.) *Pollen Chemistry & Biotechnology*; Springer International Publishing: Cham, Switzerland, 2023; ISBN 978-3-031-47562-7.
6. Kacemi, R.; Campos, M.G. Translational Research on Bee Pollen as a Source of Nutrients: A Scoping Review from Bench to Real World. *Nutrients* **2023**, *15*, 2413. [CrossRef]
7. Kostić, A.Ž.; Milinčić, D.D.; Barać, M.B.; Ali Shariati, M.; Tešić, Ž.L.; Pešić, M.B. The Application of Pollen as a Functional Food and Feed Ingredient—The Present and Perspectives. *Biomolecules* **2020**, *10*, 84. [CrossRef]
8. Li, Q.-Q.; Wang, K.; Marcucci, M.C.; Sawaya, A.C.H.F.; Hu, L.; Xue, X.-F.; Wu, L.-M.; Hu, F.-L. Nutrient-Rich Bee Pollen: A Treasure Trove of Active Natural Metabolites. *J. Funct. Foods* **2018**, *49*, 472–484. [CrossRef]
9. Kacemi, R.; Campos, M.G. Bee Pollen as a Source of Pharmaceuticals: Where Are We Now? In *Pollen Chemistry & Biotechnology*; Springer International Publishing: Cham, Switzerland, 2023; pp. 319–336.
10. Kaur, J.; Rasane, P.; Kumar, V.; Nanda, V.; Bhadariya, V.; Kaur, S.; Singh, J. Exploring the Health Benefits of Bee Pollen and Its Viability as a Functional Food Ingredient. *Rev. Agric. Sci.* **2024**, *12*, 65–78. [CrossRef]
11. Yuan, Y.; Zhong, S.; Deng, Z.; Li, G.; Zhang, J.; Li, H. Effect of Wall-Disruption on Nutrient Composition and in Vitro Digestion of Camellia and Lotus Bee Pollens. *Food Sci. Hum. Well.* **2024**, *13*, 1567–1577. [CrossRef]
12. Valentino, V.; Magliulo, R.; Farsi, D.; Cotter, P.D.; O'Sullivan, O.; Ercolini, D.; De Filippis, F. Fermented Foods, Their Microbiome and Its Potential in Boosting Human Health. *Microb. Biotechnol.* **2024**, *17*, e14428. [CrossRef]
13. Knez, E.; Kadac-Czapska, K.; Grembecka, M. Effect of Fermentation on the Nutritional Quality of the Selected Vegetables and Legumes and Their Health Effects. *Life* **2023**, *13*, 655. [CrossRef] [PubMed]
14. Tachie, C.Y.E.; Onuh, J.O.; Aryee, A.N.A. Nutritional and Potential Health Benefits of Fermented Food Proteins. *J. Sci. Food Agric.* **2024**, *104*, 1223–1233. [CrossRef] [PubMed]
15. Mayda, N.; Özkök, A.; Ecem Bayram, N.; Gerçek, Y.C.; Sorkun, K. Bee Bread and Bee Pollen of Different Plant Sources: Determination of Phenolic Content, Antioxidant Activity, Fatty Acid and Element Profiles. *J. Food Meas. Charact.* **2020**, *14*, 1795–1809. [CrossRef]

16. Bayram, N.E.; Gercek, Y.C.; Çelik, S.; Mayda, N.; Kostić, A.Ž.; Dramićanin, A.M.; Özkök, A. Phenolic and Free Amino Acid Profiles of Bee Bread and Bee Pollen with the Same Botanical Origin—Similarities and Differences. *Arab. J. Chem.* **2021**, *14*, 103004. [CrossRef]
17. Mărgăoan, R.; Stranț, M.; Varadi, A.; Topal, E.; Yücel, B.; Cornea-Cipcigan, M.; Campos, M.G.; Vodnar, D.C. Bee Collected Pollen and Bee Bread: Bioactive Constituents and Health Benefits. *Antioxidants* **2019**, *8*, 568. [CrossRef]
18. Ilie, C.-I.; Spoiala, A.; Geana, E.-I.; Chircov, C.; Ficai, A.; Ditu, L.-M.; Oprea, E. Bee Bread: A Promising Source of Bioactive Compounds with Antioxidant Properties-First Report on Some Antimicrobial Features. *Antioxidants* **2024**, *13*, 353. [CrossRef]
19. Coelho, R.M.D.; de Almeida, A.L.; do Amaral, R.Q.G.; da Mota, R.N.; de Sousa, P.H.M. Kombucha: Review. *Int. J. Gastron. Food Sci.* **2020**, *22*, 100272. [CrossRef]
20. Mora-Adames, W.I.; Fuenmayor, C.A.; Benavides-Martín, M.A.; Algecira-Enciso, N.A.; Quicazán, M.C. Bee Pollen as a Novel Substrate in Pilot-Scale Probiotic-Mediated Lactic Fermentation Processes. *LWT Food Sci. Technol.* **2021**, *141*, 110868. [CrossRef]
21. Uțoiu, E.; Matei, F.; Toma, A.; Diguță, C.; Ștefan, L.; Mănoiu, S.; Vrăjmașu, V.; Moraru, I.; Oancea, A.; Israel-Roming, F.; et al. Bee Collected Pollen with Enhanced Health Benefits, Produced by Fermentation with a Kombucha Consortium. *Nutrients* **2018**, *10*, 1365. [CrossRef]
22. Tritean, N.; Dima, Ș.O.; Trică, B.; Stoica, R.; Ghiurea, M.; Moraru, I.; Cimpean, A.; Oancea, F.; Constantinescu-Aruxandei, D. Selenium-Fortified Kombucha–Pollen Beverage by In Situ Biosynthesized Selenium Nanoparticles with High Biocompatibility and Antioxidant Activity. *Antioxidants* **2023**, *12*, 1711. [CrossRef]
23. Campos, M.G.R.; Bogdanov, S.; de Almeida-Muradian, L.B.; Szczesna, T.; Mancebo, Y.; Frigerio, C.; Ferreira, F. Pollen Composition and Standardisation of Analytical Methods. *J. Apic. Res.* **2008**, *47*, 154–161. [CrossRef]
24. Kostić, A.Ž.; Milinčić, D.D.; Nedić, N.; Gašić, U.M.; Špirović Trifunović, B.; Vojt, D.; Tešić, Ž.L.; Pešić, M.B. Phytochemical Profile and Antioxidant Properties of Bee-Collected Artichoke (*Cynara scolymus*) Pollen. *Antioxidants* **2021**, *10*, 1091. [CrossRef] [PubMed]
25. Kostić, A.Ž.; Milinčić, D.D.; Špirović Trifunović, B.; Nedić, N.; Gašić, U.M.; Tešić, Ž.L.; Stanojević, S.P.; Pešić, M.B. Monofloral Corn Poppy Bee-Collected Pollen—A Detailed Insight into Its Phytochemical Composition and Antioxidant Properties. *Antioxidants* **2023**, *12*, 1424. [CrossRef]
26. Qiao, J.; Feng, Z.; Zhang, Y.; Xiao, X.; Dong, J.; Haubruge, E.; Zhang, H. Phenolamide and Flavonoid Glycoside Profiles of 20 Types of Monofloral Bee Pollen. *Food Chem.* **2023**, *405*, 134800. [CrossRef]
27. Jiang, X.; Liu, Y.; Li, W.; Zhao, L.; Meng, F.; Wang, Y.; Tan, H.; Yang, H.; Wei, C.; Wan, X.; et al. Tissue-Specific, Development-Dependent Phenolic Compounds Accumulation Profile and Gene Expression Pattern in Tea Plant (*Camellia sinensis*). *PLoS ONE* **2013**, *8*, e62315. [CrossRef]
28. van der Hooft, J.J.J.; Akermi, M.; Ünlü, F.Y.; Mihaleva, V.; Roldan, V.G.; Bino, R.J.; de Vos, R.C.H.; Vervoort, J. Structural Annotation and Elucidation of Conjugated Phenolic Compounds in Black, Green, and White Tea Extracts. *J. Agric. Food Chem.* **2012**, *60*, 8841–8850. [CrossRef]
29. Liu, F.; Zhu, H.; Li, C.; Wang, Y.; Zhang, J.; Tang, X.; Zhang, T.; Liu, Y. A Combined Drying Process Involving Hot Air and Roasting for Improving Summer Congou Black Tea Quality. *Food Res. Int.* **2025**, *201*, 115584. [CrossRef]
30. Zhang, H.; Liu, R.; Lu, Q. Separation and Characterization of Phenolamines and Flavonoids from Rape Bee Pollen, and Comparison of Their Antioxidant Activities and Protective Effects Against Oxidative Stress. *Molecules* **2020**, *25*, 1264. [CrossRef]
31. Qiao, J.; Zhang, Y.; Haubruge, E.; Wang, K.; El-Seedi, H.R.; Dong, J.; Xu, X.; Zhang, H. New Insights into Bee Pollen: Nutrients, Phytochemicals, Functions and Wall-Disruption. *Food Res. Int.* **2024**, *178*, 113934. [CrossRef]
32. Zhang, X.; Yu, M.; Zhu, X.; Liu, R.; Lu, Q. Metabolomics Reveals That Phenolamides Are the Main Chemical Components Contributing to the Anti-Tyrosinase Activity of Bee Pollen. *Food Chem.* **2022**, *389*, 133071. [CrossRef]
33. Sknepnek, A.; Pantić, M.; Matijašević, D.; Miletić, D.; Lević, S.; Nedović, V.; Niksic, M. Novel Kombucha Beverage from Lingzhi or Reishi Medicinal Mushroom, *Ganoderma lucidum*, with Antibacterial and Antioxidant Effects. *Int. J. Med. Mushrooms* **2018**, *20*, 243–258. [CrossRef] [PubMed]
34. Bishop, P.; Pitts, E.R.; Budner, D.; Thompson-Witrick, K.A. Kombucha: Biochemical and Microbiological Impacts on the Chemical and Flavor Profile. *Food Chem. Adv.* **2022**, *1*, 100025. [CrossRef]
35. De-Melo, A.A.M.; Estevinho, L.M.; Moreira, M.M.; Delerue-Matos, C.; Freitas, A.D.S.D.; Barth, O.M.; Almeida-Muradian, L.B.D. Phenolic Profile by HPLC-MS, Biological Potential, and Nutritional Value of a Promising Food: Monofloral Bee Pollen. *J. Food Biochem.* **2018**, *42*, e12536. [CrossRef]
36. Negri, G.; Barreto, L.M.R.C.; Sper, F.L.; de Carvalho, C.; Campos, M.d.G.R. Phytochemical Analysis and Botanical Origin of *Apis mellifera* Bee Pollen from the Municipality of Canavieiras, Bahia State, Brazil. *Braz. J. Food Technol.* **2018**, *21*, e2016176. [CrossRef]
37. Kostić, A.Ž.; Milinčić, D.D.; Gašić, U.M.; Nedić, N.; Stanojević, S.P.; Tešić, Ž.L.; Pešić, M.B. Polyphenolic Profile and Antioxidant Properties of Bee-Collected Pollen from Sunflower (*Helianthus annuus* L.) Plant. *LWT Food Sci. Technol.* **2019**, *112*, 108244. [CrossRef]

38. Kostić, A.Ž.; Çobanoğlu, D.N.; Milinčić, D.D.; Felek, I.; Düzdaban, P.; Pešić, M.B. Could Phenylamides Emerge as a Powerful Chemotaxonomic Tool? A Case Study of Rapeseed Bee-Collected Pollen Originating From Serbia and Türkiye. *Chem. Biodivers.* 2025, *online ahead of print*. [CrossRef]
39. Sun, L.; Guo, Y.; Zhang, Y.; Zhuang, Y. Antioxidant and Anti-Tyrosinase Activities of Phenolic Extracts from Rape Bee Pollen and Inhibitory Melanogenesis by CAMP/MITF/TYR Pathway in B16 Mouse Melanoma Cells. *Front. Pharmacol.* **2017**, *8*, 104. [CrossRef]
40. Bridi, R.; Atala, E.; Pizarro, P.N.; Montenegro, G. Honeybee Pollen Load: Phenolic Composition and Antimicrobial Activity and Antioxidant Capacity. *J. Nat. Prod.* **2019**, *82*, 559–565. [CrossRef]
41. Marques, V.; Farah, A. Chlorogenic Acids and Related Compounds in Medicinal Plants and Infusions. *Food Chem.* **2009**, *113*, 1370–1376. [CrossRef]
42. Truchado, P.; Ferreres, F.; Tomas-Barberan, F.A. Liquid Chromatography–Tandem Mass Spectrometry Reveals the Widespread Occurrence of Flavonoid Glycosides in Honey, and Their Potential as Floral Origin Markers. *J. Chromatogr. A* **2009**, *1216*, 7241–7248. [CrossRef]
43. Mosić, M.; Trifković, J.; Vovk, I.; Gašić, U.; Tešić, Ž.; Šikoparija, B.; Milojković-Opsenica, D. Phenolic Composition Influences the Health-Promoting Potential of Bee-Pollen. *Biomolecules* **2019**, *9*, 783. [CrossRef] [PubMed]
44. Song, W.O.; Chun, O.K. Tea Is the Major Source of Flavan-3-Ol and Flavonol in the U.S. Diet. *J. Nutr.* **2008**, *138*, 1543S–1547S. [CrossRef] [PubMed]
45. De La Luz Cádiz-Gurrea, M.; Fernández-Arroyo, S.; Segura-Carretero, A. Pine Bark and Green Tea Concentrated Extracts: Antioxidant Activity and Comprehensive Characterization of Bioactive Compounds by HPLC-ESI-QTOF-MS. *Int. J. Mol. Sci.* **2014**, *15*, 20382–20402. [CrossRef]
46. Adaškevičiūtė, V.; Kaškonienė, V.; Barčauskaitė, K.; Kaškonas, P.; Maruška, A. The Impact of Fermentation on Bee Pollen Polyphenolic Compounds Composition. *Antioxidants* **2022**, *11*, 645. [CrossRef] [PubMed]
47. Aylanc, V.; Tomás, A.; Russo-Almeida, P.; Falcão, S.I.; Vilas-Boas, M. Assessment of Bioactive Compounds under Simulated Gastrointestinal Digestion of Bee Pollen and Bee Bread: Bioaccessibility and Antioxidant Activity. *Antioxidants* **2021**, *10*, 651. [CrossRef]
48. Aylanc, V.; Larbi, S.; Calhelha, R.; Barros, L.; Rezouga, F.; Rodríguez-Flores, M.S.; Seijo, M.C.; El Ghouizi, A.; Lyoussi, B.; Falcão, S.I., et al. Evaluation of Antioxidant and Anticancer Activity of Mono- and Polyfloral Moroccan Bee Pollen by Characterizing Phenolic and Volatile Compounds. *Molecules* **2023**, *28*, 835. [CrossRef]
49. Kim, S.B.; Liu, Q.; Ahn, J.H.; Jo, Y.H.; Turk, A.; Hong, I.P.; Han, S.M.; Hwang, B.Y.; Lee, M.K. Polyamine Derivatives from the Bee Pollen of *Quercus mongolica* with Tyrosinase Inhibitory Activity. *Bioorg. Chem.* **2018**, *81*, 127–133. [CrossRef]
50. Glavnik, V.; Bensa, M.; Vovk, I.; Guzelmeric, E. High-Performance Thin-Layer Chromatography-Multi-Stage Mass Spectrometry Methods for Analyses of Bee Pollen Botanically Originating from Sweet Chestnut (*Castanea sativa* Mill.). *JPC J. Planar Chrom. Mod. TLC* **2023**, *36*, 471–482. [CrossRef]
51. dos Santos, G.S.; de Almeida Veiga, A.; Carlotto, J.; Mello, R.G.; Serrato, R.V.; de Souza, L.M. Identification and Fingerprint Analysis of Novel Multi-Isomeric Lycibarbarspermidines and Lycibarbarspermines from *Lycium barbarum* L. by Liquid Chromatography with High-Resolution Mass Spectrometry (UHPLC-Orbitrap). *J. Food Compos. Anal.* **2022**, *105*, 104194. [CrossRef]
52. Gardana, C.; Del Bo', C.; Quicazán, M.C.; Corrrea, A.R.; Simonetti, P. Nutrients, Phytochemicals and Botanical Origin of Commercial Bee Pollen from Different Geographical Areas. *J. Food Compos. Anal.* **2018**, *73*, 29–38. [CrossRef]
53. Qian, D.; Xi, J.; Nie, Y.; Zou, M.; Zhang, J.; Alseekh, S.; Fernie, A.R.; Chen, J.; Huang, L. Chemical Characterization of Dicaffeoyl Polyamine Derivatives to Understand Specific Secondary Metabolites in Goji Berry. *J. Food Compos. Anal.* **2023**, *122*, 105434. [CrossRef]
54. Chesnov, S.; Bigler, L.; Hesse, M. Detection and Characterization of Natural Polyamines by High-Performance Liquid Chromatography-Atmospheric Pressure Chemical Ionization (Electrospray Ionization) Mass Spectrometry. *Eur. J. Mass Spectrom.* **2002**, *8*, 1–16. [CrossRef]
55. Ahad, H.; Jin, H.; Liu, Y.; Wang, J.; Sun, G.; Liang, X.; Akber Aisa, H. Chemical Profiling of Spermidines in Goji Berry by Strong Cation Exchange Solid-Phase Extraction (SCX-SPE) Combined with Ultrahigh-Performance Liquid Chromatography-Quadrupole Time-of-Flight Mass Spectrometry (UPLC-Q-TOF/MS/MS). *J. Chrom. B* **2020**, *1137*, 121923. [CrossRef] [PubMed]
56. Rodríguez-Flores, M.S.; Escuredo, O.; Seijo, M.C.; Rojo, S.; Vilas-Boas, M.; Falcão, S.I. Phenolic Profile of *Castanea* Bee Pollen from the Northwest of the Iberian Peninsula. *Separations* **2023**, *10*, 270. [CrossRef]
57. Luo, Y.; Jian, Y.; Liu, Y.; Jiang, S.; Muhammad, D.; Wang, W. Flavanols from Nature: A Phytochemistry and Biological Activity Review. *Molecules* **2022**, *27*, 719. [CrossRef]
58. Soobrattee, M.A.; Neergheen, V.S.; Luximon-Ramma, A.; Aruoma, O.I.; Bahorun, T. Phenolics as Potential Antioxidant Therapeutic Agents: Mechanism and Actions. *Mutat. Res./Fundam. Mol. Mech. Mutagen.* **2005**, *579*, 200–213. [CrossRef]

59. Guo, T.; Song, D.; Cheng, L.; Zhang, X. Interactions of Tea Catechins with Intestinal Microbiota and Their Implication for Human Health. *Food Sci. Biotechnol.* **2019**, *28*, 1617–1625. [CrossRef]
60. Hsu, P.-S.; Wu, T.-H.; Huang, M.-Y.; Wang, D.-Y.; Wu, M.-C. Nutritive Value of 11 Bee Pollen Samples from Major Floral Sources in Taiwan. *Foods* **2021**, *10*, 2229. [CrossRef]

antioxidants

MDPI

Article

Changes in Ascorbic Acid, Phenolic Compound Content, and Antioxidant Activity In Vitro in Bee Pollen Depending on Storage Conditions: Impact of Drying and Freezing

Rosita Stebuliauskaitė [1], Mindaugas Liaudanskas [1,2], Vaidotas Žvikas [2], Violeta Čeksterytė [3], Neringa Sutkevičienė [4], Šarūnė Sorkytė [4], Aurita Bračiulienė [1,*] and Sonata Trumbeckaitė [1,5]

1 Department of Pharmacognosy, Faculty of Pharmacy, Lithuanian University of Health Sciences, Sukilėlių av. 13, LT-50162 Kaunas, Lithuania; rosita.stebuliauskaite@lsmu.lt (R.S.); mindaugas.liaudanskas@lsmu.lt (M.L.); sonata.trumbeckaite@lsmu.lt (S.T.)

2 Institute of Pharmaceutical Technologies, Faculty of Pharmacy, Lithuanian University of Health Sciences, Sukilėlių av. 13, LT-50162 Kaunas, Lithuania; vaidotas.zvikas@lsmu.lt

3 Lithuanian Research Centre for Agriculture and Forestry, Institute of Agriculture, Instituto Ave. 1, LT-58344 Akademija, Lithuania; violeta.ceksteryte@lammc.lt

4 Animal Reproduction Laboratory, Faculty of Veterinary Medicine, Lithuanian University of Health Sciences, Tilžės Str. 18, LT-47181 Kaunas, Lithuania; neringa.sutkeviciene@lsmu.lt (N.S.); sarune.sorkyte@lsmu.lt (Š.S.)

5 Laboratory of Biochemistry, Neuroscience Institute, Lithuanian University of Health Sciences, Eivenių Str. 4, LT-50162 Kaunas, Lithuania

* Correspondence: aurita.braciuliene@lsmu.lt; Tel.: +370-62156190

Abstract: Bee pollen (BP) is a very valuable bee product, and its value depends on its proteins, lipids, amino acids, carbohydrates, vitamins, minerals, and biologically active compounds such as phenolic compounds, which may change depending on the method of pollen preparation after collection and its storage conditions. Therefore, it is very important to determine when the decline in bioactive compounds in BP occurs during storage. The purpose of this study was to evaluate the changes in the content of ascorbic acid and phenolic compounds, and to determine the antioxidant activity of BP extracts depending on their preparation method and storage conditions over a 15-month period, with assessments conducted every 3 months. Dried pollen (at +28 °C on the first day and +35 °C on the second day) and frozen (−20 °C and −80 °C) BP samples were prepared. After 3 months of storage, there was no decrease of ascorbic acid in frozen BP; however, it decreased by 20% in dried BP ($p < 0.05$). It was determined that in frozen BP, the content of total phenolic compounds decreased by 12–14% ($p < 0.05$) after 6 months, and in dried BP, it decreased by 7% ($p < 0.05$) after 3 months. The levels of flavonoids decreased by 10–17% ($p < 0.05$) in BP after 6 months. Chlorogenic and *p*-coumaric acids have been observed as the most abundant phenolic acids in BP. During storage (the 6–15-month period), the strongest antiradical and reducing activity in vitro was estimated in the frozen (−80 °C) BP, which was 1.8–3.4-fold and 2.6–3.1-fold higher, respectively, compared to the dried BP extracts. In conclusion, significant results were obtained, showing better stability of phenolic compounds and ascorbic acid during storage in frozen BP compared to dried pollen. Melisopalynological analysis revealed a polyfloral pollen mixture, with *Salix* spp. and *Brassica napus* L. predominating in all samples, comprising 34.3% and 36.8%, respectively. Among these, *Acer platanoides* L., *Malus domestica* Borkh., and *Taraxacum officinale* L. were important minor pollens present in the samples examined.

Keywords: frozen and dried bee pollen; ascorbic acid; total phenolic content; flavonoids; storage time; antioxidant activity; antiradical and reducing activity

1. Introduction

Bee pollen (BP), consisting of flower pollen grains, nectar, and salivary secretions of bees, is a highly valuable bee product. Bee pollen is known for its anti-inflammatory [1,2], antibacterial [3], antidiabetic [4–6], anticancer [7,8], and other effects. Its nutritional and medicinal value is determined by its amounts of proteins, lipids, amino acids, carbohydrates, vitamins, minerals, and various biologically active compounds such as enzymes and phenolic compounds [9,10]. Among the many biological effects on the body, antioxidant activity research has received the most attention. BP exhibits strong antioxidant activity, which is primarily ascribed to the diversity of phenolic compounds, ascorbic acid, and tocopherol [1,11–16]. In recent decades, compounds with antioxidant properties have become a target for scientific studies, as there is a constant search for new sources capable of combating oxidative stress [17,18].

The primary challenge in the use of BP in modern phytomedicine is associated with the variability in phytochemical composition, which depends on various factors, including the botanical and territorial origin of the pollen, natural conditions, and especially the preparation and storage conditions of the raw material [19,20]. To ensure the production of high-quality BP products, it is essential to guarantee not only suitable preparation conditions but also optimal storage conditions. These directly influence the phytochemical composition and biological activity of BP, as well as its potential applications as a functional food [20,21].

To maximize the retention of vitamin C, polyphenols, and antioxidant properties in bee pollen, freezing (−20 °C) is the most effective method, while drying at 40 °C results in significant degradation of these bioactive compounds. Dry pollen is usually provided to consumers in Lithuania, but its phytochemical composition and quality are highly dependent on the drying temperature [22]. Unfortunately, the drying process, especially at higher temperatures, can result in a significant reduction in the amount of biologically active compounds, including vitamins, which are particularly susceptible to high temperatures [23]. The degradation of ascorbic acid and phenolic compounds in bee pollen during storage can be influenced by several factors. Oxidation is one of the key pathways, especially for ascorbic acid, which is highly sensitive to oxygen, light, and temperature. Ascorbic acid is reversibly oxidized to dehydroascorbic acid (DHA) under light, heat, metal ions, and alkaline pH, which then irreversibly degrades to 2,3-diketogulonic acid [24]. Enzymatic activity, particularly from polyphenol oxidase, may also contribute to the degradation of phenolic compounds when enzyme inactivation is incomplete, especially in dried pollen [25]. Heat during drying may cause thermal decomposition of flavonoids and vitamin C, while even frozen samples are susceptible to slow oxidative processes over extended periods.

Pollen that is freshly collected and frozen immediately after collection is very popular in the foreign market [26]. To date, comprehensive studies systematically analyzing changes in polyphenolic compounds, vitamin C, and antioxidant activity of bee pollen during extended storage—particularly at regular 3-month intervals over a 15-month period—are lacking. Such data are essential for evaluating the suitability of bee products for consumption and their capacity to retain bioactive compounds over time. In order to introduce a preparation method for BP that allows for the production of high-quality pharmaceutical preparations, functional foods, or dietary supplements, or to consume high-quality BP in its native state, it is essential to study the differences in the phytochemical composition between dried and frozen BP and to determine the variations in the phytochemical composition of BP during storage [23,27]. Since only the highest quality BP must reach consumers, the requirements for their quality are increasing.

Therefore, the aim of our study was to evaluate the changes in the composition of phenolic compounds and ascorbic acid of bee pollen samples and to determine antiradical and reducing activity in vitro of bee pollen extracts depending on preparation methods (freezing or low temperature drying) over a 15-month storage period, with changes assessed every three months. This study was oriented toward consumer guidance, aiming to provide practical recommendations regarding how long bee pollen can be stored under typical conditions before significant degradation of key bioactive compounds begins. To date, the scientific literature lacks detailed studies that comprehensively examine these parameters over such a long storage duration, making our findings particularly novel and significant in the field of phytochemical stability.

2. Materials and Methods

2.1. Collection of Pollen and Estimation of Its Botanical Origin

Bee pollen (BP) was collected from an apiary located in Talkoniai, Pasvalys district in 2024 in April–May. The location coordinates are 55.9598° N, 24.3422° E. Pollen loads were collected with pollen traps in good weather during peak plant blooming periods of about 200 g/per day. Pure, fresh bee-collected pollen was cleaned and distributed for storage according to the research conditions described below. During the research, freshly collected and immediately frozen and dried bee pollen samples were examined.

2.2. Preparation of Bee Pollen

2.2.1. Preparation of Dried Bee Pollen Samples

Freshly collected BP samples were dried in a dryer, maintaining a temperature of +28 °C for 24 h and +35 °C on the second day for 24 h. The prepared raw material was stored in a sealed container in a dry, dark, well-ventilated room at room temperature (+19–21 °C) throughout the entire study.

2.2.2. Preparation of Frozen Bee Pollen Samples

Freshly collected BP samples were immediately frozen at −20 °C and −80 °C in freezers. They were kept in sealed containers at a constant temperature (−20 °C and −80 °C, respectively) throughout the study. The extracted raw material was immediately processed. Variations in the qualitative and quantitative composition of the BP were studied every three months. All results were recalculated to the dry weight.

2.3. Microscopic Examination of Pollen Samples

BP granules were evaluated according to morphological characteristics such as color and shape [28]. Pollen sample preparation for botanical composition analysis was performed using the melissopalynology technique described by Louveaux et al. [29]. Pollen photos taken using a Nikon Eclipse E600, model C-LP (Nikon Corporation, Tokyo, Japan) microscope were compared with the known plant pollen photos presented in a pollen catalogue [30].

Pollen expression results: About 400–500 pollen grains were counted in each sample. The frequency of pollen grains of each melliferous plant is expressed as a percentage of the total pollen sum. Pollen considered as monofloral was mainly produced from one plant species, or pollen content from one plant species was predominant (constituting more than 45%). The pollen content of other plant species was designated as follows: secondary pollen, 16–45%; important minor pollen, 3–15%; minor pollen. All analyses were carried out in triplicate. Estimated values are expressed as means ± standard deviations (SDs) and the coefficient of variance (CV).

2.4. Chemicals

All solvents, reagents, and standards used were of analytical grade. The solvents used were as follows: purified water (Millipore®, Bedford, MA, USA), ethanol 96.0% (*v*/*v*) (AB Vilnius degtinė, Vilnius, Lithuania).

The reagents used were as follows: ≥99% gallic acid monohydrate, ≥99% Folin–Ciocâlteu reagent, ≥99% sodium carbonate, ≥99% hexamethylenetetramine, ≥99% acetic acid, ≥99% aluminum chloride hexahydrate, ≥98% Trolox® (6-hydroxy-2,5,7,8-tetramethylchroman-2-carboxylic acid), ≥99% ammonium acetate, ≥98% TPTZ (2,4,6-tri-(2-pyridyl)-1,3,5-triazine), ≥37% hydrochloric acid, ≥99% sodium acetate trihydrate, ≥99% glacial acetic acid (Sigma-Aldrich®, Steinheim, Germany), ≥98% rutin standard (Extrasynthesis, Genoa, Italy), ≥98% ABTS (2,2′-azino-bis-(3-ethylbenzthiazoline-6-sulfonic acid)) (TCI Europe, Zvijndrecht, Belgium), ≥98% potassium persulfate (Alfa Aesar GmbH and Co KG, Kandel, Germany), ≥98% neocuproine, ≥98% copper (II) chloride dihydrate, ≥98% iron (III) chloride hexahydrate, ≥98%sodium 2,6-dichlorophenolindophenolate, ≥99% chloral hydrate (2,2,2-trichloroethane-1,1-diol) (Roth, Karlsruhe, Germany).

The standards of phenolic compounds used were as follows: ≥98% *p*-coumaric acid, ≥99% gallic acid, ≥98% ferulic acid, ≥98% rosmarinic acid, ≥98% chlorogenic acid, ≥98% kaempferol, ≥98% kaempferol-3-O-glucoside, ≥98% kaempferol-3-O-rutinoside, ≥98% quercetin, ≥98% hyperoside, ≥98% quercetin-3-arabinopyranoside, ≥98% rutin, ≥98% ≥98% isorhamnetin-3-glucoside, ≥98% isorhamnetin-3-rutinoside, ≥98% luteolin, ≥98% luteolin-4-O-glucoside, ≥98% luteolin-3,7-diglucoside, ≥98% phloridzin, ≥98% phloretin, ≥98% apigenin, ≥98% naringenin, ≥98% isoquercitrin, ≥98% isorhamnetin, ≥98% avicularin, ≥98% quercitrin (Sigma-Aldrich®, Steinheim, Germany).

2.5. Preparation of the Bee Pollen Extracts

2.5.1. Preparation of Ethanol Bee Pollen Extract

The methodology for the preparation of ethanol bee pollen extract was described by Kahraman et al. [31]. A total of 1 g (exact weight) of crushed, dried, and frozen (−20 °C and −80 °C) BP was weighed. The weighed amount of the raw material was transferred to labeled vials made of dark glass with the same volume, and 10 mL of 70.0% (*v*/*v*) ethanol was added. Three vials of each raw material (n = 9) were prepared and exposed to an ultrasonic bath for 50 min. Afterward, solid particles were separated from the liquid phase using a vacuum pump connected to a filtration system, with the liquid portion collected to an exact volume of 10 mL. The raw material residue on the filter was washed with a 70.0% (*v*/*v*) ethanol–water mixture. The obtained ethanol extracts were poured into dark glass sealed bottles, labeled, and stored at room temperature in a dark place until the next step. The prepared extracts were used to identify and determine the amounts of phenolic compounds and assess the antiradical and reducing activity in vitro.

2.5.2. Preparation of Aqueous Bee Pollen Extracts

The methodology for the preparation of aqueous bee pollen extract was described by Ares et al. [32]. A total of 2.5 g (exact weight) of crushed dried and frozen (−20 °C and −80 °C) BP was weighed. The weighed amount was transferred to labeled conical flasks of equal volume (100 mL), poured with 25 mL of purified water, and left for 10 min to extract three samples of each raw material (n = 9). Using a vacuum pump with a connected filtration system, the solid particles from the liquid phase to an exact volume of 25 mL were separated, washing the residue of the raw material on the filter with purified water. The obtained aqueous extracts were used for the estimation of ascorbic acid in the BP samples.

2.6. Evaluation of Total Phenolic Compounds (TPC) and Flavonoid Contents in Bee Pollen Samples

All spectrophotometric measurements were carried out using a HALO DB–20 UV–VIS spectrophotometer (Dynamica GmbH, Salzburg, Austria). Total phenolic content was determined using the Folin–Ciocâlteu method in the ethanolic extracts of BP [33]. The light absorption of a 10 mm layer of the analyzed solution was measured at a wavelength of 765 nm. The calibration curve equation was y = 0.4016x + 0.0148 (R^2 = 0.9985). The results were determined based on the standard curves of gallic acid prepared in the range of 0.03–1.0 mg/L and expressed as milligrams of gallic acid equivalent (GAE) per gram of dry sample weight (mg GAE/g DW). The total flavonoid content in the ethanolic BP extracts was estimated using the methodology described by Urbonavičiūtė et al. [34]. The absorption of the analyzed solution was recorded at a wavelength of 407 nm. The calibration curve equation was y = 0.503x − 0.0141 (R^2 = 0.9954). The results were calculated based on a rutin calibration curve and expressed as milligrams of rutin equivalent (RE) per gram of dry sample weight (mg RE/g DW).

2.7. Determination of Ascorbic Acid in Bee Pollen Samples Using the Titrimetric Method

A total of 1 mL of the prepared BP aqueous extract, 1 mL of hydrochloric acid (20 g/L solution), and 13 mL of purified water were added to a 50 mL conical flask. The resulting mixture was titrated with a 0.001 M solution of sodium 2,6-dichlorophenolindophenolate in a microburette until a pale pink color persisted for 10 s. The titration was performed three times, and the arithmetic mean was calculated. A total of 1 mL of a 0.001 M solution of sodium 2,6-dichlorophenolindophenolate corresponded to 0.000088 g of ascorbic acid. The amount of ascorbic acid for the dry raw material was calculated using the following formula:

$$X = (V \times 0.000088 \times V_1 \times 100 \times 100)/(m \times V_2 \times (100 - W))$$

X—amount of ascorbic acid, %;

V—amount of sodium 2.6-dichlorophenolindophenolate 0.001 M solution used for titration, mL;

V_1—volume of the research extract, mL;

V_2—test extracts taken for titration, volume, mL;

m—weighed raw material mass, g;

W—loss on drying, %.

2.8. Quantitative and Qualitative Determination of Phenolic Compounds in Bee Pollen Samples Using the UHPLC–MS/MS Method

To qualitatively and quantitatively evaluate the individual phenolic compounds in the BP samples, a validated UHPLC–MS/MS method was employed [35]. The separation of phenolic compounds was carried out using an Acquity H-class UPLC system (Waters, Milford, CT, USA) coupled with a triple quadrupole tandem mass spectrometer, Waters Xevo TQD (Waters, Milford, CT, USA), equipped with an electrospray ionization (ESI) source for MS/MS data acquisition. A YMC Triart C18 column (100 × 2.0 mm; 1.9 µm) (YMC Europe GmbH, Dinslaken, Germany) was used for the analysis. The column temperature was maintained at 40 °C. Gradient elution was performed with a mobile phase consisting of 0.1% formic acid aqueous solution (solvent A) and acetonitrile (solvent B) at a flow rate of 0.5 mL min^{-1}. A linear gradient profile was applied as follows for solvent A: initially 95% for 1 min, reduced to 70% over 4 min, further decreased to 50% over 7 min, and returned to 95% over 2 min. Negative electrospray ionization was used for the analysis, with the following parameters: capillary voltage, −2 kV; source temperature, 150 °C; desolvation temperature, 400 °C; desolvation gas flow, 700 L h^{-1}; cone gas flow, 20 L h^{-1}. The collision energy and cone voltage were optimized individually for each compound (Table 1).

Table 1. Collision energy and cone voltage of phenolic compounds.

Compound	Parent Ion (*m*/*z*)	Daughter Ion (*m*/*z*)	Cone Voltage, V	Collision Energy, eV
p-Coumaric acid	163	93	28	22
Gallic acid	169	51	36	30
Ferulic acid	193	134	32	18
Chlorogenic acid	353	191	32	14
Rosmarinic acid	359	161	36	16
Phloretin	273	167	42	16
Phloridzin	435	273	42	14
Apigenin	269	117	54	36
Naringenin	271	151	46	18
Avicularin	433	301	50	20
Kaempferol	285	185	50	25
Kaempferol-3-O-glucoside	447	284	54	28
Kaempferol-3-O-rutinoside	593	285	36	20
Quercetin	301	151	48	20
Quercetin-3-arabinopiranoside	433	300	56	26
Hyperoside	463	300	50	26
Rutin	609	300	70	38
Quercitrin	447	300	50	26
Isoquercitrin	463	301	52	28
Isorhamnetin	315	300	44	22
Isorhamnetin-3-glucoside	477	314	60	28
Isorhamnetin-3-O-rutinoside	623	315	70	32
Luteolin	285	133	58	36
Luteolin-4-O-glucoside	447	285	36	16
Luteolino-3.7-diglucoside	609	447	30	20

2.9. Determination of Antiradical and Reducing Activities In Vitro in Bee Pollen Extracts

The antioxidant activities of the BP extracts were established based on three different antioxidant capacity assays, namely in vitro spectrophotometric ABTS free radical scavenging and ferric (FRAP) and copper (CUPRAC) reducing activity assays using a spectrophotometer HALO DB–20 UV–VIS Dynamica (Dynamica GmbH, Salzburg, Austria).

2.9.1. Evaluation of Antiradical Activity Using the ABTS Method In Vitro

The antiradical activity was measured using a method modified from Khongkarat et al. [36]. The stock and working solutions were prepared as follows: a 2 mmol/L $ABTS^{\bullet+}$ stock solution was prepared by dissolving 0.0548 g of ABTS in 50 mL of purified water, then adding 0.0095 g of $K_2S_2O_8$. The solution was kept in the dark for 16 h to form the $ABTS^{\bullet+}$ radical, which remained stable for two days at room temperature. The working solution was obtained by diluting the stock until absorbance at 734 nm (λ) reached 0.800, measured with a 10 mm path length. Sample preparation: 10 µL of the ethanol extract was mixed with 3 mL of the working $ABTS^{\bullet+}$ solution. Samples were incubated for 30 min in the dark at room temperature, and the color change from blue-green to pale blue was observed. Absorbance at 734 nm (λ) was measured. A calibration curve (y = 0.00007x + 0.088; R^2 = 0.9932) was created using Trolox standards (125–8000 μmol/L).

2.9.2. Estimation of Reducing Activity Using the CUPRAC Method In Vitro

The method for determining reducing activity, as described by Munteanu et al. [37], was applied. Preparation of stock solution: solutions A, B, and C were mixed in a 1:1:1 ratio. Solution A: 10 mmol/L $CuCl_2 \times 2H_2O$ (0.085 g in 50 mL of purified water). Solution B: 1 mmol/L CH_3COONH_4 buffer (pH 7, 0.0038 g in 50 mL of purified water). Solution C: 7.5 mmol/L neocuproine (0.078 g in 50 mL of 96.0% ethanol). Sample preparation: 10 μL of the ethanol extract was mixed with 3 mL of freshly prepared working solution. The samples were incubated in the dark at room temperature for 30 min, with a color change observed from light blue to orange-yellow. Absorbance at 450 nm (λ) was measured using a spectrophotometer with a 10 mm path length. A blank was prepared using the working CUPRAC solution. A calibration curve (y = 0.00004x + 0.0047; R^2 = 0.9998) was constructed using Trolox standards (125–32,000 μmol/L).

2.9.3. Evaluation of Reducing Activity Using the FRAP Method In Vitro

The method for determining reducing activity, as described by Sawicki et al. [3], was applied. Preparation of working solution: solutions A, B, and C were mixed in a 1:1:10 ratio. Solution A: 20 mmol/L $FeCl_3 \times 6H_2O$ aqueous solution (0.2704 g of $FeCl_3$ dissolved in 50 mL of purified water). Solution B: 10 mmol/L TPTZ solution (0.1562 g of TPTZ dissolved in 50 mL of purified water acidified with 40 mmol/L HCl). Solution C: 300 mmol/L acetate buffer ($CH_3COONa \times 3H_2O$ solution, 0.775 g of $CH_3COONa \times 3H_2O$ mixed with 4 mL of glacial acetic acid, diluted to 250 mL with purified water). Sample preparation: 10 μL of the ethanol extract was mixed with 3 mL of the working solution. The samples were incubated in the dark at room temperature for 30 min, observing a color change from colorless to intensely blue. Absorbance was then measured at 593 nm (λ) using a spectrophotometer with a 10 mm path length. A blank was prepared using the working FRAP solution. A calibration curve (y = 0.00003x + 0.0233; R^2 = 0.9932) was constructed using Trolox standards (500–32,000 μmol/L).

2.9.4. Assessment of Antioxidant Activity In Vitro

The antioxidant activity of the BP extracts was calculated based on Trolox calibration curves and expressed as the Trolox equivalent (μmol TE/g) per absolute dry weight. The Trolox equivalent corresponds to the amount that, under the same testing conditions, exhibits the same antioxidant activity as 1 g of BP raw material [37,38]. In vitro antioxidant activity was calculated using the following formula:

$$TE = (A + b)/(a \times 1000) \times (V_{sample})/(m_{sample})$$

A—absorbance of the sample;
a—slope of the Trolox calibration curve equation;
b—deviation from the Trolox calibration curve equation;
V_{sample}—volume of BP extract, mL;
m_{sample}—weight (precise) of BP powder, g.

2.10. Statistical Analysis

To evaluate the results obtained during this research, statistical data analysis was performed using Microsoft Office Excel 2013 (Microsoft, Redmond, WA, USA) and SPSS Statistics 29 (SPSS Inc., Chicago, IL, USA). The experiments were repeated in triplicate, with the results presented as arithmetic means ± standard deviations (SDs) of replicates.

The variables were calculated and graphically presented using Microsoft Office Excel 2013, and the SPSS Statistics 29 package was used to assess whether differences between the obtained values were statistically significant. One-way analysis of variance (ANOVA) was applied using Tukey's Post Hoc test (significance level $p < 0.05$). The correlation between the content of total phenolic compounds, total flavonoids, the sum of individual phenolic compounds, ascorbic acid, and antioxidant activity (measured using ABTS, CUPRAC, and FRAP methods) was expressed by Pearson's correlation coefficient. The values of the Pearson's correlation coefficients were selected and applied based on the values provided by Šimoliūnienė R. et al. [39].

3. Results

3.1. Changes in Total Phenolic Compounds (TPC) and Flavonoid Contents in Bee Pollen Samples

In order to assess changes in the composition of phenolic compounds and ascorbic acid in the BP samples after different preparation methods (freezing at −20 °C and −80 °C, or drying at low temperature) and during storage, the total contents of these compounds and flavonoids were first measured.

At the initial stage, the highest amounts of TPC, 24.5 ± 0.7 mg GAE/g and 24.4 ± 0.4 mg GAE/g, were determined in the BP samples stored at −20 °C and −80 °C, respectively, while the lowest amount, 23.3 ± 0.4 mg GAE/g, was found in the dried BP samples. However, no statistically significant differences were found between these samples ($p > 0.05$).

The amount of TPC decreased under different storage conditions and over the storage period. During storage (after 3 months), the amount of TPC decreased when the BP samples were stored at temperatures of −20 °C and −80 °C. Meanwhile, a statistically significant (7%, $p < 0.05$) decrease in the TPC was established in the dried BP samples (Figure 1). In the next stage of the storage studies, after 6, 9, 12, and 15 months of storage at −20 °C, the amount of TPC in the BP samples decreased by 14%, 26%, 39%, and 49%, while at −80 °C, the decreases were 12%, 19%, 29%, and 34%, respectively (Figure 1). When analyzing the trends in the changes in TPC in the dried BP samples after 6, 9, 12, and 15 months of storage, the amount of TPC decreased by 10%, 27%, 46%, and 58%, respectively (Figure 1). After the completion of the storage period (after 15 months), the TPC content was 1.2–1.7-fold lower in the dried BP samples compared to that in the frozen BP samples.

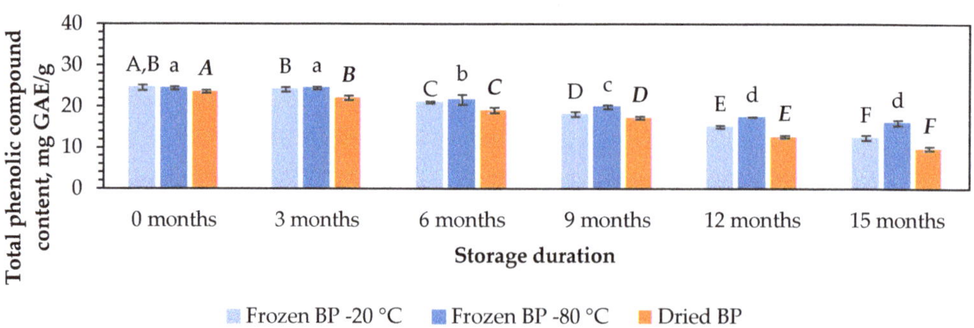

Figure 1. Variation in the amount of total phenolic compounds in bee pollen samples. Different letters indicate statistically significant differences between sample storage durations under the same conditions across different storage periods: −20 °C—uppercase letters; −80 °C—lowercase letters; dried—uppercase letters in bold italics ($p < 0.05$). Assessments were conducted every 3 months over a 15-month period.

The findings of this study revealed that the storage duration had an impact on the content of TPC in both the frozen and dried BP samples. It was found that in the BP samples stored at −20 °C and −80 °C, the content of TPC significantly decreased after 6 months of storage ($p < 0.05$). In the dried BP samples, the TPC content significantly decreased after 3 months of storage ($p < 0.05$) and gradually declined over the entire 15-month period.

The total flavonoid content in the BP samples ranged from 14.9 ± 0.5 mg RE/g to 15.7 ± 0.4 mg RE/g (Figure 2). No statistically significant changes were found between the BP samples frozen at different temperatures (−20 °C and −80 °C). The total flavonoid content in the dried BP samples was 4.5% lower than in the frozen (−20 °C) BP samples, but this difference was statistically insignificant.

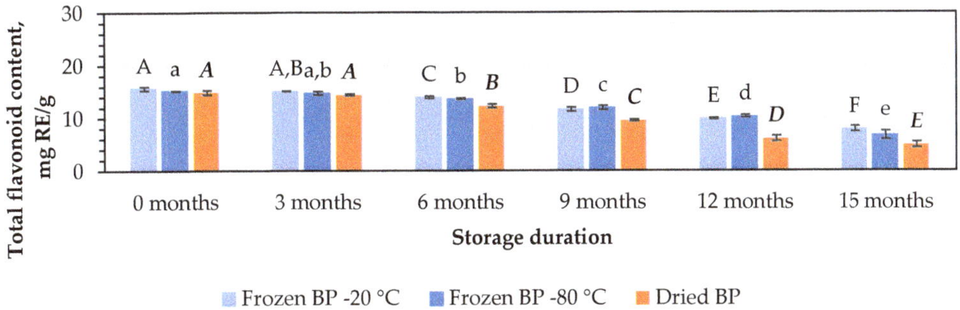

Figure 2. Variations in the total flavonoids in the bee pollen samples. Different letters indicate statistically significant differences between sample storage durations under the same conditions across different storage periods: −20 °C—uppercase letters; −80 °C—lowercase letters; dried—uppercase letters in bold italics ($p < 0.05$). Assessments were conducted every 3 months over a 15-month period.

During this study, the total flavonoid content decreased over the storage period in both the frozen and dried BP samples. After 3, 6, 9, 12, and 15 months of storage at −20 °C, the total flavonoids in the BP samples decreased by 6%, 101%, 26%, 37%, and 50%, while at −80 °C, the decreases were 3%, 10%, 21%, 32%, and 56%, respectively (Figure 2). In the dried BP samples, after 3, 6, 9, 12, and 15 months of storage, the total flavonoids diminished by 3%, 17%, 36%, 59%, and 67%, respectively (Figure 2).

The data from this study revealed that the total flavonoid content remained stable for 3 months in both the dried and fresh frozen BP samples ($p < 0.05$), followed by a steady decline over the subsequent storage period.

3.2. Changes in Ascorbic Acid Contents in Bee Pollen Samples

At the initial stage, the highest ascorbic acid content, 64.3 ± 2.3 mg/g, was found in the frozen (−20 °C) BP samples, while a lower amount, 51.3 ± 2.9 mg/g, was observed in the dried BP samples (Figure 3). No statistically significant change was found between the BP samples frozen at different temperatures (−20 °C and −80 °C), and the total ascorbic acid contents in the dried BP samples were 15% and 20%, respectively, lower than in the frozen ones ($p < 0.05$).

After 3 months of storage, the level of ascorbic acid decreased by 7% in the frozen BP samples, but no statistically significant difference was determined. In contrast, in dried BP samples, the content of ascorbic acid decreased statistically significantly by 20% ($p < 0.05$) (Figure 3). After 6, 9, 12, and 15 months of storage, the ascorbic acid content decreased most significantly in the dried BP samples, with reductions of 32%, 50%, 66%, and 79%, respectively. In the samples stored at −20 °C, the losses were also considerable—22%, 32%, 44%, and 69%—while storage at −80 °C resulted in comparatively lower decreases of 9%,

15%, 22%, and 43%, respectively (Figure 3). It was found that over the 6–15-month period, the ascorbic acid content in the dried BP samples decreased 1.8–3.4-fold more than in the frozen BP samples stored at −80 °C.

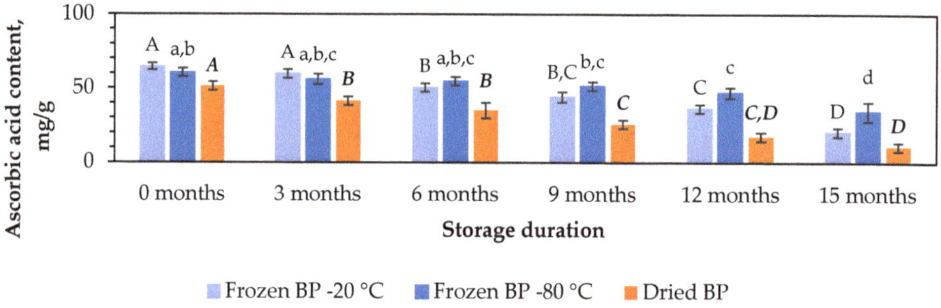

Figure 3. Variations in the amount of ascorbic acid in the bee pollen samples. Different letters indicate statistically significant differences between sample storage durations under the same conditions across different storage periods: −20 °C—uppercase letters; −80 °C—lowercase letters; dried—uppercase letters in bold italics ($p < 0.05$). Assessments were conducted every 3 months over a 15-month period.

When evaluating different storage durations, the highest amount of ascorbic acid remained when the BP samples were stored at −80 °C. A statistically significant decrease was observed only after 9 months ($p < 0.05$). Meanwhile, in the BP samples stored at −20 °C, a statistically significant decrease was observed after 6 months ($p < 0.05$), whereas in the dried samples, the decline started as early as 3 months ($p < 0.05$) and continued progressively throughout the remaining storage period.

3.3. Quantitative and Qualitative Determination of Phenolic Compounds in Bee Pollen Samples

When studying the variation in the qualitative and quantitative composition of phenolic compounds in the BP samples using ultra-efficient liquid chromatography, 25 phenolic compounds were determined (5 phenolic acids and 20 flavonoids).

After conducting the analysis of individual phenolic compounds, it was determined that in the BP samples stored at −20 °C, the highest amounts were of rutin (395.4 ± 47.6 μg/g), accounting for 30% of the sum of individual phenolic content (SIPC). High levels of *p*-coumaric acid (106.7 ± 18.3 μg/g), chlorogenic acid (106.4 ± 12.0 μg/g), and quercetin (101.3 ± 5.7 μg/g) were also identified, while the lowest amounts were found for avicularin (0.4 ± 0.04 μg/g), isorhamnetin (2.6 ± 0.1 μg/g), and quercitrin (2.6 ± 0.1 μg/g) (Figure 4).

After 15 months of BP sample storage at −20 °C, changes in phenolic compounds were determined (increasing order): isorhamnetin (+7%) > isorhamnetin-3-glucoside (+3%) > ferulic acid (−3%) > chlorogenic acid (−9%) > *p*-coumaric acid (−9%) > phloridzin (−18%) > naringenin (−28%) > hyperoside (−33%) > luteolin-3,7-diglucoside (−34%) > quercetin (−36%) > isoquercitrin (−37%) > gallic acid (−46%) > quercitrin (−48%) > rutin (−55%) > phloretin (−58%) > luteolin (−64%) > kaempferol (−75%) > quercetin-3-arabinoside (−76%) > kaempferol-3-O-glucoside (−97%) > avicularin (0%) (Figure 4). After 15 months, avicularin was no longer detected. The changes in phenolic compounds occurred gradually over time, with the exception of avicularin, kaempferol, quercetin-3-rutinoside, and kaempferol-3-O-glucoside (which showed a 1.2–2.3-fold decrease over the 6–15 month period), gallic acid, luteolin, and quercetin-3-arabinoside (which showed a 1.2–1.4-fold decrease over the 6–9 month period), and quercitrin (which showed a 1.3-fold decrease over the 9–12 month period).

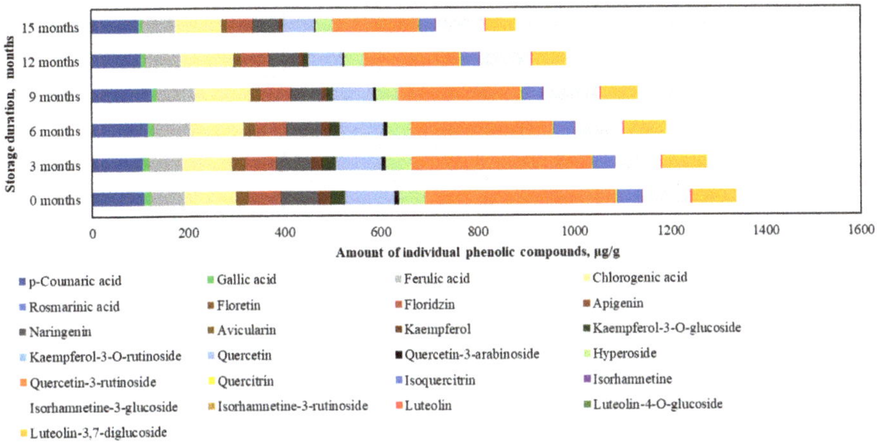

Figure 4. Variations in the contents of the individual phenolic compounds in the frozen (−20 °C) bee pollen samples.

When evaluating the changes in individual phenolic compounds in fresh frozen BP samples stored at −80 °C, it was found that the highest amounts were of rutin (405.5 ± 29.5 µg/g, accounting for 30% of SIPC) and quercetin (104.4 ± 5.3 µg/g), while the lowest amounts were of avicularin (0.4 ± 0.09 µg/g) and quercitrin (2.6 ± 0.2 µg/g) (Figure 5).

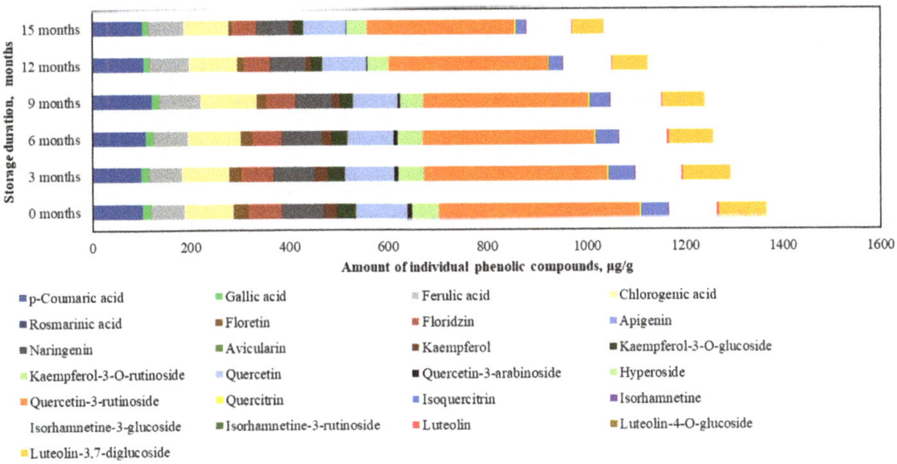

Figure 5. Variation in content of the individual phenolic compounds in frozen (−80 °C) bee pollen samples.

After 15 months of fresh frozen BP sample storage at −80 °C, changes in phenolic compounds were observed (increasing order): ferulic acid (+6%) > *p*-coumaric acid (+1%), isorhamnetin-3-glucoside (−8%) > chlorogenic acid (−9%), quercetin (−17%) > naringenin (−20%) > isorhamnetin (−21%) > hyperoside (−22%) > quercitrin (−25%) > rutin (−27%) > phloridzin (−27%) > luteolin-3,7-diglucoside (−33%) > gallic acid (−34%) > kaempferol-3-O-glucoside (−53%) > luteolin (−57%) > isoquercitrin (−63%) > kaempferol (+68%) > phloretin (−72%) > quercetin-3-arabinoside (−85%) > avicularin (−97%) (Figure 5). All identified phenolic compounds were preserved at higher or lower levels in the BP samples stored at −80 °C. After 15 months, the most significant decreases in flavonoid levels in the frozen (−80 °C) BP samples were observed for quercetin-3-arabinoside and avicularin. The changes in phenolic compounds were generally gradual, with the exception of luteolin (a significant 1.3-fold decrease after 3 months), quercetin-3-arabinoside (a significant 5.3-fold decrease during the 6–15 month period), avicularin and phloretin (a significant 1.7–14.8-fold decrease during the 9–15 month period), isoquercitrin (a significant 1.5-fold decrease during the 9–12 month period), and isorhamnetin (a 1.4-fold decrease during the 12–15 month period).

In the phenolic compound analysis of the dried BP samples, as in the frozen BP samples, the predominant compound was rutin. The highest content of rutin was found in the dried BP samples (366.0 ± 49.8 μg/g, accounting for 29% of SIPC), and the lowest amounts were detected for avicularin (0.3 ± 0.01 μg/g) and isorhamnetin (2.3 ± 0.2 μg/g) (Figure 6).

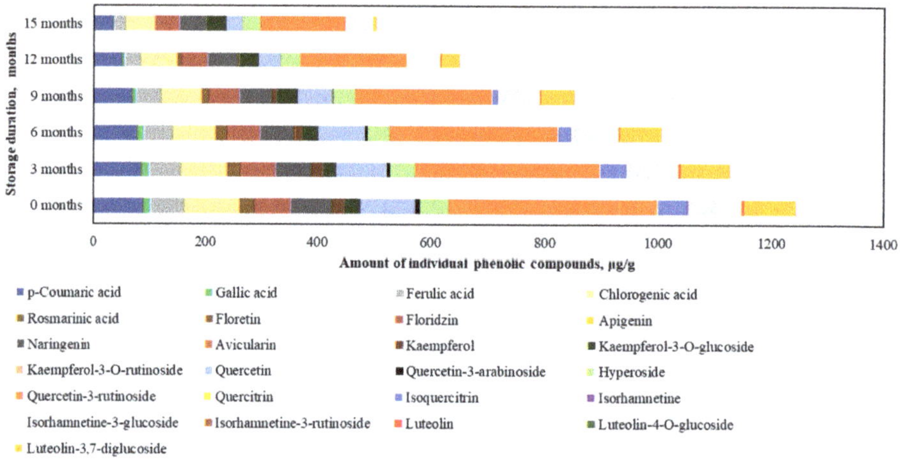

Figure 6. Variations in the contents of the individual phenolic compounds in the dried bee pollen samples.

After 15 months of storage, changes in the phenolic compounds in the dried BP samples were established (increasing order): avicularin (+36%) > kaempferol-3-O-glucoside (+31%) > naringenin (−33%) > phloridzin (−34%) > hyperoside (−41%) > chlorogenic acid (−48%) > isorhamnetin-3-glucoside (−49%) > rutin (−59%) > *p*-coumaric acid (−59%) > quercetin (−67%) > ferulic acid (−68%) > gallic acid (−84%) > quercitrin (−87%) > luteolin-3,7-diglucoside (−94%) > phloretin (−99%) > luteolin (0%) > kaempferol (0%) > isorhamnetin (0%) > isoquercitrin (0%) > quercetin-3-arabinoside (0%) (Figure 6). It is important to note that after 12 months, isorhamnetin, isoquercitrin, and quercetin-3-arabinoside were no longer detected, and after 15 months, luteolin and kaempferol were also absent. Reviewing the results of the dried BP samples, it can be concluded that, compared to the frozen BP samples, the levels of most compounds decreased significantly after 3 months. The greatest quantitative changes were observed in the analysis of kaempferol, quercetin-3-arabinoside, quercitrin, isoquercitrin, isorhamnetin, phloretin, and luteolin.

By evaluating the changes in individual phenolic acids and flavonoids under different storage conditions and durations, the highest total amount of identified phenolic compounds (9105.7 ± 170.8 μg/g) was determined in the frozen BP samples stored at −80 °C (which showed the least variation during storage), while the lowest amount (6961.5 ± 151.7 μg/g) was observed in the dried pollen samples ($p < 0.05$). The fresh frozen (−20 °C) BP samples contained 8509.3 ± 224.2 μg/g. The phenolic compounds identified in the BP samples, in decreasing order, were: rutin > chlorogenic acid > quercetin > *p*-coumaric acid > isorhamnetin-3-glucoside > luteolin-3,7-diglucoside > naringenin > ferulic acid > phloridzin > hyperoside > isorquercitrin > kaempferol-3-O-glucoside > phloretin > kaempferol > gallic acid > quercetin-3-arabinoside > luteolin > isorhamnetin > quercitrin > avicularin. When analyzing the variation in the contents of individual phenolic compounds, the amounts of phloridzin, naringenin, and hyperoside showed the least change during storage, whereas avicularin in the frozen samples decreased most significantly. The results of this study showed that the profiles of BP samples, including complex and individual phenolic compounds, varied significantly under different storage conditions and durations.

3.4. Evaluation of the Antiradical and Reducing Activity of Bee Pollen Extracts In Vitro

A review of the literature has shown that both antiradical (ABTS) and reducing (CUPRAC, FRAP) in vitro assays are used for evaluating the antioxidant activity of BP extracts. Different biologically active compounds exhibit distinct antioxidant mechanisms, which may lead to varying results. To ensure accuracy, it is essential to apply multiple methods for evaluation. Considering this possibility and recognizing that using two different methods might result in significant variations and complicate the interpretation of the data, it was decided to employ three different methodologies.

The in vitro antiradical activity values of the BP extracts ranged from 76.5 ± 9.8 µmol TE/g in the frozen (−80 °C) BP extracts to 79.1 ± 0.1 µmol TE/g in the dried BP extracts. No statistically significant differences between the values were found ($p > 0.05$).

After 3 months of storage at −20 °C and −80 °C, the antiradical activity of the BP extracts decreased by 5% and 2%, respectively, while the dried BP extracts showed a slightly higher decrease of 5%. However, these results were statistically insignificant (Figure 7). After 6, 9, 12, and 15 months of storage under different conditions, the antiradical activity of the BP extracts changed as follows: for the frozen BP samples stored at −20 °C, the activity of the BP extracts decreased by 11%, 13%, 25%, and 43%; for the frozen samples stored at −80 °C, the decreases were 6%, 7%, 17%, and 36%; and for the dried samples, the decreases were 18%, 25%, 37%, and 65%, respectively (Figure 7). The research results revealed that the antiradical activity of the frozen (−80 °C) BP extracts decreased slightly less than that of the frozen (−20 °C) and dried BP extracts. Over the 6–15-month period, the antiradical activity of the frozen (−80 °C) BP decreased 1.2–1.8-fold compared to the extracts of the frozen (−20 °C) BP and 1.8–3.4-fold less compared to the dried BP extracts.

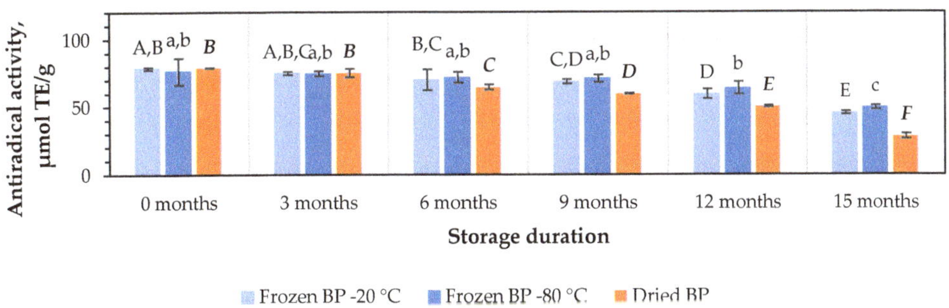

Figure 7. Variations in the antiradical activity in the bee pollen samples determined using the ABTS method in vitro. Different letters indicate statistically significant differences between sample storage durations under the same conditions across different storage periods: −20 °C—uppercase letters; −80 °C—lowercase letters; dried—uppercase letters in bold italics ($p < 0.05$). Assessments were conducted every 3 months over a 15-month period.

A statistically significant decrease in the antiradical activity of the frozen (−80 °C) BP extracts was observed only after 12 months of storage ($p < 0.05$). In the case of the frozen (−20 °C) BP extracts, a statistically significant decrease was observed after 9 months ($p < 0.05$), while in the dried BP extracts, a decrease was already noted after 6 months of storage ($p < 0.05$). Thereafter, the antiradical activity progressively decreased throughout the 15-month storage period.

Using the CUPRAC method, the values of the reducing activity capacity in the preliminary stage ranged from 179.2 ± 3.5 µmol TE/g to 195.2 ± 11.0 µmol TE/g. The strongest reducing activity (195.2 ± 11.0 µmol TE/g) was determined in the dried BP extracts, and lower activity (179.2 ± 3.5 µmol TE/g) was found for the frozen (−80 °C) BP extracts (Figure 8). No statistically significant differences between the values in the preliminary stage were found.

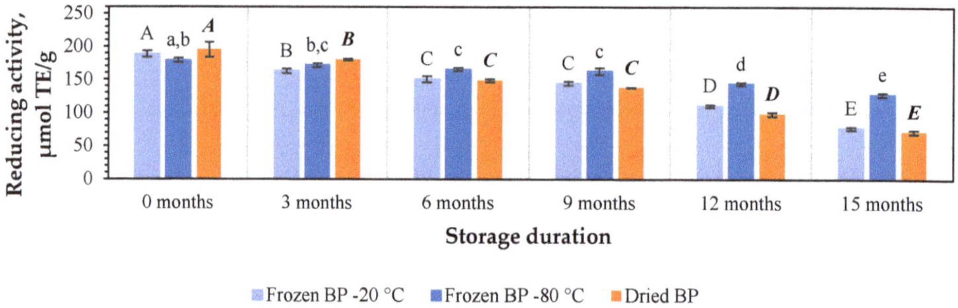

Figure 8. Variations in the reducing activity of the bee pollen samples determined using the CUPRAC method in vitro. Different letters indicate statistically significant differences between sample storage durations under the same conditions across different storage periods: −20 °C—uppercase letters; −80 °C—lowercase letters; dried—uppercase letters in bold italics ($p < 0.05$). Assessments were conducted every 3 months over a 15-month period.

Under different storage conditions for 3, 6, 9, 12, and 15 months, the reducing activity of the frozen (−20 °C) BP extracts decreased by 10%, 15%, 20%, 25%, and 36%; the frozen (−80 °C) BP extracts diminished by 7%, 10%, 13%, 17%, and 18%; and the dried BP extracts reduced by 11%, 31%, 32%, 43%, and 56%, respectively (Figure 8). It is noteworthy that after 15 months, the reducing activity of the frozen (−80 °C) BP extracts changed less and remained more stable than the frozen (−20 °C) and dried BP extracts. Over a 6–15-month period, the reducing activity of the frozen (−80 °C) BP extracts decreased 2.6–3.1-fold less compared to the dried BP extracts.

The reducing activity, determined using the CUPRAC method, of the frozen (−80 °C) BP extracts decreased statistically significantly only after 6 months of storage ($p < 0.05$). However, a statistically significant decline in the reducing activity of the frozen (−20 °C) and dried BP extracts was estimated after 3 months ($p < 0.05$). Subsequently, the reducing activity continued to decrease gradually in vitro.

The estimation of the reducing potency of the BP extracts in vitro using the FRAP method revealed that this activity ranged from 84.3 ± 3.2 µmol TE/g to 90.0 ± 2.2 µmol TE/g. Similar to the results described earlier, the strongest reducing activity (90.0 ± 2.2 µmol TE/g) was observed in the dried BP extracts, while the lowest activity (84.3 ± 3.2 µmol TE/g) was observed in the frozen (−20 °C) BP extracts (Figure 9). No statistically significant variations between the values in the preliminary stage were found.

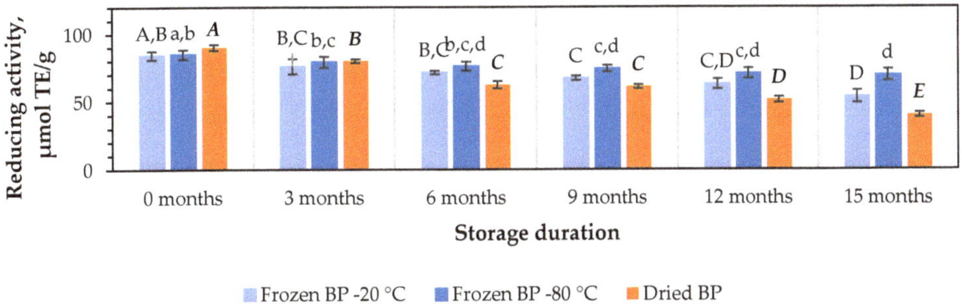

Figure 9. Variations in reducing activity of bee pollen samples determined using the FRAP method in vitro. Different letters indicate statistically significant differences between sample storage durations under the same conditions across different storage periods: −20 °C—uppercase letters; −80 °C—lowercase letters; dried—uppercase letters in bold italics ($p < 0.05$). Assessments were conducted every 3 months over a 15-month period.

The detailed changes in reducing activity during the study, conducted every three months under different storage conditions using the FRAP method, were as follows: for frozen (−20 °C) BP extracts, the reducing activity decreased by 14%, 20%, 24%, 41%, and 59%; for the frozen (−80 °C) BP extracts, it diminished by 4%, 8%, 9%, 20%, and 29%; and for the dried BP extracts, it declined by 8%, 24%, 29%, 50%, and 63% (Figure 9).

When determining the impact of the different storage conditions and durations on the reducing activity of the BP extracts using the FRAP method, it was found that after 3 months of storage, a statistically significant decline in reducing activity was found in the dried BP extracts ($p < 0.05$). In contrast, the reducing activity of the frozen BP extracts, which showed more stable activity, decreased significantly only after 9 months of storage ($p < 0.05$). The reducing activity showed a further gradual decline over 15 months in vitro.

Summarizing the results of the impact of different storage conditions (freezing and drying) and storage duration on the antioxidant capacity of the BP extracts, the strongest antiradical and reducing activity in vitro (over the 6–15-month period) was observed in the frozen (−80 °C) BP extracts compared to the dried BP extracts. During this period, the antiradical and reducing activity in vitro in the frozen BP samples was 1.8–3.4-fold and 2.6–3.1-fold higher than in the dried BP extracts, respectively.

3.5. Correlation Analysis

During this study, the strength of the correlation between the total phenolics content (TAP), sum of individual phenolic compounds (SIPC), total flavonoids (TFC), ascorbic acid (AAC), and the antioxidant activity was determined using the ABTS, FRAP, and CUPRAC methods. Under different conditions (−20 °C, −80 °C, and dried) the stored BP samples were compared and evaluated separately. The matrix created allows for a comprehensive comparison of the correlation relationships between the characteristics selected for evaluation. The intensity of the green color indicates the strength of the correlation (Figure 10).

In the frozen (−20 °C) BP samples, a statistically significant, very strong correlation was observed between the reducing activity measured using the CUPRAC method and TAP ($r = 0.969$, $p < 0.05$), TFC ($r = 0.948$, $p < 0.05$), and AAC ($r = 0.966$, $p < 0.05$). Similarly, a very strong correlation was established between the antiradical activity measured using the ABTS method and TAP ($r = 0.940$, $p < 0.05$), TFC ($r = 0.916$, $p < 0.05$), and AAC ($r = 0.940$, $p < 0.05$). In the frozen (−80 °C) BP samples, only two very strong correlations were found: between the reducing activity determined using the CUPRAC method and TAP ($r = 0.944$,

$p < 0.05$) and ACC (r = 0.911, $p < 0.05$). In the dried BP samples, four statistically significant very strong correlations were found between the reducing activity determined using the CUPRAC method and TAP (r = 0.993, $p < 0.05$), TFC (r = 0.927, $p < 0.05$), SIPC (r = 0.939, $p < 0.05$), and AAC (r = 0.963, $p < 0.05$). Three very strong correlations were also found for the FRAP method and TAP (r = 0.946, $p < 0.05$), SIPC (r = 0.964, $p < 0.05$), and AAC (r = 0.967, $p < 0.05$). Additionally, the ABTS method showed three strong correlations with TAP (r = 0.970, $p < 0.05$), SIPC (r = 0.943, $p < 0.05$), and AAC (r = 0.949, $p < 0.05$).

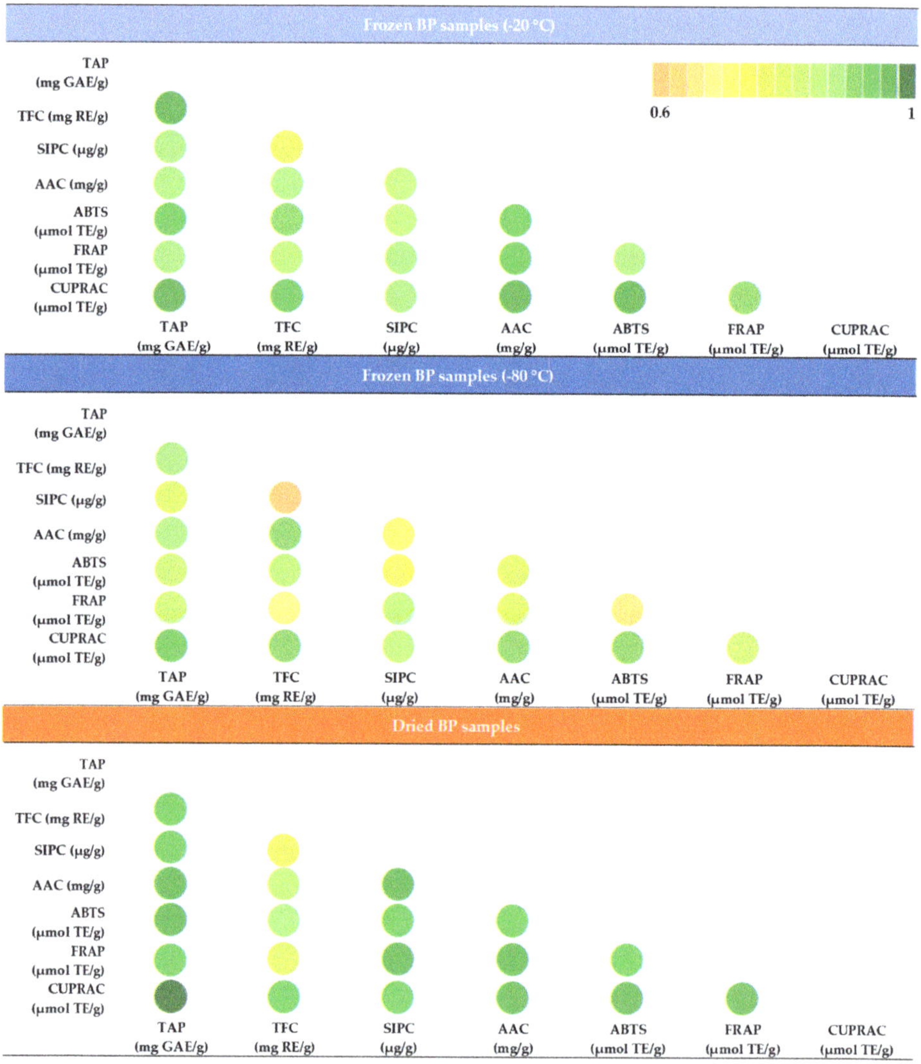

Figure 10. Correlation matrix of the quantitative characteristics of the bee pollen based on Pearson correlation coefficients; different colors indicate the strength of the correlation ($p < 0.05$). Abbreviations: TAP—total amount of phenolic compounds; TFC—total flavonoid content; SIPC—sum of individual phenolic contents; AAC—ascorbic acid content; ABTS—antiradical activity of the BP extracts determined using the ABTS method; FRAP—reducing activity of the BP extracts determined using the FRAP method; CUPRAC—reducing activity of the BP extracts determined using the CUPRAC method.

3.6. Botanical Origin of Bee Pollen

The botanical composition of the samples selected for flavonoid analysis is shown in Table 2. Melisopalynological pollen analysis showed that the pollen of *Salix* spp. and *Brassica napus* L. dominated in all the samples, constituting 34.3% and 36.8%, respectively. *Acer platanoides* L., Malus *domestica* Borkh., and *Taraxacum officinale* L. pollens accounted for 12.8%, 9.0%, and 5.9%, respectively, which are also considered as important minor pollens according to the study results.

Table 2. The botanical composition of the samples selected for flavonoid analysis.

Plant Pollen	Average, %	Variation Limits		SD	CV, %
		Min.	Max.		
Brassica napus L.	35.37	34.30	36.40	1.05	2.97
Salix spp.	36.87	36.4	37.40	0.50	1.37
Acer platanoides L.	12.83	11.0	15.40	2.29	17.84
Malus domestica Borkh.	9.03	8.10	10.40	1.21	13.39
Taraxacum officinale L.	5.93	5.40	6.70	0.68	11.47

Note: standard deviation of the mean (SD); coefficient of variance (CV).

The coefficient of variance (CV) of the botanical origin of the pollen ranged from 1.37% to 17.84% (Table 2). According to statistics, a CV value is acceptable when it ranges between 10 and 20% [40].

4. Discussion

Bee pollen is used as a valuable food source not only for bees but also for humans, as it contains proteins, lipids, amino acids, carbohydrates, vitamins, minerals, and biologically active compounds such as enzymes, phenolic compounds, etc. [20]. The composition of BP varies depending on its botanical and geographical origin, according to the region, and even according to BP collection time [41]. Phenolic compounds, specifically phenolic acids and flavonoids, are important BP compounds known to be potent antioxidants and crucial non-nutrient elements in BP. The method of collection and storage conditions of BP has a significant influence on its phytochemical composition and quality. BP is used by humans not only in its native state but also in the pharmaceutical form (tablets, pastilles etc.), as a food supplement, and in functional food due to biologically active compounds. Consequently, it is particularly essential to choose the right method of BP preparation and its storage conditions. To consumers in Lithuania, dry BP is usually provided, but its quality is highly dependent on the drying temperature. Unfortunately, after drying of pollen, especially at higher temperatures, the amount of biologically active compounds can change dramatically, i.e., decrease, and vitamins are particularly sensitive to high temperatures. Pollen that is freshly collected and frozen immediately after collection is very popular in the foreign market, especially in Romania and other countries, which have long-lasting traditions of usage of BP. Therefore, it is necessary to investigate whether dried or frozen pollen is more valuable, and what optimal storage conditions are necessary, particularly by identifying when the decline in biologically active compounds begins, in order to optimize the amount of biologically active substances in it. Phenolic compounds, particularly phenolic acids and flavonoids, are key bioactive components in BP, recognized for their strong antioxidant properties and essential role in functional foods and pharmaceuticals. While previous studies have highlighted the influence of processing and storage conditions on BP quality, there remains a critical knowledge gap regarding the precise degradation dynamics of polyphenols and vitamin C over time. Our study specifically addresses this gap by systematically investigating the long-term changes in the phytochemical profile of dried and frozen BP under different storage conditions, focusing on when significant reductions in polyphenol and vitamin C content begin. The objective of our study was not to evaluate oxygen-restricted conditions, but rather to examine the stability of bioactive compounds in BP under storage conditions that reflect common consumer and producer practices, namely drying and freezing without specialized packaging. By analyzing these changes at three-month intervals over a 15-month storage period, the first detailed timeline of antioxidant compound degradation in BP was established, offering crucial insights into its optimal preservation. Furthermore, our study provides a comprehensive analysis of the variations in the amount of individual phenolic acids and flavonoids during storage over a 15-month period, offering detailed insights into their stability and degradation dynamics over time. To the best of our knowledge, no such detailed studies have been conducted.

The main findings are as follows: (1) It was determined that during storage (over a 3–15-month period), the total phenolic compound contents began to statistically significantly decrease after 3 months in the dried BP samples, while in the frozen BP samples, this decrease occurred after 6 months ($p < 0.05$). (2) The highest ascorbic acid content remained in the frozen (−80 °C) BP samples, with a statistically significant decrease in the content observed only after 9 months. In the frozen (−20 °C) BP samples, the ascorbic acid content decreased after 6 months ($p < 0.05$), while in the dried BP samples, it decreased after 3 months of storage ($p < 0.05$). (3) The qualitative and quantitative composition of the individual phenolic compounds changed during the 3–15-month period. The content of phloridzin, naringenin, and hyperoside changed the least during storage, while avicularin decreased most significantly in the frozen BP samples (by 100%). (4) The strongest antiradical and reducing activity in vitro (over a 6–15-month period) was observed in the frozen (−80 °C) BP extracts compared to the dried BP extracts and was 1.8–3.4-fold and 2.6–3.1-fold higher, respectively.

Phenolic compounds are among the most widespread secondary metabolites produced in all plant cells [42,43]. In BP, their contents range from 3% to 5% and are highly dependent on the origin of the raw material [44]. The total amount of phenolic compounds in all the tested samples from the BP collected in Lithuania varied from 23.3 ± 0.4 mg GAE/g to 24.5 ± 0.7 mg GAE/g and was lower than that described for Anzer pollens from Turkey (44.1–124.2 mg GAE/g) [45] and pollens from Brazil (41.5–213.2 mg GAE/g) [46]. However, it was similar or higher than that described for Turkish pollen samples (7.9 mg GAE/g and 17.5 mg GAE/g) [47], Italian pollen samples (13.5–24.8 mg GAE/g) [48], Romanian pollen samples (4.4–16.4 mg GAE/g) [49], and Portuguese pollen samples (10.5–16.8 mg GAE/g) [50]. Likewise, the amounts of flavonoids in our BP samples varied from 14.9 ± 0.5 mg RE/g to 15.7 ± 0.4 mg RE/g; however, in other studies, it varied between 4.5 and 7.1 mg CAE/g, 1.8 and 4.4 mg/g QE/g [51], and 1.4 and 9.1 mg QE/g [52]. The total phenolics and total flavonoid contents of twenty-two BP samples from Portugal were determined as 12.9–19.8 mg GAE/g and 4.5–7.1 mg CAE/g, respectively [53].

The most abundant phenolic acids in BP samples are chlorogenic, gallic, ferulic, cinnamic, and caffeic acids, as well as hydroxycinnamic, *o*-coumaric and *p*-coumaric acids [44]. We also confirmed that the major phenolic acids in the BP samples were chlorogenic and *p*-coumaric acids. Flavonoids constituted the most significant group of phenolic compounds present in the BP. The main flavonols in BP are quercetin and kaempherol and their glycosides [44], which was also confirmed in our study. Straumite et al. showed that in BP samples collected in Latvia in 2022 (May and June), the total phenolic contents of polyfloral pollen samples varied from 16.1 ± 0.5 mg GAE/g to 26.1 ± 0.2 mg GAE/g, so the amounts determined were similar to those in our studies [12]. This may have been influenced by the similar geographical location, similar vegetation, climate zone, and time of collection. Meanwhile, research by other scientists (Muliuolytė et al.) shows that in BP samples collected and dried in 2022 in June–August from the apiary located in the south region of Lithuania, the total amount of phenolic compounds varied from 42.1 ± 3.1 mg GAE/g to 93.6 ± 7.3 mg GAE/g, which is higher than in the samples we studied [54]. Such differences could have been affected by differences between seasons, the conditions of extract preparation, different vegetation, weather conditions in that year, and other factors. There have been no studies on how the total amount of phenolic compounds in pollen changes during storage lasting 3 months. In that aspect, our study is new and relevant. However, Stojko et al. investigated three types of BP extracts, namely ethanol extracts, enzymatic hydrolysates from pollen, and ethanol extract of pepsin-digested BP, and showed that 12-month storage of BP extracts decreased phenolic compound concentrations in all three types of extracts, and the changes depended on the storage conditions [44]. The highest decrease in the phenolic compound concentration was in pollen extracts stored at room temperature in light, while storing at 4–8 °C in the dark was the best storage condition for all types of extracts [44]. Anjos et al. investigated the effects of the storage conditions of BP samples from Portugal (frozen at −20 °C or dried at +42 °C) on the quality of BP [27]. They found that even the botanical origin is an important factor in explaining the variation between samples, and the storage method was a highly important factor explaining changes in the amount of total phenolic compounds, total flavonoids, reducing sugars, and lipids [27]. According to Abhay et al., the drying process supports the degradation of phenolic compounds via both enzymatic and non-enzymatic oxidation of phenolic compounds [55]. According to other authors, polyphenol-oxidase enzymes are activated during drying.

Munteanu et al. studied BP samples from nine apiaries in Romania in 2019 [37]. The total amount of flavonoids in frozen BP at the initial stage was determined to be from 2.5 ± 0.3 mg RE/g to 11.8 ± 1.6 mg RE/g, and in dried samples, it was from 2.6 ± 0.04 mg RE/g to 6.9 ± 0.3 mg RE/g. Thus, the total amount of flavonoids was higher in our studied BP samples. The changes after 3 and 6 months in their frozen BP samples were similar to those found in our study, and in dried BP, twice lower amounts of flavonoids were determined, while the total amount of flavonoids in the dried pollen under our chosen conditions after 3 months did not decrease.

By analyzing and comparing the compounds identified during this research in the BP samples, the variations in the phenolic acids profiles were determined. In a study by Turkish scientists [56], caffeic acid and vanillin were detected, and almost twice lower amounts of chlorogenic acid. The abundance of the identified flavonoids in the BP samples was also more diverse. The major compounds were kaempferol, naringenin, and (−)-epicatechin, and the change after 3 months did not differ significantly, as in our studies. The phytochemical composition of BP collected in different regions differs due to different vegetation, different climatic conditions, etc. A study of dried (+30 °C) BP samples performed in Latvia in 2018 showed that the dominant compounds were rutin (428.4 ± 29.1 µg/g), quercetin (127.3 ± 6.2 µg/g), and naringenin (108.9 ± 4.9 µg/g). The results were similar to our research due to similar vegetation in Lithuania, geographical location, and climatic conditions. According to Anjos et al., under favorable storage conditions, higher quantities of biologically active compounds can be extracted from BP samples [27]. This was confirmed by our study results. The levels of *p*-coumaric acid, ferulic acid, chlorogenic acid, isorhamnetin, and isorhamnetin-3-glucoside increased when stored frozen (−20 °C and −80 °C) for 6 and 9 months, while the levels of avicularin and kaempferol-3-O-glucoside increased after 6 and 9 months. After 12 months of storage, the largest growth in the levels of these compounds was detected in dried BP samples, but later, the levels began to decrease. In a study by Anjos et al. of dried BP samples, a significant increase in avicularin (+19%) and naringenin (+11%) was observed after 9 months of storage. Our study results after

9 months showed an increase in avicularin (+26%) and kaempferol-3-O-glucoside (+31%), while naringenin levels decreased during storage (−10%), and after 15 months of storage, it they decreased by almost a third (−33%). Our results are consistent with previous studies. Yousefi et al. reported accelerated degradation of phenolic compounds in freeze-dried berries compared to frozen samples, attributing this to residual enzymatic activity and oxidative processes [57]. Sawicki et al. similarly observed increased flavonoid degradation in freeze-dried apple peels, particularly under elevated storage temperatures [58]. In the context of BP, Komosińska-Vassev et al. and Pascoal et al. demonstrated significant losses of flavonoids during storage at room temperature, which aligned with the trends we observed in our dried samples [6,25]. Furthermore, Almeida-Muradian et al. emphasized that ultra-low temperatures (−80 °C) effectively preserved polyphenolic compounds, supporting our findings regarding the superior stability of frozen BP under such conditions [59].

The amount of ascorbic acid in all the analyzed BP samples ranged from 51.3 ± 2.9 mg/g to 64.3 ± 2.3 mg/g, which is higher than the values reported in other studies, where ascorbic acid content in frozen BP samples collected in Turkey was 42.2 ± 3.4 mg/g, and in dried samples, it was 30.2 ± 2.7 mg/g [5]. Climate zone and vegetation may influence these results. Kolayli et al. investigated the variations in ascorbic acid in frozen (−20 °C) and dried (+30 °C) BP samples, storing the samples under the appropriate conditions for 12 months. After 12 months, in frozen BP samples, ascorbic acid decreased by 27%, and in dried ones, it decreased by 53% [5]. The ascorbic acid content decreased in dried BP samples after 3 months, whereas in fresh-frozen BP samples, no changes were observed after 3 months, and a statistically significant decrease was detected only after 9 months ($p < 0.05$). In fresh-frozen BP samples stored at −20 °C, the ascorbic acid content decreased after 6 months ($p < 0.05$).

We conducted a degradation kinetics analysis using a first-order kinetic model to describe the decrease in total phenolic content, total flavonoid content, and ascorbic acid in bee pollen samples stored under different conditions over 15 months. The kinetic modeling results strongly support the application of a first-order degradation model to describe the storage stability of bioactive compounds in bee pollen. The model provided both mechanistic insights and quantitative parameters: (1) Demonstrate the compound and condition-specific nature of degradation; (2) Confirm that drying accelerates the loss of all tested compounds, especially ascorbic acid and flavonoids; (3) Highlight −80 °C freezing as the most effective strategy for preserving phenolic content, flavonoids, and vitamin C (see Supplementary Material).

The data from our research revealed that the antiradical activity in vitro of the BP extracts varied from 76.5 ± 9.8 μmol TE/g to 79.1 ± 0.1 μmol TE/g. Research has shown that the storage duration can influence the variation in antiradical activity in vitro [60]. The antiradical activity of pollen extracts collected from Brazilian apiaries ranged from 37.6 ± 1.2 μmol TE/g to 70.5 ± 3.3 μmol TE/g, exhibiting weaker antiradical activity compared to the pollen extracts collected under the climatic conditions of Lithuania. After 9 months of storage, the antiradical activity in vitro of our dried BP extracts decreased by 25%, while the Brazilian BP extracts showed a decrease of 17% to 32% [60]. Previous studies have shown that the methanolic extract of BP had a higher $ABTS^{\bullet+}$ radical scavenging capacity (5.37–6.47 TE mg/g RW) compare to bee bread (4.86–5.70 TE mg/g RW) [61]. The results of our investigation demonstrated that the reducing activity in vitro of dried and frozen (−20 °C and −80 °C) BP extracts varied from 179.2 ± 3.5 μmol TE/g to 195.2 ± 11.0 μmol TE/g using the CUPRAC method, and from 84.3 ± 3.2 μmol TE/g to 90.0 ± 2.2 μmol TE/g using the FRAP method. Alimoğlu et al. determined the reducing activity in vitro of frozen (−25 °C) and dried (+30 °C) BP extracts using the CUPRAC method and evaluated the stability of the samples after 12 months [38]. The researchers established that the reducing activity of frozen BP extracts varied from 92.0 ± 8.9 μmol TE/g to 148.6 ± 11.5 μmol TE/g, while for dried BP extracts, it ranged from 89.0 ± 7.6 μmol TE/g to 150.7 ± 9.2 μmol TE/g. According to Turkish scientists, after 12 months of storage, the reducing activity in vitro of frozen BP extracts decreased by an average of 31% (in our study, the decrease was 41%), while in dried extracts, the decrease was 46% on average (in our study, the decrease was 50%). Comparing both studies, it was found that, in the initial stage, the BP extracts examined in this study exhibited stronger reducing activity in vitro; however, compared to the results obtained by Turkish scientists, they were less stable during storage [38]. Croatian scientists carried out research to compare the reducing activity in vitro of BP extracts stored under different conditions (frozen at −20 °C and dried at +30 °C) using the FRAP method [11]. In the initial stage, the reducing activity values of the BP extracts examined in this study were similar to those described by Valentić et al. in their in vitro reducing activity studies (frozen: 105.3 ± 34.1 μmol TE/g; dried: 100.2 ± 7.6 μmol TE/g). According to these researchers, after 9 months, the reducing activity in vitro of frozen BP extracts (−20 °C) was nearly twice as high as that of dried BP extracts (+30 °C) [11]. The results of the Croatian scientist's study confirm the findings of our research. During our study (over a 6–15-month period), a 2.6–3.1-fold stronger reducing activity in vitro was observed in frozen BP extracts compared to dried extracts.

A statistically significant very strong or strong positive correlation was determined between the content of total phenolic compounds, total flavonoids, the sum of individual phenolic compounds, ascorbic acid, and the in vitro antioxidant activity in all the examined BP samples. The strongest positive correlation was observed between the total phenolic compound content and the reducing activity using the CUPRAC method in dried BP samples ($r = 0.993$, $p < 0.05$), while in the frozen (−80 °C) BP samples, correlations were observed between the reducing activity using the CUPRAC method and the total phenolic compound content ($r = 0.944$, $p < 0.05$) and the ascorbic acid content ($r = 0.911$, $p < 0.05$). Habryka et al. found positive correlations between the content of total phenolic compounds and the total flavonoids ($r = 0.9992$) and phenolic acids ($r = 0.9714$) [13]. In many studies, very strong and strong positive correlations were found between the total phenolic compound content of pollen and antioxidant activity ($r = 0.930$ [62], $r = 0.950$ [63], $r = 0.870$ [46]). Therefore, in our study, the relationships described from the literature were observed.

BP is rich in antioxidants, particularly phenolic compounds, the effect of which depends on the presence of double bonds and the positioning of hydroxyl groups on the aromatic ring [64,65]. This ring structure also influences the lipophilicity of polyphenols, particularly flavonoids. Lipophilic antioxidants play a crucial role in protecting lipid membranes [66,67]. Additionally, BP serves as a source of hydrophilic antioxidants, which protect against oxidative damage to cellular structures, including the cytoplasm, organelles, and extracellular fluid. Ascorbic acid and phenolic acids, such as caffeic acid, are examples of hydrophilic antioxidants found in BP. These compounds are effective antioxidants in aqueous phases as well [68]. The antioxidative activity of phenolic acids is influenced by factors such as the number of hydroxyl groups, the arrangement of functional groups, and the steric effects they induce. Monohydroxy derivatives of benzoic acid exhibit the highest antioxidative properties when hydroxylation occurs at the meta position, while dihydroxy derivatives show strong antioxidative activity in both ortho- and meta-hydroxylation. The proximity of the COOH group to ortho-diphenolic functional groups affects the accessibility of hydrogen in the meta position, enhancing antioxidative effectiveness. Gallic acid demonstrates significant antioxidant capacity, equivalent to 3 mM Trolox, due to its three accessible hydroxyl groups. However, the presence of methyl groups at the 3-OH and 5-OH positions reduces its activity compared to trihydroxy derivatives [68]. Monohydroxy derivatives of cinnamic acids are more effective hydrogen donors than those of phenylacetic acid. Furthermore, the introduction of a second hydroxyl group in the ortho position, as seen in caffeic acid, or in the para position, as in protocatechin acid, enhances their antioxidative properties. As a result, diphenols like caffeic, chlorogenic, and protocatechin acids exhibit a stronger ability to scavenge radicals compared to monophenols such as *p*-coumaric acid. Additionally, substituting the 3-OH group with a methoxyl group in caffeic acid, as in ferulic acid, increases antioxidative activity in the lipid phase [69].

Flavonoids' chemical structure is characterized by a diphenylpropan ring system (C6-C3-C6) with a benzo-γ-pyrone skeleton. The antioxidative properties of flavonoids are influenced by the presence of a double bond between C2 and C3 in the C ring. Additionally, a carbonyl group at the C4 position enables flavonoids to scavenge hydroxyl radicals. The presence of a hydroxyl group at the C3 position in the C ring allows for the inhibition of lipid peroxidation. The ability to scavenge hydroxyl radicals increases with the number of hydroxyl groups present in the B ring, particularly at positions $3'$ and $4'$. Furthermore, hydroxyl groups at the C5 and C7 positions in the A ring, C3$'$ and C4$'$ in the B ring, and C3 in the C ring enhance the inhibition of lipid peroxidation [68,70–73]. Various flavonoid derivatives have different additions to antioxidant activity [74]. Among quercetin glycosides quercetin 3-O rutoside (rutin) is a common form [75,76]. Its sugar component consists of a glucose and rhamnose disaccharide. Some researchers suggest that the optimal level of rutin in pollen determines its biological and nutritional quality [75]. Quercetin derivates such as C(3)-OH and C(4$'$)-OH glycoside showed decreased antioxidant activity compared to its free form, quercetin aglycone (without linked sugars) [77]. However, it is claimed that supplements containing quercetin in the aglycone form are less bio-absorbed by the body than quercetin glycoside, which is often found in food [78,79]. Liangqin Xie et al. determined that the antioxidant activity of flavonoid C-glycosides is equal to or greater than that of O-glycosides. Glycosides maintain relatively stable antioxidant activity compared to their aglycones. The structural stability differences between flavonoid O-glycosides, C-glycosides, and their aglycones contribute to changes in their content during digestion, resulting in differences in their antioxidant activity. The antioxidant activity of flavonoid O-glycosides in plasma was higher than that of C-glycosides, while in urine, C-glycosides exhibited higher activity than O-glycosides, reflecting their distinct absorption and metabolism pathways [80].

Apis mellifera honeybees can detect floral scents to locate and distinguish different flowers while searching for food. Studies show that both odor intensity and quality affect bees' ability to distinguish floral scents [81]. Rapeseed is very attractive to insect pollinators due to the rich yellow color of its flowers, its special aromas, and the sugar content in its nectar [82,83]. Semi-natural habitats in the landscape have a positive effect on the diversity of pollen collected by bees [84]. Pollen from flowers of different plants has a specific size, shape, and exine surface ornamentation. The investigation of honey plant pollen is of great importance for the identification of floral resources of nectar-producing plants [28]. Monofloral and polyfloral honey is collected in Lithuania, as well as monofloral or polyfloral pollen and bee bread [85–87]. Data show that in Lithuania, beekeeping products are mainly collected from rapeseed, but monofloral honey is also collected from apple trees (*Malus domestica*, Borkh.), linden (*Tilia cordata* Mill.), caraway (*Carum carvi* L.), white clover (*Trifolium repens* L.), and willow (*Salix* spp.) [88].

Our data show that in the investigated samples, Salix spp. and *Brassica napus* L. were the dominant pollen types, comprising 34.3% and 36.8% of the total, respectively. Additionally, *Acer platanoides* L., *Malus domestica* Borkh., and *Taraxacum officinale* L. were identified as significant minor pollen contributors, accounting for 12.8%, 9.0%, and 5.9% of the composition, respectively. Baltrušaitytė, V. et al. studied the botanical composition and antioxidant activity of 35 Lithuanian honey samples and 9 samples of bee bread. The strongest (DPPH•) radical scavenging, which was greater than 80%, was found for 4 samples out of 35, including monofloral spring rape, willow, and polyfloral spring honey [89].

The biologically active compounds present in BP, such as flavonoids and phenolic acids, exhibit a broad range of biological activities, with their effects potentially varying depending on the stability of these compounds during storage. The degradation of ascorbic acid and phenolic compounds in BP during storage can be influenced by several factors, including oxygen, light, and temperature. Enzymatic activity, particularly that of polyphenol oxidase, may contribute to the degradation of phenolic compounds if enzyme inactivation is incomplete, especially in dried pollen [25]. A key pathway involves enzymatic oxidation, in which polyphenol oxidases catalyze the conversion of ortho-diphenols to highly reactive quinones. These quinones can subsequently participate in non-enzymatic polymerization reactions, forming brown pigments such as melanins. This process not only alters the visual appearance of the product but may also lead to a reduction in antioxidant activity, thereby affecting its functional quality [90]. It can also affect another mechanism, hydrolysis of flavonoid glycosides, which can occur in the presence of β-glucosidases, resulting in the release of aglycones. Flavonoid glycosides are subject to enzymatic hydrolysis, and their stability depends on both the sugar moiety and the position of glycosylation. In a study by Zhang et al., quercetin 3-O-glucoside was hydrolyzed most efficiently by β-glycosidase from propolis, while glycosides containing disaccharides such as rutin and naringin remained stable and resistant to enzymatic cleavage. These findings highlight that mono-glucosides are more susceptible to degradation, which may influence the overall bioavailability and antioxidant activity of flavonoid-rich products during processing or storage [91]. Temperature is another factor that can accelerate degradation. Heat during drying may cause thermal decomposition of flavonoids and vitamin C, while even frozen samples are susceptible to slow oxidative processes over extended periods. To maximize the retention of vitamin C, polyphenols, and antioxidant properties in BP, freezing (−20 °C) is the most effective method, while drying at 40 °C results in significant degradation of these bioactive compounds. High-pressure processing emerges as a promising alternative to preserve bioactive compounds while ensuring microbiological safety [92,93]. Freezing (−20 °C) is the most effective method to preserve vitamin C in BP, as drying at 40 °C causes significant losses, with up to 90% degradation after 12 months of storage [94]. The antioxidant potential of BP decreases over time, particularly at room temperature, with freeze-drying and high-pressure processing preserving flavonoids and polyphenols more effectively than hot air drying [92]. Fresh pollen contains higher levels of polyphenols, flavonoids, and amino acids than dried pollen, with drying at 40 °C reducing these bioactive compounds, while freeze-drying and microwave-assisted drying better preserve them [95]. Storage conditions greatly influence the stability of bioactive compounds, with freezing (−20 °C) significantly slowing polyphenol degradation, whereas room temperature storage and light exposure accelerate their loss [93,94]. Rutin and other flavonoids degrade more rapidly in hot air-dried pollen compared to freeze-dried or microwave-dried pollen, with high-pressure processing (HPP) emerging as a promising alternative for preserving these compounds [92,95]. Although our study did not investigate these mechanisms in detail, the observed trends are consistent with known degradation pathways reported in the literature. Future research should explore the stability of individual compounds and the specific mechanisms through which environmental factors influence their degradation. To maintain the stability of phenolic compounds and preserve the biological activity of BP during storage, it is essential to consider the influence of environmental factors [14,27,96]. It is recommended that BP be stored at low temperatures, such as in a refrigerator (approximately 4 °C) or in a freezer (−20 °C), in order to minimize the risk of oxidation of phenolic compounds. Exposure to higher temperatures is known to accelerate the degradation of these active components. The stability of flavonoids during storage can be influenced by both light and oxygen, with these factors potentially having a synergistic effect [97]. Scientific studies have shown that short-term exposure of apples to UV radiation stimulates the synthesis of phenolic compounds; however, prolonged exposure leads to their degradation and a decrease in their content. It has also been found that low doses of gamma ionizing radiation break glycosidic bonds of tannins and contribute to an increase in proanthocyanidin and flavan-3-ol levels. The enhanced synthesis of phenolic compounds is a response to photobiological stress induced by light radiation and environmental temperature fluctuations, during which enzymes such as superoxide dismutase, catalase, phenol peroxidase, and others are activated [98]. Flavonoid degradation can occur immediately or be delayed for varying periods depending on the molecular structure and environmental conditions. The photostability of various flavonoids depends on a 3-OH group and is a key factor in flavonoid photoreactivity. Tournaire et al. [99] highlighted that flavonoid reactivity to oxygen depends on the C-ring structure. They found that flavanones and flavanols are chemically stable in the presence of oxygen, while flavones and flavonols exhibit higher sensitivity. Furthermore, the rate of oxygen-induced degradation significantly decreases when the hydroxyl group at position 3 is absent. This can be explained by the fact that the 3-OH group activates the C2–C3 double bond in the presence of oxygen, leading to the opening of the C-ring in the flavonoid structure [100]. Additionally, pollen should be stored in dark containers or sealed bags to protect it from direct UV exposure. Moisture, which promotes enzyme activity and oxidative processes, should be avoided; therefore, BP should be kept in a dry environment to prevent biochemical reactions that may reduce its effectiveness.

In summary, new, important, and significant results were obtained, showing the value and superiority of fresh frozen pollen compared to dried pollen, since the preparation method of dried pollen dominates in Lithuania. Moreover, this study provides novel insights into the dynamic changes in individual compounds, specifically phenolic acids and flavonoids, during storage over a 15-month period with assessments conducted every three months—an aspect that, to the best of our knowledge, has not been previously detailed. The obtained results are important for both beekeepers and consumers in order to know when the amounts of antioxidants, including phenolic compounds and vitamins, start to decrease during storage. This is also very important in the pharmaceutical industry in order to use the highest quality of BP samples for the production of food supplements and pharmaceutical preparations with BP.

5. Conclusions

It was determined that during storage (over a 3–15-month period), the total amount of phenolic compounds in dried and frozen (at −20 °C and −80 °C) bee pollen gradually decreased by 7% to 58%, while the total amount of flavonoids decreased by 10% to 67% ($p < 0.05$). The highest total amounts of phenolic compounds (24.4 ± 0.4 mg GAE/g, $p < 0.05$) and flavonoids (15.2 ± 0.07 mg GAE/g, $p < 0.05$) were maintained in the fresh frozen BP samples stored at −80 °C. In the dried BP samples, the total phenolic content started to decrease after 3 months, while in the fresh frozen BP samples, it decreased after 6 months ($p < 0.05$). After UESC-MS analysis, 20 flavonoids (mainly rutin, quercetin, isorhamnetin-3-glucoside, luteolin-3,7-diglucoside, naringenin, and phloridzin) and 5 phenolic acids (mainly *p*-coumaric and chlorogenic acid) were estimated in the BP samples. It was determined that after the 3–15 month-storage period, the highest total amount of identified phenolic compounds (9105.7 ± 170.8 µg/g) was found in fresh frozen

BP samples stored at −80 °C (which showed the least variation during storage), while the lowest amount (6961.5 ± 151.8 μg/g) was observed in the dried pollen samples ($p < 0.05$). The qualitative and quantitative composition of individual phenolic compounds changed during the 3–15-month period. The levels of phloridzin, naringenin, and hyperoside changed the least during storage, while avicularin showed the most significant decrease in the frozen BP samples (decreased by 100%). When evaluating different storage conditions and durations, the highest amount of ascorbic acid remained in the frozen BP samples stored at −80 °C (a statistically significant decrease in content was observed only after 9 months, $p < 0.05$). In the frozen BP samples stored at −20 °C, the ascorbic acid content decreased after 6 months ($p < 0.05$), while in the dried BP samples, a decrease was observed after 3 months of storage ($p < 0.05$). The strongest antiradical and reducing activity in vitro (over a 6–15-month period) were established in the frozen (−80 °C) BP extracts compared to the dried pollen extracts. During storage, the antiradical and reducing activities in the frozen (−80 °C) BP pollen extracts were 1.8–3.4-fold and 2.6–3.1-fold higher compared to the dried BP pollen extracts, respectively. Melisopalynological analysis revealed a polyfloral pollen mixture, with *Salix* spp. and *Brassica napus* L. predominating in all the samples, comprising 34.3% and 36.8%, respectively. Additionally, the pollens of *Acer platanoides* L., *Malus domestica* Borkh., and *Taraxacum officinale* L. contributed significantly to the composition.

Supplementary Materials: The following supporting information can be downloaded at: https://www.mdpi.com/article/10.3390/antiox14040462/s1, Table S1. Parameters of the first-order degradation model for phenolic compounds in bee pollen samples. Figure S1. First-order degradation model of phenolic compounds in bee pollen samples. Table S2. Parameters of the first-order degradation model for flavonoids in bee pollen samples. Figure S2. First-order degradation model of flavonoids in bee pollen samples. Table S3. Parameters of the first-order degradation model for ascorbic acid in bee pollen samples. Figure S3. First-order degradation model of ascorbic acid in bee pollen samples.

Author Contributions: Conceptualization, S.T. and N.S.; methodology, M.L., Š.S., N.S., R.S. and V.Č.; software, M.L., V.Ž. and V.Č.; formal analysis, M.L., R.S., V.Č. and V.Ž.; investigation, R.S., M.L., V.Č., and Š.S.; resources, N.S. and S.T.; data curation, R.S., A.B., M.L., S.T., V.Č. and N.S.; writing—original draft preparation, S.T. and A.B.; writing—review and editing, S.T., A.B., Š.S., M.L., V.Č. and N.S.; visualization, A.B., R.S. and Š.S.; supervision, S.T., project administration S.T. and N.S. All authors have read and agreed to the published version of the manuscript.

Funding: This research was partially funded by the National Paying Agency under the Ministry of Agriculture for application "In vitro variation study of the phytochemical composition of pollen collected by bees in relation to the method of preparation and storage time", project No. 2024_BITE-KK-24-1-02475-PR001.

Institutional Review Board Statement: Not applicable.

Informed Consent Statement: Not applicable.

Data Availability Statement: The data are contained within the article.

Acknowledgments: The study was partially funded by the National Paying Agency under the Ministry of Agriculture for application "In vitro variation study of the phytochemical composition of pollen collected by bees in relation to the method of preparation and storage time", project No. 2024_BITE-KK-24-1-02475-PR001.

Conflicts of Interest: The authors declare no conflicts of interest.

References

1. Albuquerque, B.R.; Heleno, S.A.; Oliveira, M.B.; Barros, L.; Ferreira, I.C. Phenolic compounds: Current industrial applications, limitations and future challenges. *Food Funct.* **2021**, *12*, 14–29. [CrossRef] [PubMed]

2. Algethami, J.S.; El-Wahed, A.A.; Elashal, M.H.; Ahmed, H.R.; Elshafiey, E.H.; Omar, E.M.; Naggar, Y.A.; Algethami, A.F.; Shou, Q.; Alsharif, S.M.; et al. Bee pollen: Clinical trials and patent applications. *Nutrients* **2022**, *14*, 2858. [CrossRef]
3. Sawicki, T.; Starowicz, M.; Kłębukowska, L.; Hanus, P. The profile of polyphenolic compounds, contents of total phenolics and flavonoids, and antioxidant and antimicrobial properties of bee products. *Molecules* **2022**, *27*, 1301. [CrossRef] [PubMed]
4. Dias, L.G.; Tolentino, G.; Pascoal, A.; Estevinho, L.M. Effect of processing conditions on the bioactive compounds and biological properties of bee pollen. *J. Apic. Res.* **2016**, *55*, 357–365. [CrossRef]
5. Kolayli, S.; Birinci, C.; Kanbur, E.D.; Ucurum, O.; Kara, Y.; Takma, C. Comparison of biochemical and nutritional properties of bee pollen samples according to botanical differences. *Eur. Food Res. Technol.* **2024**, *250*, 799–810. [CrossRef]
6. Komosinska-Vassev, K.; Olczyk, P.; Kaźmierczak, J.; Mencner, L.; Olczyk, K. Bee pollen: Chemical composition and therapeutic application. *eCAM* **2015**, *2015*, 297425. [CrossRef]
7. Rozman, A.S.; Hashim, N.; Maringgal, B.; Abdan, K. A comprehensive review of stingless bee products: Phytochemical composition and beneficial properties of honey, propolis, and pollen. *Appl. Sci.* **2022**, *12*, 6370. [CrossRef]
8. Alfawaz, H.A.; El-Ansary, A.; Al-Ayadhi, L.; Bhat, R.S.; Hassan, W.M. Protective effects of bee pollen on multiple propionic acid-induced biochemical autistic features in a rat model. *Metabolites* **2022**, *12*, 571. [CrossRef]
9. Gerçek, Y.C.; Celik, S.; Bayram, S. Screening of plant pollen sources, polyphenolic compounds, fatty acids and antioxidant/antimicrobial activity from bee pollen. *Molecules* **2021**, *27*, 117. [CrossRef]
10. Ares, A.M.; Martín, M.T.; Toribio, L.; Bernal, J. Determination of free amino acids in bee pollen by liquid chromatography with fluorescence detection. *Food Anal. Methods* **2022**, *15*, 2172–2180. [CrossRef]
11. Prđun, S.; Svečnjak, L.; Valentić, M.; Marijanović, Z.; Jerković, I. Characterization of bee pollen: Physico-chemical properties, headspace composition and FTIR spectral profiles. *Foods* **2021**, *10*, 2103. [CrossRef] [PubMed]
12. Straumite, E.; Bartule, M.; Valdovska, A.; Kruma, Z.; Galoburda, R. Physical and microbiological characteristics and antioxidant activity of honey bee pollen. *Appl. Sci.* **2022**, *12*, 3039. [CrossRef]
13. Habryka, C.; Socha, R.; Juszczak, L. Effect of bee pollen addition on the polyphenol content, antioxidant activity, and quality parameters of honey. *Antioxidants* **2021**, *10*, 810. [CrossRef]
14. Ertosun, S.; Aylanc, V.; Falcão, S.I.; Vilas-Boas, M. Thermal Stability and Antioxidant Activity of Bioactive Compounds in Bread Enriched with Bee Pollen and Bee Bread. *Antioxidants* **2023**, *12*, 1691. [CrossRef] [PubMed]
15. Boulfous, N.; Belattar, H.; Ambra, R.; Pastore, G.; Ghorab, A. Botanical Origin, Phytochemical Profile, and Antioxidant Activity of Bee Pollen from the Mila Region, Algeria. *Antioxidants* **2025**, *14*, 291. [CrossRef]
16. Rojo, S.; Escuredo, O.; Rodríguez-Flores, M.S.; Seijo, M.C. Botanical Origin of Galician Bee Pollen (Northwest Spain) for the Characterization of Phenolic Content and Antioxidant Activity. *Foods* **2023**, *12*, 294. [CrossRef]
17. Estevinho, L.M.; Dias, T.; Anjos, O. Influence of the storage conditions (frozen vs. dried) in health-related lipid indexes and antioxidants of bee pollen. *Eur. J. Lipid Sci. Technol.* **2019**, *121*, 1800393. [CrossRef]
18. Martinello, M.; Mutinelli, F. Antioxidant activity in bee products: A review. *Antioxidants* **2021**, *10*, 71. [CrossRef]
19. Nainu, F.; Masyita, A.; Bahar, M.A.; Raihan, M.; Prova, S.R.; Mitra, S.; Emran, T.B.; Simal-Gandara, J. Pharmaceutical prospects of bee products: Special focus on anticancer, antibacterial, antiviral, and antiparasitic properties. *Antibiotics* **2021**, *10*, 822. [CrossRef]
20. El Ghouizi, A.; Bakour, M.; Laaroussi, H.; Ousaaid, D.; El Menyiy, N.; Hano, C.; Lyoussi, B. Bee pollen as functional food: Insights into its composition and therapeutic properties. *Antioxidants* **2023**, *12*, 557. [CrossRef]
21. Keskin, M.; Özkök, A. Effects of drying techniques on chemical composition and volatile constituents of bee pollen. *Czech J. Food Sci.* **2020**, *38*, 203–208. [CrossRef]
22. Isik, A.; Ozdemir, M.; doymaz, I. Effect of hot air drying on quality characteristics and physicochemical properties of bee pollen. *Food Sci. Technol.* **2019**, *39*, 224–231. [CrossRef]
23. Kayacan, S.; Sagdic, O.; Doymaz, I. Effects of hot-air and vacuum drying on drying kinetics, bioactive compounds and color of bee pollen. *J. Food Meas. Charact.* **2018**, *12*, 1274–1283. [CrossRef]
24. Yin, X.; Chen, K.; Cheng, H.; Chen, X.; Feng, S.; Song, Y.; Liang, L. Chemical Stability of Ascorbic Acid Integrated into Commercial Products: A Review on Bioactivity and Delivery Technology. *Antioxidants* **2022**, *11*, 153. [CrossRef]
25. Pascoal, A.; Rodrigues, S.; Teixeira, A.; Feás, X.; Estevinho, L.M. Biological Activities of Commercial Bee Pollens: Antimicrobial, Antimutagenic, Antioxidant and Anti-inflammatory. *Food Chem. Toxicol.* **2014**, *63*, 233–239. [CrossRef]
26. Kostić, A.Ž.; Milinčić, D.D.; Barać, M.B.; Shariati, M.A.; Tešić, Ž.L.L.; Pešić, M.B. The Application of pollen as a functional food and feed ingredient—The present and perspectives. *BioMolecules* **2020**, *10*, 84. [CrossRef] [PubMed]
27. Anjos, O.; Paula, V.; Delgado, T.; Estevinho, L. Influence of the storage conditions on the quality of bee pollen. *Žemdirbystė (Akademija)* **2019**, *106*, 87–94. [CrossRef]
28. Shubharani, R.; Roopa, P.; Sivaram, V. Pollen morphology of selected bee forage plants. *Glob. J. BioSci Biotechnol.* **2013**, *2*, 82–90.
29. Louveaux, J.; Maurizio, A.; Vorwohl, G. Methods of melissopalynology. *Bee World* **1978**, *59*, 139–153. [CrossRef]
30. Čeksterytė, V. *Electronic Catalog of Plant Pollens Found in Lithuanian Honey*; Lithuania Ministry of Agriculture: Vilnius, Lithuania, 2012.

31. Kahraman, H.A.; Tutun, H.; Kaya, M.M.; Usluer, M.S.; Tutun, S.; Yaman, C.; Keyvan, E. Ethanolic Extract of Turkish Bee Pollen and Propolis: Phenolic Composition, Anti-Radical, Anti-Proliferative, and Antibacterial Activity. *Biotechnol. Biotechnol. Equip.* **2022**, *36*, 45–56. [CrossRef]
32. Ares, A.M.; Valverde, S.; Bernal, J.L.; Nozal, M.J.; Bernal, J. Extraction and Determination of Bioactive Compounds from Bee Pollen. *J. Pharm. Biomed. Anal.* **2018**, *147*, 110–124. [CrossRef] [PubMed]
33. Bobinaitė, R.; Viškelis, P.; Venskutonis, P.R. Variation of total phenolics, anthocyanins, ellagic acid and radical scavenging capacity in various raspberry (*Rubus* spp.) cultivars. *Food Chem.* **2012**, *132*, 1495–1501. [CrossRef] [PubMed]
34. Urbonavičiūtė, A.; Jakštas, V.; Kornyšova, O.; Janulis, V.; Maruška, A. Capillary electrophoretic analysis of flavonoids in single-styled hawthorn (*Crataegus monogyna* Jacq.) ethanolic extracts. *J. Chromatogr. A.* **2006**, *1112*, 339–344. [CrossRef]
35. González-Burgos, E.; Liaudanskas, M.; Viškelis, J.; Žvikas, V.; Janulis, V.; Gómez-Serranillos, M.P. Antioxidant activity, neuroprotective properties and bioactive constituents analysis of varying polarity extracts from *Eucalyptus globulus* leaves. *J. Food Drug Anal.* **2018**, *26*, 1293–1302. [CrossRef]
36. Khongkarat, P.; Phuwapraisirisan, P.; Chanchao, C. Phytochemical content, especially spermidine derivatives, presenting antioxidant and antilipoxygenase activities in Thai bee pollens. *PeerJ* **2022**, *10*, e13506. [CrossRef] [PubMed]
37. Munteanu, I.G.; Apetrei, C. Analytical methods used in determining antioxidant activity: A review. *Int. J. Mol. Sci.* **2021**, *22*, 3380. [CrossRef]
38. Alimoglu, G.; Guzelmeric, E.; Yuksel, P.I.; Celik, C.; Deniz, I.; Yesilada, E. Monofloral and polyfloral bee pollens: Comparative evaluation of their phenolics and bioactivity profiles. *LWT* **2021**, *142*, 110973. [CrossRef]
39. Šimoliūnienė, R.; Tomkevičiūtė, J.; Jokšienė, Ž.; Šimatonienė, V.; Kriščiukaitis, A.; Šaferis, V. *Basics of Biostatistics*; Publishing House of the Lithuanian University of Health Sciences: Kaunas, Lithuania, 2015; pp. 133–142.
40. Aronhime, S.; Calcagno, C.; Jajamovich, G.H.; Dyvorne, H.A.; Robson, P.; Dieterich, D.; Fiel, I.M.; Martel-Laferriere, V.; Chatterji, M.; Rusinek, H.; et al. DCE-MRI of the liver: Effect of linear and nonlinear conversions on hepatic perfusion quantification and reproducibility. *J. Magn. Reson.* **2014**, *40*, 90–98. [CrossRef]
41. Sánchez, J.S.; Reyes, E.S.; Durán, S.S.; Gonzalez-paramas, A.M.; Estelvhino, L.; Migue, M.G.; Watson, X.X.; Fernández-López, J.; Almeida-Muradian, L.; Pak, S.; et al. *Bee Products—Chemical and Biological Properties*, 1st ed.; Springer International Publishing: Cham, Switzerland, 2017; pp. 221–259.
42. Qaderi, M.M.; Martel, A.B.; Strugnell, C.A. Environmental factors regulate plant secondary metabolites. *Plants* **2023**, *12*, 447. [CrossRef]
43. Li, Q.Q.; Wang, K.; Marcucci, M.C.; Sawaya, A.C.H.F.; Hu, L.; Xue, X.F.; Wu, L.M.; Hu, F.L. Nutrient-rich bee pollen: A treasure trove of active natural metabolites. *J. Funct. Foods* **2018**, *49*, 472–484. [CrossRef]
44. Rzepecka-Stojko, A.; Stojko, J.; Kurek-Górecka, A.; Górecki, M.; Kabała-Dzik, A.; Kubina, R.; Moździerz, A.; Buszman, E. Polyphenols from bee pollen: Structure, absorption, metabolism and biological activity. *Molecules* **2015**, *20*, 21732–21749. [CrossRef]
45. Ulusoy, E.; Kolayli, S. Phenolic composition and antioxidant properties of Anzer bee pollen. *J. Food Biochem.* **2014**, *38*, 73–82. [CrossRef]
46. Freire, K.R.L.; Lins, A.C.S.; Dórea, M.C.; Santos, F.A.R.; Camara, C.A.; Silva, T.M.S. Palynological origin, phenolic content, and antioxidant properties of honeybee-collected pollen from bahia, brazil. *Molecules* **2012**, *17*, 1652–1664. [CrossRef]
47. Kalaycıoğlu, Z.; Kaygusuz, H.; Döker, S.; Kolaylı, S.; Erim, F.B. Characterization of Turkish honeybee pollens by principal component analysis based on their individual organic acids, sugars, minerals, and antioxidant activities. *Food Sci. Technol.* **2017**, *84*, 402–408. [CrossRef]
48. Gabriele, M.; Parri, E.; Felicioli, A.; Sagona, S.; Pozzo, L.; Biondi, C.; Domenici, V.; Pucci, L. Phytochemical composition and antioxidant activity of Tuscan bee pollen of different botanic origins. *Ital. J. Food Sci.* **2015**, *27*, 248.
49. Mărghitaş, L.A.; Stanciu, O.G.; Dezmirean, D.S.; Bobiş, O.; Popescu, O.; Bogdanov, S.; Campos, M.G. In vitro antioxidant capacity of honeybee-collected pollen of selected floral origin harvested from Romania. *Food Chem.* **2009**, *115*, 878–883. [CrossRef]
50. Morais, M.; Moreira, L.; Feás, X.; Estevinho, L.M. Honeybee-collected pollen from five Portuguese Natural Parks: Palynological origin, phenolic content, antioxidant properties and antimicrobial activity. *Food Chem. Toxicol.* **2011**, *49*, 1096–1101. [CrossRef]
51. Mayda, N.; Özkök, A.; Ecem Bayram, N.; Gerçek, Y.C.; Sorkun, K. Bee bread and bee pollen of different plant sources: Determination of phenolic content, antioxidant activity, fatty acid and element profiles. *J. Food Meas. Charact.* **2020**, *14*, 1795–1809. [CrossRef]
52. Araújo, J.S.; Chambó, E.D.; Costa, M.A.P.d.C.; Cavalcante da Silva, S.M.P.; Lopes de Carvalho, C.A.; M Estevinho, L. Chemical composition and biological activities of mono- and heterofloral bee pollen of different geographical origins. *Int. J. Mol. Sci.* **2017**, *18*, 921. [CrossRef]
53. Feás, X.; Vázquez-Tato, M.P.; Estevinho, L.; Seijas, J.A.; Iglesias, A. Organic bee pollen: Botanical origin, nutritional value, bioactive compounds, antioxidant activity and microbiological quality. *Molecules* **2012**, *17*, 8359–8377. [CrossRef]

54. Muliuolytė, L.; Kaškonienė, V.; Maruška, A. The Impact of Storage Conditions and Extrahent Concentration on the Content of Bioactive Compounds in Bee Bread and Pollen. In Proceedings of the 66th International Open Readings Conference for students of Physics and Natural Sciences, Vilnius, Lithuania, 18–21 April 2023; Vilnius University Press: Vilnius, Lithuania, 2023; p. 280.
55. Abhay, S.M.; Hii, C.L.; Law, C.L.; Suzannah, S.; Djaeni, M. Effect of hot-air drying temperature on the polyphenol content and the sensory properties of cocoa beans. *Int. Food Res. J.* **2016**, *23*, 1479.
56. Kostić, A.Ž.; Gercek, Y.C.; Bayram, N.E. Phenolic Acids in Pollen. In *Pollen Chem & Biotechnol*; Springer International Publishing: Cham, Switzerland, 2024; pp. 103–125.
57. Yousefi, S.; Emam-Djomeh, Z.; Mousavi, S.M.; Oromiehie, A.R. Effect of Carrier Type and Spray Drying on the Physicochemical Properties of Pomegranate Juice Powder. *Iran. J. Chem. Chem. Eng.* **2011**, *30*, 41–49.
58. Sawicki, T.; Bączek, N.; Wiczkowski, W. Betalain Profile, Content and Antioxidant Capacity of Red Beetroot Dependent on the Genotype and Root Part. J. Funct. *Foods* **2016**, *27*, 249–261.
59. de Almeida-Muradian, L.B.; Pamplona, L.C.; Coimbra, S.; Barth, O.M. Chemical Composition and Botanical Evaluation of Dried Bee Pollen Pellets. *J. Food Compos. Anal.* **2005**, *18*, 105–111. [CrossRef]
60. Muñoz, E.; Velásquez, P.; Rodriguez, K.; Montenegro, G.; Giordano, A. Influence of *Brassica campestris* and *Galega officinalis* on antioxidant activity of bee pollen. *Rev. Bras. Farmacogn.* **2020**, *30*, 444–449. [CrossRef]
61. Čeksterytė, V.; Kurtinaitienė, B.; Venskutonis, P.R.; Pukalskas, A.; Kazernavičiūtė, R.; Balžekas, J. Evaluation of antioxidant activity and flavonoid composition in differently preserved bee products. *Czech J. Food Sci.* **2016**, *34*, 133–142.
62. Özkök, D.; Silici, S. Antioxidant activities of honeybee products and their mixtures. *Food Sci. Biotechnol.* **2017**, *26*, 201–206. [CrossRef]
63. Kaškonienė, V.; Ruočkuvienė, G.; Kaškonas, P.; Akuneca, I.; Maruška, A. Chemometric analysis of bee pollen based on volatile and phenolic compound compositions and antioxidant properties. *Food Anal. Methods* **2014**, *8*, 1150–1163. [CrossRef]
64. Gómez-Caravaca, A.M.; Gómez-Romero, M.; Arráez-Román, D.; Segura-Carretero, A.; Fernández-Gutiérrez, A. Advances in the analysis of phenolic compounds in products derived from bees. *J. Pharm. Biomed. Anal.* **2006**, *41*, 1220–1234. [CrossRef]
65. Tyszka-Czochara, M.; Knapik-Czajka, M.; Goździalska, A.; Francik, R.; Jaśkiewicz, J. Polyphenols in a diet. Some aspects of metabolism and bioavailability of phenolic compounds. *Farm. Polska* **2003**, *59*, 589–597.
66. Grajek, W. *Antioxidants in Food*, 1st ed.; WNT: Warsaw, Poland, 2007; pp. 258–259.
67. Bartosz, G. *The Second Face of Oxygen*, 2nd ed.; PWN: Warsaw, Poland, 2003; pp. 30–57.
68. Rice-Evans, C.A.; Miller, N.J.; Paganga, G. Structure–antioxidant activity relationships of flavonoids and phenolic acids. *Free Radic. Biol. Med.* **1996**, *20*, 933–956. [CrossRef] [PubMed]
69. Castellucio, C.; Paganga, G.; Melikan, N.; Bowell, G.P.; Pridham, J.; Sampson, J.; Rice-Evans, C. Antioxidant potential of intermediates in phenylopropanoid metabolism in higher plants. *FEBS Lett.* **1995**, *368*, 188–192. [CrossRef] [PubMed]
70. Iriti, M. Introduction to polyphenols, plant chemicals for human health. *Mini Rev. Med. Chem.* **2011**, *11*, 1183–1185.
71. Robak, J. Flavonoids as drugs inhibiting aggregation of blood platelets. *Farm. Polska* **1993**, *49*, 1–4.
72. Cholbi, M.R.; Paya, M.; Alcaraz, M.J. Inhibitory effects of phenolic compounds on CCl_4-induced microsomal lipid peroxidation. *Experientia* **1991**, *47*, 195–199. [CrossRef]
73. Cook, N.C.; Samman, S. Flavonoids—Chemistry, metabolism, cardioprotective effects and dietary sources. *J. Nutr. Biochem.* **1996**, *7*, 66–76. [CrossRef]
74. Raudonis, R.; Bumblauskiene, L.; Jakstas, V.; Pukalskas, A.; Janulis, V. Optimization and validation of postcolumn assay for screening of radical scavengers in herbal raw materials and herbal preparations. *J. Chromatogr. A* **2010**, *1217*, 7690–7698. [CrossRef]
75. Bonvehi, S.J.; Soliva Torrentó, M.; Centelles Lorente, E. Evaluation of polyphenolic compounds in honeybee-collected pollen produced in Spain. *J. Agric. Food Chem.* **2001**, *49*, 1848–1853. [CrossRef]
76. Čeksterytė, V.; Kazlauskas, S. Composition of flavonoids in Lithuanian honey and beebread. *Biologija* **2006**, *2*, 28–33.
77. Burda, S.; Oleszek, W. Antioxidant and antiradical activities of flavonoids. *J. Agric. Food Chem.* **2001**, *49*, 2774–2779. [CrossRef]
78. Dabeek, W.M.; Marra, M.V. Dietary quercetin and kaempferol: Bioavailability and potential cardiovascular-related bioactivity in humans. *Nutrients* **2019**, *11*, 2288. [CrossRef] [PubMed]
79. Alizadeh, S.R.; Ebrahimzadeh, M.A. Quercetin derivatives: Drug design, development, and biological activities, a review. *Eur. J. Med. Chem.* **2022**, *229*, 114068. [CrossRef]
80. Xie, L.; Deng, Z.; Zhang, J.; Dong, H.; Liu, X. Comparison of flavonoid O-glycoside, C-glycoside and their aglycones on antioxidant capacity and metabolism during in vitro digestion and in vivo. *Foods* **2022**, *11*, 882. [CrossRef]
81. Wright, G.A.; Skinner, B.D.; Smith, B.H. Ability of honeybee, *Apis mellifera*, to detect and discriminate odors of varieties of canola (*Brassica rapa* and *Brassica napus*) and snapdragon flowers (*Antirrhinum majus*). *J. Chem. Ecol.* **2002**, *28*, 721–740. [CrossRef] [PubMed]
82. Stanley, D.A.; Gunning, D.; Stout, J.C. Pollinators and pollination of oilseed rape crops (*Brassica napus* L.) in Ireland: Ecological and economic incentives for pollinator conservation. *J. Insect Conserv.* **2013**, *17*, 1181–1189. [CrossRef]
83. Abrol, D.P. Honeybees and rapeseed: A pollinator-plant interaction. *Adv. Bot. Res.* **2007**, *45*, 337–367.

84. Donkersley, P. Trees for bees. *Agric. Ecosyst. Environ.* **2019**, *270–271*, 79–83. [CrossRef]
85. Čeksterytė, V.; Borutinskaitė, V.; Matuzevičius, D.; Treigytė, G.; Navakauskas, D.; Kurtinaitienė, B.; Navakauskienė, R. Evaluation of Proteome profiles of *Salix* spp. pollen and relationship between glucose oxidase activity and pollen content in willow honey. *Balt. For.* **2019**, *25*, 83–96. [CrossRef]
86. Borutinskaitė, V.; Treigytė, G.; Matuzevičius, D.; Čeksterytė, V.; Kurtinaitienė, B.; Serackis, A.; Navakauskas, D.; Navakauskienė, R. Proteomic studies of honeybee-and manually-collected pollen. *Zemdirbystė-Agriculture* **2019**, *106*, 183–190. [CrossRef]
87. Čeksterytė, V.; Kurtinaitienė, B.; Jaškūnė, K.; Kretavičius, J. The influence of storage conditions on invertase, glucose oxidase activity and free acidity of bee bread and bee-collected pollen mixed with honey and vegetable oils. *J. Apic. Res.* **2020**, *59*, 862–875. [CrossRef]
88. Čeksterytė, V.; Kurtinaitienė, B.; Balžekas, J. Pollen diversity in honey collected from Lithuania's protected landscape areas. *EAP* **2013**, *62*, 277–282. [CrossRef]
89. Baltrušaitytė, V.; Venskutonis, P.R.; Čeksterytė, V. Radical scavenging activity of different floral origin honey and beebread phenolic extracts. *Food Chem.* **2007**, *101*, 502–514. [CrossRef]
90. Yoruk, R.; Marshall, M.R. Physicochemical properties and function of plant polyphenol oxidase: A review. *J. Food Biochem.* **2003**, *27*, 361–422. [CrossRef]
91. Zhang, Y.; Yang, R.; Zhao, W.; Hua, Y.; Kong, X.; Chen, Y. Enzymatic hydrolysis of flavonoid glycosides by propolis β-glycosidase. *Molecules* **2012**, *17*, 2259–2273.
92. Anjos, O.; Seixas, N.; Antunes, C.A.L.; Campos, M.G.; Paula, V.; Estevinho, L.M. Quality of bee pollen submitted to drying, pasteurization, and high-pressure processing—A comparative approach. *Int. Food Res.* **2023**, *170*, 112964. [CrossRef]
93. Rzepecka-Stojko, A.; Stec, M.; Kurzeja, E.; Gawrońska, E.; Pawłowska-Góral, K. The effect of storage of bee pollen extracts on polyphenol content. *Pol. J. Environ. Stud.* **2012**, *21*, 1007–1011.
94. Szczesna, T.; Rybak-Chimielewska, H.; Bornus, L. Effect of storage on variation of contents of vitamin C and A in pollen collected by bees. *Apiacta* **1991**, *2*, 1991.
95. Castagna, A.; Benelli, G.; Conte, G.; Sgherri, C.; Signorini, F.; Nicolella, C.; Ranieri, A.; Canale, A. Drying techniques and storage: Do they affect the nutritional value of bee-collected pollen? *Molecules* **2020**, *25*, 4925. [CrossRef]
96. Rodríguez-Pólit, C.; Gonzalez-Pastor, R.; Heredia-Moya, J.; Carrera-Pacheco, S.E.; Castillo-Solis, F.; Vallejo-Imbaquingo, R.; Barba-Ostria, C.; Guamán, L.P. Chemical Properties and Biological Activity of Bee Pollen. *Molecules* **2023**, *28*, 7768. [CrossRef]
97. Chaaban, H.; Ioannou, I.; Paris, C.; Charbonnel, C.; Ghoul, M. Photostability of Flavanones, Flavonols, and Flavones and Evolution of Their Antioxidant Activity. *J. Food Process Preserv.* **2016**, *40*, 13203. [CrossRef]
98. De la Rosa, L.A.; Escamilla, J.O.M.; Garcia, J.R.; Parrilla, E.A. Phenolic compounds. In *Postharvest Physiology and Biochemistry of Fruits and Vegetables*, 1st ed.; Yahia, E.M., Carrillo-López, A., Eds.; Woodhead Publishing: Cambridge, UK, 2019; pp. 253–271.
99. Tournaire, C.; Croux, S.; Maurette, M.-T.; Beck, I.; Hocquaux, M.; Braun, M.; Oliveros, E. Antioxidant Activity of Flavonoids: Efficiency of Singlet Oxygen Quenching. *J. Photochem. Photobiol. B Biol.* **1993**, *19*, 205–215. [CrossRef] [PubMed]
100. Ramešová, S.; Sokolová, R.; Degano, I.; Bulíčková, J.; Žabka, J.; Gál, M. On the Stability of the Bioactive Flavonoids Quercetin and Luteolin Under Oxygen-Free Conditions. *Anal. Bioanal. Chem.* **2012**, *402*, 975–982. [CrossRef] [PubMed]

antioxidants

MDPI

Article

Botanical Origin, Phytochemical Profile, and Antioxidant Activity of Bee Pollen from the Mila Region, Algeria

Nassiba Boulfous [1,*], Hakima Belattar [1], Roberto Ambra [2], Gianni Pastore [2] and Asma Ghorab [3]

[1] Laboratory of Natural Sciences and Materials, Institute of Natural Sciences and Life, University Center Abdelhafid Boussouf, Mila 43000, Algeria; h.belattar@centre-univ-mila.dz

[2] CREA (Council for Agricultural Research and Economics), Research Centre for Food and Nutrition, 00178 Rome, Italy; roberto.ambra@crea.gov.it (R.A.); giovanni.pastore@crea.gov.it (G.P.)

[3] Department of Vegetal Biology and Soil Sciences, Facultade de Ciencias, Universidade de Vigo, 32004 Ourense, Spain; asma.ghorab@uvigo.gal

* Correspondence: n.boulfous@centre-univ-mila.dz

Abstract: Bee pollen is a complex mixture of floral pollen, and nectar fused substances from bee saliva. It is well known for its high content of proteins, carbohydrates, lipids, vitamins, and phenolic compounds, among various other physiologically active components. Its composition varies significantly depending on its botanical sources and environmental conditions. This study investigates the relationship between the botanical origins, chemical compositions, and antioxidant activities of 15 bee pollen samples collected from different areas in the Mila region of northeastern Algeria. The botanical origins were identified using a palynological method, categorizing 11 samples as monofloral and the rest as polyfloral. The total phenolic and flavonoid contents were measured, and their antioxidant capacities were evaluated through DPPH radical scavenging assay, reducing power assay (FRAP), and total antioxidant capacity (TAC). HPLC analysis was conducted to measure 17 phenolic compounds. The data indicated that the total phenolic content (TPC) and flavonoid content (TFC) ranged from 7.72 $\pm$ 0.29 to 23.49 $\pm$ 1.48 mg GAE/g and from 1.48 $\pm$ 0.00 to 5.57 $\pm$ 0.27 mg QE/g of pollen, respectively. The variations in the concentration of bioactive compounds among samples led to significant differences in their antioxidant activities: DPPH (IC_{50}: 1.12 $\pm$ 0.15 to 0.21 $\pm$ 0.00 mg/mL), FRAP (EC_{50}: 0.06 $\pm$ 0.00 to 0.29 $\pm$ 0.00 mg/mL), and TAC (262.17 $\pm$ 3.41 to 677.14 $\pm$ 12.81 EAA mg/100 g of bee pollen), with the most active samples being monofloral types from *Cistus* type and *Brassica* type. A strong correlation was observed between TPC, TFC, and antioxidant activity. Among the 17 tested compounds, only coumaric acid, rutin, myricetin, naringenin, resveratrol, and kaempferol were detected. In conclusion, both monofloral and polyfloral bee pollen samples represent a rich source of polyphenols with significant antioxidant potential.

Keywords: bee pollen; botanical origin; phenolic contents; antioxidant activity; HPLC; Algeria

1. Introduction

Bee pollen, a widely recognized honeybee product, is valued for its biologically active compounds and use as a natural remedy. It results from bees collecting pollen from flowers and mixing it with nectar and their secretions [1]. Typically, bee pollen contains carbohydrates (40–85%), proteins (14–30%), and lipids (1–10%), along with vitamins, minerals, carotenoids, phenolics, and trace elements. Over 250 bioactive compounds have been identified in bee pollen [2].

In apitherapy, it is used for its medicinal properties, including antimicrobial, anti-inflammatory, antiviral, analgesic, immunostimulant, hepatoprotective, and antioxidant effects [2]. Studies emphasize its strong antioxidant properties, which contribute to cellular protection and overall health by reducing oxidative stress [3–7].

The quality and antioxidant power of bee pollen is influenced by factors such as the geographical location where it is collected (and the specific plants bees forage on), the time of year, weather conditions, and how the pollen is processed. Determining its botanical source is essential, and this is primarily performed through palynology, which involves microscopic analysis of pollen grains to trace their plant origins [8]. Flavonoids like isoquercetin, myricetin, tricetin, quercetin, luteolin, selagin, kaempferol, and isorhamnetin have been identified in bee pollen samples from Brazil [9].

Algeria, located in North Africa, has a Mediterranean climate and boasts a rich plant diversity with 4000 species, including 2500 in its northern region, 10% of which are endemic. Examples include *Erysimum cheiri* (Brassicaceae) and *Hypochaeris saldensis* (Asteraceae). Numerous plants, such as eucalyptus, orange trees, sunflowers, clover, French honeysuckle, rosemary, and thyme, as well as natural forests like pine, provide nectar and pollen for bees, enabling year-round honey production. The region's diverse flora and government support have fostered the growth of beekeeping [10,11]. Honey and propolis have been extensively studied by researchers [12–20]. On the other hand, only a few studies have characterized the antioxidant properties of Algerian bee pollens.

In this context, the current work is devoted to determining the botanical origins, analyzing the antioxidant capacity and polyphenolic composition using HPLC of bee pollen collected from the Mila region of northeastern Algeria. By examining these aspects, the study seeks to understand the relationships between the botanical sources of the pollen, its phenolic content, and its antioxidant properties.

2. Materials and Methods

2.1. Collection of Samples

In this study, 15 bee pollen samples from the Mila region (northeastern Algeria) were analyzed. Bee pollen samples were collected by local beekeepers using pollen traps from *Apis mellifera* between 2022 and 2023 years from different locations, and stored in the freezer at 4 °C until further analysis (Table 1, Figure 1).

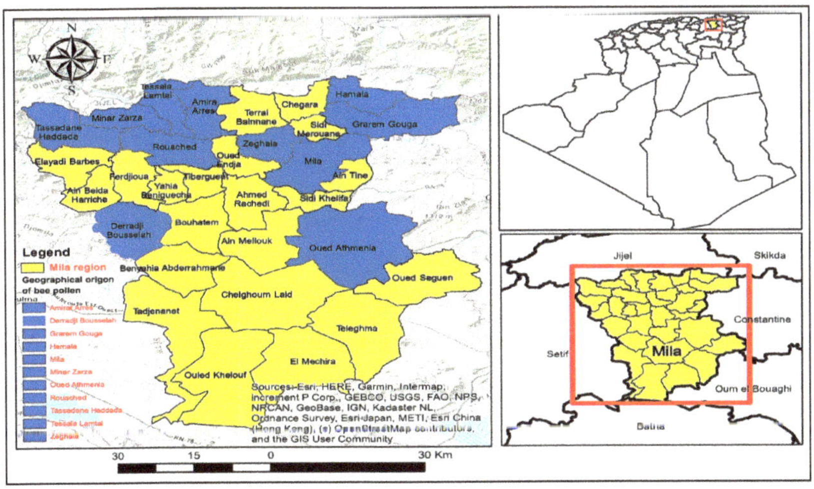

Figure 1. Geographic distribution origin by municipality (in blue) of the bee pollen samples collected the Mila region (northeastern Algeria). Created with ArcGIS 10.8.

Table 1. Geographical location and harvesting period of bee pollen samples.

Sample	Date of Harvest	Geographical Origin	Latitude (°N)	Longitude (°E)	Altitude (m)
P1	Spring 2023	Tessala Lamtai	36.4500	6.2667	600
P2	Spring 2023	Tassadane Haddada	36.3667	6.2500	700
P3	Spring 2023	Grarem Gouga	36.4000	6.5667	500
P4	Spring 2023	Derradji bousselah	36.3833	6.3333	550
P5	Spring 2023	Rouached	36.4667	6.1167	650
P6	Spring 2023	Zeghaia	36.4500	6.4000	450
P7	Spring 2023	Hamala	36.3667	6.4167	600
P8	Spring 2023	Rouached	36.4667	6.1167	650
P9	Spring 2023	Minar Zarza	36.3167	6.2000	750
P10	Summer 2023	Amira Arres	36.3000	6.2833	800
P11	Spring 2023	Tessala Lamtai	36.4500	6.2667	600
P12	Spring 2023	Oued Althmenia	36.4667	6.3667	450
P13	Spring 2023	Tassala Lemtai	36.4500	6.2667	600
P14	Spring 2023	Mila	36.4500	6.2667	710
P15	Spring 2023	Tassadane Haddada	36.3667	6.2500	700

2.2. Determination of Botanical Origin

Botanical origin was determined using a melissopalynological procedure [21]. First, the original samples were homogenized, then 1 g was weighed and placed in separate vials. Subsequently, a colorimetric separation using a white light and a black background was carried out to obtain different subsamples, which were weighted and dissolved in distilled water (15 mL). The solutions were shaken for 10 min and at 4500 rpm for 5 min centrifuged. An aliquot of 100 µL was taken from the sediment to prepare the slides. An optical microscope was used to identify the botanical origin of the different subsamples. The pollen spectra of the samples were determined considering the weight of each subsample and its botanical origin. The results were expressed in percentages.

Bee pollen samples were categorized microscopically into four sections based on the abundance of pollen grains: dominant pollen (≥45%), secondary pollen (16–45%), significant minor pollen (>3–15%), and minor pollen (<3%). The dominant pollen indicates the botanical origin of the bee pollen. However, identifying pollen grains from the same genus or family can be challenging. In such cases, only the genus or family name is specified, such as *Anthemis* sp., *Taraxacum* sp., or Asteraceae. According to Louveaux et al. [21], a sample is classified as monofloral if the percentage of a specific pollen type exceeds 45% of the total.

2.3. Preparation of Bee Pollen Extracts

Ethanolic extracts (EEPs) of the bee pollen samples were prepared according to the method of Almaraz-Abarca et al. [22], with minor modifications. The EEP was prepared by mixing 10 g of crushed bee pollen, with a precision of 0.01 g in 100 mL final volume 80% (v/v) aqueous ethanol for 24 h at room temperature. The sample was filtered under reduced atmosphere pressure using a water pump. The filtrate obtained was centrifuged at 10,000 rpm for 10 min, and then it was evaporated under reduced pressure in a rotary vacuum evaporator. The evaporated extract was dried in a laboratory incubator at 38 °C until a solid mass was obtained, and stored in the freezer at 4 °C until further analysis.

2.4. Phytochemical Analysis of Samples

2.4.1. Assessment of Total Phenol Content

The determination of the total phenolic content (TPC) was carried out based on the method developed by Singleton and Rossi [23], adapted to bee pollen. Briefly, 0.2 mL of the extract (2 mg/mL) was mixed with Folin–Ciocalteu's reagent (1 mL). After 3 min, 0.8 mL of 7.5% sodium carbonate was added to the mixture. The solution was kept in the dark at room temperature for 2 h, and the absorbance was measured at 765 nm. The total phenolic content was expressed as mg gallic acid equivalents (GAE) per g bee pollen. All measurements were performed in triplicate.

2.4.2. Assessment of Total Flavonoid Content

The total flavonoid content (TFC) was determined using a modified version of the method described by Arvouet-Grand et al. [24]. Specifically, 0.75 mL of bee pollen extract (2 mg/mL) was mixed with 0.75 mL of a 2% aluminum chloride solution. This mixture was then incubated for 15 min at room temperature. Subsequently, the absorbance of the solution was measured at 430 nm. A calibration curve was generated using quercetin as a standard, and the flavonoid content of the bee pollen extract was determined by applying the linear regression equation of the calibration curve. The results were expressed as milligrams of quercetin equivalents (QE) per gram of bee pollen. All measurements were performed in triplicate.

2.4.3. DPPH Radical Scavenging Assay

The ability of bee pollen extracts to scavenge free radicals was evaluated using the DPPH method, which measures the antioxidant's capacity to neutralize the stable 2,2-diphenyl-1-picrylhydrazyl radical using a modified method described by Tomás et al. [25]. Specifically, 0.75 mL of phenolic extracts, with concentrations ranging from 3 to 0.43 mg/mL, were mixed with 0.75 mL of DPPH solution (0.04 mg/mL). The absorbance of the resulting mixture was measured at 515 nm. The percentage of radical inhibition was calculated using the following equation:

$$\%\ \text{Inhibition} = [(A_{DPPH} - A_{Sample})/A_{DPPH}] \times 100$$

The EC_{50} value, representing the concentration of the extract required to inhibit 50% of DPPH radicals, was determined by plotting the percentage of radical inhibition against the extract concentration.

2.4.4. Reducing Power (FRAP)

The FRAP method, which relies on the reduction in the Fe^{3+}–TPTZ (ferric tripyridyltriazine) complex to its ferrous form under acidic conditions, was conducted following a previously described method [26]. A 0.25 mL aliquot of bee pollen extract (with concentrations ranging from 3 to 0.43 mg/mL) was combined with 1.25 mL of sodium phosphate buffer (pH 6.6). To this, 1.25 mL of 1% potassium ferricyanide was added, and the mixture was incubated at 50 °C for 20 min. Next, 1.25 mL of 10% trichloroacetic acid was added. The solution was then centrifuged at 3000× *g* for 10 min, and 1.25 mL of the supernatant was transferred to a new tube. Afterward, 1.25 mL of water and 0.25 mL of 0.1% ferric chloride were added, and the absorbance was measured at 700 nm. Extract concentration providing 0.5 of absorbance (EC_{50}) was calculated from the graph of absorbance against extract concentration in the solution. Ascorbic acid was used as positive control. Tests were conducted in triplicate, and the results were given as mean ± SD.

2.4.5. Total Antioxidant Capacity

The total antioxidant capacity (TAC) was measured using the ammonium molybdate colorimetric method described by Prieto et al. [27]. This method is based on the reduction of molybdenum (VI) to molybdenum (V) by antioxidants present in the extracts, leading to the formation of green phosphate/Mo (V) complexes, which can be quantitatively measured. In brief, 0.1 mL of the ethanolic bee pollen extract was combined with 1 mL of reagent solution (0.6 M sulfuric acid, 28 mM sodium phosphate, and 4 mM ammonium molybdate). The mixture was sealed and incubated in a thermal block at 95 °C for 90 min. The absorbance of the reaction mixture was then measured at 695 nm, using a blank for reference. Ascorbic acid was used as the standard for calibration, and the results were expressed as milligrams of ascorbic acid equivalent per 100 g of bee pollen.

2.5. Determination of Phenolic Compounds by HPLC

The analysis of phenolic acid compounds in ethanolic pollen extracts was performed using a Thermo Scientific Vanquish HPLC system, equipped with a Diode Array Detector (DAD) set to 280, 300, 320, and 350 nm. Spectra were continuously recorded in the range of 210–600 nm, with a data collection rate of 10.0 Hz. The system also included a Vanquish Autosampler maintained at 6 °C. A 250 × 4.6 mm C18 Luna column (Phenomenex, Torrance, CA, USA) was used, kept at 25 °C. The mobile phase consisted of 0.5% formic acid in water (phase A) and acetonitrile (phase B). The gradient used for analysis was as follows: 95% A for 1 min; 90% A for 4 min; 90% A for 5 min; 85% A for 47 min; 75% A for 6 min; 48% A for 15 min, followed by 13 min of re-equilibration with 95% A between runs. Standards were prepared at approximately 1 mg/mL, then diluted to create calibration curves (ranging from 0 to 50 μg/mL) for various phenolic acids, including gallic acid, trans-cinnamic acid, caffeic acid, syringic acid, ferulic acid, p-coumaric acid, hydroxycinnamic acid, vanillic acid, chlorogenic acid, kaempferol, epicatechin, naringenin, resveratrol, quercetin, myricetin, rutin, and catechin. The standards used for HPLC analysis were obtained from Sigma-Aldrich (Merk Life Science S.r.l., Milan, Italy). The limits of detection (LOD) and quantification (LOQ) values are provided in Table 2. The sample was injected at room temperature with a volume of 20 μL.

Table 2. Retention times (Rt), detection, and quantification the phenolic standards.

Time (min)	RT (min)	LOD (μg/mL)	LOQ (μg/mL)
Gallic Acid	7.33	0.03	0.09
Hidroxycinnamic acid	17.00	0.03	0.09
Chlorogenic acid	21.33	0.02	0.06
Catechin	22.94	0.04	0.13
Vanillic acid	25.08	0.01	0.04
Caffeic acid	27.25	0.02	0.06
Syringic acid	28.00	0.03	0.10
Epicatechin	33.00	0.04	0.13
Trans cinnamic acid	38.35	0.11	0.33
p-Coumaric acid	44.74	0.04	0.12
Ferulic acid	53.66	0.09	0.28
Rutin	65.61	0.08	0.24
Myricetin	71.64	0.02	0.06
Resveratrol	74.55	0.04	0.13

Table 2. *Cont.*

Time (min)	RT (min)	LOD (µg/mL)	LOQ (µg/mL)
Quercetin	79.10	0.01	0.04
Naringenin	84.95	0.05	0.16
Kaempferol	86.50	0.04	0.12

2.6. Statistical Analysis

All experiments were conducted in triplicate, and the results were expressed as mean ± standard deviation (SD). The data were processed using R-4.4.2.2 (2024). A one-way analysis of variance (ANOVA), followed by Tukey's multiple comparison test, was performed to determine statistical significance, with *p*-values less than 0.05 considered significant. Furthermore, Pearson's correlation coefficients were calculated to evaluate the relationships between the analyzed parameters.

3. Results

3.1. Botanical Originof Bee Pollen

The results of the palynological analysis of bee pollen samples are summarized in Table 3 and Figure 2.

Table 3. Palynological analysis of bee pollen samples.

Sample	Predominant Pollen Grains (>45%)	Secondary Pollen Grains (16–45%)	Important Minor Pollen (3–15%)	Minor Pollen (<3%)	Classification
P1	*Cistus* type [a] (63%)	*Brassica* type [b] (16%)	*Trifolium repens* [j] (11%), *Onobruchis* sp. [j] (7%)	*Anthemis* type [e], *Castanea sativa* [i], *Centaurea* sp. [e], Fabaceae [j], Rosaceae [k]	Monofloral
P2	*Cistus* type [a] (82%)	/	*Olea europaea* [l] (14%)	*Quercus* sp. [i]	Monofloral
P3	*Cistus* type [a] (65%)	*Erica* sp. [c] (25%)	*Quercus* sp. [i] (7%)	*Oxalis* sp. [m]	Monofloral
P4	*Brassica* type [b] (82%)	/	*Meliotus* type [j] (15%)	Rosaceae [k], *Prunus* sp. [k], *Cistus* type [a], *Carduus* type [e], Apiaceae [d]	Monofloral
P5	/	*Brassica* type [b] (39%), *vitex* sp. [g] (41%)	*Melilotus* type [j] (10%), *Erica* sp. [c] (7%)	*Cistus* type [a]	Polyfloral
P6	*Brassica* type [b] (50%)	*Erica* sp. [c] (36%)	*Rubus* sp. [k] (6%), *Aster* type [e] (4%), *Taraxacum* type [e] (4%)	*Capparis spinosa* [n], *Cistus* type [a]	Monofloral
P7	*Brassica* type [b] (55%)	*Cistus* type [a] (29%),	*Melilotus* type [j] (8%), *Lotus* sp. [j] (5%)	*Rubus* sp. [k], *Anthemis* type [e]	Monofloral
P8	/	*Brassica* type [b] (43%), *vitex* sp. [g] (42%)	*Melilotus* type [j] (7%), *Erica* sp. [c] (5%)	*Cistus* type [a]	Polyfloral
P9	*Pimpinella anisum* [d] (49%)	/	*Hedysarum coronarium* [j] (10%), *Taraxacum* type [e] (15%), *Anthemis* type [e] (12%), *Quercus* sp. [i] (11%)	*Rubus* sp. [k], *Eucalyptus* sp., *Carduus* type [e], *Trifolium* type [j], *Cistus* type [a]	Monofloral
P10	*Erica* sp. [c] (53%)	*Melilotus* sp. [j] (32%)	Asreraceae [e] (7%), *Brassica* type [b] (5%),	Apiaceae [d], Rosaceae [k], *Carduus* type [e], *Asparagus* type [p], *Alluim* sp. [h], *Rubus* sp. [k]	Monofloral
P11	*Cistus* type [a] (47%)	Liliaceae (44%)	*Carduus* type [e] (9%)	/	Monofloral
P12	*Brassica* type [b] (71%)	Apiaceae [d] (26%)	/	*Carduus* type [e]	Monofloral

Table 3. *Cont.*

Sample	Predominant Pollen Grains (>45%)	Secondary Pollen Grains (16–45%)	Important Minor Pollen (3–15%)	Minor Pollen (<3%)	Classification
P13	*Aster* type [e] (52%)	*Cistus* type [a] (23%)	*Erica* sp. [c] (12%), *Quercus* sp. [i] (10%)	*Asparagus* type [p], *Anthemis* type [e], *Taraxacum* type [e], Fabaceae [j]	Monofloral
P14	/	Oleaceae [l] (39%), *Tamarix* sp. [f] (29%)	*Quercus* sp. [i] (8%), Lamiaceae [g] (9%), *Brassica* type [b] (5%), *Centaurea* sp. [e] (7%)	*Carduus* type [e], *Taraxacum* type [e], *Plantago* sp. [q], *Trifolium pratanse* [j], *Eucalyptus* sp. [o], *Genista* type [j], *Melilotus* sp. [j]	Polyfloral
P15	/	*Artemesia* sp. [e] (37%), *Brassica* type [b] (21%), *Helianthus annuus* [e] (41%)	*Alluim* sp. [h] (4%)	*Trifolium repens* [j], *Cistus* type [a], Rosaceae [k], *Tamarix* sp. [f], Chenopodiaceae, Apiaceae [d]	Polyfloral

Different letters (a–q) indicate the families: [a] Cistaceae, [b] Brassicaceae, [c] Ericaceae, [d] Apiaceae, [e] Asteraceae, [f] Tamaricaceae, [g] Lamiaceae, [h] Alliaceae, [i] Fagaceae, [j] Fabaceae, [k] Rosaceae, [l] Oleaceae, [m] Oxalidaceae, [n] Capparaceae, [o] Myrtaceae, [p] Asparagaceae, and [q] Plantaginaceae.

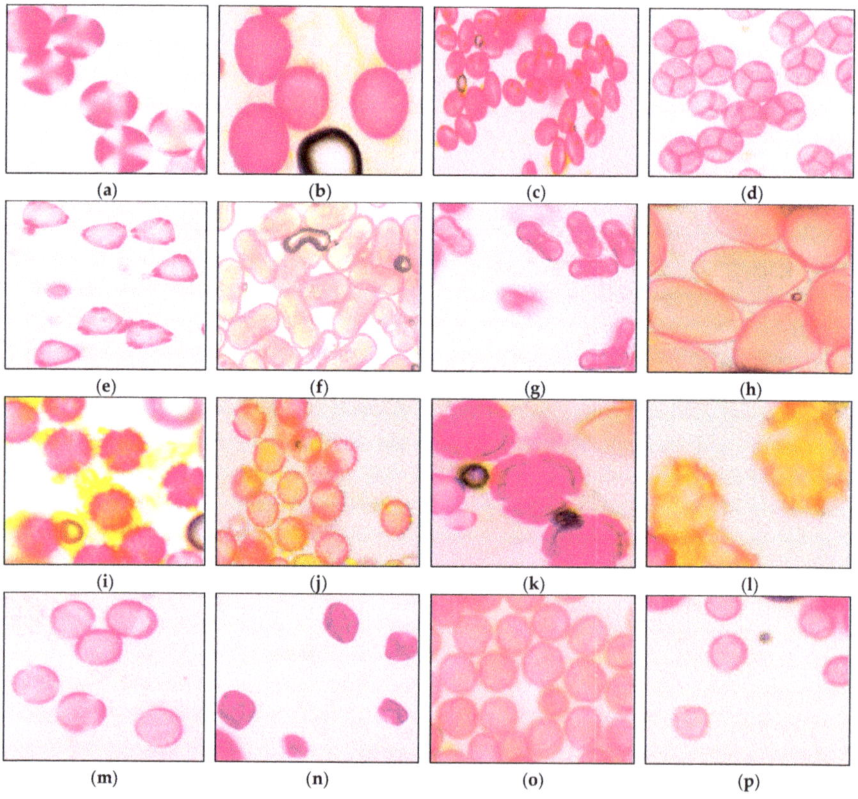

Figure 2. Microscopic appearance of the different pollen types found in Algerian pollen samples. **(a)** *Brassica* type. **(b)** *Cistus* type. **(c)** *Hedysarum coronarium*. **(d)** *Erica* sp. **(e)** *Eucalyptus* sp. **(f)** Apiaceae. **(g)** *Pimpinella anisum*. **(h)** Liliaceae. **(i)** *Aster* type. **(j)** *Helianthus annuus*. **(k)** *Carduus* type. **(l)** *Taraxacum* type. **(m)** *Tamarix* sp. **(n)** *Rubus* sp. **(o)** *Quercus* sp. **(p)** *Olea europaea*.

Among the analyzed samples, 30 pollen types from 17 families were identified. The most diverse families were Asteraceae, with 7 types, and Fabaceae, with 6 types. The remaining families were represented by only one or two types. The most frequently occurring pollen types were *Cistus*, *Brassica*, and various Asteraceae, appearing in nearly all samples.

The analysis classified eleven samples as monofloral and four samples as polyfloral. Samples P5, P8, P14, and P15 exhibited pollen from various botanical origins at different relative frequencies, as shown in Table 3. The monofloral samples were identified as originating from the following types: *Cistus* (P1, P2, P3, and P11), *Brassica* (P4, P6, P7, and P12), *Pimpinella anisum* (P9), *Erica* sp., (P10), and *Aster* (P13).

3.2. Concentration of Total Phenols, Flavonoids, and Antioxidant Capacity of Bee Pollen Extracts

Table 4 provides a summary of the total phenolic content, flavonoid levels, and antioxidant capacity of bee pollen.

Table 4. Antioxidant content and activities of the ethanolic extract of bee pollen.

Sample	Total Phenolic Content (TPC) (mgGAE/g)	Total Flavonoid Content (TFC) (mg QE/g)	DPPH Radical Scavenging Assay (IC_{50}) (mg/mL)	Total Antioxidant Capacity (TAC) (EAA mg/100 g)	Reducing Power (FRAP) (EC_{50}) (mg/mL)
P1	12.37 ± 0.19 ^{b,c}	2.80 ± 0.33 ^{c,d}	0.26 ± 0.01 ^{a,b}	401.52 ± 1.13 ^{d,e}	0.12 ± 0.02 ^{c,d}
P2	23.49 ± 1.48 ^{i}	2.99 ± 0.02 ^{d}	0.34 ± 0.04 ^{a,b,c}	465.56 ± 25.63 ^{g}	0.09 ± 0.00 ^{a,b}
P3	19.69 ± 0.69 ^{i}	2.95 ± 0.07 ^{d}	0.52 ± 0.01 ^{d,e,f}	262.17 ± 3.41 ^{a}	0.06 ± 0.00 ^{a}
P4	19.05 ± 1.90 ^{h,i}	1.70 ± 0.05 ^{a,b}	0.25 ± 0.03 ^{a,b}	612.54 ± 5.12 ^{i}	0.09 ± 0.00 ^{a,b}
P5	15.91 ± 0.64 ^{e,f,g}	4.74 ± 0.24 ^{f,g}	0.67 ± 0.05 ^{f}	672.36 ± 3.41 ^{j}	0.10 ± 0.00 ^{b,c}
P6	13.69 ± 0.50 ^{c,d,e}	2.43 ± 0.05 ^{c,d}	0.42 ± 0.00 ^{b,c,d,e}	246.56 ± 10.25 ^{a}	0.13 ± 0.00 ^{c,d}
P7	17.09 ± 0.32 ^{g,h}	3.92 ± 0.24 ^{e}	0.50 ± 0.04 ^{b,c,d,e,f}	595.96 ± 6.83 ^{i}	0.12 ± 0.00 ^{a,b}
P8	17.18 ± 0.27 ^{f,g}	1.77 ± 0.21 ^{a,b}	0.38 ± 0.01 ^{a,b,c}	373.26 ± 5.12 ^{c,d}	0.13 ± 0.01 ^{c,d}
P9	14.70 ± 0.37 ^{c,d,e,f}	5.57 ± 0.27 ^{h}	0.53 ± 0.00 ^{d,e,f}	551.92 ± 1.13 ^{h}	0.12 ± 0.00 ^{b,c,d}
P10	14.87 ± 1.02 ^{d,e,f}	4.28 ± 0.31 ^{e,f}	0.58 ± 0.05 ^{e,f}	592.03 ± 8.54 ^{i}	0.13 ± 0.01 ^{d}
P11	8.28 ± 0.67 ^{a}	2.94 ± 0.10 ^{d}	1.12 ± 0.15 ^{g}	354.46 ± 6.83 ^{b,c}	0.29 ± 0.00 ^{f}
P12	13.42 ± 0.30 ^{c,d}	3.74 ± 0.33 ^{e}	0.21 ± 0.00 ^{a}	677.14 ± 12.81 ^{j}	0.07 ± 0.00 ^{a}
P13	7.72 ± 0.29 ^{a}	1.48 ± 0.00 ^{a}	0.67 ± 0.07 ^{f}	421.12 ± 8.54 ^{e,f}	0.20 ± 0.00 ^{e}
P14	11.05 ± 0.08 ^{b}	4.98 ± 0.09 ^{g,h}	0.63 ± 0.01 ^{f}	433.08 ± 17.09 ^{f}	0.15 ± 0.01 ^{d}
P15	8.52 ± 0.08 ^{a}	2.28 ± 0.03 ^{b,c}	0.99 ± 0.08 ^{g}	330.19 ± 5.98 ^{b}	0.18 ± 0.01 ^{e}
Average	14.32	3.26	0.54	476.24	0.12

We compared differences between the resulting TPC, TFC, DPPH, TAC, and FRAP of all BPEs using one-way ANOVA and Tukey's multiple comparisons tests. Within the same column, different letters (a–j) indicate significant differences ($p < 0.05$).

The TPC of the tested samples ranged from 7.72 ± 0.29 to 23.49 ± 1.48 mg GAE/g. The highest TPC was recorded in sample P2, followed by P3 and P4, with values of 19.69 ± 0.69 GAE/g and 19.05 ± 1.90 GAE/g, respectively. The lowest TPC was observed in sample P13.

The TFC ranged from 1.48 ± 0.00 to 5.57 ± 0.27 mg QE/g of pollen. Sample P9 had the highest flavonoid content (5.57 ± 0.27 mg QE/g of pollen), followed closely by samples P14, P5, and P10, with values of 4.98 ± 0.09, 4.74 ± 0.24, and 4.28 ± 0.31 mg QE/g, respectively. The lowest TFC was found in sample P13, accordingly with the lowest. Interestingly, a contrasting trend was observed in samples P2, P3, and P4, where higher TPC was accompanied by lower TFC.

Different assays were employed to evaluate the antioxidant potential of bee pollen extracts, including the DPPH free radical scavenging activity, the reducing power assay (FRAP), and the molybdate ion reduction assay. The results are shown in Table 4.

DPPH radical scavenging activities ranged from IC_{50} 1.12 ± 0.15 mg/mL to IC_{50} 0.21 ± 0.00 mg/mL. The lowest IC_{50} value indicates high radical scavenging activity. The monofloral samples (*Brassica* type) P12 and P4 exhibited the highest antioxidant activity (0.21 mg/mL, 0.25 mg/mL, respectively) together with *Cistus* type monofloral bee pollen (P1), followed by P2 (0.34 ± 0.04 mg/mL), P8 (0.38 ± 0.01 mg/mL), P6 (0.42 ± 0.00 mg/mL), and P7 (0.50 ± 0.04 mg/mL).

The reducing power method revealed that sample P3 demonstrated the highest inhibition (EC_{50} 0.06 ± 0.00 mg/mL), while P11 showed the lowest inhibition (EC_{50} 0.29 ± 0.00 mg/mL). The antioxidant capacity of the samples decreased in the following order: P13 > P15 > P14 > P6, P8, P10 > P1, P7, P9 > P5 > P4, P2 > P12.

The total antioxidant capacity (TAC) values ranged from 262.17 ± 3.41 to 677.14 ± 12.81 EAA mg/100 g. Sample P12 exhibited the highest antioxidant activity (677.14 ± 12.81 EAA mg/100 g), followed by P5 (672.36 ± 3.41 EAA mg/100 g), P4 (612.54 ± 5.12 EAA mg/100 g), and P7 (595.96 ± 6.83 EAA mg/100 g). Conversely, P6 and P2, representing monofloral *Brassica* and *Cistus* types, respectively, showed the lowest activity. Significant differences were found between the mean values for all the variables analyzed in the bee pollen samples ($p < 0.05$).

3.3. Phenolic Compounds by HPLC Analysis

In this study, phenolic compounds in ethanolic bee pollen extracts were analyzed using Thermo Scientific Vanquish HPLC. Figure 3A,B, illustrate chromatograms for phenolic standards and the phenolic profile of bee pollen extracts, respectively, using sample P9 as an example. Table 5 outlines the phenolic compounds (μg/g) identified in the extracts.

Table 5. HPLC analysis of phenolic compounds of bee pollens (μg/g).

Sample	p-Coumaric Acid	Rutin	Myricetin	Resveratrol	Naringenin	Kaempferol
P1	Not detected	124.8	995.4	186.2	177.1	37.0
P2	Not detected	103.8	532.1	59.6	84.5	39.5
P3	Not detected	488.1	2081.7	377.7	7.6	62.0
P4	89.0	48.1	255.1	291.2	161.0	347.6
P5	77.6	38.3	36.1	453.4	488.7	11.5
P6	40.9	613.1	89.9	255.2	112.1	5.4
P7	Not detected	187.0	977.2	69.7	102.1	4.6
P8	23.2	12.9	56.7	160.6	625.9	90.8
P9	32.8	1052.7	2062.1	48.4	1418.1	172.4
P10	64.8	23.8	52.8	156.3	220.8	21.8
P11	Not detected	32.2	10.5	8.4	705.9	6.6
P12	130.3	496.3	883.4	413.1	205.6	67.0
P13	3.9	71.1	163.2	122.7	192.5	17.2
P14	Not detected	361.2	696.2	149.3	193.3	22.7
P15	Not detected	11.4	26.7	149.2	97.8	13.7
Average	31.3	303.3	700.1	191.5	354.6	53.5

Out of 17 tested molecules, only p-coumaric acid, rutin, myricetin, naringenin, resveratrol and kaempferol were detected. Notably, high levels of rutin (11.4–1052.7 μg/g), myricetin (10.5–1418.1 μg/g), and naringenin (7.6–1418.1 μg/g) were found. p-coumaric Acid was absent in samples P1, P2, P3, P7, P11, P14, and P15. Sample P9 had the highest amounts of rutin (1052.7 μg/g) and naringenin (1418.1 μg/g) and the second-highest level of myricetin (1062.1 μg/g), correlating with its high flavonoid content (5.57 ± 0.27 mg/100 g).

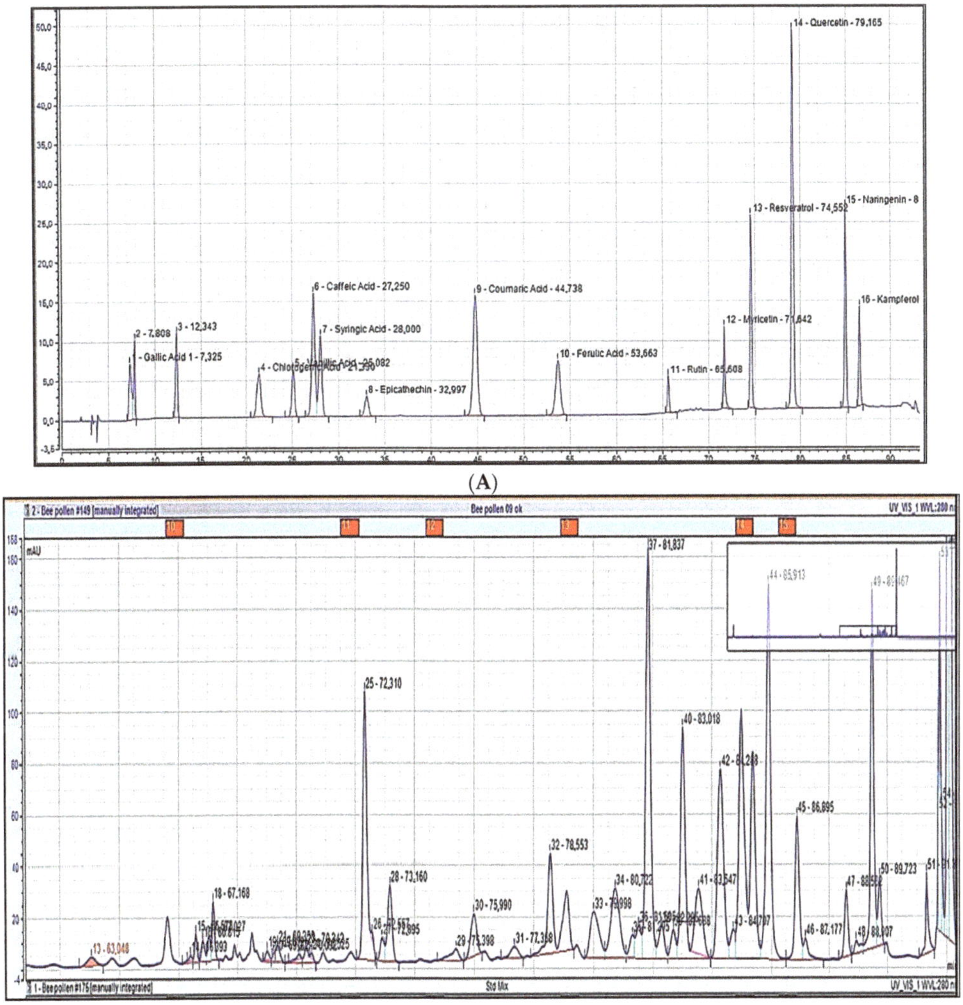

Figure 3. Chromatogram of the standards (**A**). As an example, the chromatogram of sample 9, zoomed with RT > 40 min (**B**).

3.4. Correlation Among Phenolic Compounds and Antioxidant Activity

Using Pearson's correlation (Figure 4), this study identified a strong positive correlation between total phenolic content (TPC) and kaempferol (r = 0.368), TPC and myricetin (r = 0.298), total flavonoid content (TFC), and total antioxidant capacity (TAC) (r = 0.472), TFC and rutin (r = 0.482), TFC and myricetin (r = 0.417), TFC and naringenin (r = 0.428). These correlations are explained by the fact that rutin, myricetin, and naringenin belong to the flavonoid family. Additionally, DPPH radical scavenging activity correlated positively with FRAP. A positive correlation was also observed between TAC and specific phenolic compounds, such as coumaric acid and kaempferol. There was further evidence of relationships between coumaric acid with resveratrol and kaempferol, as well as between rutin with myricetin and naringenin. However, a negative correlation was observed between TPC and both DPPH and FRAP (r = −0.594 and r = −0.734, respectively), which was attributed to the EC_{50} expression of these assays, where lower values indicate higher activity. Negative correlations were also noted between DPPH and TAC, resveratrol, and kaempferol, as well as between FRAP and TAC, coumaric acid, myricetin, and resveratrol. Among phenolic compounds, only naringenin and resveratrol showed a weak negative correlation. Pollen sample P12, belonging to the *Brassica* type with average TPC and TFC levels, showed the highest activity in both DPPH and TAC assays and ranked second in FRAP. Conversely, monofloral bee pollen P13 (*Aster* type) with low TPC and TFC exhibited the lowest FRAP activity.

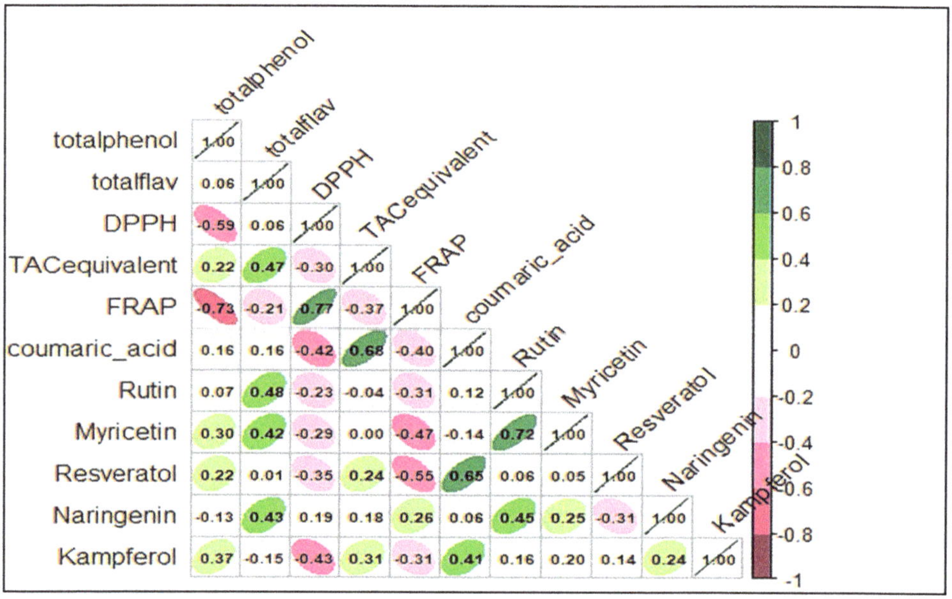

Figure 4. Pearson Correlation shown by the parameters studied in bee pollen samples from different botanical origins.

4. Discussion

Bee pollen is a highly nutritious natural product, rich in antioxidants, carbohydrates, proteins, and essential amino. It also contains unsaturated fatty acids, vitamins, and minerals. Due to its nutritional and bioactive properties, bee pollen is widely used in dietary supplements, functional foods, and nutraceuticals for its health-promoting benefits. Its bioactive compounds, especially phenolic acids and flavonoids, contribute to their biological activities. Since pollen is derived from various plant species, its quality, safety, and characterization are determined by its botanical and geographical origins [28–30].

The botanical analysis of the bee pollen samples from this research highlights the rich floral diversity of the Mila region (northeastern Algeria), influenced by its Mediterranean climate, orography, and human activity. Most identified pollen types originate from spontaneous species with high melliferous potential, supporting the sustainable growth of beekeeping in the region [10,16,31]. With the exception of the Brassicaceae family, all the mentioned families are recognized as excellent pollen sources. However, most species within the Brassicaceae and Asteraceae families are primarily classified as nectar-producing. The Fabaceae family is known for its strong nectar and pollen production potential [32]. The dominant pollen types in the samples were *Cistus* and *Brassica*, suggesting that honeybees in the Mila region primarily collect these types for sustenance. Almost all samples were monofloral (11 samples). Carpes et al. [33] highlighted that the nutritional quality of certain pollen types makes bees favor a single floral source. Similarly, Aličić et al. [34] noted that bees typically visit the same plant species when collecting pollen, resulting in predominantly monofloral pollen with minor contributions from other species. There is limited data on the characterization of Algerian pollen in the literature. Bakchiche et al. [35] investigated pollen from the Laghouat region and identified taxa from families such as Asteraceae, Betulaceae, Boraginaceae, Brassicaceae, Cistaceae, Ericaceae, Fabaceae, Liliaceae, Myrtaceae, Salicaceae, and Rosaceae, with Brassicaceae being dominant. Mokhtari et al. [11] analyzed eight bee pollen samples from different Algerian regions and classified them as polyfloral, identifying 40 pollen types across 22 botanical families. The most diverse families included Asteraceae (8 types), Brassicaceae (4 types), and Cistaceae, Fabaceae, and Fagaceae (3 types each). These pollen types were also noted in other Algerian bee products, especially honey [16,20,36,37]. Similarly to our findings, El Ghouizi et al. [38] attributed the botanical origin of eight monofloral bee pollen samples to the Apiaceae, Fabaceae, Resedaceae, Rosaceae, and Lamiaceae families. Significant variation in pollen composition was observed, likely due to the diversity of environmental flora at different altitudes where beehives are located. Furthermore, bee pollen samples collected from the same or nearby areas can reflect different plant sources. This variation may depend on the dominant plant species in the vicinity of the hives and the foraging preferences of honeybees [26].

Several studies across various countries have investigated the total bioactive components of bee pollen. The TPC and TFC of Algerian bee pollen were similar to those found in samples from Marocco by El Ghouizi et al. [38], who reported TPC in monofloral bee pollen ranging between 8.070 $\pm$ 1.037 mg GAE/g and 32.387 $\pm$ 0.148 mg GAE/g, and TFC between 0.202 $\pm$ 0.044 mg QE/g and 6.30 $\pm$ 0.37 mg QE/g. The TPC values in our samples are consistent also with the results reported by Ilie et al. [39] for Romanian bee pollen. However, our TFC values are notably higher. In addition, we obtained higher levels of TPC and TFC when compared with the findings of Aylanc et al. [26]. A recent study in Algeria by Hemmami et al. [40] showed TPC values ranging from 3.79 to 9.17 mg GAE/g and TFC values between 2.07 and 5.50 mg QE/g. Compared to other studies involving bee pollen samples from Algeria, Brazil, Northwest Spain, Turkey, and India [5,7,11,18,28,41], our extracts exhibited lower amounts of total phenols and flavonoids. A high phenolic content does not necessarily correspond to elevated flavonoid levels, a finding consistent with previous research [26]. This discrepancy was observed in certain samples.

Bee pollen exhibits strong antioxidant potential due to its rich polyphenolic content. Various methods are used to assess its antioxidant activity, as different compounds require specific treatments and yield varying responses [29]. According to the literature, our DPPH radical scavenging assay results are consistent with those reported for Brazilian bee pollen, where the highest antioxidant activity was observed in the ethanolic extract of *Brassica napus* pollen [42]. However, our findings exceed the previously documented range of 7.77 to 1.68 mg/mL [6]. In addition, the methanolic extract of coriander bee pollen from India was revealed as a potent free radicals inhibitor (93.75 $\pm$ 0.05%) due to a minimum IC_{50} value (0.006 $\pm$ 0.21 mg/mL) [43]. The IC_{50} values of Greek pollen [4] and pollen from Portugal and New Zealand [44] were 0.181 $\pm$ 1.70 mg/mL and 0.04–0.5 mg/mL, respectively, which are within the range of our Algerian bee pollen samples from several regions. When comparing the FRAP assay results with those of other researchers, we observe notable similarities and differences, providing valuable insights into antioxidant capacity variations. Studies evaluating the antioxidant capacity of bee pollen extracts from various floral regions in Morocco [38] reported EC_{50} values ranging from 0.133 $\pm$ 0.036 to 0.790 $\pm$ 0.175 mg/mL. Similarly, analyses of ethanolic pollen extracts from apiaries in Bahia, Brazil [41] showed antioxidant activity values between 1.45 ± 0.03 and 8.77 ± 0.23 mg/mL. These findings suggest that the antioxidant activity of Moroccan and Brazilian bee pollen is generally lower than that of Algerian samples. The antioxidant capacity of bee pollen extracts has been extensively evaluated using the FRAP method, though results often differ in their expression. Alimoglu et al. [5] investigated 11 bee pollen samples from Turkey, reporting FRAP values between 3.42 ± 0.07 mg Trolox/g and 16.03 ± 0.25 mg Trolox/g. Similarly, Ulusoy and Kolayli [45] analyzed 13 Anzer bee pollen samples, and Karkar et al. [46], examined 10 samples from Turkey, finding FRAP values ranging from 11.77 ± 0.63 mmol Trolox/g to 4.75 ± 0.02 mmol Trolox/g and from 1.53 ± 0.06 mg Trolox/g to 4.75 ± 0.02 mg Trolox/g, respectively. Additionally, coriander bee pollen extracts demonstrated a wide range of FRAP values, from 32.54 ± 0.57 to 103.98 ± 0.82 mmol Fe^{2+}/g, with the highest values [7]. These variations highlight the diverse antioxidant potential of bee pollen across different geographical and botanical origins. Limited data exists on total antioxidant capacity using the phosphomolybdenum method. In this study, pollen samples exhibited lower TAC compared to Moroccan samples (3.98 ± 0.16 mg EAA/g to 9.69 ± 0.34 mg EAA/g) [38] and Algerian samples (808.2–3311 mg/100 g) [40], but higher values than those reported by Rebiai and Lanez [47]. Significant variations in TPC, TFC, and antioxidant capacities of bee pollen have been observed worldwide. These differences are influenced by factors such as the choice of standards (e.g., gallic acid, pinocembrin, or chlorogenic acid), the extraction solvent (e.g., methanol, ethanol, or dichloromethane), and the extraction method. Additionally, the botanical and geographical origins of bee pollen, along with environmental factors like soil type, climate conditions, and beekeeping practices, play a crucial role in determining its composition [30,38,46].

Bee pollen has garnered increased attention for its phenolic profile, which has been fairly well reported in the literature [48]. In this study, the phenolic compounds identified in Algerian bee pollen samples align with the findings of Bayram et al. [49], who reported rutin as the dominant compound in Turkish bee pollen. Additionally, resveratrol and kaempferol were present in considerable amounts, whereas coumaric acid was detected in lower concentrations and was absent in samples P1, P2, P3, P7, P14, and P15. However, research on Algerian bee pollen remains scarce, highlighting the need for further studies. Hemmami et al. [40], analyzed 13 Algerian bee pollen, identifying high levels of rutin (6751.996 ± 168.799 μg/g), quercetin (20,749.05 ± 518.726 μg/g), and gallic acid (648.558 ± 12.971 μg/g). All 13 Hemmami and coworker's samples contained gallic acid, p-coumaric acid, and quercetin. However, different from our results specific compounds, i.e., vanillin, chlorogenic acid, naringin, vanillic acid, rutin, and caffeic acid, were completely lacking in several samples, and only one sample contained all nine studied phenolic compounds. Aylanc et al. [26] detected caffeic acid and p-coumaric acid only in one Moroccan sample, with concentrations of 150 μg/g and 200 μg/g, respectively. Similarly, Kahraman et al. [30] found kaempferol (3264.5 μg/g) in Turkish bee pollen but did not identify epicatechin, syringic acid, vanillic acid, or catechin. For instance, Karkar et al. [46] did not detect gallic acid in ethanolic extracts of Turkish chestnut bee pollen but found low levels of syringic acid, kaempferol, and myricetin. Alimoglu et al. [5] reported different profiles using ethyl acetate and dichloromethane extracts, demonstrating that solvent selection significantly affects phenolic extraction. Thus, bee pollen extracts contain phenolic compounds, whose profiles differ markedly, as clearly documented in the literature.

Numerous studies have explored the relationship between phenolic content and the antioxidant capacity of bee pollen [3,50,51], and recent findings highlight a strong positive association [26,28,29,47]. Limmongkon et al. [52] reported a significant correlation (r = 0.773) between the DPPH and FRAP assays, both of which measure the electron transfer potential of antioxidants. Our results are consistent with this observation, further supporting the link between phenolic compounds and their antioxidant efficacy.

Certain bee pollen samples present higher TPC values and lower DPPH IC_{50}. The presence of specific phenolic compounds with high radical-scavenging efficiency may explain this relationship. This trend is expected, as phenolic compounds are known for their ability to scavenge free radicals [53]. The strong correlations may be due to the presence of flavonols such as rutin, kaempferol, and myricetin derivatives. Their planar structures, influenced by hydroxyl groups at position 3, facilitate radical capture through conjugation and electron delocalization. Moreover, the high number of hydroxyl groups contributes significantly to their potency as electron scavengers. Despite this, several studies have concluded that antioxidant capacity is not necessarily linked to total phenolic content [46,54–56]. In this study, the sample with the highest TPC (P2) showed the lowest TAC. This indicates that the presence of specific phenolic compounds, rather than their concentration, may play a more critical role. Synergistic and antagonistic effects among phenolic compounds and interactions with other phytochemicals in the extract further influence antioxidant activity [52]. Environmental factors, such as soil conditions and plant physiology, also impact free radical reactions and the antioxidant capacity of bee pollen [34,41]. Furthermore, the collection period significantly affects antioxidant activity, with higher values observed during UV-intense periods, particularly from early to late summer [3]. Consequently, proper management practices and the timing of collection are crucial for optimizing the chemical and functional properties of bee pollen.

While this study provides valuable insights into the botanical origin, phytochemical profile, and antioxidant activity of bee pollen from the Mila region, certain limitations should be acknowledged. The sample size, though representative, could be expanded to improve the generalizability of the findings. Additionally, while three antioxidant assays (DPPH, FRAP, and TAC) were employed, incorporating a broader range of assays would provide a more comprehensive evaluation of the antioxidant potential. Further, the HPLC analysis, which identified key phenolic compounds, could be extended to include a wider range of bioactive compounds for a more detailed phytochemical characterization. Despite these limitations, the findings of this study contribute to the growing body of knowledge on natural antioxidants and highlight the need for further investigations to overcome these challenges.

5. Conclusions

This study investigated bee pollen samples from the Mila region in northeastern Algeria. Bee pollen is recognized as a valuable functional food due to its high levels of phenolics and strong antioxidant properties. These compounds are known to offer various health benefits. The study aimed to determine the botanical origin of the pollen samples, analyze their antioxidant capacity using different methods (DPPH, FRAP, and TAC), and characterize their phenolic composition using HPLC. The results showed significant variations in phenolic content and antioxidant activity among pollen samples from different plant sources. Phenolic compounds were identified as the primary contributors to the antioxidant properties of the pollen. HPLC analysis revealed the presence of six phenolic compounds across all samples, with rutin, myricetin, and naringenin being the most abundant. Moreover, the results suggest that bee pollen from the Mila region could contribute to human health as a powerful natural antioxidant and should be applied to various fields of food and pharmaceutical industries. However, the study emphasizes the need and opportunity for further research, increasing such as the number of samples analyzed, expanding the research to other regions of Algeria, and exploring other research approaches to gain a deeper understanding of the therapeutic and economic value of Algerian bee pollen.

Author Contributions: N.B. and H.B. conceived and designed the methodology. N.B., R.A. and G.P. carried out the experiments. A.G. determined the botanical origin of bee pollen. N.B. and H.B. analyzed and interpreted data. N.B. writing—original draft preparation. H.B., R.A., G.P. and A.G. writing—review and editing. H.B. supervision. All authors have read and agreed to the published version of the manuscript.

Funding: This research received no external funding.

Institutional Review Board Statement: Not applicable.

Informed Consent Statement: Not applicable.

Data Availability Statement: Data is contained within the article.

Acknowledgments: The authors would like to thank the beekeepers and for their collaboration in this study and CREA (Council for Agricultural Research and Economics), Research Centre for Food and Nutrition, 00178 Rome, Italy; under the administration of Fausta Natella for HPLC analysis.

Conflicts of Interest: The authors declare no conflicts of interest.

References

1. Rupanar, S.V.; Satpute, R.; Gadhave, P.; Kakade, N. Bee Pollen beyond Nutrition: An Integrated Review of Its Chemical, Pharmacological, and Biological Properties. *J. Drug Res. Ayurvedic Sci.* **2024**, *9*, 211. [CrossRef]
2. Anjum, S.I.; Ullah, A.; Gohar, F.; Raza, G.; Khan, M.I.; Hameed, M.; Ali, A.; Chen, C.-C.; Tlak Gajger, I. Bee Pollen as a Food and Feed Supplement and a Therapeutic Remedy: Recent Trends in Nanotechnology. *Front. Nutr.* **2024**, *11*, 1371672. [CrossRef] [PubMed]
3. LeBlanc, B.W.; Davis, O.K.; Boue, S.; DeLucca, A.; Deeby, T. Antioxidant Activity of Sonoran Desert Bee Pollen. *Food Chem.* **2009**, *115*, 1299–1305. [CrossRef]
4. Graikou, K.; Kapeta, S.; Aligiannis, N.; Sotiroudis, G.; Chondrogianni, N.; Gonos, E.; Chinou, I. Chemical Analysis of Greek Pollen—Antioxidant, Antimicrobial and Proteasome Activation Properties. *Chem. Cent. J.* **2011**, *5*, 33. [CrossRef]
5. Alimoglu, G.; Guzelmeric, E.; Yuksel, P.I.; Celik, C.; Deniz, I.; Yesilada, E. Monofloral and Polyfloral Bee Pollens: Comparative Evaluation of Their Phenolics and Bioactivity Profiles. *LWT* **2021**, *142*, 110973. [CrossRef]
6. Soares De Arruda, V.A.; Vieria Dos Santos, A.; Figueiredo Sampaio, D.; Da Silva Araújo, E.; De Castro Peixoto, A.L.; Estevinho, L.M.; De Almeida-Muradian, L.B. Brazilian Bee Pollen: Phenolic Content, Antioxidant Properties and Antimicrobial Activity. *J. Apic. Res.* **2021**, *60*, 775–783. [CrossRef]
7. Thakur, M.; Nanda, V. Screening of Indian Bee Pollen Based on Antioxidant Properties and Polyphenolic Composition Using UHPLC-DAD-MS/MS: A Multivariate Analysis and ANN Based Approach. *Food Res. Int.* **2021**, *140*, 110041. [CrossRef]
8. Komosinska-Vassev, K.; Olczyk, P.; Kaźmierczak, J.; Mencner, L.; Olczyk, K. Bee Pollen: Chemical Composition and Therapeutic Application. *Evid. -Based Complement. Altern. Med.* **2015**, *2015*, 1–6. [CrossRef]
9. Freire, K.R.L.; Lins, A.C.S.; Dórea, M.C.; Santos, F.A.R.; Camara, C.A.; Silva, T.M.S. Palynological Origin, Phenolic Content, and Antioxidant Properties of Honeybee-Collected Pollen from Bahia, Brazil. *Molecules* **2012**, *17*, 1652–1664. [CrossRef] [PubMed]
10. Tamali, H.S.; Özkırım, A. Beekeeping Activities in Turkey and Algeria. *Mellifera* **2019**, *19*, 30–40.
11. Mokhtarı, M.B.; El Ouar, I.; Zeghına, I.; Tartouga, M.A.; Ghorab, A.; Bahrı, L.; Bensouıcı, C. Palynological Analysis, Phenolic Components and Anti-Inflammatory Activity of some Bee Pollens Collected from the Northeast Region of Algeria. *Uludağ Arıcılık Derg.* **2022**, *22*, 45–58. [CrossRef]
12. Ayad, A.S.; Hébert, M.P.A.; Doiron, J.A.; Loucif-Ayad, W.; Daas, T.; Smagghe, G.; Alburaki, M.; Barnett, D.A.; Touaibia, M.; Surette, M.E. Algerian Propolis from Distinct Geographical Locations: Chemical Profiles, Antioxidant Capacity, Cytotoxicity and Inhibition of 5-Lipoxygenase Product Biosynthesis. *Chem. Biodivers.* **2024**, *21*, e202301758. [CrossRef]
13. Piccinelli, A.L.; Mencherini, T.; Celano, R.; Mouhoubi, Z.; Tamendjari, A.; Aquino, R.P.; Rastrelli, L. Chemical Composition and Antioxidant Activity of Algerian Propolis. *J. Agric. Food Chem.* **2013**, *61*, 5080–5088. [CrossRef] [PubMed]
14. Boufadi, Y.M.; Soubhye, J.; Riazi, A.; Rousseau, A.; Vanhaeverbeek, M.; Nève, J.; Boudjeltia, K.Z.; Van Antwerpen, P. Characterization and Antioxidant Properties of Six Algerian Propolis Extracts: Ethyl Acetate Extracts Inhibit Myeloperoxidase Activity. *Int. J. Mol. Sci.* **2014**, *15*, 2327–2345. [CrossRef]
15. Bouhala, A.; Ouchemoukh, S.; Moussi, A.; Beldjoudi, S. Altitude Effect on the Properties of Honeys from the Region of Jijel (Algeria). *Pol. J. Food Nutr. Sci.* **2020**, *70*, 169–178. [CrossRef]
16. Ghorab, A.; Rodríguez-Flores, M.S.; Nakib, R.; Escuredo, O.; Haderbache, L.; Bekdouche, F.; Seijo, M.C. Sensorial, Melissopalynological and Physico-Chemical Characteristics of Honey from Babors Kabylia's Region (Algeria). *Foods* **2021**, *10*, 225. [CrossRef] [PubMed]

17. Hosoya, T.; Tsuchiya, I.; Ohta, T.; Benhanifia, M.; Kumazawa, S. Composition of Algerian Propolis, Plant Origin, and Its Antiangiogenic Activity In Vitro. *Molecules* **2021**, *26*, 6510. [CrossRef] [PubMed]
18. Rebiai, A.; Ben Seghir, B.; Hemmami, H.; Zeghoud, S.; Belfar, M.L.; Kouadri, I. Determination of Some Phenolic Acids in Algerian Propolis. *Ovidius Univ. Ann. Chem.* **2021**, *32*, 120–124. [CrossRef]
19. Soltani, E.K.; Zaim, K.; Mokhnache, K.; Haichour, N.; Mezaache-Aichour, S.; Charef, N.; Zerroug, M.M. Polyphenol Contents, Antioxidant and Antibacterial Activities of Aqueous Algerian Propolis Extracts. *Phytothérapie* **2021**, *19*, 408–415. [CrossRef]
20. Zerrouk, S.; Bahloul, R. Palynological and Physicochemical Properties of Multifloral Honey Produced in Some Regions of Algeria. *J. Apic. Res.* **2023**, *62*, 345–354. [CrossRef]
21. Louveaux, J.; Maurizio, A.; Vorwohl, G. Methods of Melissopalynology. *Bee World* **1978**, *59*, 139–157. [CrossRef]
22. Almaraz-Abarca, N.; Campos, M.d.G.; Ávila-Reyes, J.A.; Naranjo-Jiménez, N.; Herrera-Corral, J.; González-Valdez, L.S. Variability of Antioxidant Activity among Honeybee-Collected Pollen of Different Botanical Origin. *Interciencia* **2004**, *29*, 574–578.
23. Singleton, V.L.; Rossi, J.A. Colorimetry of Total Phenolics with Phosphomolybdic-Phosphotungstic Acid Reagents. *Am. J. Enol. Vitic.* **1965**, *16*, 144–158. [CrossRef]
24. Arvouet-Grand, A.; Vennat, B.; Pourrat, A.; Legret, P. Standardization of propolis extract and identification of principal constituents. *J. Pharm. Belg.* **1994**, *49*, 462–468. [PubMed]
25. Tomás, A.; Falcão, S.I.; Russo-Almeida, P.; Vilas-Boas, M. Potentialities of Beebread as a Food Supplement and Source of Nutraceuticals: Botanical Origin, Nutritional Composition and Antioxidant Activity. *J. Apic. Res.* **2017**, *56*, 219–230. [CrossRef]
26. Aylanc, V.; Falcão, S.I.; Ertosun, S.; Vilas-Boas, M. From the Hive to the Table: Nutrition Value, Digestibility and Bioavailability of the Dietary Phytochemicals Present in the Bee Pollen and Bee Bread. *Trends Food Sci. Technol.* **2021**, *109*, 464–481. [CrossRef]
27. Prieto, P.; Pineda, M.; Aguilar, M. Spectrophotometric Quantitation of Antioxidant Capacity through the Formation of a Phosphomolybdenum Complex: Specific Application to the Determination of Vitamin E. *Anal. Biochem.* **1999**, *269*, 337–341. [CrossRef] [PubMed]
28. Rojo, S.; Escuredo, O.; Rodríguez-Flores, M.S.; Seijo, M.C. Botanical Origin of Galician Bee Pollen (Northwest Spain) for the Characterization of Phenolic Content and Antioxidant Activity. *Foods* **2023**, *12*, 294. [CrossRef]
29. Rzepecka-Stojko, A.; Stojko, J.; Kurek-Górecka, A.; Górecki, M.; Kabała-Dzik, A.; Kubina, R.; Moździerz, A.; Buszman, E. Polyphenols from Bee Pollen: Structure, Absorption, Metabolism and Biological Activity. *Molecules* **2015**, *20*, 21732–21749. [CrossRef]
30. Kahraman, H.A.; Tutun, H.; Kaya, M.M.; Usluer, M.S.; Tutun, S.; Yaman, C.; Sevin, S.; Keyvan, E. Ethanolic Extract of Turkish Bee Pollen and Propolis: Phenolic Composition, Antiradical, Antiproliferative and Antibacterial Activities. *Biotechnol. Biotechnol. Equip.* **2022**, *36*, 45–56. [CrossRef]
31. Zerrouk, S.; Seijo, M.C.; Boughediri, L.; Escuredo, O.; Rodríguez-Flores, M.S. Palynological Characterisation of Algerian Honeys According to Their Geographical and Botanical Origin. *Grana* **2014**, *53*, 147–158. [CrossRef]
32. Ghorab, A.; Mesbah, M.; Nakib, R.; Kabli, N.; Bekdouche, F.; Seijo, M. Input to the Knowledge of the Melliferous Plants Diversity in Babors Kabylia's Region (NorthEast of Algeria). *Livest. Res. Rural Dev.* **2021**, *33*, 1–11.
33. Carpes, S.T.; Cabral, I.S.; Luz, C.F.P.; Capeletti, J.P.; Alencar, S.M.; Masson, M.L. Palynological and Physicochemical Characterization of Apis Mellifera L. Bee Pollen in the Southern Region of Brazil. *J. Food Agric. Environ.* **2009**, *7*, 667–673.
34. Aličić, D.; Šubarić, D.; Jašić, M.; Pašalić, H.; Ačkar, Đ. Antioxidant Properties of Pollen. *Hrana U Zdr. I Boles. Znan. -Stručni Časopis Za Nutr. I Dijetetiku* **2014**, *3*, 6–12.
35. Bakchiche, B.; Temizer, İ.K.; Güder, A.; Çelemli, Ö.G.; Yegin, S.Ç.; Bardaweel, S.; Ghareeb, M. Chemical Composition and Biological Activities of Honeybee Products From Algeria. *J. Apple Biotechnol. Rep.* **2020**, *7*, 109498. [CrossRef]
36. Makhloufi, C.; Kerkvliet, J.D.; D'albore, G.R.; Choukri, A.; Samar, R. Characterization of Algerian Honeys by Palynological and Physico-Chemical Methods. *Apidologie* **2010**, *41*, 509–521. [CrossRef]
37. Homrani, M.; Escuredo, O.; Rodríguez-Flores, M.S.; Fatiha, D.; Mohammed, B.; Homrani, A.; Seijo, M.C. Botanical Origin, Pollen Profile, and Physicochemical Properties of Algerian Honey from Different Bioclimatic Areas. *Foods* **2020**, *9*, 938. [CrossRef] [PubMed]
38. El Ghouizi, A.; El Menyiy, N.; Bakour, M.; Badiaa, L. Moroccan Monofloral Bee Pollen: Botanical Origin, Physicochemical Characterization, and Antioxidant Activities. *J. Food Qual.* **2021**, *2021*, 1–10. [CrossRef]
39. Ilie, C.-I.; Oprea, E.; Geana, E.-I.; Spoiala, A.; Buleandra, M.; Gradisteanu Pircalabioru, G.; Badea, I.A.; Ficai, D.; Andronescu, E.; Ficai, A.; et al. Bee Pollen Extracts: Chemical Composition, Antioxidant Properties, and Effect on the Growth of Selected Probiotic and Pathogenic Bacteria. *Antioxidants* **2022**, *11*, 959. [CrossRef] [PubMed]
40. Hemmami, H.; Ben Seghir, B.; Ben Ali, M.; Rebiai, A.; Zeghoud, S.; Brahmia, F. Phenolic Profile and Antioxidant Activity of Bee Pollen Extracts from Different Regions of Algeria. *Ovidius Univ. Ann. Chem.* **2020**, *31*, 93–98. [CrossRef]
41. Araújo, J.S.; Chambó, E.D.; de Carvalho Costa, M.A.P.; Cavalcante da Silva, S.M.P.; Lopes de Carvalho, C.A.; Estevinho, L.M. Chemical Composition and Biological Activities of Mono- and Heterofloral Bee Pollen of Different Geographical Origins. *Int. J. Mol. Sci.* **2017**, *18*, 921. [CrossRef] [PubMed]

42. Fatrcová-Šramková, K.; Nôžková, J.; Kačániová, M.; Máriássyová, M.; Rovná, K.; Stričík, M. Antioxidant and Antimicrobial Properties of Monofloral Bee Pollen. *J. Environ. Sci. Health Part B* **2013**, *48*, 133–138. [CrossRef] [PubMed]
43. Gercek, Y.C.; Celik, S.; Bayram, S. Screening of Plant Pollen Sources, Polyphenolic Compounds, Fatty Acids and Antioxidant/Antimicrobial Activity from Bee Pollen. *Molecules* **2021**, *27*, 117. [CrossRef] [PubMed]
44. Campos, M.G.; Webby, R.F.; Markham, K.R.; Mitchell, K.A.; Da Cunha, A.P. Age-Induced Diminution of Free Radical Scavenging Capacity in Bee Pollens and the Contribution of Constituent Flavonoids. *J. Agric. Food Chem.* **2003**, *51*, 742–745. [CrossRef]
45. Ulusoy, E.; Kolayli, S. Phenolic Composition and Antioxidant Properties of Anzer Bee Pollen: Phenolics and Antioxidant Properties of Anzer Bee Pollen. *J. Food Biochem.* **2014**, *38*, 73–82. [CrossRef]
46. Karkar, B.; Şahin, S.; Güneş, M.E. Evaluation of Antioxidant Properties and Determination of Phenolic and Carotenoid Profiles of Chestnut Bee Pollen Collected from Turkey. *J. Apic. Res.* **2021**, *60*, 765–774. [CrossRef]
47. Rebiai, A.; Lanez, T. Chemical Composition and Antioxidant Activity of Apis Mellifera Bee Pollen from Northwest Algeria. *J. Fundam. Appl. Sci.* **2012**, *4*, 156–163. [CrossRef]
48. Mosić, M.; Trifković, J.; Vovk, I.; Gašić, U.; Tešić, Ž.; Šikoparija, B.; Milojković-Opsenica, D. Phenolic Composition Influences the Health-Promoting Potential of Bee-Pollen. *Biomolecules* **2019**, *9*, 783. [CrossRef] [PubMed]
49. Bayram, N.E.; Gercek, Y.C.; Çelik, S.; Mayda, N.; Kostić, A.Ž.; Dramićanin, A.M.; Özkök, A. Phenolic and Free Amino Acid Profiles of Bee Bread and Bee Pollen with the Same Botanical Origin—Similarities and Differences. *Arab. J. Chem.* **2021**, *14*, 103004. [CrossRef]
50. Mohdaly, A.A.A.; Mahmoud, A.A.; Roby, M.H.H.; Smetanska, I.; Ramadan, M.F. Phenolic Extract from Propolis and Bee Pollen: Composition, Antioxidant and Antibacterial Activities: Phenolic Extract from Propolis and Bee Pollen. *J. Food Biochem.* **2015**, *39*, 538–547. [CrossRef]
51. Pukalskas, A.; Kazernavičiutė, R.; Balžekas, J. Evaluation of Antioxidant Activity and Flavonoid Composition in Differently Preserved Bee Products. *Czech J. Food Sci.* **2016**, *34*, 133–142. [CrossRef]
52. Limmongkon, A.; Janhom, P.; Amthong, A.; Kawpanuk, M.; Nopprang, P.; Poohadsuan, J.; Somboon, T.; Saijeen, S.; Surangkul, D.; Srikummool, M.; et al. Antioxidant Activity, Total Phenolic, and Resveratrol Content in Five Cultivars of Peanut Sprouts. *Asian Pac. J. Trop. Biomed.* **2017**, *7*, 332–338. [CrossRef]
53. Denisow, B.; Denisow-Pietrzyk, M. Biological and Therapeutic Properties of Bee Pollen: A Review. *J. Sci. Food Agric.* **2016**, *96*, 4303–4309. [CrossRef] [PubMed]
54. Leja, M.; Mareczek, A.; Wyżgolik, G.; Klepacz-Baniak, J.; Czekońska, K. Antioxidative Properties of Bee Pollen in Selected Plant Species. *Food Chem.* **2007**, *100*, 237–240. [CrossRef]
55. Mărghitaş, L.A.; Stanciu, O.G.; Dezmirean, D.S.; Bobiş, O.; Popescu, O.; Bogdanov, S.; Campos, M.G. In Vitro Antioxidant Capacity of Honeybee-Collected Pollen of Selected Floral Origin Harvested from Romania. *Food Chem.* **2009**, *115*, 878–883. [CrossRef]
56. Morais, M.; Moreira, L.; Feás, X.; Estevinho, L.M. Honeybee-Collected Pollen from Five Portuguese Natural Parks: Palynological Origin, Phenolic Content, Antioxidant Properties and Antimicrobial Activity. *Food Chem. Toxicol.* **2011**, *49*, 1096–1101. [CrossRef] [PubMed]

antioxidants

MDPI

Article

Phytochemical and Bioactivity Evaluation of Bee Pollen and Androecia of *Castanea*, *Salix*, and *Quercus* Species

Nisa Beril Sen [1], Irena Vovk [2,*,†], Hasan Kırmızıbekmez [1] and Etil Guzelmeric [1,*,†]

1 Department of Pharmacognosy, Faculty of Pharmacy, Yeditepe University, Kayisdagi Cad., Atasehir, 34755 Istanbul, Türkiye; nisaberil.sen@yeditepe.edu.tr (N.B.S.); hkirmizibekmez@yeditepe.edu.tr (H.K.)

2 Laboratory for Food Chemistry, National Institute of Chemistry, Hajdrihova 19, SI-1000 Ljubljana, Slovenia

* Correspondence: irena.vovk@ki.si (I.V.); etil.ariburnu@yeditepe.edu.tr (E.G.); Tel.: +386-1476-0341 (I.V.); +90-216-578-0558 (E.G.)

† These authors contributed equally to this work.

Abstract: Qualitative and quantitative differences in the chemical composition between bee pollen originated from *Castanea sativa* (Türkiye and Slovenia), *Salix* spp. (Türkiye and Slovenia), and *Quercus* spp. (Türkiye) and androecia of *Castanea sativa*, *Salix alba*, and *Quercus pubescens* (apetalous trees) were evaluated for the first time by new high-performance thin-layer chromatography (HPTLC) and ultra-performance liquid chromatography (UPLC) methods using marker compounds. N^1,N^5,N^{10}-tricaffeoylspermidine was isolated, and its structure was elucidated by nuclear magnetic resonance (NMR) and high-resolution mass spectrometry (HRMS). It was the main and the marker compound common to bee pollen (≈3–41 mg/g) and androecia (≈3–6 mg/g) samples. To the best of our knowledge, this is the first report of the identification of N^1,N^5,N^{10}-tricaffeoylspermidine in bee pollen originated from *Salix* spp. and androecia of *C. sativa*, *S. alba*, and *Q. pubescens*. The botanical origins of bee pollen were determined via phytochemical profiling using HPTLC-image analyses showing that bee pollen from the same botanical source had almost identical profiles regardless of collection location, geographical differences, and the bee race. *In vitro* tests and HPTLC-effect-directed analyses (EDAs) were performed to assess antioxidant and xanthine oxidase (XO) inhibitory activities of bee pollen, androecia, and N^1,N^5,N^{10}-tricaffeoylspermidine. HPTLC EDA combined with image analyses was used for comparing the activities of bee pollen, androecia, N^1,N^5,N^{10}-tricaffeoylspermidine, and also other marker compounds (quercetin, myricitrin, hyperoside, quercitrin, isoquercitrin, and rutin). The remarkable bioactivity of N^1,N^5,N^{10}-tricaffeoylspermidine was for the first time evaluated by HPTLC-EDA and *in vitro* tests. This is the first study performing HPTLC-XO inhibitory activity analyses on the HPTLC NH_2 F_{254S} plates. Further bioactivity studies on botanically and chemically well-characterized bee pollen samples are needed to aid in the use of bee pollen-containing supplements in the prevention and treatment of diseases.

Keywords: bee pollen; N^1,N^5,N^{10}-tricaffeoylspermidine; HPTLC; UPLC; HPTLC-effect-directed analyses; antioxidant activity; xanthine oxidase inhibitory activity; chemical profiling; image analysis

1. Introduction

A complete flower consists of four main parts: green-colored sepals, colorful petals, and male (androecium) and female (gynoecium) reproductive organs. The apetalous flower is a type of incomplete flower without petals that serves as a valuable food source for bees as bees directly come into contact with the anthers (a part of androecium) containing pollen sacs. When electrically charged bees come close to these flowers, their bodies are covered by pollen grains due to the electrostatic field. These pollen grains are converted into pellets by honeybees using their saliva, flower nectar, honey, etc. The final product is known as bee pollen. During this transformation, bee pollen grains may have morphologically and chemically different characteristic features from flower pollen [1]. The main bioactive compounds of bee pollen are phenolic acids (e.g., chlorogenic acid and *p*-coumaric acid), flavonoids (e.g., quercetin, kaempferol, and their glycoside derivatives), and phenolamides, which are formed by conjunction of hydroxycinnamic acid derivatives (e.g., caffeic acid, *p*-coumaric acid, and ferulic acid) with putrescine, spermidine, or spermine [2–6]. The presence and quantity of these compounds strongly depend on the botanical source of the bee pollen. It is, therefore, relevant to characterize the main chemical composition of bee pollen, which is linked to its therapeutic properties such as antioxidant, anti-inflammatory, anticariogenic, antibacterial, hepatoprotective, anti-atherosclerotic, and immune enhancing potential [2].

A fraction with a molecular weight less than 1000 obtained from an aqueous extract of *Cistus ladaniferus* bee pollen was found to be responsible for stimulating bone formation and inhibiting osteoclastic bone resorption [7]. The lack of knowledge of the chemical components responsible for the targeted bioactivity causes difficulties in the application of bee pollen supplements in phytotherapy. Standardization over bioactive component(s) is a key parameter for achieving reproducible pharmacological activity. Another challenge for the therapeutic use of food supplements containing bee pollen is that the botanical origin of bee pollen is not well defined. This is because palynological analysis applied to determine the botanical source of bee pollen using a microscope is sometimes insufficient for the identification of species with similar pollen grains; in this case, identification is accomplished either by giving the genus name (*Cistus* spp., *Papaver* spp., or *Salix* spp.) or the family name (Asteraceae, Brassicaceae, Boraginaceae, etc.). There are many species in a genus or a family; therefore, their chemical composition will be different. One study showed that the anti-inflammatory activity of an ethanolic extract of *Cistus* spp. originated bee pollen collected in Spain contained quercetin, kaempferol, and their derivatives [8]. In Spain, various *Cistus* spp. are grown such as *Cistus ladanifer* L., *Cistus albidus* L., *Cistus salviifolius* L., and *Cistus clusii* and their chemical composition may differ from each other [9]. Thus, there is a need for alternative methods that can be used for identifying the botanical source of bee pollen.

High-performance thin-layer chromatography (HPTLC) enables simultaneous analysis of bee pollen and its flower source under the same conditions, leading to the determination of the botanical origin of bee pollen. In a recent study, HPTLC profiling of flower and bee pollen of *Hedera helix* led to the determination of the botanical origin of bee pollen samples. Additionally, when hydroalcoholic bee pollen extracts were standardized using common markers such as quercetin-3-*O*-β-glucopyranosyl-(1→2)-β-galactopyranoside, platanoside, and afzelin found both in the *H. helix* flower and bee pollen, they exhibited antioxidant and xanthine oxidase (XO) inhibitory activities [10]. Further bioactivity studies on botanically and chemically well-characterized bee pollen samples are needed to aid in the use of bee pollen-containing supplements in the prevention and treatment of diseases.

Castanea sativa (sweet chestnut), *Salix* spp. (willow), and *Quercus* spp. (oak) are examples of trees having apetalous flowers. Their inflorescence, known as catkins, contains male flowers responsible for releasing pollen grains. In Türkiye and Slovenia, these trees grow all over the county, providing distinctive sources to honeybees for bee pollen production. Apparently, there are only a few phytochemical and bioactivity studies about *Castanea sativa* [11,12], *Salix* spp. [13–18], and *Quercus* spp. flower pollen [19]. The methanol extract of *C. sativa* flower pollen collected from nine populations showed high antibacterial activity, especially against *Micrococcus luteus*, methicillin resistant *Staphylococcus aureus*, and *Streptococcus aureus* [11]. Among those extracts, the one with the highest total phenolic content (TPC) value showed the highest 2,2-diphenyl-1-picrylhydrazyl (DPPH) radical scavenging activity [11]. Anti-urease activity was reported for aqueous and methanolic extracts of *C. sativa* blossoms in which rutin was found to be the dominant phenolic compound [12]. There are no bioactivity studies for *S. alba* flowers; although, some other *Salix* species displayed antioxidant, hepatoprotective, neuroprotective, antibacterial, and anticancer activities. *Salix* spp. flowers' extracts in methanol (*S. aegyptiaca*, *S. matsudana*, *S. babylonica*, *S. excelsa*, and *S. acmophylla*), ethanol (*S. caprea* and *S. aegyptiaca*), cyclohexane (*S. aegyptiaca*), butanol (*S. aegyptiaca*), and water (*S. aegyptiaca*) showed antioxidant activity by various assays such as DPPH, superoxide hydrogen peroxide, and nitric oxide scavenging assays [14,15]. The ethanol extract of *S. caprea* flowers significantly suppressed hepatic lipid peroxidation (LPO) which indicated its hepatoprotective effect [14]. The acetone-soluble fraction of *S. caprea* flowers was shown to be an effective antioxidant and chemopreventive agent against phorbol ester-induced tumor promotion *in vivo* [13]. The anti-inflammatory and antimicrobial activities of the methanolic extract of *S. tetrasperma* flowers, with dominant rutin, kaempferide 3-*O*-glucoside, trichocarposide, coumaroylquinic acid, and salicin were investigated, resulting in the extract significantly reducing the expression of several pro-inflammatory biomarkers in brainstem and sciatic nerve tissue, as well inhibiting *P. aeruginosa* PAO1 growth and biofilm formation [17,18]. In addition, the ethanol extract of *S. subserrata* flowers significantly reduced the elevated serum levels of intracellular liver enzymes as well as liver biomarkers in carbon tetrachloride (CCl_4)-induced liver damage *in vivo*, demonstrating its hepatoprotective effect [16]. There is only one phytochemical study on *Quercus* spp. flower pollen. Accordingly, hydroxycinnamic acid amides were isolated from *Quercus dentata* flower pollen, and their structures were elucidated as N^1,N^5,N^{10}-tricaffeoylspermidine, N^1-*p*-coumaroyl-N^5,N^{10}-dicaffeoylspermidine, N^1,N^{10}-di-*p*-coumaroyl-N^5-caffeoylspermidine, and N^1,N^5,N^{10}-tri-*p*-coumaroylspermidine. In view of the chemical composition and the health beneficial effects of the flowers of these species' bee pollen, which is botanically derived from these species, has recently been the focus of scientific studies.

Uric acid is formed by an enzyme called xanthine oxidase (XO) by catalyzing the oxidation of hypoxanthine to xanthine to uric acid. During this reaction, reactive oxygen species (ROS) are also produced. Excessive uric acid production, along with ROS, can trigger inflammation that leads to many diseases. Prolonged use of allopurinol, an XO inhibitor, may make individuals more prone to the development of kidney and liver diseases [20]. Therefore, research on compounds derived from natural sources as bee products is of interest to eliminate the risk factors. There are various *in vitro* studies evaluating the xanthine oxidase inhibitory activities of extracts from natural sources. The advantage of HPTLC-effect-directed analysis (HPTLC-EDA) in evaluating XO inhibitory activities of extracts is that, besides providing information on bioactivity, HPTLC-EDA also provides information on the active compound(s) separated on the chromatogram as demonstrated for quercetin-3-*O*-β-glucopyranosyl-(1→2)-β-galactopyranoside and afzelin from bee pollen originated from *H. helix* [10].

The aim of this study was to compare chemical and bioactivity profiles of extracts from *C. sativa*, *Salix* spp., and *Quercus* spp. bee pollen (samples from Türkiye and Slovenia) and androecia, containing anthers composed of pollen sacs of *C. sativa*, *S. alba*, and *Q. pubescens* which were considered to be possible bee pollen sources. The steps of this study were (1) HPTLC profiling of androecia and bee pollen samples; (2) isolation of the marker compound (N^1,N^5,N^{10}-tricaffeoylspermidine, found in all samples) by column chromatography and structural elucidation by NMR and HRMS; (3) development and validation of the UPLC method to quantify the main components in hydroalcoholic extracts of androecia and bee pollen; (4) evaluation of the antioxidant and XO enzyme inhibition activities of all extracts and N^1,N^5,N^{10}-tricaffeoylspermidine by spectroscopic methods; (5) evaluation of the antioxidant and XO enzyme inhibition activities of hydroalcoholic extracts from androecia and bee pollen, N^1,N^5,N^{10}-tricaffeoylspermidine, as well as other marker compounds for androecia and bee pollen by HPTLC-EDA combined with image analysis.

2. Materials and Methods

2.1. Chemicals

All chemicals used were at least of analytical grade. Methanol (HPLC grade and analytical grade), ethanol, acetonitrile (HPLC grade), ethyl acetate, glacial acetic acid, formic acid (98–100%), and dimethyl sulfoxide (DMSO) were acquired from Sigma-Aldrich (Steinheim, Germany). Xanthine oxidase (XO), 2,2′-azino-bis(3-ethylbenzothiazoline-6-sulfonic acid) diammonium salt (ABTS), 2,2-diphenyl-1-picrylhydrazyl (DPPH·), 2,4,6-tri(2-pyridyl)-s-triazine (TPTZ), sephadex LH-20, and *p*-anisaldehyde were acquired from Sigma-Aldrich (Steinheim, Germany). Sodium acetate, di-sodium hydrogen phosphate heptahydrate, *o*-phosphoric acid (85%), polyethylene glycol 400 (PEG 400), polyamide (for column chromatography), and copper (II) sulfate pentahydrate were from Merck (Darmstadt, Germany). Sodium phosphate monobasic dihydrate, sulfuric acid (95–97%), and hydrochloric acid were from Riedel-de Haen (Seelze, Germany). Ethylenediaminetetraacetic acid (EDTA), iron (III) chloride, and ammonium acetate were obtained from Fluka (Steinheim, Germany). Nitro blue tetrazolium chloride (NBT) was acquired from Cayman Chemical Company (Ann Arbor, MI, USA), and 2-aminoethyl diphenylborinate (NP) was from Alfa Aesar (Karlsruhe, Germany). Standards of rutin, isoquercitrin, hyperoside, myricitrin, quercitrin, quercetin, and trolox ((±)-6-hydroxy-2,5,7,8-tetramethylchromane-2-carboxylic acid) (97%) were purchased from Sigma-Aldrich (Steinheim, Germany). Allopurinol (98%) was acquired from Acros Organics (UK). Neocuproine was obtained from Sigma-Aldrich. Xanthine (99%) was obtained from Alfa Aesar (Kandel, Germany). Distilled water was used for extractions and all analyses except for UPLC analyses, for which ultrapure water was obtained by means of the Simplicity UV purification system (Millipore, Darmstadt, Germany).

2.2. Preparation of Standard Solutions

All standard solutions of rutin, isoquercitrin, hyperoside, myricitrin, quercitrin, quercetin, and N^1,N^5,N^{10}-tricaffeoylspermidine (isolated compound) were prepared in methanol. Standard stock solutions were prepared at concentrations of 200 μg/mL for HPTLC analyses and 350 μg/mL for UPLC analyses. A standard mixture (MIX) for HPTLC analyses was prepared by mixing equal volumes (100 μL) of all seven standard solutions (200 μg/mL). Stock solutions for UPLC analyses (350 μg/mL) were further diluted with methanol to prepare seven working solutions for each standard in a concentration range of 3.5–350 μg/mL. The seven working standard solutions were mixed in the order of the lowest to the highest concentration to prepare standard mixtures for the calibration curves (0.5–50 μg/mL).

2.3. Samples

Bee pollen samples were obtained from professional beekeepers who placed their beehives in various regions of Türkiye and Slovenia. Bee pollen samples were encoded after palynological analysis was applied according to the standard methodology [21]. Bee pollen samples were classified according to Barth [22] as dominant pollen (>45%), secondary pollen (15–45%), important minor pollen (3–15%), and minor pollen (<3%).

Codes, botanical sources with the proportion of dominant pollen grains, and the collection sites of the bee pollen samples are presented in Table 1.

Table 1. Botanical origins and locations of collected bee pollen samples.

Bee Pollen Samples	Sample Codes	Botanical Sources (Dominant Pollen (%))	Locations
C-bee pollen	C1	*Castanea sativa* (85.8%)	İzmit (Türkiye)
	C2	*Castanea sativa* (97.0%)	Artvin (Türkiye)
	C3	*Castanea sativa* (98.0%)	Ptuj (Slovenia)
S-bee pollen	S1	*Salix* spp. (52.8%)	Ordu (Türkiye)
	S2	*Salix* spp. (89.8%)	Giresun (Türkiye)
	S3	*Salix* spp. (45.1%)	Çanakkale (Türkiye)
	S4	*Salix* spp. (56.7%)	Begunje na Gorenjskem (Slovenia)
	S5	*Salix* spp. (62.1%)	Žirovnica (Slovenia)
	S6	*Salix* spp. (63.5%)	Kranj (Slovenia)
	S7	*Salix* spp. (87.8%)	Ljubljana (Slovenia)
Q-bee pollen	Q1	*Quercus* spp. (53.8%)	Ankara (Türkiye)

Bee pollen samples that originated from *Castanea sativa* were encoded as "C" and numbered from 1 to 3, samples originated from *Salix* spp. are encoded as "S" and numbered from 1 to 7, and one sample that originated from *Quercus* spp. is encoded as "Q1" (Table 1). The terminology used in this paper (C-bee pollen, S-bee pollen, and Q-bee pollen) is presented in Table 1.

The *Castanea sativa* Mill. (sweet chestnut), *Salix alba* L. (white willow), and *Quercus pubescens* Willd. (pubescent oak) samples used in this study were selected and collected according to the locations and plant diversity around the beehives. The male flowers (catkins) for *C. sativa* were collected from the forest in İzmit (Türkiye) (40°39′39.0″ N 30°08′01.2″ E) from the same collection site as the bee pollen coded as C1. *S. alba* catkins were collected in Çanakkale (Türkiye) (39°31′33.6″ N 26°16′45.7″ E), closely located to the bee pollen samples encoded as S3. Additionally, *Q. pubescens* is mainly grown in the forests surrounding Ankara [23], where the sample Q1 was obtained. As a representative sample, *Q. pubescens* catkins were collected from Çanakkale (Türkiye) (39°30′35.8″ N 26°17′00.2″ E). These samples were identified by Assoc. Prof. Dr. Etil Guzelmeric. The voucher specimens of *C. sativa* (YEF23009), *S. alba* (YEF24008), and *Q. pubescens* (YEF24009) were deposited at the Herbarium of the Faculty of Pharmacy, Yeditepe University, İstanbul, Türkiye. Finally, plant parts named androecia, containing anthers composed of pollen sacs, were separated from the catkins of *C. sativa*, *S. alba*, and *Q. pubescens*. All androecia samples and bee pollen samples were kept at −20 °C before being used for chemical composition and bioactivity studies.

2.4. Preparation of Sample Test Solutions

Bee pollen and androecia samples (5 g) were extracted with 80% $ethanol_{(aq)}$ (50 mL) in an ultrasonic bath (30 min). The solutions obtained were filtered, and the liquid part was evaporated under reduced pressure at 40 °C to obtain hydroalcoholic extracts. Hydroalcoholic extracts (250 mg) were dissolved in methanol (5 mL) using a sonicator and filtered through 0.45 µm hydrophilic-regenerated cellulose (RC) membrane filters (Minisart, Sartorius Stedim Biotech, Goettingen, Germany) to obtain stock sample test solutions (50 mg/mL). Stock sample test solutions diluted with methanol were used for HPTLC analyses (20 mg/mL) and bioactivity analyses (0.25–2.5 mg/mL for free radical-scavenging activity with 2,2-diphenyl-1-picrylhydrazyl (DPPH); 2,2′-azino-bis-3-ethylbenzothiazoline-6-sulfonic acid (ABTS) assays; ferric reducing antioxidant power (FRAP) assays; and cupric reducing antioxidant capacity (CUPRAC) assays; 0.025–1 mg/mL for xanthine oxidase (XO) inhibitory activity (superoxide radical scavenging activity) assay). Stock sample test solutions filtered through 0.2 μm GHP polypropylene membrane filters (Pall Corporation, Ann Arbor, MI, USA) and diluted with methanol (0.25–2.5 mg/mL) were used for UPLC analyses.

2.5. Isolation and Structure Elucidation of Marker Compounds from the Bee Pollen Sample

The bee pollen sample (C1; 50 g) was extracted with 80% ethanol$_{(aq)}$ (500 mL) for 30 min in an ultrasonic bath. After filtration, the solvent was evaporated under vacuum at 40 °C to yield hydroalcoholic extract (19.38 g), which was suspended in water (50 mL) and partitioned with ethyl acetate (50 mL × 3). The ethyl acetate fraction (1.99 g) was applied to polyamide column chromatography (CC) (12 g) and eluted with a water–methanol mixture (20–100% methanol) to give 8 main fractions (Fr. A–H; Figure 1). Fraction D (619 mg) was applied to the Sephadex LH-20 (95 g) CC with 600 mL of methanol to yield 8 fractions (Fr. D_{1-8}). The targeted compound was purified from Fr. D_8 (34 mg) by medium pressure liquid chromatography (MPLC, BÜCHI Labortechnik, Flawil, Switzerland) using a C_{18} MPLC column (30 g) and methanol–water (10–60% methanol) (Figure 1). The flow rate was 8 mL/min, and the run time was approximately 4 h. All isolation steps were performed at room temperature. The structure of the compound was elucidated by NMR (1D) and HRMS methods.

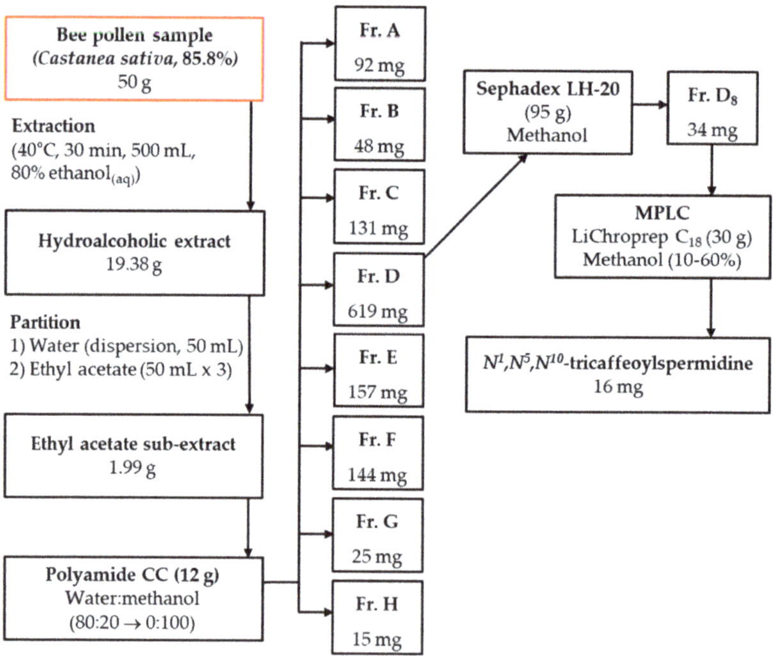

Figure 1. The isolation scheme of the targeted compound (N^1,N^5,N^{10}-tricaffeoylspermidine).

2.6. HPTLC Analyses

HPTLC analyses were performed on 20 cm × 10 cm glass-backed HPTLC silica gel 60 NH_2 F_{254S} plates (Merck, Art. No. 1.13192). Solutions were applied to the plate as 8 mm bands, 8 mm from the bottom of the plate and 15 mm from the left edge, with a semi-automatic applicator Linomat 5 (Camag, Muttenz, Switzerland) equipped with a 100 μL Hamilton syringe. Three different sets of solutions were applied to the plates:

(1) N^1,N^5,N^{10}-tricaffeoylspermidine (200 μg/mL, 2 μL) and stock sample test solutions of bee pollen (20 mg/mL, 5 μL); (2) standard solutions (200 μg/mL, 2 μL: rutin, isoquercitrin, hyperoside, myricitrin, quercitrin, quercetin, and N^1,N^5,N^{10}-tricaffeoylspermidine) and stock sample test solutions of androecia (20 mg/mL, 5 μL); (3) MIX (28.5 μg/mL, 8 μL) and stock sample test solutions of bee pollen and androecia samples (20 mg/mL, 5 μL). The plate was developed up to 8 cm in a saturated (20 min) twin-trough chamber (20 cm × 10 cm, Camag) with a developing solvent system containing ethyl acetate–formic acid–water (35:4:4, *v*/*v*/*v*).

The same derivatization procedure was used for analyses of all three sets of solutions. After drying the plate in a stream of cold air, the plate was heated on a TLC plate heater (Camag) at 100 °C for 3 min and immersed in an NP derivatization reagent, and after drying, it was also immersed in PEG reagent [24] by a Chromatogram Immersion Device III (Camag) for 3 s. An additional anisaldehyde–sulfuric acid derivatization reagent was used only for analyses of the 3rd set of solutions. In this case, the developed and dried plate was dipped for 2 s into anisaldehyde detection reagent by means of a Chromatogram Immersion Device III. Anisaldehyde (ANIS) detection reagent was prepared by mixing glacial acetic acid (20 mL) and methanol (170 mL). During cooling with cold water, 10 mL of sulfuric acid was added in a dropwise manner, and, subsequently, anisaldehyde (1 mL) was added to the mixture [25].

Documentation of the plate images was performed using a Visualiser (Camag) and the following illumination conditions: (1) 254 nm—before derivatization; (2) 366 nm—before derivatization and after derivatization with anisaldehyde or NP reagents, as well as after enhancement of the fluorescence with PEG; (3) white light (transmission mode)—after derivatization with anisaldehyde and after application of PEG reagent. The winCATS program was used to operate all the instruments (Camag, Version 128 1.4.8.2031).

2.7. UPLC Analysis

UPLC analysis was performed using the Nexera UHPLC series (Shimadzu, Kyoto, Japan), consisting of a quaternary pump, an autosampler, a thermostatted column compartment, and a photodiode array (PDA) detector. The UPLC system was operated by LabSolutions software (version 5.111). Analyses were carried out on a Poroshell 120 EC-C_{18} column (3 mm × 150 mm I.D., 2.7 μm particle size, Agilent (USA)). The column temperature was set to 25 °C. Mobile phase A [*o*-phosphoric acid–water (0.1:99.9, *v*/*v*)] and mobile phase B (acetonitrile) were degassed and filtered through 0.2 μm filter before analyses. The following gradient elution with different flow rates was applied: 17–17.2% B (0–2 min, 0.5 mL/min), 17.2–17.3% B (2–10 min, 0.5 mL/min), 17.3–20% B (10–11 min, 0.5 mL/min), 20–28% B (11–15 min, 0.5 mL/min), 28–40% B (15–25 min, 0.275 mL/min), 40–70% B (25–26 min, 0.2 mL/min), 70% B (26–27 min, 0.5 mL/min), and 70–17% B (27–28 min, 0.5 mL/min). The injection volume was 2 μL. Three acquisition wavelengths were used for quantitative analyses: 256 nm for rutin, hyperoside, isoquercitrin, quercitrin, and quercetin; 260 nm for myricitrin; 320 nm for N^1,N^5,N^{10}-tricaffeoylspermidine. This newly developed UPLC method was validated according to the International Conference on Harmonisation (ICH) 1995 guidelines [26] and was used for the quantification of rutin, hyperoside, isoquercitrin, quercitrin, quercetin, myricitrin, and N^1,N^5,N^{10}-tricaffeoylspermidine in androecia and bee pollen sample test solutions.

2.8. In Vitro Bioactivity Analyses

In vitro antioxidant activity (DPPH, FRAP, CUPRAC, and ABTS) and xanthine oxidase (XO) inhibitory (superoxide radical scavenging) activity were analyzed for bee pollen and androecia sample test solutions as well as for the isolated N^1,N^5,N^{10}-tricaffeoylspermidine.

The concentration ranges of bee pollen and androecia sample test solutions used for antioxidant activity assays and XO inhibitory activity assays were in the range 0.25–2.5 mg/mL and 0.025–1 mg/mL, respectively. The concentrations of N^1,N^5,N^{10}-tricaffeoylspermidine were 50 μg/mL for antioxidant activity assays and 10–400 μg/mL for XO inhibitory activity assays. The antioxidant activities obtained were presented as mg of trolox equivalents (TE) per g of hydroalcoholic extract (mg TE/g extract) for samples and mg TE/g for N^1,N^5,N^{10}-tricaffeoylspermidine. Half-maximal inhibitory concentration values (IC_{50}) were calculated for XO inhibitory activity. Inhibition values (%) were calculated based on absorbances (A) measured against blank solutions by using the following formula:

$$\text{Inhibition (\%)}: (1 - (A_{\text{Sample}} - A_{\text{Sample Blank}})/(A_{\text{Control}} - A_{\text{Control Blank}})) \times 100 \quad (1)$$

The blanks, controls (substrate and enzyme solutions), and control blank (substrate solution) are described in detail in the following subsections.

2.8.1. DPPH• Assay

After adding equal volumes (20 μL) of the diluted sample test solutions, trolox standard solutions (3.125–100 μg/mL), and methanol (as a blank) separately into wells of a 96-well microplate, DPPH solution (0.1 mM, 280 μL) was added, respectively, into each well. Incubation in the dark at room temperature for 30 min was followed by reading the absorbance at 530 nm [10].

2.8.2. FRAP Assay

Equal volumes (20 μL) of diluted sample test solutions, trolox standard solutions (3.125–100 μg/mL), and water (blank) were added to separate wells of a 96-well microplate. A freshly prepared FRAP solution (280 μL of a mixture of iron (III) chloride solution (2×10^{-2} M), TPTZ solution (1×10^{-2} M) and sodium acetate buffer (pH 3.6) solution, 1:1:10, *v/v/v*) was added into each well. The absorbance was read at 593 nm after 6 min [27].

2.8.3. CUPRAC Assay

Copper (II) sulfate pentahydrate (10 mM, 85 μL), neocuproine (7.5 mM, 85 μL), ammonium acetate buffer solution (85 μL, pH 7), and water (51 μL) were added into separate wells of a 96-well microplate. This was followed by the addition of either diluted sample test solutions (43 μL) or trolox standard solutions (3.125–200 μg/mL) or water (blank) into each well. After incubation at room temperature for 30 min, the absorbance was read at 450 nm [28].

2.8.4. ABTS Assay

As in our previous study [10], a slightly modified ABTS assay [29] was applied. Before adding ABTS reagent (280 μL) into each well of a 96-well microplate plate, equal volumes (20 μL) of the diluted sample test solutions, trolox standard solutions (6.25–100 μg/mL), or methanol (blank) were added. After incubation at room temperature for 6 min, the absorbance was read at 734 nm.

2.8.5. Xanthine Oxidase Inhibitory Activity

Equal volumes (20 μL) of diluted sample test solutions and allopurinol (10–200 μg/mL, used as a positive control) solution were placed into separate wells of a 96-well microplate. Addition of the substrate solution (a mixture of 0.4 mM xanthine and 0.24 mM NBT, 80 μL) was followed by the addition of XO solution (50 mU/mL in sodium phosphate buffer (pH 7.5), 80 μL). Hydrochloric acid (0.6 M, 80 μL) was used to terminate the reaction after

20 min of incubation at 37 °C. The absorbance was measured at 560 nm against a blank solution containing all reagents except XO [30].

2.9. HPTLC-Effect-Directed Analyses (EDAs)

2.9.1. HPTLC-DPPH•

HPTLC-DPPH• assays were performed on two HPTLC NH_2 F_{254S} plates. One of the following two sets of solutions was applied to either of the plates: (1) N^1,N^5,N^{10}-tricaffeoylspermidine (200 µg/mL, 2 µL) and sample test solutions of bee pollen (20 mg/mL, 5 µL); (2) standard solutions (200 µg/mL, 2 µL: rutin, isoquercitrin, hyperoside, myricitrin, quercitrin, quercetin, and N^1,N^5,N^{10}-tricaffeoylspermidine) and androecia sample test solutions (20 mg/mL, 5 µL). Other application conditions as well as developing and drying (after development) conditions were the same as those described in Section 2.6. HPTLC analyses. After development and drying, the plates were dipped into DPPH solution (0.1% methanolic solution) for 3 s using a Chromatogram Immersion Device III (Camag). After drying in the air in the dark for 30 min, the plates were documented immediately (t = 0 min) and after 5, 10, 15, 20, 30, 45, 60, 90, and 120 min under white light in transmission mode using a Visualiser (Camag). Between intervals, the plates were stored in the dark. Antioxidant compounds were detected as yellow-colored bands on the purple background.

2.9.2. HPTLC-Xanthine Oxidase Inhibitory Activity

HPTLC-XO inhibitory activity was performed on two silica gel NH_2 F_{254S} plates. One of the two sets of solutions was applied to separate plates: (1) N^1,N^5,N^{10}-tricaffeoylspermidine (1 mg/mL, 5 µL) and sample test solutions of bee pollen (50 mg/mL, 5 µL); (2) standard solutions (1 mg/mL, 5 µL: rutin, isoquercitrin, hyperoside, myricitrin, quercitrin, quercetin, and N^1,N^5,N^{10}-tricaffeoylspermidine) and androecia sample test solutions (50 mg/mL, 5 µL). Other application conditions, as well as developing and drying (after development) conditions, are described in Section 2.6. HPTLC analyses. After development, allopurinol (1 mg/mL, 15 µL), which was used as the positive control, was applied to the HPTLC plates. The plates were then dipped into a derivatization chamber containing a mixture of phosphate buffer solution (pH 7.6), EDTA (1 mM), NBT (1 mM), and XO (0.1 U/mL) using a Chromatogram Immersion Device III (Camag). After incubating at 37 °C for 30 min in the dark and drying in the air, the plates were then immersed into a phosphate buffer solution containing xanthine (1.5 mM) and incubated for 30 min at 37 °C in the dark. The plates were documented immediately (t = 0 min) and after 5, 10, 15, 20, 30, 45, 60, 90, and 120 min under white light in transmission mode using a Visualiser (Camag). Between intervals, the plates were stored in the dark. The compounds having XO inhibitory activity were then detected as white/yellow zones on a purple background [10].

2.10. Image Analyses

Images of the HPTLC plates documented as described in Section 2.6 HPTLC analyses and Section 2.9. HPTLC-Effect-Directed Analyses (EDAs) were converted to a different format using winCATS software (Camag, Version 1.4.9.2001) and transformed to video-densitograms in absorption mode or fluorescence mode using VideoScan TLC/HPTLC Evaluation Software (Version 1.02.00) (Camag). Absorption mode was applied only in the case of the image of the plate documented at white light after derivatization with anisaldehyde reagent. Fluorescence mode was applied in all other cases where the images of the plates were documented: (1) at 366 nm after derivatization with anisaldehyde reagent, as well as after enhancement of the fluorescence with PEG that followed derivatization with NP reagent; (2) at white light after HPTLC-DPPH analyses. Videodensitograms obtained after applying NP/PEG or anisaldehyde reagents were used to generate the profile comparison for comparing chromatographic fingerprint profiles of sample test solutions prepared from bee pollen and androecia samples. Additionally, videodensitograms obtained after HPTLC-DPPH analyses were also used for the integration and calculation of peak areas. The total peak areas obtained for each of the tracks were used for a comparison of the analyzed androecia and bee pollen samples and standard compounds.

2.11. Statistical Analyses

Assays for bioactivity and quantitative analyses were performed in triplicates. Microsoft Excel 2013 was used to calculate the average values of the replicates and standard deviations (SD), stated as the average value ± SD. One-way analysis of variance (ANOVA) with Tukey's test, performed by Minitab 17, was applied to evaluate statistically significant differences between the means of bioactivity test results ($p < 0.05$).

3. Results and Discussion

3.1. Isolation and Structural Elucidation of N^1,N^5,N^{10}-Tricaffeoylspermidine from Chestnut Bee Pollen

Polyamines (putrescine, spermidine, and spermine) and their derivatives conjugated to hydroxycinnamic acids were found in some bee pollen samples [2–6,31–33]. After isolation of N^1,N^5,N^{10}-tricaffeoylspermidine from chestnut bee pollen, its structure was elucidated according to the HRMS (Figure S1), ^{13}C, and ^{1}H NMR data (Table 2; Figures S2 and S3). In this study, N^1,N^5,N^{10}-tricaffeoylspermidine was found as a common compound in bee pollen and androecia samples, and it was found for the first time in bee pollen samples originating from *Salix* spp. In previous studies, N^1,N^5,N^{10}-tricaffeoylspermidine was identified in bee pollen originating from *Castanea sativa, Quercus mongolica, Phellodendron chinense, Fagopyrum esculentum, Prunus armeniaca, Camellia sinensis,* and *Cocos nucifera* [5,34,35]. N^1,N^5,N^{10}-tricaffeoylspermidine was found in male but not in female *Salix purpurea* catkins [36]. N^1,N^5,N^{10}-tricaffeoylspermidine was identified in the pollen of *Quercus* species such as *Q. rubra, Q. coccinea, Q. dentata, Q. cerris, Q. macrocarpa, Q. palustris, Q. velutina, Q. bicolor, Q. prinus, Q. robur,* and *Q. phellos* [19,37]. To the best of our knowledge, this is the first report of the identification of N^1,N^5,N^{10}-tricaffeoylspermidine in *C. sativa, S. alba,* and *Q. pubescens* androecia.

3.2. HPTLC Chemical Profiling

A new HPTLC method was developed and applied for evaluating chemical profiles of bee pollen samples (with the dominant pollen of *C. sativa, Salix* spp., and *Quercus* spp.) and from androecia (*C. sativa, S. alba,* and *Q. pubescens*) samples. The fingerprints were investigated on HPTLC NH_2 F_{254S} plates before (at 254 nm and at 366 nm) and after derivatization with NP detection reagent (at 366 nm), followed by enhancement and stabilization of fluorescence with PEG detection reagent (at 366 nm and at white light), and after derivatization with anisaldehyde reagent (at 366 nm and at white light). Use of the amino plates provided a better chromatographic separation, in particular of the spermidine derivatives. N^1,N^5,N^{10}-tricaffeoylspermidine was well separated from other spermidine derivatives on the HPTLC amino plate but not on the HPTLC silica gel plate [33].

3.2.1. HPTLC Chemical Profiling of Bee Pollen Samples

Differences in the chemical fingerprints of the bee pollen samples (20 mg/mL, 5 μL) of different botanical origin were first compared with N^1,N^5,N^{10}-tricaffeoylspermidine (200 μg/mL, 2 μL) after the separation on the HPTLC NH_2 F_{254S} plate (Figure 2). N^1,N^5,N^{10}-tricaffeoylspermidine (isolated in this study) at $R_F \approx 0.67$ was detected as a dark grey zone at 254 nm (Figure 2A, track 1) and a blue zone at 366 nm (Figure 2B, track 1) before derivatization. This R_F zone became much more intense after derivatization with NP reagent (light-blue fluorescent zone at 366 nm (Figure 2C, track 1)) and even more intense after the enhancement of fluorescence with PEG reagent (light-blue fluorescent zones at 366 nm (Figure 2D, track 1)) but was not visible at white light (Figure 2E, track 1). Zones at the same R_F with the same colors (Figure 2A–D, tracks 2–12), except at white light, detected in bee pollen samples indicated the presence of N^1,N^5,N^{10}-tricaffeoylspermidine. Among bee pollen samples, the most intensive zones of N^1,N^5,N^{10}-tricaffeoylspermidine were observed for all C-bee pollen (Figure 2, tracks 2–4).

Table 2. The structure and NMR data (^{13}C (100 MHz) and ^{1}H (400 MHz)) of N^1,N^5,N^{10}-tricaffeoylspermidine (DMSO).

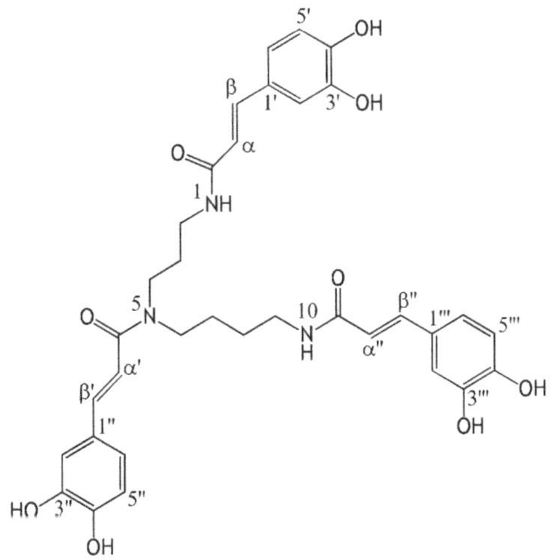

$C_{34}H_{37}N_3O_9$
Mol. Wt.: 631.7

Position	δ_C(ppm)	δ_H(ppm), J (Hz)
2	36.7	3.47 m
3	27.3	1.42–1.72 m
4	45.9	3.19 †
6	47.5	3.19 †
7	27.0 †	1.42–1.72 m
8	27.0 †	1.42–1.72 m
9	37.0	3.36 m
Tricaffeoyl (N^1,N^5,N^{10})		
$\alpha/\alpha'/\alpha''$	116.1	6.33 d, (15.9)
$\beta/\beta'/\beta''$	146.5	7.24 d, (15.9)
$1'/1''/1'''$	126.3	-
$2'/2''/2'''$	113.9	6.94 †
$3'/3''/3'''$	139.7	-
$4'/4''/4'''$	142.7	-
$5'/5''/5'''$	116.1	6.70–6.79 †
$6'/6''/6'''$	120.8	6.70–6.79 †
3-CO	165.9	-

† Overlapped signals.

After development, the same band pattern (R_F zones at ≈ 0.87, 0.85, 0.81, 0.76, 0.74, 0.67, 0.60, 0.53, 0.25, 0.22, 0.19, 0.14, and 0.06) was observed at 254 nm and 366 nm for all bee pollen samples (Figure 2A,B, tracks 2, 3, and 4) with *C. sativa* as the dominant pollen (from 85.8% to 98%) as confirmed by the palynological analyses (Table 1). After use of NP/PEG reagents, the same pattern with the most intense zones (R_F ≈ 0.60, 0.67, 0.74, and

0.81) was observed at 366 nm for all bee pollen samples from different geographical regions and even countries (Figure 2D, track 2—İzmit (Türkiye), track 3—Artvin (Türkiye), and track 4—Ptuj (Slovenia)). Additionally, before derivatization, blue zones ($R_F \approx 0.35$) were observed at 366 nm only for C-bee pollen (Figure 2B, tracks 2, 3, and 4).

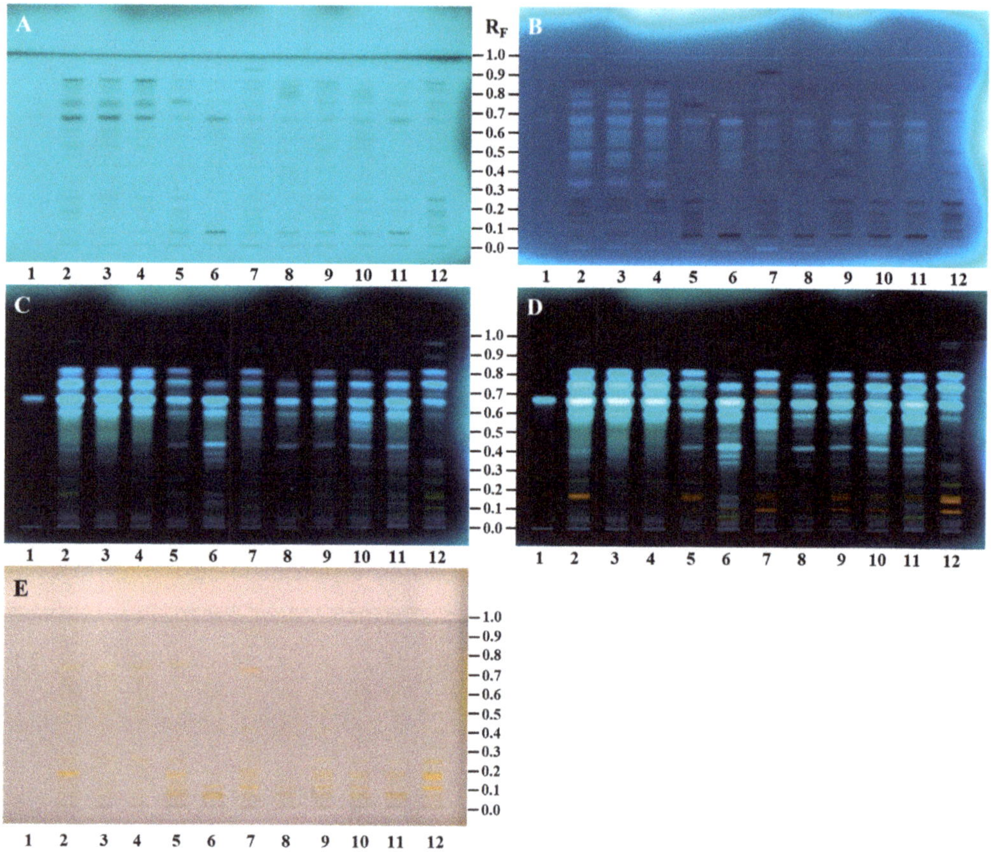

Figure 2. HPTLC NH_2 F_{254S} plate developed with ethyl acetate–formic acid–water (35:4:4, *v*/*v*/*v*), documented before (at 254 nm (**A**) and 366 nm (**B**)) and after derivatization with NP reagent (at 366 nm (**C**)), followed by enhancement and stabilization of the zones with PEG reagent (at 366 nm (**D**) and white light (**E**)). Applications per band: track 1: N^1,N^5,N^{10}-tricaffeoylspermidine (0.4 µg); track 2–4: C-bee pollen (C1–C3, 0.1 mg); track 5–11: S-bee pollen (S1–S7, 0.1 mg); track 12: Q-bee pollen (Q1, 0.1 mg).

In addition to the light-blue fluorescent zone of N^1,N^5,N^{10}-tricaffeoylspermidine ($R_F \approx 0.67$), the pattern of light-blue fluorescent zones was observed at $R_F \approx 0.60$, 0.74, and 0.81 in all bee pollen samples at 366 nm after derivatization with NP reagent (Figure 2C, tracks 2–12). However, the colors of the zones at $R_F \approx 0.81$ in the tracks of two *Salix* spp. samples (Figure 2C,D, tracks 6 and 8) were a shade darker compared to the colors of the zones at the same R_F in the tracks of the other nine samples (Figure 2C,D, tracks 2–5, 7, and 9–12). All these zones ($R_F \approx 0.60$, 0.74, and 0.81) became much more intense at 366 nm after the application of PEG reagent. These zones, together with the zone of N^1,N^5,N^{10}-tricaffeoylspermidine ($R_F \approx 0.67$), were found to be more intense in the C-bee pollen compared to other bee pollen samples (Figure 2D). Only C-bee pollen showed the pattern of three zones at R_F values $\approx$ 0.19, 0.22, and 0.25 at 254 nm (dark grey zones,

Figure 2A, tracks 2–4) and 366 nm (black zones, Figure 2B, tracks 2–4) after development and at 366 nm after using NP (green zones, Figure 2C, tracks 2–4) and PEG reagents (green zones, Figure 2D, tracks 2–4). At the same conditions (366 nm, NP/PEG reagents), the green zone at $R_F \approx 0.22$ was detected also in Q-bee pollen (Figure 2D, track 12) but not in S-bee pollen (Figure 2D, tracks 5–11). Additionally, the green zones at $R_F \approx 0.25$ (Figure 2C,D) were observed at 366 nm in all bee pollen samples except S2 (Figure 2C,D, track 6) and S4 (Figure 2C,D, track 8). Orange zones were detected at $R_F \approx 0.25$ with white light in S-bee pollen (Figure 2E, tracks 5–11) and Q-bee pollen (Figure 2E, track 12) but not in C-bee pollen (Figure 2E, tracks 2–4). Green zones at $R_F \approx 0.07$ were detected at 366 nm after application of PEG reagent in all tracks of *Salix* spp. (Figure 2D, tracks 5–11) and *C. sativa* (Figure 2D, tracks 2–4) but were not detected in *Quercus* spp. (Figure 2D, track 12). At the same R_F ($\approx$0.07), brown zones were observed at white light after application of PEG reagent in S-bee pollen (Figure 2E, tracks 5–11) but were not detected in C-bee pollen (Figure 2E, tracks 2–4) and Q-bee pollen (Figure 2E, track 12).

After derivatization with NP reagent, a pattern of three light-blue fluorescent zones at R_F values $\approx$ 0.35, 0.38, and 0.43 (Figure 2C, tracks 5–11) was observed at 366 nm only for S-bee pollen (Figure 2C, tracks 5–11). The intensity of these zones further increased after the application of PEG reagent (Figure 2D, tracks 5–11). Additionally, one green- ($R_F \approx 0.92$), one red- ($R_F \approx 0.72$), and two orange-colored ($R_F \approx 0.41$ and 0.31) zones were detected at 366 nm after using PEG reagent in only one (S3; Figure 2D, track 7) of the seven S-bee pollen samples (Figure 2D, tracks 5–11). These four zones were detected as orange with white light after application of PEG reagent (Figure 2E, track 7). Zones at $R_F \approx 0.11$, colored yellow after using NP reagent and orange after using PEG reagent, were detected at 366 nm in S-bee pollen (Figure 2D, tracks 5–11); *C. sativa* (Figure 2D, track 2); and *Quercus* spp. (Figure 2D, track 12). These zones were also orange at white light after the application of PEG reagent in all bee pollen samples from *Salix* spp. (Figure 2E, tracks 5–11) and *Quercus* spp. (Figure 2E, track 12) but not in C-bee pollen (Figure 2E, tracks 2–4). Zones at $R_F \approx 0.16$ and 0.19, colored yellow after derivatization with NP reagent and orange after application of PEG reagent, were detected at 366 nm in S-bee pollen (Figure 2C,D, tracks 5, 7, 9–11), C-bee pollen (Figure 2C,D, track 2), and Q-bee pollen (Figure 2C,D, track 12). These zones were also orange at white light after the application of PEG reagent (Figure 2E, tracks 2, 5, 7, and 9–12). After the application of PEG reagent, orange-brown zones ($R_F \approx 0.05$; Figure 2E, tracks 5–11)) and yellow zones ($R_F \approx 0.74$, 0.52; Figure 2E, tracks 2–4) were detected with white light in S-bee pollen and C-bee pollen, respectively.

In general, bee pollen samples originating from the same botanical source as *C. sativa* (Figure 2, tracks 2–4) and *Salix* spp. (Figure 2, tracks 5–11) showed similar HPTLC profiles. However, due to different percentages of the dominant botanical source pollen (85.8–98.0% for C-bee pollen and 45.1–89.8% for S-bee pollen), HPTLC profiles of bee pollen from the same botanical source showed different additional zones (Figure 2) related to pollen of other botanical sources present in the samples.

3.2.2. HPTLC Chemical Profiling of Androecia Samples

Differences in the chemical fingerprints of the androecia samples (20 mg/mL, 5 μL) of *C. sativa*, *S. alba*, and *Q. pubescens* were compared and evaluated with the seven compounds investigated (standards of rutin, hyperoside, isoquercitrin, myricitrin, quercitrin, quercetin, and N^1,N^5,N^{10}-tricaffeoylspermidine) (200 μg/mL, 2 μL) after the separation on the HPTLC NH_2 F_{254S} plate (Figure 3). In this way, the potential of these seven compounds as identification markers for distinguishing among the three floral sources was evaluated. After development, all seven compounds (rutin: $R_F \approx 0.19$, track 1; hyperoside: $R_F \approx 0.42$, track 2; isoquercitrin: $R_F \approx 0.43$, track 3; myricitrin: $R_F \approx 0.47$, track 4; quercitrin: $R_F \approx 0.60$, track 5; N^1,N^5,N^{10}-tricaffeoylspermidine: $R_F \approx 0.67$, track 6; quercetin: $R_F \approx 0.93$, track 7) were detected at 254 nm as grey zones (Figure 3A). Although the amounts (0.4 µg) of these compounds on the plate were the same, the zone of N^1,N^5,N^{10}-tricaffeoylspermidine was less intense (Figure 3A, track 6) than the zones of the six flavonoids (Figure 3A, tracks 1–5, 7). At 366 nm (Figure 3B), the zones for five of these compounds (rutin (track 1), hyperoside (track 2), isoquercitrin (track 3), myricitrin (track 4), and quercitrin (track 5)) were dark blue, while the zones of N^1,N^5,N^{10}-tricaffeoylspermidine (track 6) and quercetin (track 7) were light blue and light green, respectively. After derivatization with NP, six compounds (Figure 3C, tracks 1–5 and 7) were detected as yellow zones at 366 nm, while N^1,N^5,N^{10}-tricaffeoylspermidine (Figure 3C, track 6) was detected as a light-blue fluorescent zone. Similarly, after application of PEG reagent, six compounds (Figure 3D, tracks 1–5 and 7) were detected as orange zones at 366 nm and at white light, but N^1,N^5,N^{10}-tricaffeoylspermidine (Figure 3D, track 6) was detected as a light-blue fluorescent zone at 366 nm but was not detected with white light (Figure 3E, track 6).

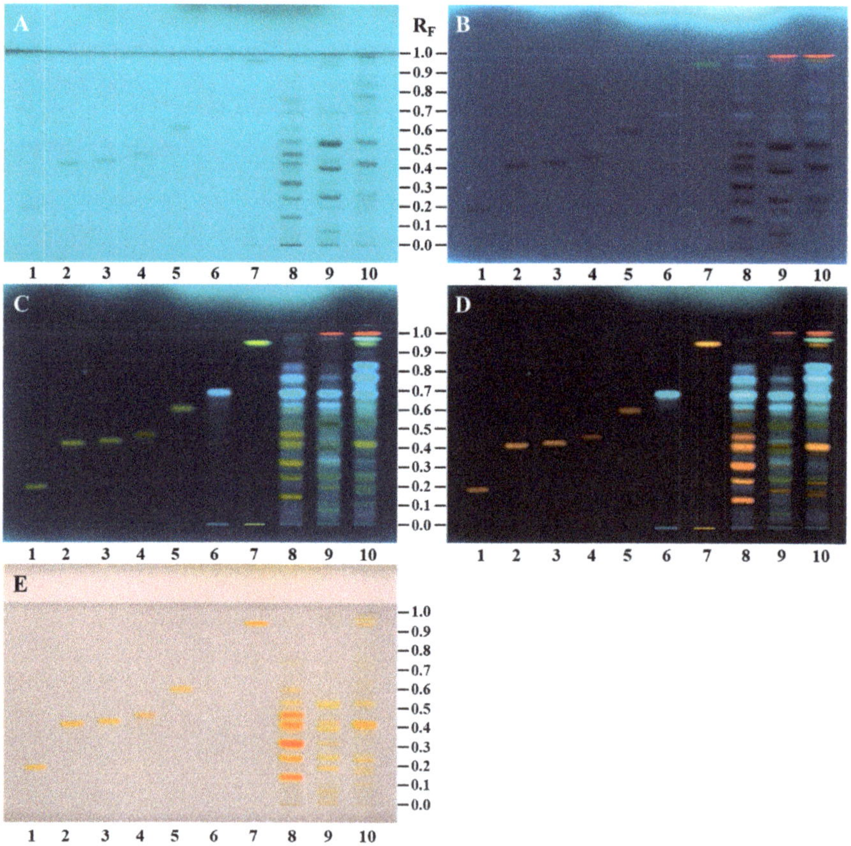

Figure 3. HPTLC NH_2 F_{254S} plate developed with ethyl acetate–formic acid–water (35:4:4, *v*/*v*/*v*), documented before (at 254 nm (**A**) and 366 nm (**B**)) and after derivatization with NP reagent (**C**), followed by enhancement and stabilization of the zones with PEG reagent (at 366 nm (**D**) and white light (**E**)). Applications per band: track 1: rutin (0.4 µg); track 2: hyperoside (0.4 µg); track 3: isoquercitrin (0.4 µg); track 4: myricitrin (0.4 µg); track 5: quercitrin (0.4 µg); track 6: N^1,N^5,N^{10}-tricaffeoylspermidine (0.4 µg); track 7: quercetin (0.4 µg); track 8: *C. sativa* androecia (0.1 mg); track 9: *S. alba* androecia (0.1 mg); track 10: *Q. pubescens* androecia (0.1 mg).

The presence of an N^1,N^5,N^{10}-tricaffeoylspermidine zone at $R_F \approx 0.67$ was observed as one of the most intense zones for all androecia samples (Figure 3, tracks 8–10). The zone at the $R_F \approx 0.19$ of rutin (Figure 3, track 1) was detected in androecia *S. alba* (Figure 3, track 9) and *Q. pubescens* (Figure 3, track 10) but not in *C. sativa*. The zone at the $R_F \approx 0.42$ of hyperoside (Figure 3, track 2) was detected in *C. sativa* (Figure 3, track 8) and *Q. pubescens* (Figure 3, track 10) androecia. The zone at the $R_F \approx 0.43$ of isoquercitrin (Figure 3, track 3) was detected in all androecia samples (*C. sativa* (Figure 3, track 8), *S. alba* (Figure 3, track 9), and *Q. pubescens* (Figure 3, track 10). Zones of myricitrin ($R_F \approx 0.47$, Figure 3, track 4) and quercitrin ($R_F \approx 0.60$) (Figure 3, track 5) were detected only in *C. sativa* androecia (Figure 3, track 8). The zone at $R_F \approx 0.93$ of quercetin (Figure 3, track 7) was detected only in *Q. pubescens* androecia (Figure 3, track 10). Differently colored zones were detected at the same R_F ($\approx$0.93) in *C. sativa* androecia (Figure 3, track 8) after development (Figure 3A: grey at 254 nm; Figure 3B: light blue at 366 nm) and after using NP reagent (Figure 3C: yellow at 366 nm) and PEG reagent (Figure 3D: orange at 366 nm). At the same R_F ($\approx$0.93) the zone was detected also in *S. alba* androecia at 254 nm (grey zone, (Figure 3A, track 9)) and 366 nm (red zone, (Figure 3B, track 9)) but only after development. The zones at $R_F \approx 0.86$ were detected in *C. sativa* and *Q. pubescens* androecia at 254 nm (grey, Figure 3A, tracks 8 and 10) and 366 nm (violet, Figure 3B, tracks 8 and 10) only after development. The zones at $R_F \approx 0.96$ were detected after development at 254 nm only in *Q. pubescens* androecia (grey zone (Figure 3A, track 10)) but at 366 nm in *C. sativa* (red zone (Figure 3B, track 8)), *S. alba* (red zone (Figure 3B, track 9)), and *Q. pubescens* androecia (yellow zone (Figure 3B, track 10)). After using NP reagent, the zone at the same R_F ($\approx$0.96) was detected at 366 nm in *C. sativa* (green-blue zone, Figure 3C, track 8), *S. alba* (orange zone, Figure 3C, track 9), and *Q. pubescens* androecia (green-blue zone, Figure 3C, track 10). The green-blue (after NP reagent (Figure 3C, tracks 8 and 10)) or blue (after PEG reagent (Figure 3D, tracks 8 and 10)) colored zones were detected at 366 nm in the *C. sativa* and *Q. pubescens* androecia. At 366 nm, the zone at the same R_F in the *S. alba* androecia (Figure 3C, track 9) was yellow-orange or yellow after using NP and PEG reagents (Figure 3D, track 9), respectively. The same R_F zone in *S. alba* and *Q. pubescens* androecia was yellow at white light after using PEG reagent. After development, zones at $R_F \approx 0.77$ were detected in *C. sativa* (light blue at 366 nm, (Figure 3B, track 8)) and *Q. pubescens* androecia (grey at 254 nm (Figure 3A, track 10) and light blue at 366 nm (Figure 3B, track 10)). Several other zones were also detected in *C. sativa*, *S. alba*, and *Q. pubescens* androecia after development (at 254 and 366 nm) as well as after using NP (at 366 nm) and PEG reagents (at 366 nm and at white light). Apart from the light-blue fluorescent zone of N^1,N^5,N^{10}-tricaffeoylspermidine ($R_F \approx 0.67$), three additional light-blue fluorescent zones at $R_F \approx 0.60$, 0.74, and 0.81 were detected at 366 nm after using NP and were common to all androecia samples (Figure 3C, tracks 8–10).

Additional blue fluorescent zones at $R_F \approx 0.43$ and 0.35 (Figure 3D, track 9) were observed for *Salix alba* androecia at 366 nm after using NP/PEG reagents. Dark blue zones at $R_F \approx 0.35$ and 0.21 (Figure 3C,D, track 8) were detected in *C. sativa* androecia at 366 nm after using NP reagent. These zones became more intense after using PEG reagent. The zones at $R_F \approx 0.25$ were detected in *S. alba* and *Q. pubescens* androecia (Figure 3, tracks 9 and 10) after development (at 254 nm—Figure 3A: dark grey and grey, respectively; dark blue at 366 nm Figure 3B) and after using NP (green at 366 nm—Figure 3C) and PEG reagents (yellow at white light—Figure 3E and at 366 nm—Figure 3D).

The zones at $R_F \approx 0.24$ were detected in *C. sativa* and *Q. pubescens* androecia (Figure 3, tracks 8 and 10) as dark grey at 254 nm (Figure 3A) and dark blue at 366 nm (Figure 3B) after development, as orange with white light (Figure 3E), and at 366 nm (Figure 3D) after using PEG reagent. The zone at $R_F \approx 0.16$ detected only in *Q. pubescens* androecia (Figure 3, track 10) was grey at 254 nm (Figure 3A) and 366 nm (Figure 3B) after development, yellow at 366 nm after using NP reagent (Figure 3C), and orange at 366 nm (Figure 3D) and with white light (Figure 3E) after using PEG reagent. Light-blue zones ($R_F \approx 0.15$ and 0.35) detected at 366 nm in *S. alba* androecia (Figure 3B, track 9) after development became more intense after using NP (Figure 3C) and PEG reagents (Figure 3D). The zone at $R_F \approx 0.15$ that was also observed in *C. sativa* androecia (Figure 3, track 8) was dark grey at 254 nm (Figure 3A) and dark blue at 366 nm (Figure 3B) after development and orange at 366 nm (Figure 3D) and at white light (Figure 3E) after using PEG reagent.

Zones at $R_F \approx 0.32$ were detected only in *C. sativa* and *S. alba* androecia (Figure 3, tracks 8 and 9) after development (grey at 254 nm (Figure 3A) and dark blue at 366 nm (Figure 3B)) and after derivatization with NP reagent (orange at 366 nm (Figure 3C)). The application of PEG reagent resulted in more intense orange zones at 366 nm (Figure 3D), which were detected as orange with white light for *C. sativa* (Figure 3E, track 8) or yellow for *S. alba* (Figure 3E, track 9) androecia. The zone at $R_F \approx 0.38$ of the track of *C. sativa* androecia (Figure 3, track 8) was detected only after development (grey zone at 254 nm (Figure 3A) and dark grey zone at 366 nm (Figure 3B)) and after using NP reagent (yellow zone at 366 nm (Figure 3C)), but it was not detected at 366 nm after using PEG reagent because of the interference of another highly intensive zone. An additional zone at $R_F \approx 0.37$ was observed for *Q. pubescens* androecia (Figure 3, track 10) at 366 nm (orange zone, Figure 3D) and at white light (yellow zone, Figure 3E) after using PEG reagent.

Zones at $R_F \approx 0.39$ and 0.52 were found in *S. alba* androecia (Figure 3, track 9) at 254 nm (dark grey, Figure 3A) and 366 nm (dark blue, Figure 3B) after development and after using NP (black-green at 366 nm Figure 3C) and PEG (green at 366 nm (Figure 3D) and yellow at white light Figure 3E) reagents. The zones at $R_F \approx 0.53$ were detected in *C. sativa* and *Q. pubescens* androecia after development (dark grey at 254 nm (Figure 3A, tracks 8 and 10) and dark blue at 366 nm (Figure 3B, tracks 8 and 10)) and after derivatization with NP reagent (yellow at 366 nm (Figure 3C, tracks 8 and 10)). After using PEG reagent, yellow zones ($R_F \approx 0.53$) were detected at white light for *C. sativa* and *Q. pubescens* androecia (Figure 3E, tracks 8 and 10) and at 366 nm for *Q. pubescens* androecia (Figure 3D, track 10), but orange zones were detected at 366 nm at the same R_F value for *C. sativa* androecia (Figure 3D, track 8). After derivatization with NP reagent, light-blue zones at $R_F \approx 0.58$ and 0.71 were detected at 366 nm in *S. alba* (Figure 3C, track 9) and *Q. pubescens* androecia (Figure 3C, track 10), and using PEG reagent resulted in enhancement of the light-blue fluorescence of these zones. Zones at $R_F \approx 0.11$ were detected in *S. alba* and *Q. pubescens* androecia (Figure 3, tracks 9 and 10) after development (grey at 254 nm (Figure 3A) and dark blue at 366 nm (Figure 3B)). After using NP reagent, the yellow zones ($R_F \approx 0.11$) were observed at 366 nm for *S. alba* and *Q. pubescens* androecia (Figure 3C, tracks 9 and 10). After using PEG reagent, this R_F zone was observed for *S. alba* and *Q. pubescens* androecia as yellow and orange at 366 nm (Figure 3D, track 9) and at white light (Figure 3E, track 10), respectively. The zone at $R_F \approx 0.07$ was observed only in androecia of *S. alba* (Figure 3, track 9) after development (dark grey at 254 nm (Figure 3A) and dark blue at 366 nm (Figure 3B)) and after using NP (yellow at 366 nm (Figure 3C)) and PEG reagents (green at 366 nm (Figure 3D) and brown orange at white light (Figure 3E)). This is the first study comparing the chemical composition of androecia belonging to apetalous trees including *C. sativa*, *S. alba*, and *Q. pubescens*.

3.2.3. Comparisons of HPTLC Fingerprints of Bee Pollen and Androecia of *C. sativa*, *S. alba*, and *Q. pubescens*

HPTLC is a powerful analytical technique enabling the detection of various compounds at the same time using different derivatization reagents. In general, NP/PEG reagents are used for detecting phenolic compounds whereas an anisaldehyde reagent is not so specific but it is a more universal reagent used for detecting terpenic compounds and phenolic compounds as catechins [25,38].

Bee pollen is a processed product of flower grains. Even if processing causes some differences in the morphological and chemical properties of the flower pollen, there is a strong relationship between bee pollen and its botanical source in terms of chemical composition. HPTLC fingerprints of bee pollen and *C. sativa*, *S. alba*, and *Q. pubescens* androecia were compared under 366 nm after application of NP/PEG and anisaldehyde (Figure 4).

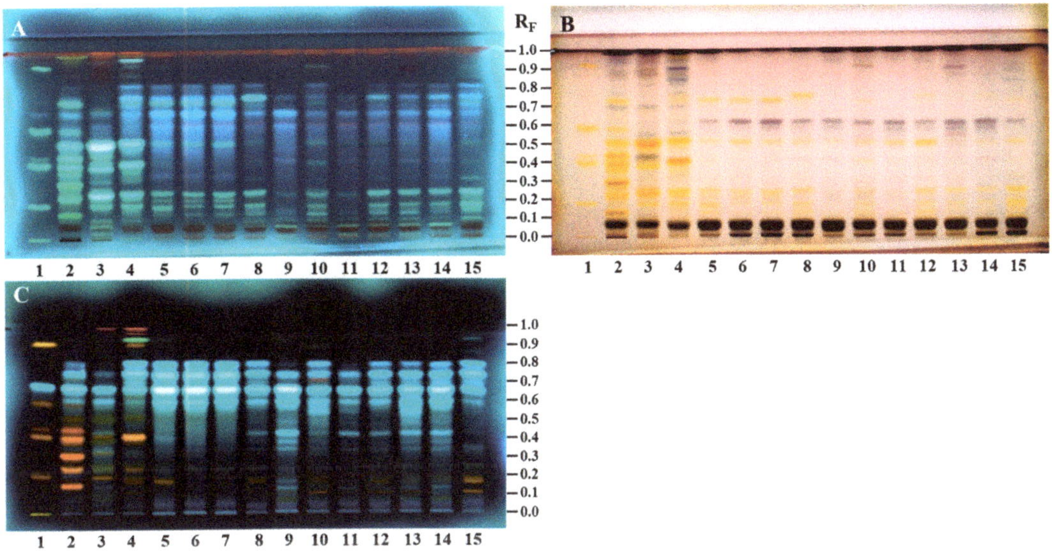

Figure 4. HPTLC NH_2 F_{254S} plate developed with ethyl acetate–formic acid–water (35:4:4, $v/v/v$), documented after derivatization with anisaldehyde/H_2SO_4 reagent (at 366 nm (**A**) and white light (**B**)) and after derivatization with NP/PEG reagents (at 366 nm (**C**)). Applications per band: track 1: Standard MIX; track 2: *C. sativa* androecia (0.1 mg); track 3: *S. alba* androecia (0.1 mg); track 4: *Q. pubescens* androecia (0.1 mg); tracks 5–7: C-bee pollen (C1–C3, 0.1 mg); tracks 8–14: S-bee pollen (S1–S7, 0.1 mg); track 15: Q-bee pollen (Q1, 0.1 mg).

All fingerprints were also analyzed under white light after post-chromatographic derivatization with an anisaldehyde reagent. Detailed information for the compounds found in bee pollen and androecia samples before and after using NP/PEG reagent is given in Section 3.2.1. HPTLC Chemical Profiling of Bee Pollen Samples and Section 3.2.2. HPTLC Chemical Profiling of Androecia Samples. This section focuses mainly on the chromatographic zones of compounds important for bee pollen of the same botanical origin and identification of their botanical sources. This is the first study to evaluate chemical composition differences between *C. sativa* (Türkiye and Slovenia), *Salix* spp. (Türkiye and Slovenia), and *Quercus* spp. (Türkiye) bee pollen samples and *C. sativa*, *S. alba*, and *Q. pubescens* androecia to determine the botanical origin of bee pollen.

HPTLC fingerprints of bee pollen and *C. sativa*, *S. alba*, and *Q. pubescens* androecia obtained after application of NP/PEG (at 366 nm) and anisaldehyde (ANIS, at 366 nm and white light) are given in Figure 4. Additionally, the common chromatographic zones found in the bee pollen samples and related botanical sources (Figure 4) are summarized in Table 3. For all bee pollen and androecia samples, three common light-blue zones (at $R_F \approx 0.60$, 0.67 (N^1,N^5,N^{10}-tricaffeoylspermidine), and 0.74) were detected at 366 nm after using NP/PEG reagents (Figure 4C, Table 3). Using anisaldehyde reagent, two additional common zones (at $R_F \approx 0.07$ black-brown at 366 nm (Figure 4A, Table 3) and brown under white light (Figure 4B, Table 3) and at $R_F \approx 0.61$ violet color zone only under white light (Figure 4B, Table 3)) were detected in all bee pollen and androecia samples. Apart from common characteristic zones for all bee pollen and androecia samples, anisaldehyde derivatization enabled the detection of compounds that could be used as markers for identification. Accordingly, the following two zones were found only Q-bee pollen and *Q. pubescens* androecia: green-blue zone at $R_F \approx 0.93$ (at 366 nm, Figure 4A, Table 3) and violet at $R_F \approx 0.91$ (under white light, Figure 4B, Table 3). The following three characteristic zones were detected in S-bee pollen and *S. alba* androecia using NP/PEG at 366 nm (Figure 4C, Table 3): green zone at $R_F \approx 0.07$ and light-blue zones at $R_F \approx 0.35$ and 0.43.

Table 3. Common chromatographic zones in C-bee pollen (C1–C3), S-bee pollen (S1–S7), Q-bee pollen (Q1), and androecia of *C. sativa* (C-a), *S. alba* (S-a), and *Q. pubescens* (Q-a).

		Samples													
Reagent	**R_F values/Color**	**C-a**	**C1**	**C2**	**C3**	**S-a**	**S1**	**S2**	**S3**	**S4**	**S5**	**S6**	**S7**	**Q-a**	**Q1**
NP/PEG (366 nm)	0.07/Green	-	-	-	-	+	+	+	+	+	+	+	+	-	-
	0.11/Orange	-	+	-	-	-	+	+	+	+	+	+	+	+	+
	0.16/Orange	-	+	-	-	-	+	-	+	-	+	+	+	+	+
	0.19/Orange	-	-	-	-	+	+	-	+	-	+	+	+	+	+
	0.22/Green	-	+	+	+	-	-	-	-	-	-	-	-	-	-
	0.25/Green	-	+	+	+	+	+	-	+	-	+	+	+	+	+
	0.35 /Light blue	-	-	-	-	+	+	+	+	+	+	+	+	-	-
	0.38/Light blue	-	-	-	-	-	+	+	+	+	+	+	+	-	-
	0.43/Light blue	-	-	-	-	+	+	+	+	+	+	+	+	-	-
	0.6/Light blue	+	+	+	+	+	+	+	+	+	+	+	+	+	+
	0.67/Light blue (N^1,N^5,N^{10}-tricaffeoylspermidine)	+	+	+	+	+	+	+	+	+	+	+	+	+	+
	0.70/Light blue	-	-	-	-	-	+	-	-	-	+	+	+	+	+
	0.74/Light blue	+	+	+	+	+	+	+	+	+	+	+	+	+	+
	0.81/Light blue	+	+	+	+	+	+	-	+	-	+	+	+	+	+
ANIS (white light)	0.07/Brown	+	+	+	+	+	+	+	+	+	+	+	+	+	+
	0.61/Violet	+	+	+	+	+	+	+	+	+	+	+	+	+	+
	0.72/Yellow	+	+	+	+	-	-	-	-	-	-	-	-	+	+
	0.77/Blue	+	+	+	+	-	-	-	-	-	-	-	-	+	+
	0.84/Blue	-	-	-	-	-	+	-	-	-	+	+	-	+	+
	0.89/Red-Brown	-	-	-	-	+	-	-	+	-	-	-	-	-	-
	0.91/Violet	-	-	-	-	-	-	-	-	-	-	-	-	+	+
ANIS (366 nm)	0.07/Black-Brown	+	+	+	+	+	+	+	+	+	+	+	+	+	+
	0.11/Green	-	-	-	-	+	+	+	+	+	+	+	+	+	+
	0.19/Green	-	+	+	+	+	+	-	+	-	+	+	+	+	+
	0.22/Green	+	+	+	+	-	-	-	-	-	-	-	-	+	+
	0.25/Green	-	+	+	+	+	+	-	+	-	+	+	+	+	+
	0.52/Green-Blue	+	+	+	+	+	+	-	+	-	+	+	-	+	+
	0.61/Blue	-	+	+	+	+	-	+	-	-	-	-	+	-	-
	0.63/Red	-	+	+	+	+	+	+	+	+	+	+	+	-	+
	0.67/Blue (N^1,N^5,N^{10}-tricaffeoylspermidine)	+	+	+	+	+	+	+	+	+	+	+	+	+	+
	0.74/Green	+	+	+	+	-	-	-	-	-	-	-	-	+	+
	0.76/Blue	+	+	+	+	-	+	+	+	+	+	+	+	+	+
	0.81/Blue	+	+	+	+	-	+	-	+	-	+	+	+	+	+
	0.93/Green-Blue	-	-	-	-	-	-	-	-	-	-	-	-	+	+

(+) Detected; (-) Not detected.

Comparisons were performed using HPTLC fingerprints based on chromatograms and videodensitogram profiles obtained from chromatograms by image analysis in fluorescence mode. Bee pollen samples from Türkiye and Slovenia having the same botanical origin with the highest percentage of dominant pollen grains (Table 1) were compared, and their profiles obtained at 366 nm after using NP/PEG were compared with *C. sativa*, *S. alba*, and *Q. pubescens* androecia profiles obtained under the same conditions (Figure 5). The profiles of C-bee pollen from Türkiye and Slovenia showed almost identical patterns (Figure 5A), and, additionally, they had similar profiles as their botanical source *C. sativa* (Figure 5D), especially at R_F values between ≈ 0.55 and 0.85. S-bee pollen from Türkiye and Slovenia showed almost identical HPTLC patterns (Figure 5B). The pattern of the three characteristic zones at $R_F \approx 0.3$ to 0.5 was only observed in S-bee pollen (Figure 5B) but not in C-bee pollen (Figure 5A) and Q-bee pollen (Figure 5C). HPTLC fingerprints of *S. alba* androecia and S-bee pollen (Figure 5E) were nearly identical. Q-bee pollen (Figure 5C) and *Q. pubescens* androecia (Figure 5F) had very similar HPTLC profiles.

Among the botanical sources studied (*C. sativa*, *S. alba*, and *Q. pubescens*), *C. sativa* is the only species in the *Castanea* genus found in both countries (Türkiye and Slovenia). In addition to *S. alba* and *Q. pubescens*, other species belonging to *Salix* spp. and *Quercus* spp. should be studied to obtain more information about the botanical sources of bee pollen.

HPTLC fingerprints based on chromatograms at 366 nm after using anisaldehyde (Figure 6A) and NP/PEG (Figure 6B) reagents and videodensitogram profiles obtained from them by image analysis in fluorescence mode enabled the distinction between bee pollen samples from different botanical sources. Figure 6A,B show a comparison of the profiles of C3 and S2 bee pollen samples having different botanical sources with the highest dominant pollen grains (Table 1) and Q1 bee pollen after using anisaldehyde and NP/PEG at 366 nm, respectively. It is evident that different derivatization reagents provide a wide range of information about zones for compounds which can be used as identification markers.

3.3. *UPLC Analyses*

A newly developed UPLC-PDA method was first validated using the parameters as follows: specificity, linearity, recovery, intraday and interday precision, limit of detection (LOD), and limit of quantification (LOQ). This method was used for the quantification of the compounds N^1,N^5,N^{10}-tricaffeoylspermidine, rutin, myricitrin, hyperoside, isoquercitrin, quercitrin, and quercetin in bee pollen and androecia samples.

3.3.1. UPLC Method Validation

Specificity

The compounds investigated were identified by comparing the retention times (t_R) and UV spectra of the compounds evaluated in bee pollen and androecia samples with the t_R and UV spectra of compounds in the standard mixture (Figure 7). Retention times of the compounds investigated were found as follows: rutin ($t_R \approx 6.60$); myricitrin ($t_R \approx 7.05$); hyperoside ($t_R \approx 7.56$); isoquercitrin ($t_R \approx 8.13$); quercitrin ($t_R \approx 13.85$); N^1,N^5,N^{10}-tricaffeoylspermidine ($t_R \approx 20.15$); and quercetin ($t_R \approx 24.01$) (Figure 7). Since specificity is the ability of the method to measure only the compound of interest without being influenced by other components that may be present in the sample, the evaluation of this parameter was based on whether or not peaks of the respective standards are detected in the blank chromatogram. As a result of this evaluation, chromatographic peaks belonging to N^1,N^5,N^{10}-tricaffeoylspermidine, rutin, myricitrin, hyperoside, isoquercitrin, quercitrin, and quercetin were not detected in the blank chromatogram, which confirmed the specificity of the method.

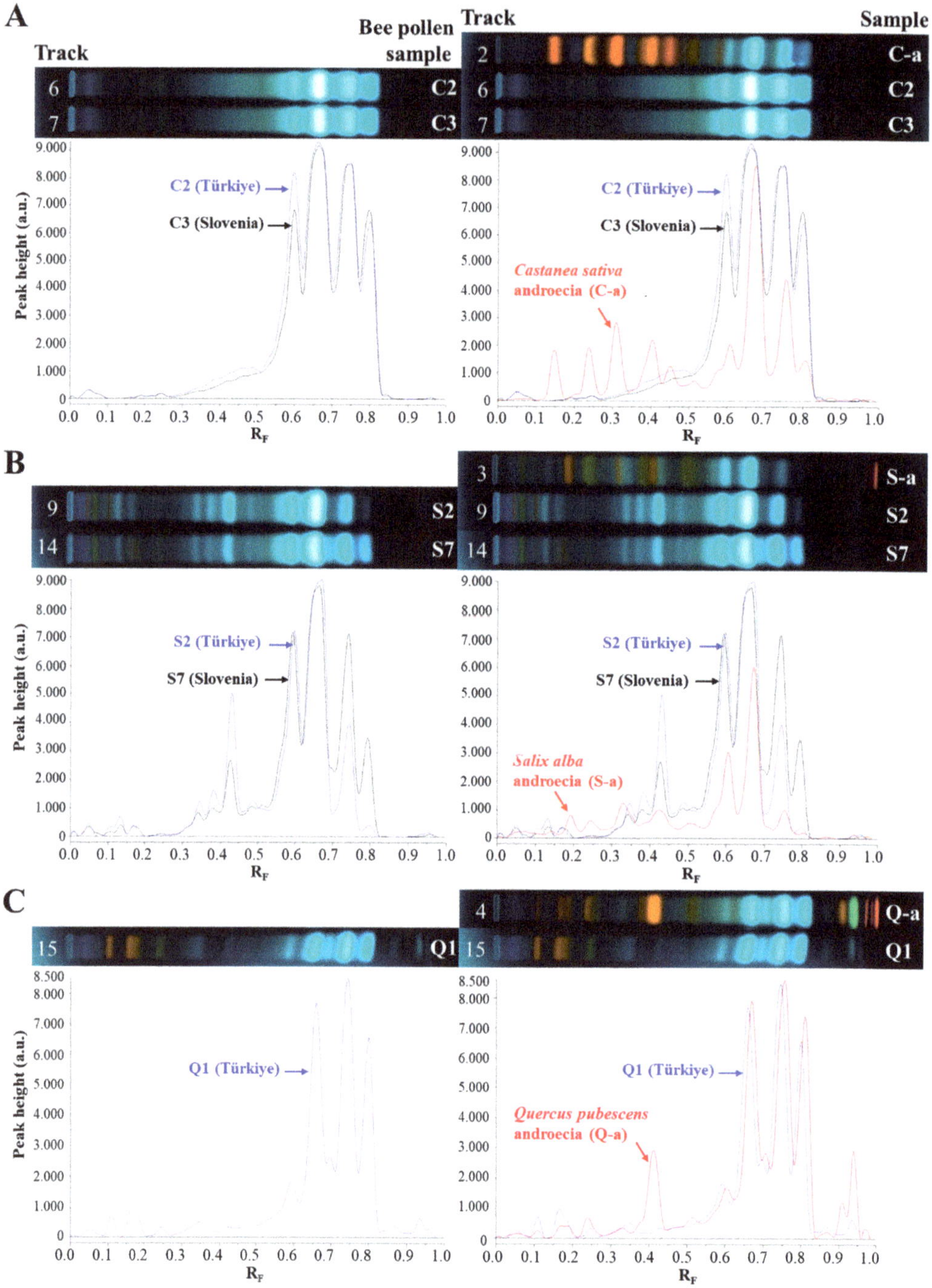

Figure 5. HPTLC fingerprints based on chromatograms at 366 nm after NP/PEG and videodensitogram profiles obtained from chromatograms (Figure 4) by image analysis in fluorescence mode. (**A**) C-bee pollen (C2 and C3), (**B**) S-bee pollen (S2 and S7), (**C**) Q-bee pollen (Q1), (**D**) C2 and C3 bee pollen and *C. sativa* androecia (C-a), (**E**) S2 and S7 bee pollen and *S. alba* androecia (S-a), (**F**) Q1 bee pollen and *Q. pubescens* androecia (Q-a). Application amounts of all samples per band: 0.1 mg.

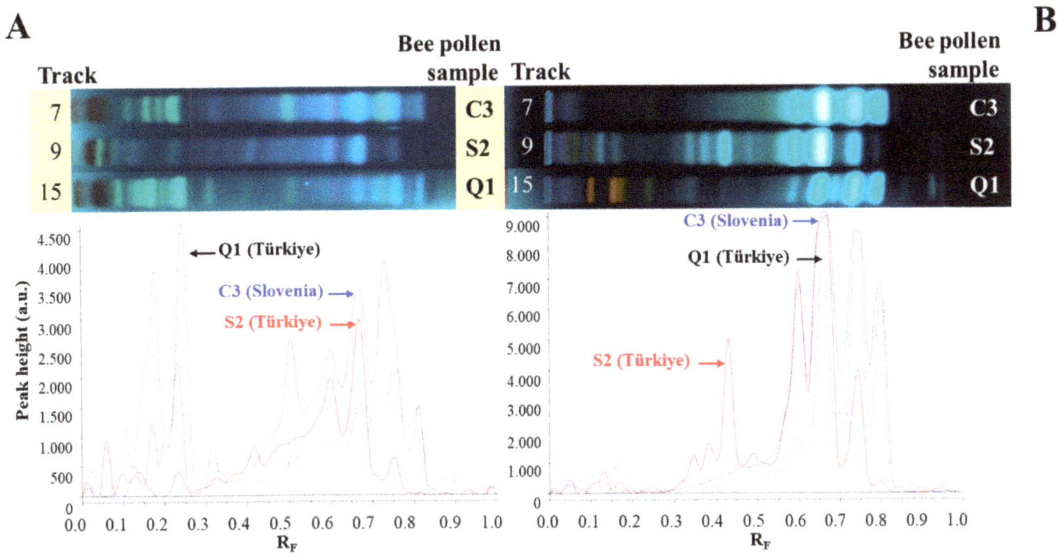

Figure 6. HPTLC fingerprints of C-bee pollen (C3), S-bee pollen (S2), and Q-bee pollen (Q1) based on chromatograms at 366 nm after using anisaldehyde (**A**) and NP/PEG (**B**) reagents and videodensitogram profiles obtained from chromatograms (Figure 4A,C) by image analysis in fluorescence mode. Application amounts of all samples per band: 0.1 mg.

Linearity of the Calibration Curve, Limit of Detection (LOD), and Limit of Quantification (LOQ)

Calibration curves of the investigated standards were obtained by plotting the average peak areas obtained after analyzing standard solutions with seven different concentrations ranging from 0.5 to 50 µg/mL in triplicate. The coefficient of determination values (r^2) for calibration curves of each of the standards were over 0.99, showing a good linear correlation (Table 4). LOD and LOQ values were determined using the equations 3 × (SD/S) and 10 × (SD/S), respectively (Table 4).

Table 4. Linearity, LOD, and LOQ data for the standards investigated.

Standards	Linearity Range (µg/mL)	r^2	S *	Intercept	SD **	LOD (µg/mL)	LOQ (µg/mL)
N[1],*N*[5],*N*[10]-tricaffeoylspermidine	0.5–50	0.9986	17,251	15,374.333	331.482	0.058	0.192
Rutin		0.9996	7,345	427.920	126.484	0.052	0.172
Myricitrin		0.9999	9,531	1,833.133	176.365	0.056	0.185
Hyperoside		0.9997	11,143	3258.767	460.255	0.124	0.413
Isoquercitrin		0.9989	9,942	825.153	342.088	0.103	0.344
Quercitrin		0.9985	16,156	9,244.433	666.800	0.124	0.413
Quercetin		0.9997	31,160	23,892.333	154.500	0.015	0.050

* S: slope; ** SD: standard deviation of intercept.

Precision

Intraday and interday precision is the degree of agreement between repeated measurements of the same sample with the developed method and were determined by analyzing 5 µg/mL of all standards at different times of the same day and on two different days, respectively. The results were expressed with relative standard deviation (RSD) values (Table 5). The RSD values were found in the range of 0.069–0.891 for intraday precision and 0.015–0.678 for interday precision. As recommended in [39], precision results should be ≤5%. The values obtained (Table 5) meet these criteria, indicating the sensitivity of the method.

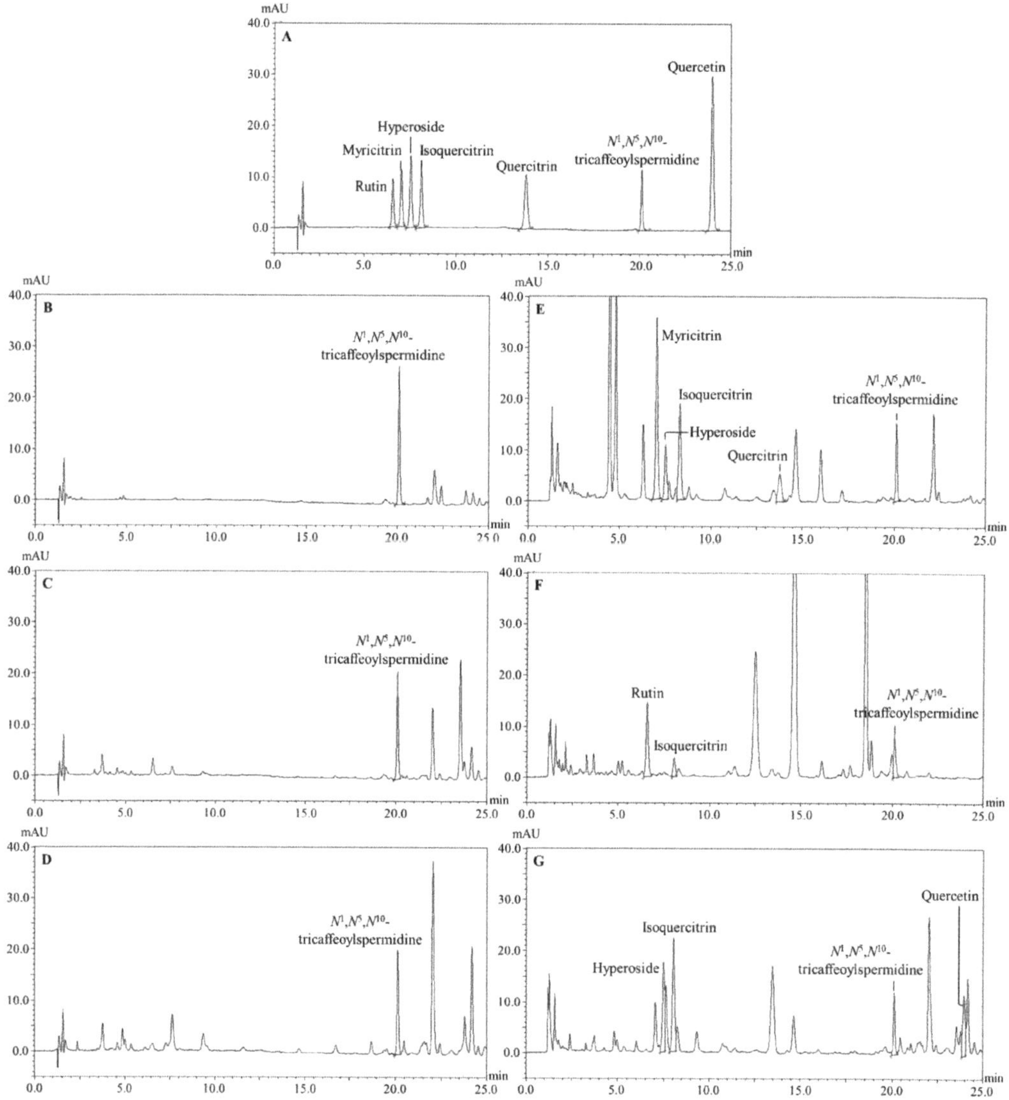

Figure 7. UPLC chromatograms of compounds investigated (**A**), C-bee pollen—C1 (**B**), S-bee pollen—S1 (**C**), Q-bee pollen—Q1 (**D**), *C. sativa* androecia (**E**), *S. alba* androecia (**F**), and *Q. pubescens* androecia (**G**) at 256 nm (**A**,**E**–**G**) and 320 nm (**B**–**D**).

Table 5. Intraday and interday precision data.

Standards (5 μg/mL)	Intraday Precision		Interday Precision	
	Average Concentration (μg/mL ± SD) (n = 3)	RSD (%) (n = 3)	Average Concentration (μg/mL ± SD) (n = 3)	RSD (%) (n = 3)
N^1,N^5,N^{10}-tricaffeoylspermidine	4.914 ± 0.023	0.470	4.906 ± 0.025	0.512
	4.862 ± 0.026	0.528	4.991 ± 0.012	0.239
	4.906 ± 0.025	0.512		
Rutin	5.030 ± 0.045	0.891	5.104 ± 0.026	0.512
	5.002 ± 0.041	0.825	5.091 ± 0.018	0.348
	4.983 ± 0.018	0.358		
Myricitrin	4.822 ± 0.015	0.318	4.845 ± 0.033	0.678
	4.855 ± 0.024	0.489	5.062 ± 0.001	0.015
	4.905 ± 0.034	0.684		
Hyperoside	4.799 ± 0.027	0.573	4.929 ± 0.030	0.616
	4.838 ± 0.028	0.585	5.025 ± 0.016	0.325
	4.823 ± 0.013	0.279		
Isoquercitrin	5.191 ± 0.032	0.622	5.198 ± 0.032	0.615
	5.211 ± 0.026	0.501	5.284 ± 0.015	0.281
	5.178 ± 0.045	0.877		
Quercitrin	4.823 ± 0.012	0.245	4.813 ± 0.021	0.427
	4.754 ± 0.029	0.614	4.877 ± 0.012	0.253
	4.834 ± 0.005	0.103		
Quercetin	4.895 ± 0.003	0.069	4.904 ± 0.025	0.515
	4.880 ± 0.021	0.429	4.895 ± 0.015	0.298
	4.928 ± 0.013	0.256		

Accuracy

Accuracy of the developed method was evaluated by the percentage recovery and RSD values of the found amounts of standards at three concentrations as 3, 6, and 12 μg/mL using the calibration curves. The results obtained were evaluated as recovery.

The results were found to be 93.9–98.4% for N^1,N^5,N^{10}-tricaffeoylspermidine, 98.4–103.4% for rutin, 95.5–99.0% for myricitrin, 96.8–102.9% for hyperoside, 98.2–101.2% for isoquercitrin, 102.0–105.4% for quercitrin, and 94.2–98.7% for quercetin (Table 6). As a consequence, the accuracy of the developed method was determined based on the recovery percentages and was found to be within a recommended acceptable range of 80–120% [40].

3.3.2. Quantitative Analyses

The UPLC method developed and validated in this study was used for the quantitation of the main compounds in both bee pollen and androecia samples. The results are hereby presented in Table 7 as mg of the compound investigated per hydroalcoholic bee pollen and androecia extracts (mg/g).

N^1,N^5,N^{10}-tricaffeoylspermidine was the main compound in bee pollen samples (from Türkiye and Slovenia; ranging between ≈3 mg/g and 41 mg/g) (Table 7). Among the bee pollen samples, the highest content of N^1,N^5,N^{10}-tricaffeoylspermidine was found in C-bee pollen (ranging between ≈26 mg/g and 41 mg/g) (Table 7). In addition to N^1,N^5,N^{10}-tricaffeoylspermidine, other spermidine derivatives such as N^1,N^5,N^{10}-tricoumaroylspermidine, N^1,N^{10}-dicoumaroyl-N^5-caffeoylspermidine, N^1-coumaroyl-

N^5,N^{10}-dicaffeoylspermidine, and N^1-feruloyl-N^5,N^{10}-dicaffeoylspermidine were found in bee pollen samples originating from *C. sativa* [33].

Table 6. Recovery of standards investigated.

Standards	Theoretical Value (μg/mL) (n = 3)	Amount Found (μg/mL ± SD) (n = 3)	Recovery (%) (n = 3)	RSD (%) (n = 3)
N^1,N^5,N^{10}-tricaffeoylspermidine	3	2.816 ± 0.005	93.9	0.189
	6	5.786 ± 0.029	96.4	0.494
	12	11.814 ± 0.012	98.4	0.099
Rutin	3	3.094 ± 0.014	103.1	0.447
	6	5.902 ± 0.028	98.4	0.478
	12	12.404 ± 0.022	103.4	0.176
Myricitrin	3	2.866 ± 0.003	95.5	0.114
	6	5.821 ± 0.008	97.0	0.141
	12	11.879 ± 0.041	99.0	0.348
Hyperoside	3	2.904 ± 0.013	96.8	0.464
	6	6.075 ± 0.021	101.2	0.347
	12	12.351 ± 0.008	102.9	0.063
Isoquercitrin	3	2.947 ± 0.019	98.2	0.648
	6	6.067 ± 0.032	101.1	0.535
	12	12.139 ± 0.045	101.2	0.373
Quercitrin	3	3.161 ± 0.017	105.4	0.546
	6	6.191 ± 0.040	103.2	0.653
	12	12.242 ± 0.009	102.0	0.077
Quercetin	3	2.962 ± 0.005	98.7	0.172
	6	5.655 ± 0.008	94.2	0.142
	12	11.618 ± 0.015	96.8	0.126

Table 7. Contents of compounds investigated in bee pollen and androecia samples.

Samples	N^1,N^5,N^{10}-tricaffeoylspermidine	Rutin	Myricitrin	Hyperoside	Isoquercitrin	Quercitrin	Quercetin
Bee Pollen			**mg/g ± SD (n = 3)**				
C1	40.96 ± 0.22 ^{a}	N.d.	N.d.	N.d.	N.d.	N.d.	N.d.
C2	27.96 ± 0.14 ^{b}	N.d.	N.d.	N.d.	N.d.	N.d.	N.d.
C3	25.57 ± 0.16 ^{c}	N.d.	N.d.	N.d.	N.d.	N.d.	N.d.
S1	6.54 ± 0.02 ^{g}	N.d.	N.d.	N.d.	N.d.	N.d.	N.d.
S2	21.24 ± 0.07 ^{d}	N.d.	N.d.	N.d.	N.d.	N.d.	N.d.
S3	4.86 ± 0.01 ^{k}	N.d.	N.d.	N.d.	N.d.	N.d.	N.d.
S4	5.54 ± 0.04 ^{i}	N.d.	N.d.	N.d.	N.d.	N.d.	N.d.
S5	6.20 ± 0.03 ^{h}	N.d.	N.d.	N.d.	N.d.	N.d.	N.d.
S6	8.56 ± 0.11 ^{f}	N.d.	N.d.	N.d.	N.d.	N.d.	N.d.
S7	14.17 ± 0.09 ^{e}	N.d.	N.d.	N.d.	N.d.	N.d.	N.d.
Q1	3.15 ± 0.01 ^{m}	0.95 ± 0.02 ^{a}	N.d.	N.d.	N.d.	N.d.	N.d.
Androecia			**mg/g ± SD (n = 3)**				
Castanea sativa	6.19 ± 0.05 ^{h}	N.d.	16.14 ± 0.08	4.14 ± 0.07 ^{b}	8.76 ± 0.13 ^{b}	2.55 ± 0.02	N.d.
Salix alba	3.48 ± 0.01 ^{l}	0.59 ± 0.01 ^{b}	N.d.	N.d.	1.64 ± 0.03 ^{c}	N.d.	N.d.
Quercus pubescens	5.19 ± 0.05 ^{j}	0.47 ± 0.06 ^{c}	N.d.	6.96 ± 0.06 ^{a}	10.79 ± 0.04 ^{a}	N.d.	2.46 ± 0.02

N.d.: Not detected. Different letters "a–m" in the same column indicate statistically significant differences ($p \leq 0.05$).

As in this study, N^1,N^5,N^{10}-tricaffeoylspermidine was reported as the dominant compound in ethanolic extracts of *C. sativa*-originated bee pollen samples from Italy [41] and the northwest Iberian Peninsula (Galicia and North of Portugal) [42]. Previous studies [41,42], as well as this study, indicate that the main compound(s) found in bee pollen are independent of the collection sites and periods (years) but dependent on the botanical origin of the bee pollen samples. Based on the results obtained in these studies, food supplements containing bee pollen originating from *C. sativa* could be standardized over N^1,N^5,N^{10}-tricaffeoylspermidine.

N^1,N^5,N^{10}-tricaffeoylspermidine was quantified in S-bee pollen (ranging from ≈5.5 to 21 mg/g) and Q-bee pollen (≈3 mg/g) (Table 7). Quantitative data obtained for N^1,N^5,N^{10}-tricaffeoylspermidine in S-bee pollen (Table 7) and the highest intensity of the zones ($R_F \approx 0.67$) in HPTLC chromatograms (Figure 2D, tracks 5–11) confirmed that N^1,N^5,N^{10}-tricaffeoylspermidine was the major compound, suggesting that S-bee pollen-containing dietary supplements could also be standardized via N^1,N^5,N^{10}-tricaffeoylspermidine. In the case of Q-bee pollen, more samples should be investigated, and other main compounds at R_F 0.74 and 0.81 should be identified (Figure 2D, track 11). Parallel to this study, spermidine derivatives were identified in bee pollen originating from *Q. mongolica* [34]. N^1,N^5,N^{10}-tricaffeoylspermidine was also found to be a common compound in the androecia of *C. sativa* (≈6 mg/g), *S. alba* (≈3.5 mg/g), and *Q. pubescens* (≈5 mg/g) (Table 7). To the best of our knowledge, this is the first study identifying and quantifying N^1,N^5,N^{10}-tricaffeoylspermidine in the androecia of *C. sativa*, *S. alba*, and *Q. pubescens*. N^1,N^5,N^{10}-tricaffeoylspermidine and other spermidine derivatives such as N^1-*p*-coumaroyl-N^5,N^{10}-dicaffeoylspermidine, N^1,N^{10}-di-*p*-coumaroyl-N^5-caffeoylspermidine, and N^1,N^5,N^{10}-tri-*p*-coumaroylspermidine were identified in the flower pollen of *Q. dentata*, but their quantity was not reported [19]. Rutin (quercetin-3-*O*-rutinoside) was found in the androecia of *S. alba* (0.6 mg/g) and *Q. pubescens* (0.5 mg/g) but not in *C. sativa* (Table 7). To the best of our knowledge, this is the first study identifying rutin in the androecia of *S. alba* and *Q. pubescens*. Rutin was also identified in *S. aegyptiaca* catkins [15]. In this study, rutin was found in C-bee pollen and S-bee pollen by HPTLC (Figure 2) but was quantified by UPLC only in Q-bee pollen (≈1 mg/g) (Table 7), possibly because different concentrations were used for HPTLC and UPLC analyses. In another study, rutin (≈0.18 mg/g) was detected in a methanolic extract of bee pollen originating from *C. sativa* [43]. Myricitrin (myricetin-3-*O*-rhamnoside) was quantified in the androecia of *C. sativa* (≈16 mg/g), but it was not detected in other androecia samples (Table 7). To the best of our knowledge, this is the first study identifying myricitrin in the androecia of *C. sativa*. Myricetin derivatives such as myricetin-3-*O*-glucuronide (≈0.36 mg/g) and myricetin-3-*O*-glucoside (≈0.38 mg/g) were determined in the hydroalcoholic extract of *C. sativa* catkins collected in Portugal, but myricitrin was not detected [44]. Myricetin-3-*O*-glucuronide (*judia* cultivar ≈ 1 mg/g, *longal* cultivar ≈ 0.1 mg/g) and myricetin-3-*O*-glucoside (*judia* cultivar ≈ 0.6 mg/g, *longal* cultivar ≈ 0.1 mg/g) were found in extracts of lyophilized *C. sativa* flowers (both cultivars) prepared with infusion and decoction techniques [45]. Hyperoside (quercetin-3-*O*-galactoside) was found in the androecia of *C. sativa* (≈4 mg/g) and *Q. pubescens* (≈7 mg/g) but was not detected in the androecia of *S. alba* (Table 7). To the best of our knowledge, this is the first study identifying hyperoside in the androecia of *C. sativa* and *Q. pubescens*. Isoquercitrin was found to be a common component in the analyzed androecia samples, with the highest content in *Q. pubescens* (≈11 mg/g) and the lowest content in *S. alba* (≈2 mg/g) (Table 7). To the best of our knowledge, this is the first study identifying isoquercitrin in the androecia of *Q. pubescens* and *S. alba*. Isoquercitrin content (≈9 mg/g) in the androecia of *C. sativa* (Table 7) was much higher than the contents reported for the samples collected in Portugal: ≈0.9 mg/g in the hydroalcoholic extract of *C. sativa* catkins [44]; ≈2 mg/g in extracts prepared with a decoction technique from lyophilized *C. sativa* flowers (*judia* cultivar) [45]. Isoquercitrin was not detected in bee pollen samples (Table 7). However, it was found in the ethyl acetate sub-extract of bee pollen originating from *S. alba* (≈0.7 mg/g) [46]. Quercitrin (quercetin-3-*O*-rhamnoside) was only found in *C. sativa* androecia (≈3 mg/g) (Table 7), and it was reported also in a hydroalcoholic extract of *C. sativa* catkins from Portugal but in a much lower amount (0.1 mg/g) [44]. Quercetin was only quantified in the androecia of *Q. pubescens* (≈2.5 mg/g) but was not detected in other samples (Table 7). To the best of our knowledge, this is the first study identifying and quantifying quercetin in *Q. pubescens* androecia. Although quercetin was detected in *Quercus* spp. bee pollen by HPTLC (Figure 2), it was not quantified by UPLC (Table 7), which could be a consequence of different concentrations used for HPTLC and UPLC analyses.

The variations in chemical composition could be because of growth conditions such as frequency of the rain fall, soil pH, and naturally grown and cultivar trees. In addition, the plant parts used in all these studies such as the androecia and flowers (male or female) may differ in chemical composition. Some studies [44,45] did not specify the type of flowers (male or female), therefore, the terminology from these studies (catkins [15,45], flowers [44]) was used throughout this discussion.

3.4. *In Vitro Bioactivity Analyses*

3.4.1. Antioxidant Activity Determined by DPPH, FRAP, ABTS, and CUPRAC Assays

The antioxidant activities of bee pollen, androecia samples, and N^1,N^5,N^{10}-tricaffeoylspermidine were evaluated using DPPH, FRAP, ABTS, and CUPRAC assays. N^1,N^5,N^{10}-tricaffeoylspermidine showed the highest antioxidant activity. This is the first report to evaluate the antioxidant activity of N^1,N^5,N^{10}-tricaffeoylspermidine. Among all the bee pollen samples, the samples with high content of N^1,N^5,N^{10}-tricaffeoylspermidine (Table 7) as C-bee pollen containing 85–98% of *C. sativa* flower pollen (Table 1) showed higher antioxidant activity than the S-bee pollen and Q-bee pollen (Table 8). When the bee pollen samples from *Salix* spp. were compared to each other, S2 and S7 (containing 89.8% and 87.8% *Salix* spp. flower pollen, respectively (Table 1)), having higher N^1,N^5,N^{10}-tricaffeoylspermidine content (Table 7), exerted higher antioxidant activity than the other samples (Table 8). Q-bee pollen with 53.8% of *Quercus* spp. flower pollen (Table 1) and the lowest N^1,N^5,N^{10}-tricaffeoylspermidine content (Table 7) showed the lowest antioxidant activity (Table 8).

All antioxidant assays (DPPH, FRAP, CUPRAC, and ABTS) proved N^1,N^5,N^{10}-tricaffeoylspermidine (Table 8) to be a more potent antioxidant than phenolic acids (chlorogenic acid and 3,5-dicaffeoylquinic acid) and flavonoids (afzelin, platanoside, and quercetin-3-*O*-β-glucopyranosyl-(1→2)-β-galactopyranoside) [10]. Similar findings about phenolamines (di-*p*-coumaroyl spermidine, di-*p*-coumaroyl hydroxyferuloyl spermine, tri-*p*-coumaroyl spermidine, etc.) having higher antioxidant activity than flavonoids (quercetin and kaempferol derivatives) were also reported in another study [4]. In addition, it was reported that phenolamides (di-*p*-coumaroylspermidine, tri-*p*-coumaroylspermidine, and tetra-*p*-coumaroylspermine) had a positive correlation with the antioxidant activity of rape bee pollen [32].

As reported in the literature, bee pollen originating from *C. sativa*, *Salix* spp., and *Quercus* spp. were extracted using different extraction procedures, and the antioxidant activity of the extracts was evaluated using different methods. Other studies of bee pollen samples originating from *C. sativa* reported values of 17.08–38.56 mg TE/g by ABTS assay [47] and 20.24 mg TE/g by FRAP assay [48]. For bee pollen originating from *Salix* spp., values of 18.66 mg TE/g, 75.64 mg TE/g, and 9.92 TE/g were reported, respectively, by DPPH, CUPRAC, and FRAP assays [46].

Table 8. *In vitro* antioxidant activity of N^1,N^5,N^{10}-tricaffeoylspermidine, bee pollen, and androecia samples determined by DPPH, FRAP, ABTS, and CUPRAC assays.

Standard and Samples	DPPH	FRAP	ABTS	CUPRAC
		mg TE/g (n = 3)		
N^1,N^5,N^{10}-tricaffeoylspermidine	966.28 ± 28.72 ^{a}	768.97 ± 11.44 ^{a}	1047.65 ± 35.65 ^{a}	610.35 ± 15.69 ^{a}
		mg TE/g extract (n = 3)		
Bee Pollen				
C1	42.04 ± 0.79 ^{cde}	28.34 ± 1.33 ^{d}	141.94 ± 1.08 ^{cd}	47.94 ± 2.27 ^{cd}
C2	46.37 ± 0.55 ^{cd}	38.75 ± 0.66 ^{c}	154.65 ± 2.06 ^{c}	45.09 ± 1.52 ^{d}
C3	41.76 ± 0.60 ^{cde}	18.40 ± 0.28 ^{e}	125.79 ± 1.15 ^{d}	41.54 ± 1.28 ^{d}
S1	18.09 ± 0.68 ^{f}	13.81 ± 0.12 ^{ef}	69.39 ± 0.63 ^{ef}	24.08 ± 0.33 ^{e}
S2	23.28 ± 0.37 ^{ef}	14.90 ± 0.59 ^{e}	77.33 ± 0.56 ^{e}	24.67 ± 1.00 ^{e}
S3	16.44 ± 0.35 ^{f}	12.05 ± 0.35 ^{ef}	63.63 ± 1.27 ^{ef}	25.46 ± 0.65 ^{e}
S4	8.88 ± 0.60 ^{f}	5.35 ± 0.12 ^{f}	47.66 ± 1.56 ^{f}	21.33 ± 0.80 ^{e}
S5	12.46 ± 0.28 ^{f}	10.01 ± 0.88 ^{ef}	60.46 ± 0.34 ^{ef}	23.50 ± 0.73 ^{e}
S6	16.69 ± 0.38 ^{f}	11.76 ± 0.34 ^{ef}	66.25 ± 0.95 ^{ef}	24.10 ± 0.22 ^{e}
S7	24.59 ± 0.62 ^{def}	12.18 ± 0.60 ^{ef}	71.79 ± 0.58 ^{ef}	26.03 ± 1.14 ^{e}
Q1	16.41 ± 1.03 ^{f}	10.37 ± 0.21 ^{ef}	63.53 ± 0.80 ^{ef}	27.05 ± 0.38 ^{e}
Androecia				
Castanea sativa	165.98 ± 3.14 ^{b}	111.92 ± 1.52 ^{b}	309.08 ± 3.91 ^{b}	177.41 ± 2.54 ^{b}
Salix alba	25.13 ± 0.20 ^{def}	27.79 ± 0.75 ^{d}	120.98 ± 1.01 ^{d}	47.99 ± 1.21 ^{cd}
Quercus pubescens	51.56 ± 1.21 ^{c}	39.22 ± 0.67 ^{c}	147.05 ± 1.14 ^{cd}	60.47 ± 1.58 ^{c}

Different letters "a–f" in the same column indicate statistically significant differences ($p \leq 0.05$).

In *C. sativa*, *S. alba*, and *Q. pubescens* androecia samples, not only N^1,N^5,N^{10}-tricaffeoylspermidine but also other compounds may contribute to the antioxidant activity. Among the androecia samples, *C. sativa* had the highest antioxidant activity (Table 8).

3.4.2. Xanthine Oxidase Inhibitory Activity

The XO inhibitory activity was evaluated in N^1,N^5,N^{10}-tricaffeoylspermidine, bee pollen, and androecia samples using allopurinol as a positive control (IC_{50} = 3.07 µg/mL). A significant XO inhibitory activity was obtained for N^1,N^5,N^{10}-tricaffeoylspermidine (IC_{50} = 2.98 µg/mL) (Table 9). This is the first study evaluating the XO inhibitory activity of N^1,N^5,N^{10}-tricaffeoylspermidine. Among bee pollen samples, C-bee pollen had the highest XO inhibitory activity (with IC_{50} values ranging from 26.18 to 31.52 µg/mL (Table 9)) that could be related to N^1,N^5,N^{10}-tricaffeoylspermidine contents, which were higher than in other bee pollen samples (Table 7). Similarly, among five bee pollen fractions, the fraction with coumaroylspermidines showed higher XO inhibitory activity, suggesting that spermidines can be considered as the components responsible for this activity [49]. Among androecia samples, the highest XO inhibitory activity was observed for *C. sativa* (IC_{50} = 15.97 µg/mL), the next highest was for *Q. pubescens* (IC_{50} values 20.83 µg/mL), and the lowest was for *S. alba* (IC_{50} = 23.74 µg/mL) (Table 9). In addition to N^1,N^5,N^{10}-tricaffeoylspermidine, other phenolic compounds may contribute to the XO inhibitory activity.

3.5. HPTLC-Effect-Directed Analyses

3.5.1. HPTLC-DPPH Analyses of Antioxidant Activity

In addition to the *in vitro* antioxidant analyses using spectrophotometry, HPTLC-DPPH analyses of the investigated compounds, bee pollen, and androecia samples were performed to obtain additional information on the antioxidant activity of compounds separated on the HPTLC plate. Figure 8 shows the HPTLC plates with yellow-colored chromatographic zones which indicate compounds that have antioxidant activity; the images of the plates were captured under white light after 30 min. Zones at the R_F of N^1,N^5,N^{10}-

tricaffeoylspermidine were more intense in C-bee pollen than in other bee pollen samples (Figure 8A). C-bee pollen had higher contents of N^1,N^5,N^{10}-tricaffeoylspermidine than S-bee pollen and Q-bee pollen (Table 6). C-bee pollen had higher antioxidant potential than S-bee pollen and Q-bee pollen (Figure 9A). These results were found to be in parallel with *in vitro* assays (Table 7). As illustrated in Figure 9A, not only N^1,N^5,N^{10}-tricaffeoylspermidine but apparently also other compounds separated on the chromatogram contribute to the antioxidant effect. The total peak areas of the yellow zones showed that C-bee pollen had the highest antioxidant activity (Figure 9A).

Table 9. XO inhibitory activity of allopurinol, N^1,N^5,N^{10}-tricaffeoylspermidine, bee pollen, and androecia samples.

Standards and Samples	XO Inhibitory Activity
	IC$_{50}$ (μg/mL ± SD) (n = 3)
Allopurinol (Positive Control)	3.07 ± 0.05 ^{a}
N^1,N^5,N^{10}-tricaffeoylspermidine	2.98 ± 0.19 ^{a}
Bee Pollen	
C1	26.18 ± 0.99 ^{de}
C2	29.30 ± 1.94 ^{ef}
C3	31.52 ± 3.48 ^{fg}
S1	54.29 ± 0.70 ^{j}
S2	34.36 ± 1.72 ^{g}
S3	63.77 ± 2.39 ^{kl}
S4	62.79 ± 0.91 ^{k}
S5	49.60 ± 2.19 ^{ij}
S6	45.05 ± 0.41 ^{hi}
S7	40.78 ± 1.80 ^{h}
Q1	67.75 ± 2.44 ^{l}
Androecia	
Castanea sativa	15.97 ± 0.21 ^{b}
Salix alba	23.74 ± 0.93 ^{cd}
Quercus pubescens	20.83 ± 0.68 ^{bc}

Different letters "a–l" in the same column indicate statistically significant differences ($p \leq 0.05$).

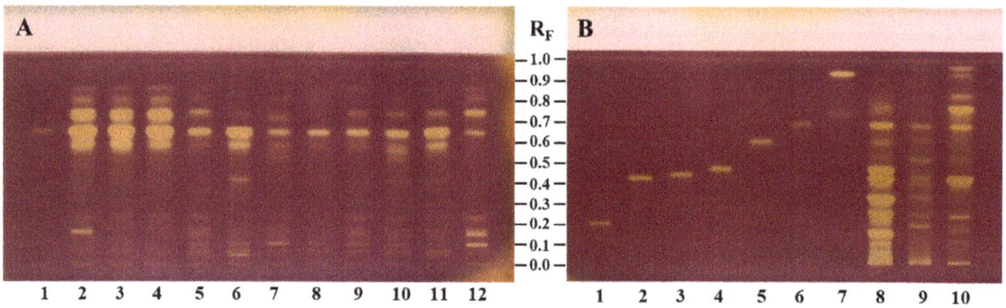

Figure 8. HPTLC silica gel NH$_2$ F$_{254S}$ plate developed with ethyl acetate–formic acid–water (35:4:4, $v/v/v$), documented at white light after detection with DPPH reagent. Applications per band: (**A**) track 1: N^1,N^5,N^{10}-tricaffeoylspermidine (0.4 μg); track 2–4: C-bee pollen (C1–C3, 0.1 mg); track 5–11: S-bee pollen (S1–S7, 0.1 mg); track 12: Q-bee pollen (Q1, 0.1 mg). (**B**) track 1: rutin (0.4 μg); track 2: hyperoside (0.4 μg); track 3: isoquercitrin (0.4 μg); track 4: myricitrin (0.4 μg); track 5: quercitrin (0.4 μg); track 6: N^1,N^5,N^{10}-tricaffeoylspermidine (0.4 μg); track 7: quercetin (0.4 μg); track 8: *C. sativa* androecia (0.1 mg); track 9: *S. alba* androecia (0.1 mg); track 10: *Q. pubescens* androecia (0.1 mg).

Evaluation of N^1,N^5,N^{10}-tricaffeoylspermidine and six other compounds (rutin, hyperoside, isoquercitrin, myricitrin, quercitrin, and quercetin, all having the same 0.4 μg application amounts) demonstrated that the intensity of the yellow zones belonging to flavonoids was higher than N^1,N^5,N^{10}-tricaffeoylspermidine (Figure 8B).

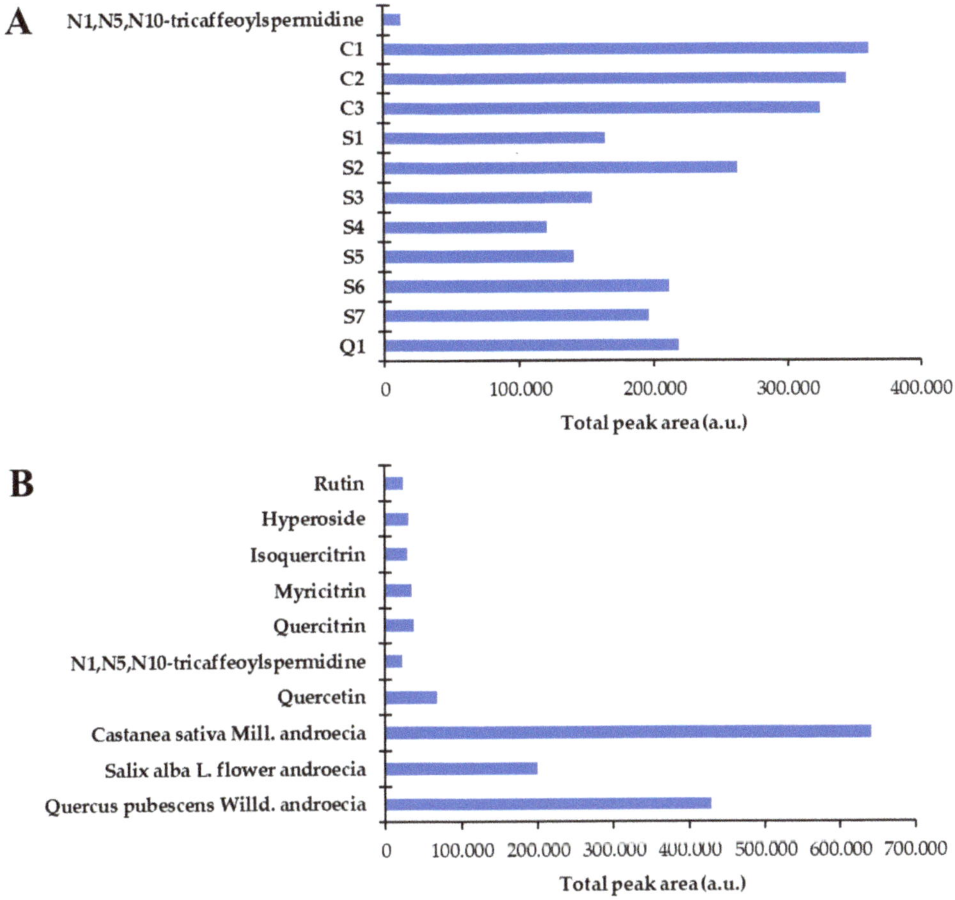

Figure 9. Comparison of total peak areas of all yellow bands detected in the videodensitograms obtained by image analysis of HPTLC-DPPH chromatograms (Figure 8A,B) in fluorescent mode. (**A**) Total peak areas of N^1,N^5,N^{10}-tricaffeoylspermidine and bee pollen samples. (**B**) Total peak areas of standards investigated and androecia samples.

In androecia samples (Figure 8B, tracks 8–10), several yellow zones were observed showing the antioxidant activity contributions of the compounds investigated as well as several other yet unidentified compounds. In particular, these intense yellow zones were observed from the application position at $R_F \approx 0.5$ in *C. sativa* androecia, showing a high contribution to the antioxidant activity (Figure 8B, track 8). Based on the total peak areas of the yellow zones, *C. sativa* androecia had the highest antioxidant activity (Figure 9B).

3.5.2. HPTLC-XO Inhibitory Activity

To supplement the results of the spectrophotometric XO inhibitory activity assay of the whole hydroalcoholic extract of each of the bee pollen and androecia samples and N^1,N^5,N^{10}-tricaffeoylspermidine (Table 9), the XO inhibitory activity of the samples and standards was analyzed by an HPTLC-XO assay providing more information about the XO inhibitors separated on the HPTLC plate. The XO inhibitors were detected as white/yellow zones on a purple background (Figures 10 and 11). This is the first study performing HPTLC-XO inhibitory activity analysis on the HPTLC NH_2 F_{254S} plates (Figures 10 and 11).

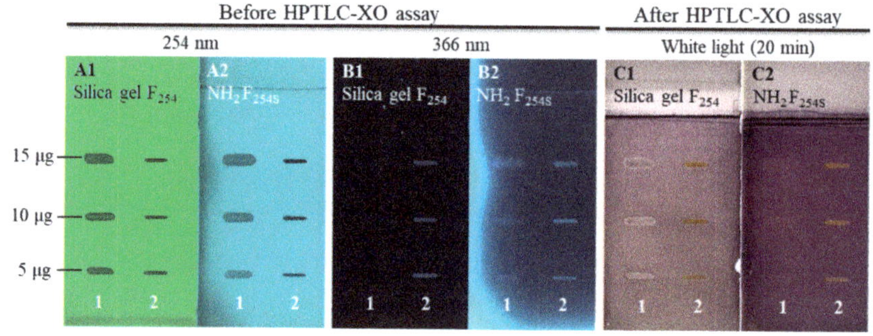

Figure 10. Comparison of HPTLC-XO assay on HPTLC silica gel F_{254} and HPTLC NH_2 F_{254S} plates using allopurinol (positive control, track 1) and N^1,N^5,N^{10}-tricaffeoylspermidine (track 2) applied on the plates after development with ethyl acetate–formic acid–water (35:4:4, *v*/*v*/*v*). Plates were documented before HPTLC-XO assay at 254 nm (**A1**,**A2**) and 366 nm (**B1**,**B2**) and under white light (20 min) after the HPTLC-XO assay (**C1**,**C2**).

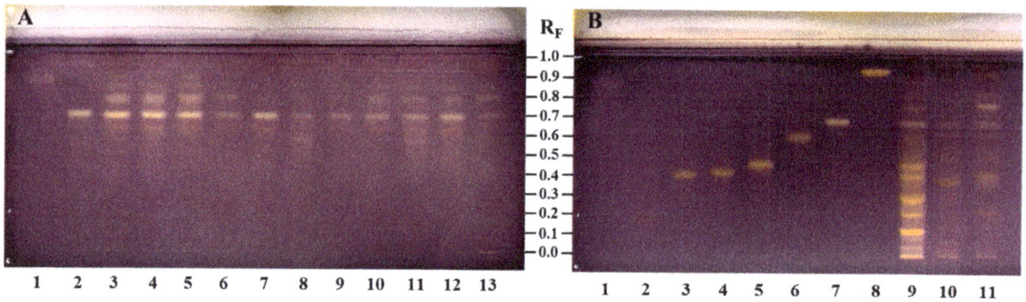

Figure 11. HPTLC NH_2 F_{254S} plate developed with ethyl acetate–formic acid–water (35:4:4, *v*/*v*/*v*), documented under white light after HPTLC-XO assay. Applications per band: (**A**) track 1: allopurinol (15 µg); track 2: N^1,N^5,N^{10}-tricaffeoylspermidine (5 µg); tracks 3–5: C-bee pollen (C1–C3, 0.25 mg); tracks 6–12: S-bee pollen (S1–S7, 0.25 mg); track 13: Q-bee pollen (Q1, 0.25 mg). (**B**) track 1: allopurinol (15 µg); track 2: rutin (5 µg); track 3: hyperoside (5 µg); track 4: isoquercitrin (5 µg); track 5: myricitrin (5 µg); track 6: quercitrin (5 µg); track 7: N^1,N^5,N^{10}-tricaffeoylspermidine (5 µg); track 8: quercetin (5 µg); track 9: *C. sativa* androecia (0.25 mg); track 10: *S. alba* androecia (0.25 mg); track 11: *Q. pubescens* androecia (0.25 mg).

To understand the applicability of the HPTLC-XO inhibitory activity on HPTLC NH_2 F_{254S} plates, a comparative analysis was performed with HPTLC silica gel F_{254} plates. Both types of plates were first developed with ethyl acetate–formic acid–water (35:4:4, *v*/*v*/*v*). After development, the plates were dried in a stream of cold air, followed by separate applications of 5 µg, 10 µg, and 15 µg of allopurinol (positive control) (Figure 10, track 1) and N^1,N^5,N^{10}-tricaffeoylspermidine (Figure 10, track 2), followed by documentation of the images of the plates at 254 nm (Figure 10A1,A2) and 366 nm (Figure 10B1,B2). The HPTLC-XO inhibitory activity assay was then performed as described in the Section 2.9.2. HPTLC-xanthine oxidase inhibitory activity and plates were documented immediately (t = 0 min) and after 5, 10, 15, 20, 30, 45, 60, 90, and 120 min under white light (Figure S4). Between intervals, the plates were stored in the dark. Figure 10C1,C2 show the plates under white light (20 min) after HPTLC-XO assay.

Several differences were observed between HPTLC-XO inhibitory activity assays on silica and amino plates during the 120 min time interval. It is evident that the stationary phase influences the detectability of both compounds on the background which becomes darker in the time interval used (0 to 120 min) (Figure S4). The background was pale purple for the silica gel plate and dark purple for the amino plate at the beginning of the documentation (t = 0 min) (Figure S4). At the beginning (t = 0 min), it was easier to detect the white zone with the lowest amount of allopurinol (5 μg) on the amino plate than on the silica gel plate. However, after 15 min, the same amount of allopurinol (5 μg) was easier to detect on the silica gel plate than on the amino plate because the background became darker (Figure S4, track 1). N^1,N^5,N^{10}-tricaffeoylspermidine was observed as a yellow zone on silica gel and amino plates (Figure S4, track 2). Intensive yellow bands for N^1,N^5,N^{10}-tricaffeoylspermidine and additional light frames around the N^1,N^5,N^{10}-tricaffeoylspermidine zone were observed from 15 to 120 min on the silica gel plate. On the amino plate, a yellow N^1,N^5,N^{10}-tricaffeoylspermidine zone was clearly observed at the beginning, and its intensity was increasing during the time intervals studied (Figure S4, track 2).

Based on these observations, HPTLC-XO inhibitory activity results for allopurinol (positive control), bee pollen, androecia samples, N^1,N^5,N^{10}-tricaffeoylspermidine, rutin, hyperoside, isoquercitrin, myricitrin, quercitrin, and quercetin were given after 20 min (Figure 11). The yellow-white zone of N^1,N^5,N^{10}-tricaffeoylspermidine (at $R_F \approx 0.67$, Figure 11A, track 2 and Figure 11B, track 7) was detected in all bee pollen (Figure 11A, tracks 3–13) and androecia samples (Figure 11B, tracks 9–11). Among bee pollen samples, C-bee pollen (Figure 11A, tracks 3–5) showed the highest number of yellow-white zones. Additionally, the C-bee pollen (Figure 11A, tracks 3–5) gave more intense yellow-white zones at $R_F \approx 0.67$ (N^1,N^5,N^{10}-tricaffeoylspermidine) and between $R_F \approx 0.74$ and ≈ 0.85 (Figure 11A, tracks 3–5) than S-bee pollen (Figure 11A, tracks 6–12) and Q-bee pollen (Figure 11A, tracks 13). Among the S-bee pollen, S2 and S7, with higher of N^1,N^5,N^{10}-tricaffeoylspermidine content (Table 7), showed more intense yellow-white zones (Figure 11A, tracks 7 and 12) than other samples. This is in agreement with the results of the spectrophotometric assay (Table 9). Among the standards investigated, flavonoids were observed as yellow zones (Figure 11B, tracks 2–6 and 8). Among the androecia samples, *C. sativa* androecia (Figure 11B, track 9) showed the highest number of yellow zones compared to *S. alba* (Figure 11B, track 10) and *Q. pubescens* androecia (Figure 11B, track 11), which is in agreement with the results of the spectrophotometric assay (Table 9).

4. Conclusions

This is the first study comparing the phytochemical composition and bioactivity profiles of bee pollen originating from *C. sativa*, *Salix* spp., and *Quercus* spp. from Türkiye and Slovenia and the androecia of *C. sativa*, *S. alba*, and *Q. pubescens*. Comparisons were based on newly developed qualitative HPTLC and quantitative UPLC methods. N^1,N^5,N^{10}-tricaffeoylspermidine, the compound common to bee pollen in this study, was found to have the highest content in C-bee pollen. In addition, N^1,N^5,N^{10}-tricaffeoylspermidine was for the first time identified and quantified in S-bee pollen, where it was found in relatively high content levels. N^1,N^5,N^{10}-tricaffeoylspermidine was for the first time identified and quantified in *C. sativa*, *S. alba*, and *Q. pubescens* androecia, where it was also found as a common compound. Among the C-, S-, and Q-bee pollen, bee pollen having the same botanical origin showed identical chemical fingerprint patterns by HPTLC image analysis showing that bee pollen composition is not related to the collection location, geographical differences, and the bee race. HPTLC profiles of bee pollen samples showed similarities and differences at specified R_F ranges when compared with profiles of the relevant botanical sources. The similarities helped with the identification of the botanical sources, but attention should be drawn to the reasons for the differences. The reason for these differences may be the morphological and chemical changes that flower pollen undergoes during pellet formation. Furthermore, since *C. sativa* is the only species in the genus *Castanea* found in both countries (Türkiye and Slovenia), other *Salix* and *Quercus* species should also be investigated in terms of chemical composition to find out if they could be a potential pollen grain source.

The antioxidant activity and XO inhibition of bee pollen, androecia, and N^1,N^5,N^{10}-tricaffeoylspermidine were evaluated by *in vitro* tests and HPTLC-EDA. HPTLC-EDA, which provided the detection of bioactive zones on the plate, was combined with image analysis that enabled the evaluation of bee pollen and androecia together with N^1,N^5,N^{10}-tricaffeoylspermidine, quercetin, myricitrin, hyperoside, quercitrin, isoquercitrin, and rutin. N^1,N^5,N^{10}-tricaffeoylspermidine had remarkable antioxidant and XO inhibitory activities, which were evaluated for the first time by HPTLC-EDA and *in vitro* tests. This is the first report on performing an HPTLC-XO inhibitory activity assay on the HPTLC NH_2 F_{254S} plates and evaluating their suitability related to commonly used HPTLC silica gel F_{254} plates.

The high content and bioactivity of N^1,N^5,N^{10}-tricaffeoylspermidine demonstrate its potential to be used as a marker compound for the standardization of bee pollen extracts. Overall, botanically identified and standardized bee pollen extracts should be used in food supplements to ensure repeatable pharmacological activity.

Supplementary Materials: The following supporting information can be downloaded at https://www.mdpi.com/article/10.3390/antiox14010040/s1: Figure S1: HRMS spectrum of N^1,N^5,N^{10}-tricaffeoylspermidine; Figure S2: 1H NMR spectrum of N^1,N^5,N^{10}-tricaffeoylspermidine (400 MHz, DMSO); Figure S3: ^{13}C NMR spectrum of N^1,N^5,N^{10}-tricaffeoylspermidine (100 MHz, DMSO); Figure S4: Influence of the type of stationary phase (HPTLC silica gel F_{254} and HPTLC NH_2 F_{254S} plates) and time on the detection of XO inhibitors. Allopurinol (positive control, track 1) and N^1,N^5,N^{10}-tricaffeoylspermidine (track 2) were applied on the plates after development with ethyl acetate–formic acid–water (35:4:4, *v*/*v*/*v*). Plates were documented before HPTLC-XO assay (at 254 and 366 nm) and after HPTLC-XO assay (under white light) at different time intervals (0–120 min).

Author Contributions: Conceptualization, E.G. and I.V.; methodology, N.B.S., E.G., I.V. and H.K.; validation, N.B.S. and E.G.; formal analysis, N.B.S., E.G., I.V. and H.K.; investigation, N.B.S., E.G., I.V. and H.K.; resources, E.G. and I.V.; data curation, N.B.S., E.G., I.V. and H.K.; writing—original draft preparation, N.B.S., E.G. and I.V.; writing—review and editing, N.B.S., E.G., I.V. and H.K.; visualization, N.B.S., E.G., I.V. and H.K.; supervision, E.G., I.V. and H.K.; project administration, E.G. and I.V.; funding acquisition, E.G. and I.V. All authors have read and agreed to the published version of this manuscript.

Funding: This study was supported by Yeditepe University Research Projects and Scientific Activities Support (YAP; Grant No: YAP-US-SAB-22026), the Scientific and Technological Research Council of Türkiye (TÜBİTAK; Project No: 119N569), and the Slovenian Research and Innovation Agency (ARIS, research core funding No. P1-0005 and the bilateral project BI-TR/20-23-004). The APC was funded by the National Institute of Chemistry (core funding No. P1-0005, ARIS).

Institutional Review Board Statement: Not applicable.

Informed Consent Statement: Not applicable.

Data Availability Statement: The data presented in this study are available upon request from the corresponding authors.

Acknowledgments: The authors acknowledge the Slovenian Beekeepers' Association (Čebelarska zveza Slovenije) and the beekeepers who generously provided bee pollen samples from Türkiye and Slovenia.

Conflicts of Interest: The authors declare no conflicts of interest. The funders had no role in the design of this study, in the collection, analyses, or interpretation of data, in the writing of this manuscript, or in the decision to publish the results.

References

1. Guzelmeric, E. Highlights on Bee Pollen. *Future Postharvest Food* **2024**, *1*, 263–265. [CrossRef]
2. Rzepecka-Stojko, A.; Stojko, J.; Kurek-Górecka, A.; Górecki, M.; Kabała-Dzik, A.; Kubina, R.; Moździerz, A.; Buszman, E.; Iriti, M. Polyphenols from Bee Pollen: Structure, Absorption, Metabolism and Biological Activity. *Molecules* **2015**, *20*, 21732–21749. [CrossRef] [PubMed]
3. Vukašinović, E.L.; Kebert, M.; Radišić, P.; Đorđievski, S.; Čelić, T.V.; Pihler, I.; Kojić, D.; Purać, J. Polyamine Profiling in Honey Bee Products: A Contribution to Functional Nutrition. *J. Food Compos. Anal.* **2024**, *125*, 105856. [CrossRef]
4. Zhang, H.; Liu, R.; Lu, Q. Separation and Characterization of Phenolamines and Flavonoids from Rape Bee Pollen, and Comparison of Their Antioxidant Activities and Protective Effects Against Oxidative Stress. *Molecules* **2020**, *25*, 1264. [CrossRef] [PubMed]
5. Qiao, J.; Feng, Z.; Zhang, Y.; Xiao, X.; Dong, J.; Haubruge, E.; Zhang, H. Phenolamide and Flavonoid Glycoside Profiles of 20 Types of Monofloral Bee Pollen. *Food Chem.* **2023**, *405*, 134800. [CrossRef]
6. Kostić, A.Ž.; Milinčić, D.D.; Špirović Trifunović, B.; Nedić, N.; Gašić, U.M.; Tešić, Ž.L.; Stanojević, S.P.; Pešić, M.B. Monofloral Corn Poppy Bee-Collected Pollen-A Detailed Insight into Its Phytochemical Composition and Antioxidant Properties. *Antioxidants* **2023**, *12*, 1424. [CrossRef]
7. Hamamoto, R.; Ishiyama, K.; Hashimoto, K.; Yamaguchi, M. Characterization of the Active Component in Bee Pollen *Cistus ladaniferus* Extract in Stimulating Bone Calcification and in Inhibiting Bone Resorption *in Vitro*. *J. Health Sci.* **2006**, *52*, 607–612. [CrossRef]
8. Maruyama, H.; Sakamoto, T.; Araki, Y.; Hara, H. Anti-Inflammatory Effect of Bee Pollen Ethanol Extract from *Cistus* sp. of Spanish on Carrageenan-Induced Rat Hind Paw Edema. *BMC Complement. Altern. Med.* **2010**, *10*, 30. [CrossRef]
9. Morales-Soto, A.; Oruna-Concha, M.J.; Elmore, J.S.; Barrajón-Catalán, E.; Micol, V.; Roldán, C.; Segura-Carretero, A. Volatile Profile of Spanish Cistus Plants as Sources of Antimicrobials for Industrial Applications. *Ind. Crops Prod.* **2015**, *74*, 425–433. [CrossRef]
10. Sen, N.B.; Guzelmeric, E.; Vovk, I.; Glavnik, V.; Kırmızıbekmez, H.; Yesilada, E. Phytochemical and Bioactivity Studies on *Hedera helix* L. (Ivy) Flower Pollen and Ivy Bee Pollen. *Antioxidants* **2023**, *12*, 1394. [CrossRef] [PubMed]
11. Avşar, C.; Özler, H.; Berber, I.; Civek, S. Phenolic Composition, Antimicrobial and Antioxidant Activity of *Castanea sativa* Mill. Pollen Grains from Black Sea Region of Turkey. *Int. Food Res. J.* **2016**, *23*, 1711–1716.
12. Şahin, H.; Kolaylı, S.; Kara, Y.; Can, Z.; Güler, H.İ.; Ozkok, A.; Serdar, Ü. A Study on Recognizing the Value of Chestnut (*Castanea sativa*) Blossom Waste. *Tarım Bilim. Derg.* **2024**, *30*, 79–89. [CrossRef]
13. Sultana, S.; Saleem, M. Salix Caprea Inhibits Skin Carcinogenesis in Murine Skin: Inhibition of Oxidative Stress, Ornithine Decarboxylase Activity and DNA Synthesis. *J. Ethnopharmacol.* **2004**, *91*, 267–276. [CrossRef]
14. Alam, M.S.; Kaur, G.; Jabbar, Z.; Javed, K.; Athar, M. Evaluation of Antioxidant Activity of Salix Caprea Flowers. *Phyther. Res.* **2006**, *20*, 479–483. [CrossRef]
15. Enayat, S.; Banerjee, S. Comparative Antioxidant Activity of Extracts from Leaves, Bark and Catkins of *Salix aegyptiaca* sp. *Food Chem.* **2009**, *116*, 23–28. [CrossRef]
16. Wahid, A.; Hamed, A.N.; Eltahir, H.M.; Abouzied, M.M. Hepatoprotective Activity of Ethanolic Extract of *Salix subserrata* against CCl4-Induced Chronic Hepatotoxicity in Rats. *BMC Complement. Altern. Med.* **2016**, *16*, 263. [CrossRef] [PubMed]
17. Sobeh, M.; Mahmoud, M.F.; Rezq, S.; Alsemeh, A.E.; Sabry, O.M.; Mostafa, I.; Abdelfattah, M.A.O.; Ait El-Allem, K.; El-Shazly, A.M.; Yasri, A.; et al. *Salix tetrasperma* Roxb. Extract Alleviates Neuropathic Pain in Rats via Modulation of the NF-KB/TNF-α/NOX/INOS Pathway. *Antioxidants* **2019**, *8*, 482. [CrossRef] [PubMed]
18. Mostafa, I.; Abbas, H.A.; Ashour, M.L.; Yasri, A.; El-Shazly, A.M.; Wink, M.; Sobeh, M. Polyphenols from *Salix tetrasperma* Impair Virulence and Inhibit Quorum Sensing of *Pseudomonas aeruginosa*. *Molecules* **2020**, *25*, 1341. [CrossRef] [PubMed]
19. Bokern, M.; Witte, L.; Wray, V.; Nimtz, M.; Meurer-Grimes, B. Trisubstituted Hydroxycinnamic Acid Spermidines from *Quercus dentata* Pollen. *Phytochemistry* **1995**, *39*, 1371–1375. [CrossRef]
20. Ryu, H.J.; Song, R.; Kim, H.W.; Kim, J.H.; Lee, E.Y.; Lee, Y.J.; Song, Y.W.; Lee, E.B. Clinical Risk Factors for Adverse Events in Allopurinol Users. *J. Clin. Pharmacol.* **2013**, *53*, 211–216. [CrossRef] [PubMed]

21. Guzelmeric, E.; Ristivojević, P.; Trifković, J.; Dastan, T.; Yilmaz, O.; Cengiz, O.; Yesilada, E. Authentication of Turkish Propolis through HPTLC Fingerprints Combined with Multivariate Analysis and Palynological Data and Their Comparative Antioxidant Activity. *LWT* **2018**, *87*, 23–32. [CrossRef]
22. Barth, O.M. Pollen Analysis of Brazilian Propolis. *Grana* **1998**, *37*, 97–101. [CrossRef]
23. Uyar, G.; Çetin, B. The Moss Flora of Ankara-Kızılcahamam Soğuksu National Park. *Turk. J. Botany* **2001**, *25*, 261–273.
24. Reich, E.; Schibli, A. *High-Performance Thin-Layer Chromatography for the Analysis of Medicinal Plants*; Thieme: New York, NY, USA, 2007.
25. Jork, H.; Funk, W.; Fischer, W.R.; Wimmer, H. *Thin-Layer Chromatography, Thin-Layer Chromatography: Reagents and Detection Methods: Volume 1a: Physical and Chemical Detection Methods: Fundamentals; Reagents I*; Thin-Layer Chromatography; Wiley-VCH: Weinheim, Germany, 1989; ISBN 9783527278343.
26. International Conference on Harmonisation ICH Harmonised Tripartite Guideline, Validation of Analytical Procedures: Text and Methodology Q2(R1) 2005. Available online: https://database.ich.org/sites/default/files/Q2(R1)%20Guideline.pdf (accessed on 25 January 2024).
27. Berker, K.I.; Gülü, K.; Demirata, B.; Apak, R. A Novel Antioxidant Assay of Ferric Reducing Capacity Measurement Using Ferrozine as the Colour Forming Complexation Reagent. *Anal. Methods* **2010**, *2*, 1770–1778. [CrossRef]
28. Apak, R.; Güçlü, K.; Özyürek, M.; Çelik, S.E. Mechanism of Antioxidant Capacity Assays and the CUPRAC (Cupric Ion Reducing Antioxidant Capacity) Assay. *Microchim. Acta* **2008**, *160*, 413–419. [CrossRef]
29. Ockun, M.A.; Baranauskaite, J.; Uner, B.; Kan, Y.; Kırmızıbekmez, H. Preparation, Characterization and Evaluation of Liposomal-Freeze Dried Anthocyanin-Enriched *Vaccinium arctostaphylos* L. Fruit Extract Incorporated into Fast Dissolving Oral Films. *J. Drug Deliv. Sci. Technol.* **2022**, *72*, 103428. [CrossRef]
30. Guzelmeric, E.; Ugurlu, P.; Celik, C.; Sen, N.B.; Helvacıoglu, S.; Charehsaz, M.; Erdogan, M.; Ockun, M.A.; Kırmızıbekmez, H.; Aydın, A.; et al. *Myrtus communis* L. (Myrtle) Plant Parts: Comparative Assessment of Their Chemical Compositions and Antioxidant, Anticancer, and Antimutagenic Activities. *South Afr. J. Bot.* **2022**, *150*, 711–720. [CrossRef]
31. Bassard, J.-E.; Ullmann, P.; Bernier, F.; Werck-Reichhart, D. Phenolamides: Bridging Polyamines to the Phenolic Metabolism. *Phytochemistry* **2010**, *71*, 1808–1824. [CrossRef] [PubMed]
32. Zhang, H.; Zhu, X.; Huang, Q.; Zhang, L.; Liu, X.; Liu, R.; Lu, Q. Antioxidant and Anti-Inflammatory Activities of Rape Bee Pollen after Fermentation and Their Correlation with Chemical Components by Ultra-Performance Liquid Chromatography-Quadrupole Time of Flight Mass Spectrometry-Based Untargeted Metabolomics. *Food Chem.* **2023**, *409*, 135342. [CrossRef]
33. Glavnik, V.; Bensa, M.; Vovk, I.; Guzelmeric, E. High-Performance Thin-Layer Chromatography-multi-Stage Mass Spectrometry Methods for Analyses of Bee Pollen Botanically Originating from Sweet Chestnut (*Castanea sativa Mill*). *JPC J. Planar Chromatogr. Mod. TLC* **2023**, *36*, 471–482. [CrossRef]
34. Kim, S.B.; Liu, Q.; Ahn, J.H.; Jo, Y.H.; Turk, A.; Hong, I.P.; Han, S.M.; Hwang, B.Y.; Lee, M.K. Polyamine Derivatives from the Bee Pollen of *Quercus mongolica* with Tyrosinase Inhibitory Activity. *Bioorg. Chem.* **2018**, *81*, 127–133. [CrossRef] [PubMed]
35. Negri, G.; Barreto, L.M.R.C.; Sper, F.L.; de Carvalho, C.; Campos, M. das G.R. Phytochemical Analysis and Botanical Origin of Apis Mellifera Bee Pollen from the Municipality of Canavieiras, Bahia State, Brazil. *Braz. J. Food Technol.* **2018**, *21*, e2016176. [CrossRef]
36. Keefover-Ring, K.; Carlson, C.H.; Hyden, B.; Azeem, M.; Smart, L.B. Genetic Mapping of Sexually Dimorphic Volatile and Nonvolatile Floral Secondary Chemistry of a Dioecious Willow. *J. Exp. Bot.* **2022**, *73*, 6352–6366. [CrossRef]
37. Meurer-Grimes, B. New Evidence for the Systematic Significance of Acylated Spermidines and Flavonoids in Pollen of Higher Hamamelidae. *Brittonia* **1995**, *47*, 130–142. [CrossRef]
38. Vovk, I.; Simonovska, B.; Vuorela, P.; Vuorela, H. Optimization of Separation of (+)-Catechin and (−)-Epicatechin on Cellulose TLC Plates. *J. Planar Chromatogr. Mod. TLC* **2002**, *15*, 433–436. [CrossRef]
39. Ng, B.; Quinete, N.; Gardinali, P.R. Assessing Accuracy, Precision and Selectivity Using Quality Controls for Non-Targeted Analysis. *Sci. Total Environ.* **2020**, *713*, 136568. [CrossRef]
40. Ilieva, D.; Surleva, A.; Murariu, M.; Drochioiu, G.; Abdullah, M.M.A.B. Evaluation of ICP-OES Method for Heavy Metal and Metalloids Determination in Sterile Dump Material. *Solid State Phenom.* **2018**, *273*, 159–166. [CrossRef]
41. Nasti, R.; Orlandini, S.; Furlanetto, S.; Casale, M.; Daci, A.; Hajdari, A.; Meneghetti, F.; Villa, S.; Mori, M.; Beretta, G. An Analytical Investigation of Hydroxylated Cinnamoyl Polyamines as Biomarkers of Commercial Bee Pollen Botanical Origin. *Int. J. Food Sci. Technol.* **2022**, *57*, 7787–7796. [CrossRef]
42. Rodríguez-Flores, M.S.; Escuredo, O.; Seijo, M.C.; Rojo, S.; Vilas-Boas, M.; Falcão, S.I. Phenolic Profile of Castanea Bee Pollen from the Northwest of the Iberian Peninsula. *Separations* **2023**, *10*, 270. [CrossRef]
43. Sonmez, E.; Kekecoglu, M.; Sahin, H.; Bozdeveci, A.; Karaoglu, S.A. Comparing the Biological Properties and Chemical Profiling of Chestnut Bee Pollen and Bee Bread Collected from Anatolia. *Braz. J. Microbiol.* **2023**, *54*, 2307–2317. [CrossRef] [PubMed]

44. Barros, L.; Alves, C.T.; Dueñas, M.; Silva, S.; Oliveira, R.; Carvalho, A.M.; Henriques, M.; Santos-Buelga, C.; Ferreira, I.C.F.R. Characterization of Phenolic Compounds in Wild Medicinal Flowers from Portugal by HPLC–DAD–ESI/MS and Evaluation of Antifungal Properties. *Ind. Crops Prod.* **2013**, *44*, 104–110. [CrossRef]
45. Carocho, M.; Barros, L.; Bento, A.; Santos-Buelga, C.; Morales, P.; Ferreira, I.C.F.R. Castanea Sativa Mill. Flowers amongst the Most Powerful Antioxidant Matrices: A Phytochemical Approach in Decoctions and Infusions. *Biomed. Res. Int.* **2014**, *2014*, 232956. [CrossRef]
46. Alimoglu, G.; Guzelmeric, E.; Yuksel, P.I.; Celik, C.; Deniz, I.; Yesilada, E. Monofloral and Polyfloral Bee Pollens: Comparative Evaluation of Their Phenolics and Bioactivity Profiles. *LWT* **2021**, *142*, 110973. [CrossRef]
47. Karkar, B.; Şahin, S.; Güneş, M.E. Evaluation of Antioxidant Properties and Determination of Phenolic and Carotenoid Profiles of Chestnut Bee Pollen Collected from Turkey. *J. Apic. Res.* **2021**, *60*, 765–774. [CrossRef]
48. Karkar, B.; Şahin, S.; Güneş, M.E. Antioxidative Effect of Turkish Chestnut Bee Pollen on Dna Oxidation System and Its Phenolic Compounds. *Gida J. Food* **2018**, *43*, 34–42. [CrossRef]
49. Wang, R.; Su, G.; Wang, L.; Xia, Q.; Liu, R.; Lu, Q.; Zhang, J. Identification and Mechanism of Effective Components from Rape (*Brassica napus* L.) Bee Pollen on Serum Uric Acid Level and Xanthine Oxidase Activity. *J. Funct. Foods* **2018**, *47*, 241–251. [CrossRef]

antioxidants

MDPI

Article

The Consumption of Beeswax Alcohol (BWA, Raydel®) Improved Zebrafish Motion and Swimming Endurance by Protecting the Brain and Liver from Oxidative Stress Induced by 24 Weeks of Supplementation with High-Cholesterol and D-Galactose Diets: A Comparative Analysis Between BWA and Coenzyme Q_{10}

Kyung-Hyun Cho *, Yunki Lee, Ashutosh Bahuguna, Sang Hyuk Lee, Chae-Eun Yang, Ji-Eun Kim and Hyo-Shin Kwon

Raydel Research Institute, Medical Innovation Complex, Daegu 41061, Republic of Korea
* Correspondence: chok@raydel.co.kr; Tel.: +82-53-964-1990; Fax: +82-53-965-1992

Abstract: The prolonged consumption of D-galactose (Gal) has been associated with severe damage in the liver and brain via exacerbation of oxidative stress, non-enzymatic glycation, and the aging process. The current study was initiated for a comparative assessment of beeswax alcohol (BWA, final 0.5% and 1.0% *w/w*) and coenzyme Q_{10} (CoQ_{10}, final 0.5% and 1.0% *w/w*) against high-cholesterol (HC, final 4%, *w/w*) and -galactose (Gal, final 30%, *w/w*)-induced adverse events in zebrafish during 24 weeks of consumption. The survivability of zebrafish decreased to 82.1% due to HC+Gal exposure, but this was substantially improved (91.0%) with the consumption of 0.5% and 1.0% BWA. In contrast, no protective effect of CoQ_{10} consumption (1.0%) was observed on the survivability of zebrafish. Nevertheless, both BWA and CoQ_{10} displayed a significant ($p < 0.001$) preventive effect against HC+Gal-induced body weight enhancement. The HC+Gal-induced cognitive changes, marked by staggered and confused swimming behavior, and retarded swimming speed and motion patterns (restricted to the bottom of the tank), were efficiently restored by BWA. A significantly higher residence time in the upper half of the tank, 3.1-and 4.5-fold reduced latency time along with 3.5-fold and 4.1-fold higher swimming distance, was logged in the 0.5% and 1.0% BWA groups, respectively, than the zebrafish that consumed HC+Gal. In addition, BWA effectively enhanced plasma ferric ion reduction (FRA) and paraoxonase (PON) activity and alleviated the total cholesterol (TC), triglyceride (TG), and blood glucose levels disrupted by the consumption of HC+Gal. Also, the HC+Gal-alleviated plasma high-density lipoprotein-cholesterol (HDL-C) was 2.6-fold ($p < 0.001$) enhanced in the group that consumed 1.0% BWA, which was significantly 1.5-fold ($p < 0.001$) better than the effect of 1.0% CoQ_{10}. Similarly, BWA displayed a superior impact over CoQ_{10} to mitigate HC+Gal-induced plasma AST and ALT levels, hepatic IL-6 production, generation of oxidized species, cellular senescence, and fatty liver changes. Moreover, BWA protects the brain against HC+Gal-induced oxidative stress, apoptosis, and myelin sheath degeneration. Conclusively, compared to CoQ_{10}, BWA efficiently can the HC+Gal-impaired brain and liver functionality to subside and improves the dyslipidemia and cognitive behavior of zebrafish.

Keywords: beeswax alcohol (BWA); behavioral analysis; brain; coenzyme Q_{10} (CoQ_{10}); D-galactose; senescence; swimming distance; myelin

1. Introduction

The long-term consumption of D-galactose (Gal) has been widely acknowledged as a contributor to aging in major organs, including the liver, heart, brain, and skin [1–4], primarily through non-enzymatic glycation and elevated oxidative stress [5,6]. In a rodent model, Gal consumption reduced longevity by exacerbating cognitive dysfunction, neurodegeneration, and liver inflammation [7,8]. The pathogenesis of galactose-induced cataracts, vision loss, and blindness due to opacity of the lens is associated with the production and accumulation of reactive oxygen species (ROS), protein denaturation, and cellular damage [9,10]. In addition, Gal-induced aging is associated with dyslipidemia, high serum cholesterol, and triglycerides with low HDL-C, which are the hallmarks of metabolic aging and fatty liver disease [11]. Because high cholesterol (HC) also induces cellular senescence and aging via an increase in ROS production, the co-consumption of HC and Gal might exacerbate the pro-oxidative and pro-inflammatory cascade to cause aging [12,13].

Beeswax alcohol (BWA), a mixture of six long-chain aliphatic alcohols (tetracosanol, hexacosanol, octacosanol, triacontanol, dotriacontanol, and tetratriacontanol) obtained from beeswax (*Apis mellifera*) [14], has been well documented to ameliorate gastrointestinal disorders, peptic ulcers, and osteoarthritis via the protection of lipid and protein in the cell membrane with potent antioxidant and anti-inflammatory activity [15–17]. Due to its many health benefits, the Korean Ministry of Food and Drug Safety has approved BWA as a functional food ingredient to improve gastrointestinal and joint health [18].

CoQ_{10} is an intracellular antioxidant that prevents cellular senescence and dysfunction from oxidative stress to treat cardiomyopathy [19]. A meta-analysis of randomized controlled trials revealed that CoQ_{10} supplementations, 60–300 mg/day, among patients with coronary artery disease was also associated with an increase in serum HDL-C, superoxide dismutase (SOD), and catalase activity and decreased malondialdehyde levels [20]. A comparative analysis revealed that BWA exhibited a greater antioxidant effect than vitamin C [21] and coenzyme Q_{10} (CoQ_{10}) [22] through its ability to inhibit LDL oxidation [21,22] and enhance HDL-associated paraoxonase (PON-1) activity [22].

Recently, we reported that 20 weeks of dietary supplementation of BWA effectively countered the high-cholesterol diet-induced dyslipidemia and protected the vital organs [23]. In a recent study, 12 weeks of BWA supplementation effectively mitigated HC (final 10%, *w*/*w*)+Gal (final 10%, *w*/*w*)-induced dyslipidemia, liver damage, inflammation, and impairment of kidney and reproductive organ function in zebrafish [24]. Moreover, the consumption of BWA ameliorated the dyslipidemia, production of ROS, and cellular senescence actuated by the consumption of HC+Gal. Despite the beneficial effect of BWA, the study [24] was limited only to the short-duration (12 weeks) consumption of HC+Gal with a relatively lower amount of Gal, which was noted to be insufficient in causing severe detrimental effects in the brain and cognitive changes. Given this, the present investigation aimed to compare the in vivo efficacy of BWA (final 0.5% and 1.0%, *w*/*w*) and CoQ_{10} (final 0.5% and 1.0%, *w*/*w*) against long-term (24 weeks) consumption of a high amount of Gal (final 30%, *w*/*w*) and cholesterol (final 4%, *w*/*w*), which induced severe detrimental effects in the brain, aging, and cognitive impairments in the zebrafish (*Danio rerio*).

2. Materials and Methods

2.1. Materials

The beeswax alcohol (BWA) was complementarily provided by Raydel® Pty, Ltd. (Thornleigh, NSW, Australia). The BWA (>86% purity) was extracted from the block beeswax of honeybee (*Apis mellifera* mainly mellifera linage). The extracted BWA contained a mixture of six long-chain aliphatic alcohols (LCAA): C24, C26, C28, C30, C32, and C34. Detailed specifications of the BWA (batch#:330020123) and the certificate of analysis are listed in Supplementary Table S1. D-galactose (Gal, Cat # G5388), coenzyme Q_{10} (CoQ_{10}, Cat # 303-98-0), and 5-bromo-4-chloro-3-indolyl β D-galactopyranoside (X-gal, Cat # B4252) were purchased from Sigma-Aldrich (St. Louis, MO, USA). All the other chemicals and reagents were of analytical grade and used as supplied. The myelin specific primary antibody was in-house produced in rabbits and was complementarily provided by Prof. Cheol-Hee Kim (Laboratory of Developmental Genetics, Department of Biology, Chungnam National University, Daejeon 34134, Republic of Korea). The secondary antibody Goat anti-rabbit IgG (AlexaFluor™ 405, A-31556) was purchased from Thermo Fischer (Waltham, MA, USA).

2.2. Formulation of Different Diets

The normal tetrabit (ND), a regular diet of zebrafish, was mixed with cholesterol (final 4%, *w*/*w*) to make a high-cholesterol (HC) diet. The HC was blended with galactose (Gal, final 30%, *w*/*w*) to make an HC-supplemented galactose diet (HC+Gal). HC+Gal was individually amalgamated with BWA (final 0.5% or 1.0%, *w*/*w*) or CoQ_{10} (final 0.5% or 1.0%, *w*/*w*) to make four distinct HC+Gal formulations enriched with BWA and CoQ_{10} and abbreviated as HC+Gal+0.5% BWA or 1.0% BWA and HC+Gal+0.5% CoQ_{10} or 1.0% CoQ_{10}, respectively. Table 1 depicts the specific amount of tetrabit, galactose, cholesterol, BWA, and CoQ_{10} used to prepare the different zebrafish diets.

Table 1. Seven distinct zebrafish diet compositions.

Diet (mg)	ND	HC	HC+30% Gal	HC+30% Gal+0.5% BWA	HC+30% Gal+1.0% BWA	HC+30% Gal+0.5% CoQ_{10}	HC+30% Gal+1.0% CoQ_{10}
Tetrabits [1]	10	9.6	6.6	6.55	6.5	6.55	6.5
Galactose	0	0	3	3	3	3	3
Cholesterol	0	0.4	0.4	0.4	0.4	0.4	0.4
BWA	0	0	0	0.05	0.1	0	0
CoQ_{10}	0	0	0	0	0	0.05	0.1
Total (mg)	10	10	10	10	10	10	10

ND, normal diet; HC, high cholesterol; BWA, beeswax alcohol; CoQ_{10}, coenzymeQ_{10}. [1] Tetrabits, a brand-name zebrafish diet, was purchased from Tetrabit Gmbh, Melle, Germany (containing 47.5% crude protein, 6.5% crude fat, 2.0% crude fiber, 10.5% crude ash, vitamin A (29,770 IU/kg), vitamin D3 (1860 IU/kg), vitamin E (200 mg/kg), and vitamin C (137 mg/kg)).

The selection of 30% Gal was based on the preliminary screening study where Gal (10–50%) was supplemented in the ND and HC and fed separately to the zebrafish (*Danio rerio*). The study outcome revealed no substantial changes in the zebrafish mortality, weight, total cholesterol, glucose level, and cognitive behavior in the ND+Gal (10–50%) groups compared to the ND (control) group. However, in the HC+Gal groups, substantial adverse effects were observed with an increasing concentration of Gal (20–50%). Notably, severe zebrafish mortality was observed at HC+Gal > 30% (i.e., 40% and 50%). Therefore, 30% Gal with HC was selected as a relevant dietary composition to induce damaging effects on the zebrafish.

2.3. Zebrafish Husbandry and Consumption of Different Diets

Zebrafish (*Danio rerio*) AB strains were maintained at 28 °C water temperature in an alternative 14 h (light) and 10 h (dark) photoperiod following the standard guidelines of Animal Care Committee and Used approved by Raydel Research Institute (RRI, approval code RRI-23-007, approval date 27 July 2023).

Adult zebrafish (~18 weeks, n = 392) were randomly segregated and maintained with a normal diet (ND, n = 56) and a high-cholesterol diet (HC, n = 336). Following four weeks of consumption, the ND-fed zebrafish (Group I) were further maintained only on the ND diet for the next 24 weeks, while the HC-fed zebrafish were further segregated into five groups (n = 56/group). The zebrafish in Group II were maintained in the HC diet only, while the zebrafish in Groups III and IV were fed with HC+Gal+0.5% BWA and HC+Gal+1.0% BWA, respectively. The zebrafish in Groups V and VI were fed with HC+Gal+0.5% CoQ_{10} and HC+Gal+1.0% CoQ_{10}, respectively. The experimental layout involves feeding zebrafish with different diets is depicted in Figure 1. The zebrafish in the different groups (I–VI) were fed the designated diet (10 mg/zebrafish) twice a day (morning and evening) until 24 weeks. The food consumption was measured by using the following formula: [(given food amount − remaining food amount)/given food amount] × 100. The feed efficiency was quantified by measuring the feed conversion ratio using the following equation: [weight gain (g)/feed consumed (g)] × 100. The feed efficiency was calculated at 24 weeks of consumption of the respective diets.

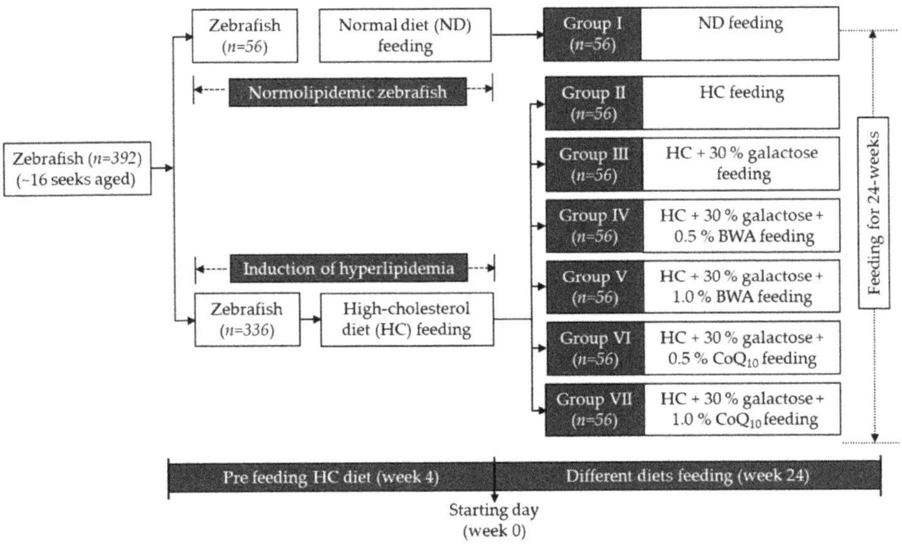

Figure 1. The experimental layout involves feeding zebrafish with different diets. The abbreviations: ND represent normal diet; HC represents high-cholesterol diet (final 4%, *wt/wt*); and BWA and CoQ_{10} represent beeswax alcohol and coenzyme Q_{10}, respectively.

The mortality of zebrafish across all the groups was periodically measured until 24 weeks of the experimental period. For the body weight analysis, zebrafish were anaesthetized by submerging them in a 0.1% solution of 2-phenoxyethanol. The anaesthetized zebrafish were surface-dried using filter paper, and body weight was analyzed using an electronic balance (Ohaus, Parsippany-Troy Hills, NJ, USA).

2.4. Behavioral Analysis of Zebrafish

The behavioral analysis of zebrafish across all the groups was analyzed using an open field tank following video recording. For the behavioral analysis, the filled water area was divided into two halves using horizontal lines in the tank. The divided water section was named the lower and upper half [as depicted in Supplementary Figure S1]. The zebrafish from the different groups were transferred into the respective tanks (divided into two halves). Following 1 h of acclimatization, the zebrafish's swimming movement was recorded (1 min). The video of swimming movement was analyzed using "Tracker video analysis and modelling software" (https://physlets.org/tracker, assessed on 15 August 2024) to quantify total swimming distance, swimming trajectory, swimming speed, total time spent in the tank's upper and lower halves, and latency time.

2.5. Collection of Blood and Organs

After 24 weeks of consumption of different diets, the zebrafish from the respective groups were sacrificed using hypothermic shock [25]. Immediately, blood was collected in ethylenediamine-tetraacetic acid (EDTA) rinsed tubes and, further, processed for the quantification of total cholesterol (TC), triglycerides (TGs), high-density lipoprotein-cholesterol (HDL-C), and the hepatic function biomarkers aspartate aminotransferase (AST) and alanine aminotransferase (ALT) using commercial kits following the recommended methodology suggested by the manufacturer. A detailed procedure for the blood analysis is provided in Supplementary Material Method S1. The organs (liver and brain) from the zebrafish were surgically removed and preserved in 10% formalin for further histological analyses.

2.6. Estimation of Blood Glucose, Ferric Ion Reduction (FRA), and Paraoxonase (PON) Activity

Blood from the different groups was analyzed using an electronic blood glucose meter (AccuCheck, Roche, Basel, Switzerland) to estimate glucose, and the results are depicted as glucose mg/dL. The ferric ion reduction ability (FRA) of plasma was examined using the method described earlier [26]. Briefly, 20 μL of plasma (0.1 mg equivalent protein) was mixed with 180 μL of FRA reagent [prepared by blending 10 mM 2,4,6-tripridyl-S triazin (1.25 mL) and 20 mM ferric chloride dissolved in 10 mL of 0.2 M acetate buffer (pH 3.6)]. The content was incubated for 60 min at room temperature, followed by an absorbance assessment at 593 nm.

Plasma paraoxonase (PON) activity was determined by the earlier described spectroscopic method with slight modification [26]. In brief, a 40 μL plasma (0.1 mg equivalent protein) sample was mixed with 160 μL of 0.55 M paraoxon-ethyl [dissolved in 90 mM Tris-HCl, 3.6 mM NaCl, and 90 mM $CaCl_2$]. After 60 min of incubation at 25 °C, absorbance at 415 nm was recorded, and PON activity was quantified as μU/L/min using a molar absorbance coefficient (ε) of 17,000 $M^{-1}cm^{-1}$ for the formed product *p*-nitrophenol.

2.7. Histological Analysis

The tissue sections (liver and brain) were individually dehydrated using ethanol and then embedded in a paraffin block. A thin tissue section (7 μm thick) was obtained from the cryo-sectioning using a microtome (Leica CM 1510S, Nussloch, Germany). The sliced tissue section was processed for hematoxylin and eosin (H&E) staining [27] to examine the changes in the liver and brain.

The senescence in the tissue sections of the liver and brain was analyzed with an SA-β-gal assay [28]. Briefly, the tissue section (7 μm) was fixed with 4% paraformaldehyde, followed by washing with PBS twice and the subsequent addition of X-gal (250 μL, 1 mg/mL). Following 16 h of incubation at room temperature in the dark, the stained section was washed three times with PBS and visualized under a microscope. Oil red O (ORO) staining [25] was performed to observe the fatty liver changes in the liver tissue. In brief, 250 μL of ORO (3 mg/mL) was poured over the tissue section (7 μm). Following 10 min of incubation, the section was washed with water and visualized under a microscope.

2.8. Immunohistology (IHC)

For the detection of hepatic interleukin (IL)-6, an immunohistochemical analysis was performed using the target-specific primary antibodies [29]. For the IL-6 detection, the hepatic section (7 μm) was incubated overnight with IL-6 specific antibody (ab9324, Abcam, London, UK) following IHC development with a secondary antibody containing the EnVision + System-HRP polymer kit (Code K4001, Dako, Glostrup, Denmark).

The myelin sheath in the brain was detected by immunohistochemical analysis using a myelin-specific antibody. For the myelin sheath detection, the zebrafish myelin-specific primary antibody from rabbit (200× diluted) was spread over the brain tissue section (7 μm thickness). Following overnight incubation at 4 °C, the tissue section was washed three times with PBS, followed by the subsequent addition of goat anti-rabbit IgG (AlexaFluor™ 405, A-31556, Thermo Fischer). After 4 h of incubation in the dark at 4 °C, the tissue section was washed three times with PBS and visualized under a fluorescent microscope at the excitation (Ex) and emission (Em) wavelengths of 401 and 422 nm, respectively.

2.9. Fluorescent Staining for Reactive Oxygen Species (ROS) Production and Apoptosis

Dihydroethidium (DHE) [30] and acridine orange (AO) [31] fluorescent staining was performed to detect ROS and the extent of apoptosis. In brief, 250 μL of DHE (30 mM) and AO (30 mg/mL) was poured on the tissue section (7 μm) and incubated in the dark for 30 min. The stained section was rinsed three times with PBS and visualized under a fluorescent microscope at 585 nm (Ex)/615 nm (Em) and 505 nm (Ex)/535 nm (Em) for the detection of DHE and AO fluorescent stained areas, respectively.

2.10. Statistical Analysis

The results in the graphs are represented as mean ± SEM obtained from the triplicate experiments. The statistical significance ($p < 0.05$) between the groups was established using a one-way analysis of variance (ANOVA) following Tukey's post hoc analysis, while the pairwise assessment was examined by T-test using the Statistical Package for Social Science software program (version 29.0, Chicago, IL, USA).

3. Results

3.1. Change in Survivability and Body Weight During Galactose Consumption

The Kaplan–Meier survival analysis, followed by a log-rank test, revealed a significant difference (log-rank: $\chi^2 = 13.5$, $p = 0.035$) in zebrafish solvability among the different groups. As represented in Figure 2A, the highest survival probability (1.0) was maintained in the ND control group throughout the 24 weeks of consumption. In contrast, the HC group experienced a substantial drop in survival probability (0.88), which declined to 0.81 in the HC+Gal group over the 24 weeks of consumption. The HC+Gal-induced mortality of zebrafish was substantially prevented by the consumption of BWA, as reflected by the 0.89 survival probability in the group that consumed BWA 0.5% and 1.0% for 24 weeks. In contrast to BWA, the CoQ_{10} consumption showed a lower effect on HC+Gal-induced zebrafish mortality, even more in the group that consumed 1.0% CoQ_{10}; 0.81 survival probability was observed, which is analogous to the survival probability observed in the HC+Gal group.

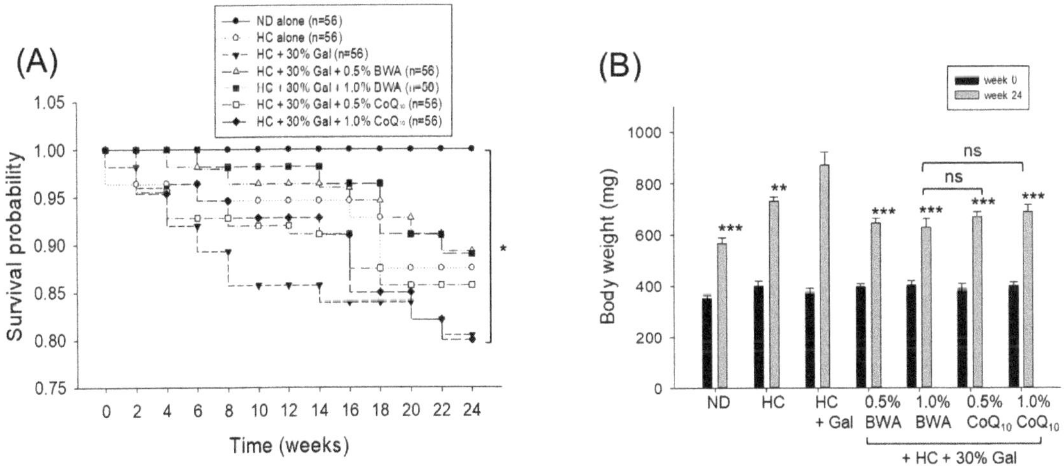

Figure 2. Zebrafish survivability and body weight across the different groups. (**A**) Survival probability across different groups during 24 weeks of consumption of the respective diets. The Kaplan–Meier survival analysis followed by log-rank test was used to determine the statistical difference in the survival probability curve. *, $p < 0.05$ (log-rank: $\chi^2 = 13.5$). (**B**) Body weight of zebrafish at the beginning (week 0) and after 24 weeks of consumption of the different diets. ND represents the normal diet; HC represents the high-cholesterol (final 4%, w/w) diet; HC+Gal represents the high-cholesterol (final 4%, w/w) with galactose (final 30%, w/w) diet; and HC+Gal+BWA (0.5%/1.0%) or CoQ_{10} (0.5%/1.0%) represents the high-cholesterol + galactose diet supplemented with beeswax alcohol (final 0.5% or 1.0% w/w) or coenzyme Q_{10} (final 0.5% or 1.0% w/w), respectively. ** ($p < 0.01$) and *** ($p < 0.001$) depict the statistical difference in the body weight measured at 24 weeks of consumption of different diets with respect to the HC+Gal group. "ns" represents the non-significant difference between the specified group concerning the HC+Gal+1.0% BWA group.

Among all the groups, ~95–100% food consumption was noticed within 30 min of feed supply (Table 2). The feed efficiency ratio among the groups was in the range of 6.3–14.6%. The highest feed efficiency was seen in the group that consumed HC+Gal (14.6%), while the least was in the group that consumed ND (6.3%), followed by the 1.0% BWA (6.8%), 0.5% BWA (7.4%), 0.5% CoQ_{10} (8.3%), and 1.0% CoQ_{10} (8.6%) groups (Table 2).

Table 2. Food consumption and food efficiency among the seven distinct zebrafish groups.

Dietary Efficiency	ND	HC	HC+30% Gal	HC+30% Gal+0.5% BWA	HC+30% Gal+1.0% BWA	HC+30% Gal+0.5% CoQ_{10}	HC+30% Gal+1.0% CoQ_{10}
Food consumption (%) [1]	95–100	95–100	95–100	95–100	95–100	95–100	95–100
Food efficiency (%) [2]	6.3	9.8	14.6	7.4	6.8	8.3	8.6

ND, normal diet; HC, high cholesterol; BWA, beeswax alcohol; CoQ_{10}, coenzymeQ_{10}. [1] The food consumption was calculated by [(given food amount − remaining food amount)/given food amount] × 100. [2] The feed efficiency was calculated at 24 weeks of consumption of the respective diets using [weight gain (g)/feed consumed (g)] × 100.

As illustrated in Figure 2B, after 24 weeks, the HC+Gal group exhibited the most pronounced body weight gain (870.6 ± 51.7 mg), approximately 2.3-fold higher from week 0 (377.4 ± 12.7 mg), followed by the HC alone group (731.5 ± 16.4 mg), which displayed around a 1.8-fold increase from the baseline (week 0, 398.6 ± 20.1 mg). The HC+Gal-mediated BW enhancement is significantly ($p < 0.001$) controlled by the consumption of BWA and CoQ_{10}.

3.2. Swimming Behavioral Analysis of Zebrafish

A substantial variation in swimming behavior concerning the swimming distance and swimming pattern was noticed across the different groups (Supplementary Videos S1–S7). The heat map of the swimming trajectories (obtained at 22 weeks of consumption) showed a regular swimming pattern of zebrafish (free movement across the tank) in the ND and BWA groups (Figure 3A). Similarly, the zebrafish in the HC and CoQ_{10} groups displayed free-swimming movements; however, the occasional thigmotaxic behavior (time spent at the bottom and side of the tank) was also noticed. In contrast, the HC+Gal group exhibited a marked cessation of swimming behavior, characterized by pronounced bottom-dwelling and thigmotaxic tendencies.

The swimming behavioral studies at 24 weeks of consumption revealed a free exploration of zebrafish movement spanning in the horizontal and vertical directions across the tank (Figure 3B). A noticeable maximum swimming time in the upper half (65.7%), the least latency time (3.1 s, towards the top phase), and the maximum swimming distance (4.3 m) was noticed in the zebrafish of the ND group (Figure 3C–F). A severe burst with clumsy swimming that is highly restricted to the bottom side was noticed in the HC+Gal group. Contrary to the ND group, a significant 3.2-fold delayed latency time, 59-fold reduced swimming time in the upper half, and 4.6-fold reduced total swimming distance were noticed in the HC+Gal group. The HC+Gal-induced severity was effectively curtailed by the consumption of BWA at both the tested concentrations (0.5 and 1.0%). As seen in Figure 3B–D, free-swimming movement across the tank with a significant 3.0-fold ($p < 0.05$) and 4.5-fold ($p < 0.01$) better latency time was observed in the 0.5% and 1.0% BWA groups compared to the HC+Gal group. Likewise, the stay-up time in the top half was significantly, 33.9-fold ($p < 0.001$) and 42.6-fold ($p < 0.001$), higher, and 3.6-fold ($p < 0.001$) and 4.2-fold ($p < 0.001$) higher total swimming distances were observed in the 0.5% and 1.0% BWA groups than the HC+Gal group (Figure 3E). The consumption of CoQ_{10} also displayed a substantial preventive effect against the HC+Gal-induced swimming impairment of zebrafish. However, when compared with BWA, precisely at 1.0%, a significant 1.2-fold ($p < 0.01$) and 1.4-fold ($p < 0.001$) lower stay time in the top half and total swimming distance, respectively, was noticed in the 1.0% CoQ_{10} group.

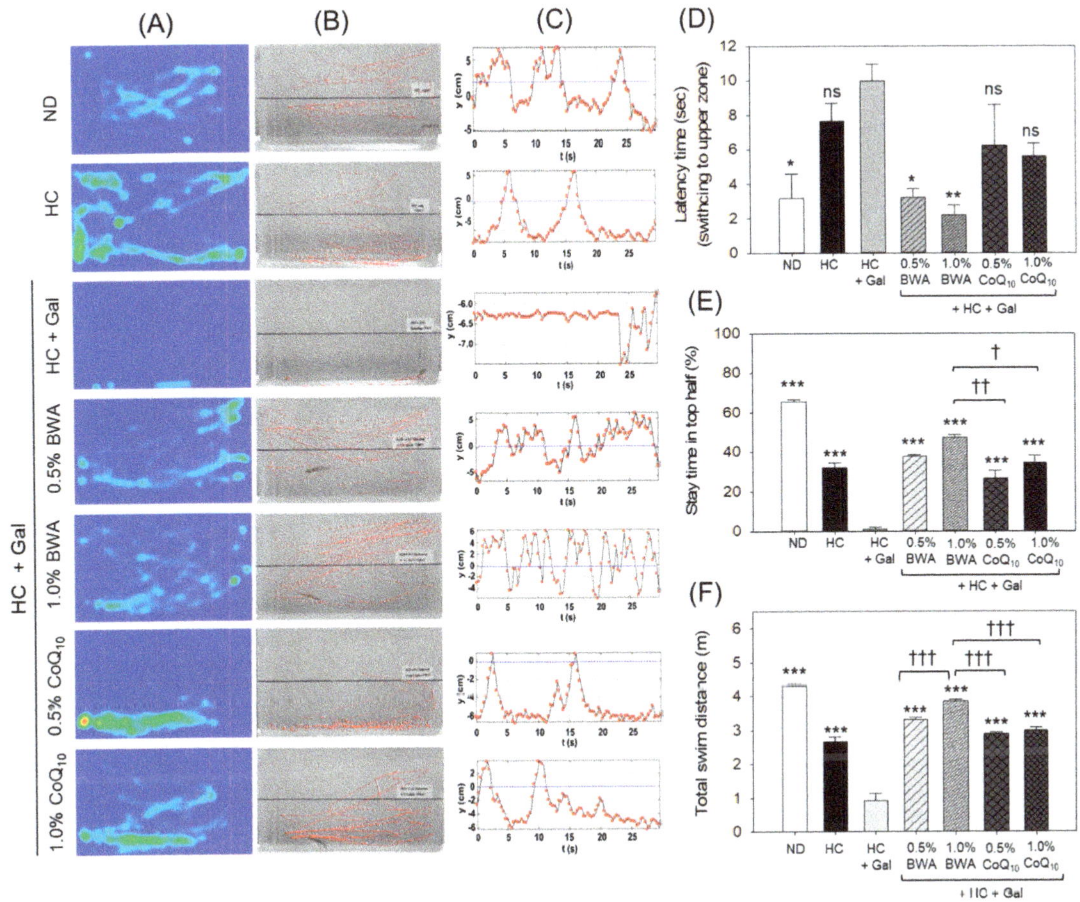

Figure 3. A representative image of the swimming trajectories and swimming behavior of zebrafish across different groups. (**A**) Heat map of zebrafish swimming trajectory for 60 s following 22 weeks consumption of different diets. (**B**) Swimming trajectory of zebrafish motion (60 s) across different groups after 24 weeks of consumption of different diets. (**C**) The vertical movement trajectories of zebrafish across different groups. The blue dotted line inside the images represents the middle line segregating the upper and lower halves of the water tank. (**D**) The latency time of movement (represents the average time zebrafish took to transition from the lower to the upper half) in 60 s. (**E**) Time of stay in the top half of the tank and (**F**) total swimming distance during 60 s. ND represents the normal diet; HC represents the high-cholesterol (final 4%, *w*/*w*) diet; HC+Gal represents the high-cholesterol (final 4%, *w*/*w*) with galactose (final 30%, *w*/*w*) diet; and HC+Gal+BWA (0.5%/1.0%) or CoQ_{10} (0.5%/1.0%) represents the high-cholesterol + galactose diet supplemented with beeswax alcohol (final 0.5% or 1.0% *w*/*w*) or coenzyme Q_{10} (final 0.5% or 1.0% *w*/*w*), respectively. * ($p < 0.01$), ** ($p < 0.01$) and *** ($p < 0.001$) depict the statistical difference with respect to the HC+Gal group, while † ($p < 0.05$), †† ($p < 0.01$), and ††† ($p < 0.01$) depict the statistical difference with respect to the HC+Gal+1.0% BWA group. "ns" represents the non-significant difference between the specified group concerning the HC+Gal group.

3.3. Lipid Profile, Glucose Level, and Antioxidant Status of Blood

As shown in Figure 4A,B, the ND group showed the lowest plasma TC and TG levels, while the HC group showed 1.7-fold and 2.0-fold elevations in TC and TG levels to those of the ND group. Furthermore, the HC+Gal group showed the highest TC and TG levels, which were significantly, 2.0- ($p < 0.001$) and 2.5-fold ($p < 0.001$), higher than in the ND group and 1.2- ($p < 0.001$) and 1.3-fold ($p < 0.001$) higher than in the HC group. However, the co-consumption of BWA resulted in a remarkable decrease in TC and TG levels; the 1.0% BWA group showed the lowest levels of TC and TG, i.e., 50% and 54% less than the HC+Gal group. Similarly, the CoQ_{10} displayed 31–35% less TC and 18–30% lower TG levels than the HC+Gal group.

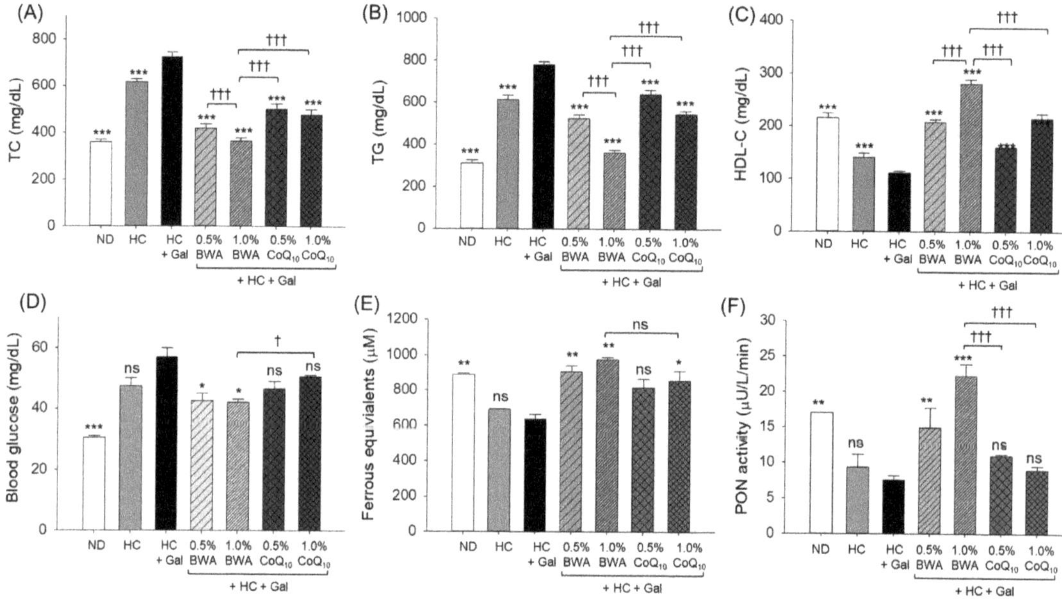

Figure 4. Blood lipid profile, blood glucose level, and antioxidant status of zebrafish (in different groups) consuming the specified diets for 24 weeks. (**A**) Total cholesterol (TC), (**B**) triglycerides (TG), (**C**) high-density lipoprotein cholesterol (HDL-C), (**D**) blood glucose level, (**E**) ferric ion reduction assay (FRA), and (**F**) paraoxonase (PON) activity. ND represents the normal diet; HC represents the high-cholesterol (final 4%, *w*/*w*) diet; HC+Gal represents the high-cholesterol (final 4%, *w*/*w*) with galactose (final 30%, *w*/*w*) diet; and HC+Gal+BWA (0.5%/1.0%) or CoQ_{10} (0.5%/1.0%) represents the high-cholesterol + galactose diet supplemented with beeswax alcohol (final 0.5% or 1.0% *w*/*w*) or coenzyme Q_{10} (final 0.5% or 1.0% *w*/*w*), respectively. * ($p < 0.05$), ** ($p < 0.01$) and *** ($p < 0.001$) depicts the statistical difference with respect to the HC+Gal group, while † ($p < 0.001$) and ††† ($p < 0.001$) depicts the statistical difference with respect to the HC+Gal+1.0% BWA group. "ns" represents the non-significant difference between the specified group concerning the HC+Gal group or HC+Gal+1.0% BWA group.

As shown in Figure 4C, the HC+Gal group showed the lowest HDL-C level, which was 49% lower than the ND group and 22% lower than the HC-consumed group, respectively. The supplementation of both BWA and CoQ_{10} displayed a substantial effect on the elevation of HDL-C disturbed by the consumption of HC+Gal. However, the BWA at 1.0% consumption displayed the most promising results by elevating the HDL-C level by 1.8-fold ($p < 0.001$) and 1.4-fold ($p < 0.001$) more than those of 0.5% and 1.0% CoQ_{10}, respectively.

A significant elevation in the blood glucose level was observed in the HC+Gal group (57.0 mg/dL) than the ND (30.5 mg/dL) and HC (47.5 mg/dL) groups (Figure 4D). Consumption of both BWA and CoQ_{10} effectively reduced the HC+Gal-elevated blood glucose level, as reflected by the 42.0 mg/dL and 50.5 mg/dL blood glucose in the 1.0% BWA and 1.0% CoQ_{10} groups.

The plasma antioxidant status of zebrafish among the different groups was determined by an FRA assay (Figure 4E). The HC+Gal group showed a lower (636.4 μM, $p < 0.01$) FRA value than the ND group (888.6 μM), which was significantly elevated by 42% ($p < 0.01$) and 53% ($p < 0.01$) in response to the 0.5% and 1.0% BWA consumption groups, respectively. Similarly, a 34.4% ($p < 0.05$) higher FRA value was noticed in the 1.0% CoQ_{10} group than in the HC+Gal group.

A significantly compromised plasma PON activity (7.5 μU/L/min) was noticed in the HC+Gal group than the PON activity (17.1 μU/L/min) observed in the ND group (Figure 4F). The consumption of BWA displayed a dose-dependent effect and elevated PON activity by 2.0-fold ($p < 0.01$) and 3.0-fold ($p < 0.001$) in response to 0.5% and 1.0% BWA than the HC+Gal group. No significant effect on the elevation of PON activity was observed in response to the consumption of CoQ_{10}.

3.4. Morphological Changes in Brain and Liver

Figure 5A–C depicts the morphology of the whole zebrafish, brain, and liver after 24 weeks of consumption of different diets. A negative effect of HC was noticed on the brain/body weight that became worse in the presence of galactose (HC+Gal), reflected by an 8% lower brain/bodyweight in the HC+Gal group than the HC group (Figure 5B,D). Notably, the most substantial protective effect was observed in the 1.0% BWA group, which displayed a significant elevation, 23%, 9%, and 12%, in the brain/body weight compared to the HC+Gal, and 0.5% and 1.0% CoQ_{10} groups, respectively. Besides the brain/body weight, the critical morphology analysis revealed that BWA consumption improved brain size and prevented shrinkage caused by HC+Gal, particularly in the cerebellum (marked by the red arrow).

As shown in Figure 5C,E, HC+Gal consumption displayed a substantial elevation in liver size and enhancement of liver/body weight, which was ~14% ($p < 0.001$) higher than the liver/body weight of the ND group. The co-consumption of BWA at both 0.5% and 1.0% effectively prevented HC+Gal-provoked hepatomegaly and decreased the liver/body weight by 25% ($p < 0.01$) and 32% ($p < 0.001$), respectively. Likewise, 0.5% and 1.0% CoQ_{10} also substantially prevented HC+Gal-induced hepatomegaly, evidenced by the 23% and 25% alleviated liver/body weight, respectively, compared to the HC+Gal group.

3.5. Histological Analysis of Brain Tissue

The histology analysis examined by H&E staining revealed a normal brain morphology with well-organized optic tectum (TeO) without hyperemia, with the least vacuolation, and with mononuclear cells with a clear zone in the ND group (Figure 6A,B). As opposed to the ND group, the HC+Gal-consumed group displayed an elevation in vacuolation (indicated by the blue arrow) and number of mononuclear cells with a clear zone (indicated by the red arrow) in the TeO and periventricular grey zone (PGZ) of the brain with the occasional hyperemia (indicated by the green arrow). The consumption of BWA prevented the HC+Gal-induced adverse changes. Precisely, BWA at 1.0% effectively reversed the HC+Gal-induced vacuolation and formation of mononuclear cells with a clear zone. Unlike BWA, CoQ_{10} at both the tested concentrations displayed less effect in managing the HC+Gal-induced impairment in the TeO region of brain tissue.

The immunohistochemical (IHC) staining revealed a ~15% lower myelin-specific fluorescent intensity in the HC-consumed group than in the ND group (Figure 6C,D). Particularly, substantial demyelination was noticed around the valvular cerebella (Val) (indicated by a brown arrow) and in the lateral recess of the diencephalic ventricle (LR) region of the brain section (indicated by a yellow arrow). The consumption of HC+Gal displayed a severe decline in myelin-specific fluorescent intensity, which was substantially, 2.4-fold ($p < 0.001$) and 2.0-fold ($p < 0.01$), lower than the fluorescent intensity observed in the control and HC groups, respectively, suggesting an adverse effect of HC+Gal on the myelin sheath. BWA consumption significantly prevents HC+Gal-induced demyelination, reflected by ~2.2-fold ($p < 0.01$) and 2.0-fold ($p < 0.01$) higher myelin-specific fluorescent intensity in the 0.5% and 1.0% BWA group than the fluorescent intensity observed in the HC+Gal group. Contrary to BWA, CoQ_{10} consumption displayed a non-significant effect in mitigating HC+Gal-provoked demyelination.

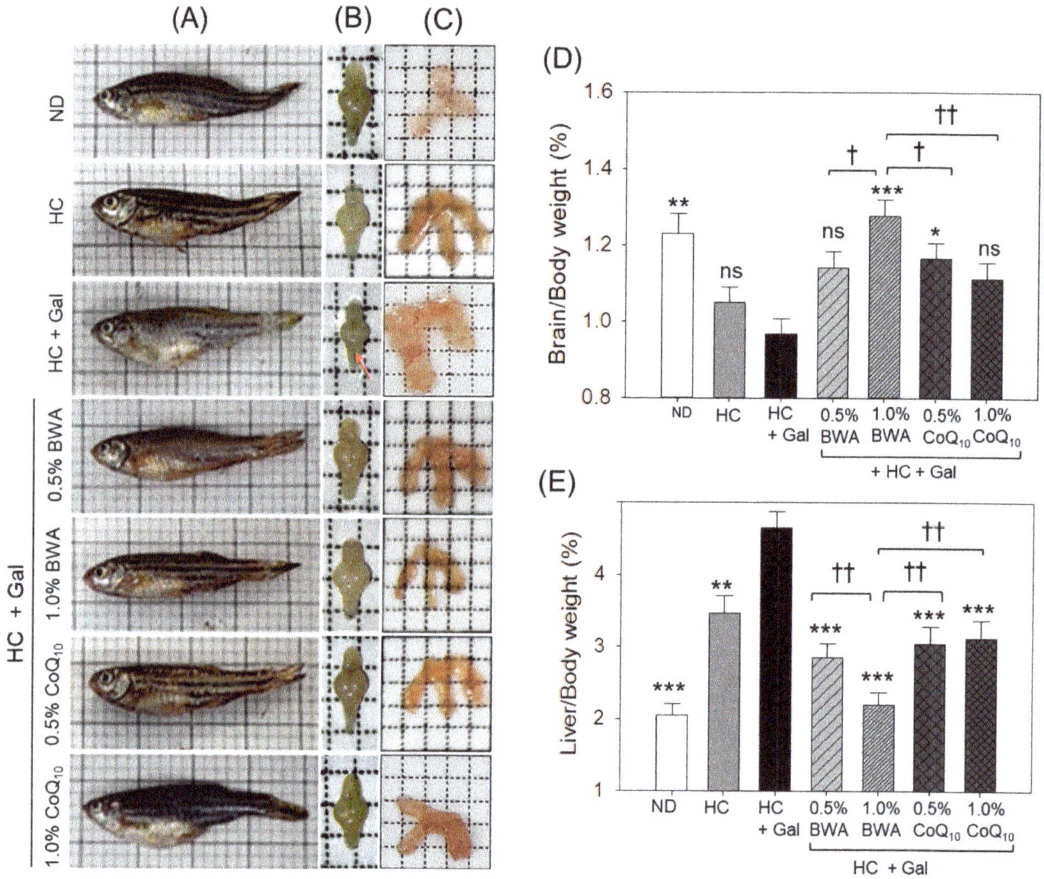

Figure 5. Pictorial view of representative morphological images of (**A**) zebrafish and (**B**) whole brain, with the red arrow depicting shrinkage of the cerebellum of the brain; (**C**) liver segregated from the different groups at 24 weeks of consumption of the designated diets. Percentage average weight (n = 15) of (**D**) brain and (**E**) liver with respect to body weight across the different groups consuming specified diets. ND represents the normal diet; HC represents the high-cholesterol (final 4%, *w*/*w*) diet; HC+Gal represents the high-cholesterol (final 4%, *w*/*w*) with galactose (final 30%, *w*/*w*) diet; and HC+Gal+BWA (0.5%/1.0%) or CoQ_{10} (0.5%/1.0%) represents the high-cholesterol + galactose diet supplemented with beeswax alcohol (final 0.5% or 1.0% *w*/*w*) or coenzyme Q_{10} (final 0.5% or 1.0% *w*/*w*), respectively. * ($p < 0.01$), ** ($p < 0.01$), and *** ($p < 0.001$) depict the statistical difference with respect to the HC+Gal group, while † ($p < 0.05$) and †† ($p < 0.01$) depict the statistical difference with respect to the HC+Gal+1.0% BWA group. "ns" represents the non-significant difference between the specified group concerning the HC+Gal group.

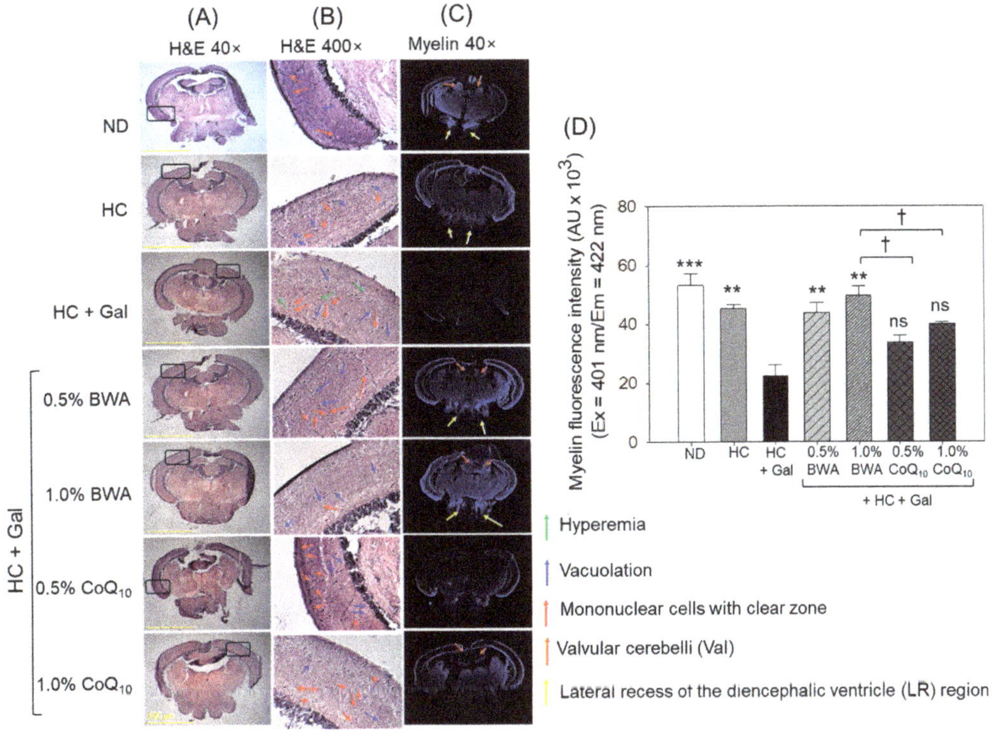

Figure 6. Brain histology of zebrafish at 24 weeks of consumption of the specified diets. (**A**) Images of hematoxylin and eosin (H&E) staining captured at 40× magnification, [scale bar, 0.5 mm]. (**B**) The digitally magnified images of the section inside the black box (as mentioned in (**A**)). (**C**) Immunohistochemical (IHC) images of myelin sheath in the brain section captured at 40× magnification. The yellow arrows highlight the myelin sheath in the lateral recess of the diencephalic ventricle (LR) region of the brain. (**D**) Image J software (version 1.53, https://imagej.net/ij; accessed on 16 June 2023) quantification of myelin fluorescent intensity. ND represents the normal diet; HC represents the high-cholesterol (final 4%, *w*/*w*) diet; HC+Gal represents the high-cholesterol (final 4%, *w*/*w*) with galactose (final 30%, *w*/*w*) diet; and HC+Gal+BWA (0.5%/1.0%) or CoQ_{10} (0.5%/1.0%) represents the high-cholesterol + galactose diet supplemented with beeswax alcohol (final 0.5% or 1.0% *w*/*w*) or coenzyme Q_{10} (final 0.5% or 1.0% *w*/*w*), respectively. ** ($p < 0.01$), and *** ($p < 0.001$) depict the statistical difference with respect to the HC+Gal group, while † ($p < 0.05$) depict the statistical difference with respect to the HC+Gal+1.0% BWA group. "ns" represents the non-significant difference between the specified group concerning the HC+Gal group.

3.6. Reactive Oxygen Species (ROS) Production, Apoptosis, and Senescence in Brain

As shown in Figure 7, the ND-alone group showed the least red fluorescence (ROS production), green fluorescence (apoptosis), and SA-β-gal-stained areas (senescence), while the HC+Gal group showed the most ROS production, apoptosis, and senescence. In the HC+Gal group, prominent red and green fluorescence was detected across the PGZ and TeO (as indicated by the yellow arrow) that mainly concentrated adjacent to the valvular cerebella (Val) (as indicated by the white arrow) (Figure 7A,B). The most vigorous blue intensity (interchanged with red), corresponding to SA-β-gal-positive cells, were detected around the vascular lacuna of area postrema (Vas) and below the tectal ventricle (TeV) region (as indicated by the red arrow) (Figure 7D–F). The 0.5% and 1.0% BWA consumption groups displayed substantial, 1.5-fold ($p < 0.001$) and 1.6-fold ($p < 0.001$), reductions in DHE and 1.8-fold ($p < 0.001$) and 2.0-fold ($p < 0.001$) decreased AO fluorescent intensity compared to the respective fluorescent intensities that appeared in the HC+Gal group (Figure 7G,H). Similarly, the 0.5% and 1.0% BWA consumption group efficiently prevented the HC+Gal-induced cellular senescence manifested by a significantly, 3.3-fold ($p < 0.001$) and 4.6-fold ($p < 0.001$), reduced SA-β-gal-stained area, respectively, than the HC+Gal group (Figure 7I). On the other hand, no beneficial effect of CoQ_{10} consumption (at 0.5% and 1.0%) was observed in curtailing HC+Gal-induced ROS production. Likewise, CoQ_{10} consumption at 0.5% was found ineffective in preventing HC+Gal-induced apoptosis; however, a significant anti-apoptotic effect reflected by a 1.4-fold ($p < 0.01$)-reduced AO fluorescent intensity compared to the HC+Gal group was noticed in response to 1.0% CoQ_{10} consumption. Interestingly, CoQ_{10} at the tested concentrations of 0.5% and 1.0% effectively prevents HC+Gal-induced cellular senescence (Figure 7I); however, the effect is inferior ($p < 0.05$) compared to the 1.0% BWA group.

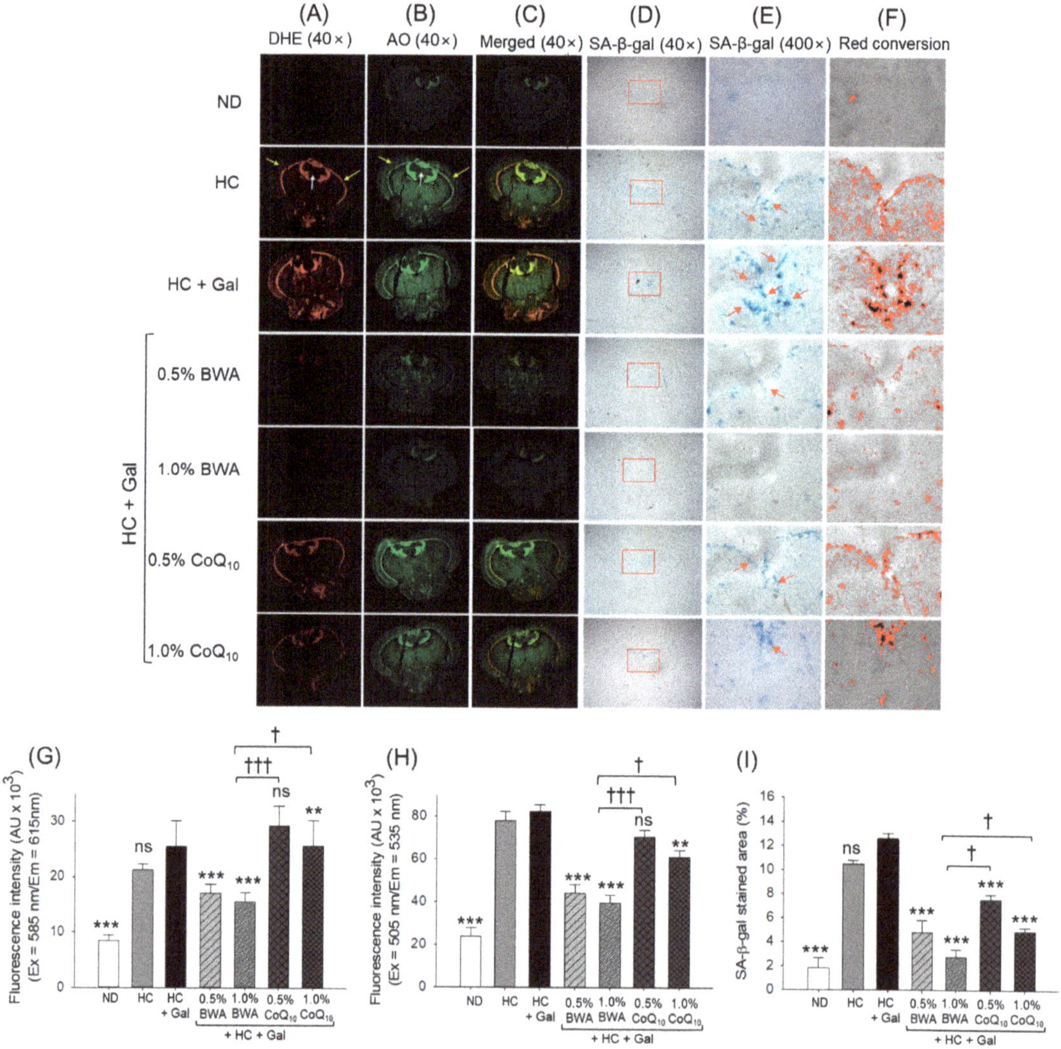

Figure 7. Fluorescent staining: (**A**) dihydroethidium (DHE), (**B**) acridine orange (AO), (**C**) merged images of DHE and AO captured at 40× magnification. (**D**) Cellular senescence-associated-β-galactose

(SA-β-gal) staining at 40× magnification. (**E**) The 400× magnified images of the SA-β-gal-stained section inside the red box (as mentioned in (**D**)). The yellow arrow represents the periventricular gray zone of the optic tectum. The white arrow depicts the valvular cerebelli while the red arrow highlights the SA-β-gal positive cells around the vascular lacuna of the postrema and below the tectal ventricle region. (**F**) Red conversion of the (400× magnified) blue-colored SA-β-gal-positive cells. The red conversion was performed with Image J software (version 1.53) at the blue color threshold value (0–120) to enhance the visibility of the SA-β-gal-stained area. (**G**) and (**H**) Quantification of DHE and AO fluorescent intensity, respectively. (**I**) Quantification of the SA-β-gal-stained area. ND represents the normal diet; HC represents the high-cholesterol (final 4%, *w*/*w*) diet; HC+Gal represents the high-cholesterol (final 4%, *w*/*w*) with galactose (final 30%, *w*/*w*) diet; and HC+Gal+BWA (0.5%/1.0%) or CoQ_{10} (0.5%/1.0%) represents the high-cholesterol + galactose diet supplemented with beeswax alcohol (final 0.5% or 1.0% *w*/*w*) or coenzyme Q_{10} (final 0.5% or 1.0% *w*/*w*), respectively. ** ($p < 0.01$) and *** ($p < 0.001$) depict the statistical difference with respect to the HC+Gal group, while † ($p < 0.05$) and ††† ($p < 0.001$) depict the statistical difference with respect to the HC+Gal+1.0% BWA group. "ns" represents the non-significant difference between the specified group concerning the HC+Gal group.

3.7. H&E Staining and Interleukin (IL)-6 Production in Liver Tissue

As shown in Figure 8A, H&E staining revealed that the HC+Gal group showed the highest neutrophil infiltration and H&E-stained area in the hepatic tissue. Co-consumption of BWA at 0.5% and 1.0% efficiently lowered the H&E-stained area around 46% ($p < 0.001$) and 65% ($p < 0.001$), and neutrophil counts by 60% ($p < 0.01$) and 64% ($p < 0.01$), respectively, compared to the HC+Gal group (Figure 8A,E). Contrary to BWA, co-consumption of CoQ_{10} at both the tested concentrations (0.5% and 1.0%) did not have a preventive effect against HC+Gal-induced hepatic damage. Notably, lipid droplets (indicated by the blue arrow) were highly prevalent in the HC+Gal group, which was substantially reduced with the consumption of 1.0% BWA.

Immunohistochemical (IHC) staining revealed a 2.6-fold higher IL-6 production in the HC-consumed group than in the control ND group (Figure 8C–F). The production of IL-6 induced by HC was significantly elevated by 1.5-fold ($p < 0.001$) when combined with galactose (HC+Gal), highlighting the detrimental role of galactose in exacerbating HC-induced inflammation (Figure 8F). The HC+Gal-induced IL-6 production was significantly, 2.1-fold ($p < 0.001$) and 2.9-fold ($p < 0.001$), reduced in the 0.5% and 1.0% BWA groups. Similarly, a 1.4-fold ($p < 0.001$) and 1.2-fold ($p < 0.001$) lower IL-6 production was observed in the 0.5% and 1.0% CoQ_{10} groups compared to the HC+Gal group. Notably, 1.0% BWA demonstrated significantly greater efficacy, inhibiting IL-6 production, reflected by the ~2.0-fold ($p < 0.001$) lower IL-6-stained area than the 0.5% and 1.0% CoQ_{10} groups, underscoring the superior activity of BWA over CoQ_{10}.

3.8. Fatty Liver Change, ROS Production, Apoptosis, and Cellular Senescence

The HC+Gal group exhibited the most pronounced increase in dark red intensity from oil red O-staining, DHE, and AO fluorescence, which were 19-fold ($p < 0.001$), 4.4-fold ($p < 0.001$), and 5.8-fold ($p < 0.001$) higher, respectively, compared to the ND-alone group (Figure 9). The 1.0% BWA group displayed remarkable decreases of 16.6-fold ($p < 0.001$), 3.5-fold ($p < 0.001$), and 4.0-fold ($p < 0.001$) in the fatty liver changes (Figure 9A,E), ROS production (Figure 9B,F), and apoptosis (Figure 9C,G) actuated by the consumption of HC+Gal. In comparison, the CoQ_{10} groups did not show notable decreases in ROS production and apoptosis in both consumed doses of 0.5% and 1.0%. Contrary to ROS and apoptosis, the consumption of 0.5% and 1.0% CoQ_{10} effectively reverted the HC+Gal-induced fatty liver changes, reflected by 4-fold ($p < 0.001$) and 2.9-fold ($p < 0.001$) reductions in ORO-stained areas, respectively.

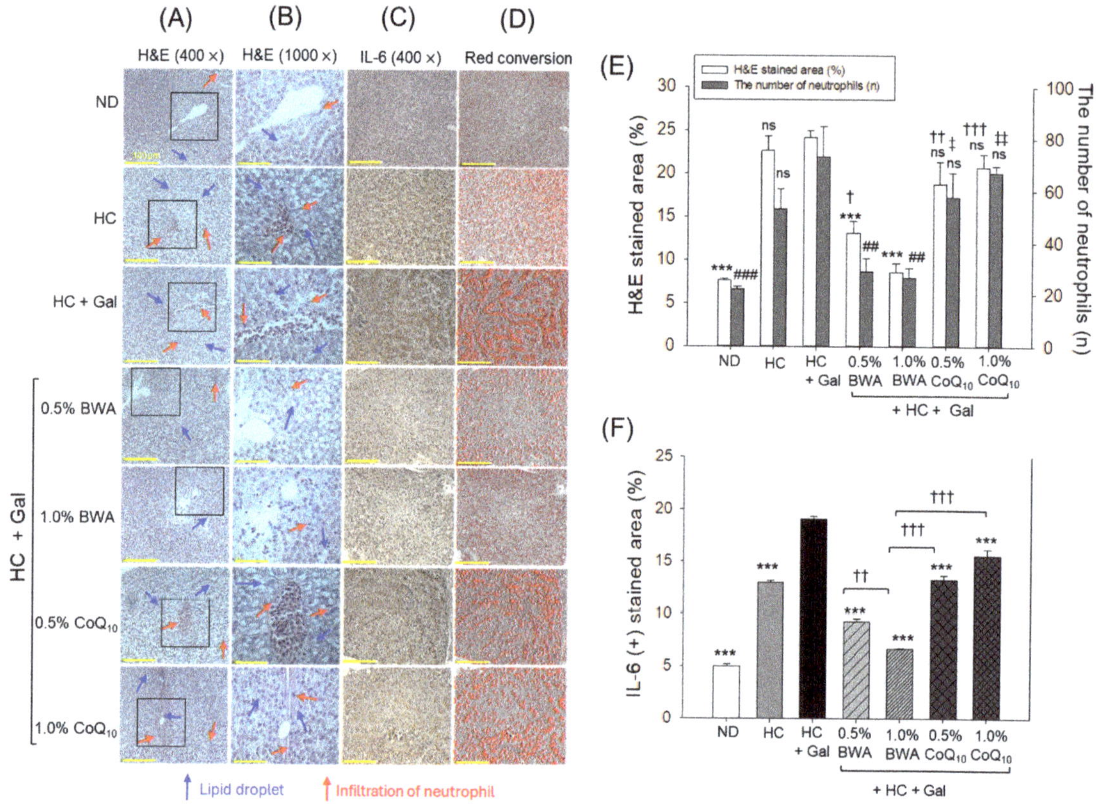

Figure 8. Hepatic histology of zebrafish across different groups consuming the specified diets for 24 weeks. (**A**) Hematoxylin and eosin (H&E) staining at 400× magnifications. (**B**) A 1000× magnified view of the H&E-stained section inside the black box (as mentioned in (**A**)). The red and blue arrows indicate neutrophil and lipid accumulation. (**C**) Interleukin (IL)-6-stained area examined by immunohistochemistry (IHC). (**D**) Brown-colored (IL-6 stained) area interchanged with red at the brown color threshold value 20–120 to enhance visibility. [Scale bar, 0.1 mm]. (**E**) Quantification of H&E-stained area and the number of neutrophils. (**F**) Percentage quantification of the IL-6-stained area. ND represents the normal diet; HC represents the high-cholesterol (final 4%, *w*/*w*) diet; HC+Gal represents the high-cholesterol (final 4%, *w*/*w*) with galactose (final 30%, *w*/*w*) diet; and HC+Gal+BWA (0.5%/1.0%) or CoQ_{10} (0.5%/1.0%) represents the high-cholesterol + galactose diet supplemented with beeswax alcohol (final 0.5% or 1.0% *w*/*w*) or coenzyme Q_{10} (final 0.5% or 1.0% *w*/*w*), respectively. *** ($p < 0.001$) (for H&E stained and IL-6-stained area), ## ($p < 0.01$), and ### ($p < 0.001$) (for neutrophil counts) depict the statistical difference with respect to the HC+Gal group. † ($p < 0.05$), †† ($p < 0.01$), and ††† ($p < 0.001$) (for H&E stained and IL-6-stained area), and ‡ ($p < 0.05$) and ‡‡ ($p < 0.001$) depict the statistical difference with respect to the HC+Gal+1.0% BWA group, respectively. "ns" represents the non-significant difference between the specified group concerning the HC+Gal group.

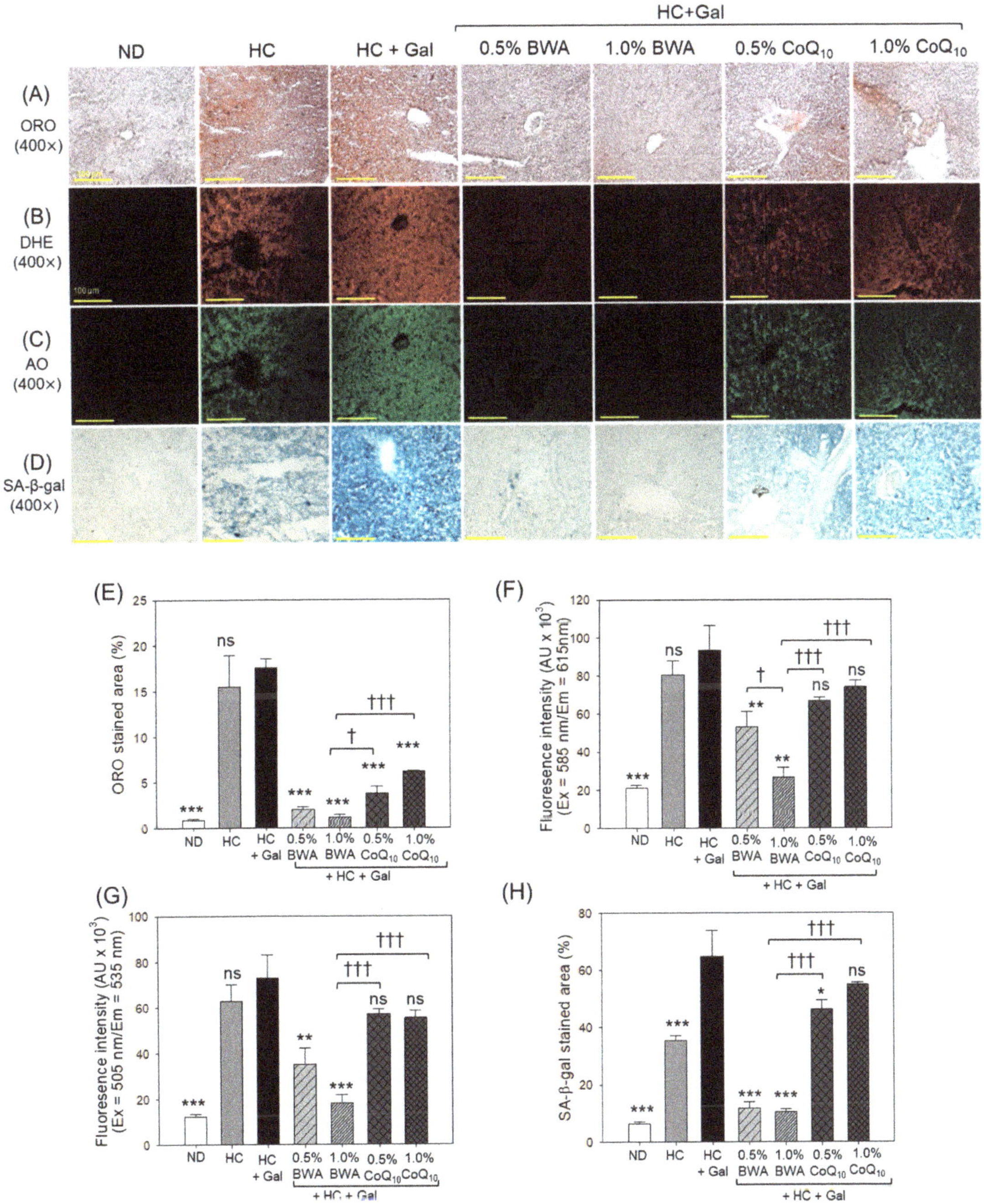

Figure 9. Evaluation of (**A**) oil red O (ORO) staining, (**B**,**C**) dihydroethidium (DHE) and acridine orange (AO) fluorescent staining, and (**D**) senescence associated-βgalactose (SA-β-gal) staining in hepatic tissue following 24 weeks of consumption of the specified diets. (**E**) Quantification of the ORO-stained area. (**F**) and (**G**) DHE and AO fluorescent intensity quantification, respectively. (**H**) Quantification of SA-β-gal-stained areas. ND rpresents the normal diet; HC represents the high-cholesterol (final 4%, *w*/*w*)

diet; HC+Gal represents the high-cholesterol (final 4%, *w*/*w*) with galactose (final 30%, *w*/*w*) diet; and HC+Gal+BWA (0.5%/1.0%) or CoQ_{10} (0.5%/1.0%) represents the high-cholesterol+galactose diet supplemented with beeswax alcohol (final 0.5% or 1.0% *w*/*w*) or coenzyme Q_{10} (final 0.5% or 1.0% *w*/*w*), respectively. * ($p < 0.05$), ** ($p < 0.01$), and *** ($p < 0.001$) depict the statistical difference with respect to the HC+Gal group. † ($p < 0.05$) and ††† ($p < 0.001$) depict the statistical difference with respect to the HC+Gal+1.0% BWA group, respectively. "ns" represents the non-significant difference between the specified group concerning the HC+Gal group.

The SA-β-gal staining revealed higher senescence in the HC+Gal group that accounts for 64.8% SA-β-gal-positive cells, which was significantly 10-fold ($p < 0.001$) and 1.8-fold ($p < 0.001$) higher than the SA-β-gal-positive cells that occurred in the ND and HC-alone groups, respectively (Figure 9D,H). HC+Gal-induced cellular senescence was significantly prevented by the consumption of BWA, as reflected by the 5.5-fold ($p < 0.001$) and 6.1-fold ($p < 0.001$) lower SA-β-gal-positive cells in the hepatic section from the 0.5% and 1.0% BWA groups, respectively, compared to the HC+Gal group. Unlike BWA, the consumption of 1.0% CoQ_{10} displayed a non-significant effect in preventing HC+Gal-induced cellular senescence.

3.9. Plasma Parameters of Hepatic Damage

As shown in Figure 10, the plasma hepatic function biomarkers AST and ALT levels were significantly, 40% ($p < 0.001$) and 34% ($p < 0.001$), higher in the HC group compared to the ND group. The co-consumption of galactose with HC (HC+Gal) substantially elevated the HC-induced AST and ALT levels by ~10% ($p < 0.01$). The intake of both BWA and CoQ_{10} was found to be efficient in preventing HC+Gal-induced AST and ALT levels. Significantly, 25% ($p < 0.01$) and 30% ($p < 0.01$), lower AST (Figure 10A) and ALT (Figure 10B) levels were noticed in the 1.0% BWA groups than in the HC+Gal group, which was significantly, 12% ($p < 0.01$) and 21% ($p < 0.001$), better than the effect exerted by 1.0% CoQ_{10}.

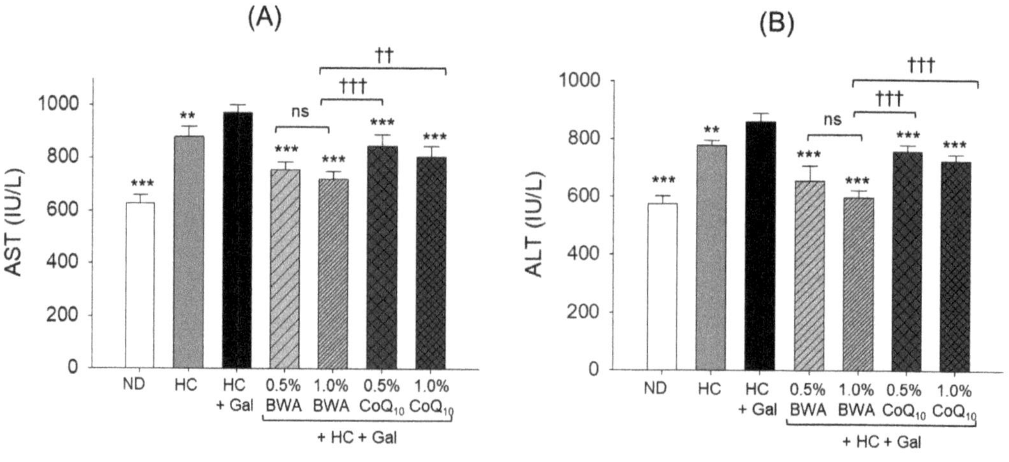

Figure 10. Plasma hepatic function biomarkers—(**A**) aspartate aminotransferase (AST) and (**B**) alanine aminotransferase (ALT)—in different groups consuming the specified diets for 24 weeks. ND represents the normal diet; HC represents the high-cholesterol (final 4%, *w*/*w*) diet; HC+Gal represents the high-cholesterol (final 4%, *w*/*w*) with galactose (final 30%, *w*/*w*) diet; and HC+Gal+BWA (0.5%/1.0%) or CoQ_{10} (0.5%/1.0%) represents the high-cholesterol + galactose diet supplemented with beeswax alcohol (final 0.5% or 1.0% *w*/*w*) or coenzyme Q_{10} (final 0.5% or 1.0% *w*/*w*), respectively. ** ($p < 0.01$) and *** ($p < 0.001$) depict the statistical difference with respect to the HC+Gal group, while †† ($p < 0.01$) and ††† ($p < 0.001$) depict the statistical difference with respect to the HC+Gal+1.0% BWA group. "ns" represents the non-significant difference between the specified group concerning the HC+Gal group.

4. Discussion

Various animal studies describe the detrimental effects of high cholesterol consumption [32–34]. Similarly, the consumption of high amounts of sugar associated with obesity, type 2 diabetes, metabolic syndrome, and cardiovascular diseases has been documented [35]. This study revealed that the high consumption of Gal (40%) can cross the placenta and impair brain development of the fetus [36]. The excessive intake of reducing sugars also causes the non-enzymatic glycation of protein that successively produces distinct toxicogenic advanced glycation end products (AGEs) [37]. The high intake of both cholesterol and carbohydrates serves as an excellent dietary model for causing obesity, hepatic damage, and insulin resistance. It also resembles the pathology and progression of disease similar to humans [38].

The outcome highlights the higher zebrafish mortality and body weight gain in the HC+Gal group, compared to the HC group, testifying to the additive effect of Gal on the augmentation of HC-provoked mortality and body weight gain. HC+Gal-induced zebrafish mortality and body weight gain were substantially prevented by the consumption of BWA, which was substantially better than the effect exerted by CoQ_{10}, attesting to the higher efficacy of BWA over CoQ_{10} in mitigating HC+Gal-induced mortality and obesity.

The adverse effect of high Gal on cellular antioxidants (SOD, catalase, and GSH-Px) [39], generation of free radicals, and induction of oxidative stress [40], which leads to cognitive disorder and brain aging has been established [40–43]. Consistently, we have also noticed a substantial adverse effect of HC+Gal on cognitive impairment and brain damage that was substantially restored by the consumption of BWA and CoQ_{10}. However, compared to CoQ_{10}, a much higher effect of BWA was observed at both the tested concentrations (0.5 and 1.0%). The higher impact of BWA over CoQ_{10} might be attributed to its substantially better antioxidant activity than CoQ_{10} [22], which also modulates the cellular antioxidants SOD, catalase, and GSH-Px [44,45].

Gal accumulation in the brain provokes glycation, leads to the generation of AGEs [40,46], and causes brain damage and cognitive change [47,48]. Therefore, a substance that can inhibit AGE formation substantially affects brain health and cognitive changes. The effect of BWA on preventing protein glycation has been recognized [21,22]; moreover, BWA, compared to CoQ_{10}, displayed a much more superior activity that protects zebrafish embryos against carboxymethyllysine (CML) (a common AGE)-induced toxicity [22], underscoring the higher efficacy of BWA over CoQ_{10} in meeting CML-induced challenges.

The adverse effect of Gal has also been noticed to disrupt myelin [49], which is an important multicellular membrane of the central nervous system (CNS) and associated with cognitive processes associated with the brain [50]. The consumption of BWA significantly mitigated demyelination in response to HC+Gal. We speculate that the substantially high myelin content in the BWA group is due to its notable inhibitory impact on cellular ROS, which has been recognized among the main culprits that provoke demyelination [51] and the subsequent brain aging and cognitive changes [52].

Also, the findings of plasma FRA and PON activity support the antioxidant nature of BWA and align with earlier studies documenting the impact of BWA on the plasma oxidative variables and the modulatory effect on cellular antioxidants [53,54]. The significant influence of BWA in enhancing the activity of HDL-associated antioxidant PON underscores its role in improving HDL functionality, a critical particle with diverse beneficial properties. The results are aligned with previous in vitro findings, where BWA was documented to augment PON-1 activity, while no effect was observed for CoQ_{10} [22], strengthening the superior antioxidant modulatory effect of BWA over CoQ_{10}.

Similarly, a higher potency of BWA than CoQ_{10} was observed in countering HC+Gal-induced dyslipidemia. The impactful role of CoQ_{10} against dyslipidemia has been described in animal and human studies [55]. On the contrary, the effect of BWA on dyslipidemia has been described in a limited number of studies. In one of our earlier studies, BWA effectively countered dyslipidemia induced by a high-cholesterol diet [23]. The high prevalence of long chain aliphatic alcohol (LCAA) like hexacosanol, triacontanol, and octacosanol in BWA are the key contributors responsible for the protective effect of BWA against HC+Gal-induced dyslipidemia. The assumption is based on earlier studies documenting the influence of hexacosanol, triacontanol, and octacosanol on cholesterol biosynthesis [56–59]. The pivotal role of hexacosanol in activating the AMP-activated protein kinase (AMPK) and subsequent inhibition of 3-hydroxy-methylglutaryl-coenzyme A reductase (HMG-CoA), a rate-limiting enzyme for cholesterol biosynthesis, has been noticed as the key event in managing cholesterol biosynthesis [57]. Moreover, a substantial modulatory effect of hexacosanol on the nuclear translocation of sterol regulatory element binding protein-2 (SREBP-2), which is an important transcriptional regulatory molecule for the expression of HMG-CoA, has been recognized as an essential event to regulate cholesterol biosynthesis [57]. A similar effect of octacosanol has been documented to inhibit cholesterol biosynthesis [58,59].

The role of galactose, high-fat diets, and oxidative stress in elevating blood glucose levels has been well documented, mainly due to their association with insulin resistance or impaired insulin secretion [60,61]. Herein, the HC+Gal group exhibited elevated blood glucose levels, which were significantly mitigated by supplementation with BWA. This effect of BWA can be attributed to its potent antioxidant properties, which help to counteract oxidative stress elevated by HC+Gal. Additionally, numerous studies have highlighted the ability of BWA to ameliorate insulin resistance [62], which consequently impacts blood glucose levels. These findings align with previous research demonstrating policosanol's blood glucose-lowering effect in zebrafish and human subjects [63].

BWA also displayed a substantial protective effect against HC+Gal-induced hepatic damage that is substantially better than the effect of CoQ_{10}. The results are aligned with those in an earlier published article, which suggests that the 6-month consumption of BWA cures fatty liver changes and insulin resistance in patients with non-alcoholic fatty liver disease (NAFLD) [62]. The better effect of BWA than CoQ_{10} can be highlighted by the higher antioxidant potential of BWA than CoQ_{10} [22], which substantially reduces oxidative stress in various organs [23]. Consistent with the hepatic histology results, significantly lower plasma hepatic function biomarker (AST and ALT) levels were observed in the BWA than CoQ_{10} group. The reduced hepatic function biomarkers (AST and ALT) in BWA can be linked with its effect on diminished TC and heightened HDL-C levels, as a negative correlation has been documented between the hepatic function biomarkers, and elevated HDL-C [64] and diminished TC [65]. The effect of hexacosanol (a BWA component) on the induction of autophagy [57] may also contribute to preventing liver damage, as autophagy has been recognized to modulate hepatic lipid accumulation and fatty liver changes [57,66].

The higher IL-6 levels in the HC+Gal group can be linked with heightened ROS production, as different studies established the inductive effect of oxidative stress on the inflammatory pathway [67]. Similar to the present study's outcomes, many earlier studies depicted BWA's role in preventing inflammatory disorders in mice [68] and humans [17], which supports the current findings. The lower extent of apoptosis in response to BWA than that of CoQ_{10} is due to the higher antioxidant activity of BWA, which effectively limits HC+Gal-induced ROS generation and leads to the inhibition of oxidative stress, which has been accredited as a main provocative agent in inducing apoptosis [69]. Likewise, oxidative stress has been recognized as the main provocatory of cellular senescence [70,71]. Compared to CoQ_{10}, BWA demonstrates a higher impact on mitigating HC+Gal-induced oxidative stress, resulting in a more pronounced reduction in cellular senescence. As summarized in Figure 11, the antioxidant and anti-inflammatory nature of BWA is a key contributor towards its diverse functionality in preventing HC+Gal-induced senescence, oxidative stress, fatty liver changes, and brain impairment.

As BWA showed substantial hepatoprotective activity and restored the fatty liver changes induced by the consumption of HC+Gal, it has a positive impact on brain health and cognitive changes. This notion is supported by earlier studies documenting the liver–brain axis [72]. The hepatic encephalopathy caused by liver dysfunction leads to the accumulation of toxic substances in the brain and is characterized by cognitive changes [73].

Many reports documented the association of NAFLD with structural changes in the brain and enhanced risk of dementia [74]. Even an excessive accumulation of fat in the liver alters liver functionality and changes the composition of fat-derived molecules in the blood, which subsequently triggers inflammation in the blood–brain barrier and provokes the entry and accumulation of toxic substances and inflammatory cells into the brain [74], leading to several detrimental effects including cognitive changes.

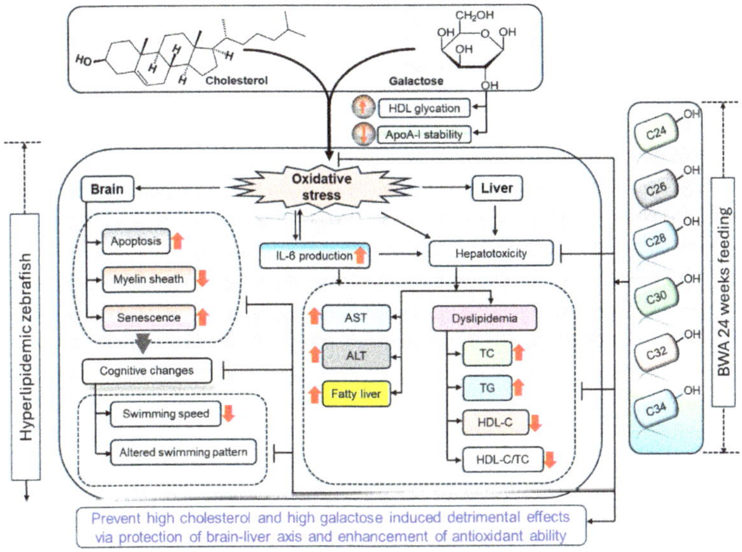

Figure 11. A summary of the protective effect exerted by beeswax alcohol (BWA) consumption to mitigate high-cholesterol + galactose (HC+Gal)-induced adversity in hyperlipidemic zebrafish.

5. Conclusions

This 24-week consumption study displayed the remarkable functionality of BWA in preventing HC+Gal-induced adverse events in the blood, brain, and liver of zebrafish. Compared to CoQ_{10}, BWA displayed a much higher effect in restoring HC+Gal-induced dyslipidemia and altered hepatic function biomarkers, blood glucose levels, and plasma FRA and PON activities. Also, a substantial effect of BWA was noticed in preventing myelin sheath degeneration in the brain and in restoring swimming behavior in zebrafish altered by the consumption of HC+Gal. The results outlined the functional superiority of BWA over CoQ_{10} and endorsed its dietary supplementation to counter high-cholesterol and galactose-induced adverse events and the restoration of cognitive changes.

Supplementary Materials: The following supporting information can be downloaded at https://www.mdpi.com/article/10.3390/antiox13121488/s1, Supplementary Table S1: Certificate of analysis and composition of beeswax alcohol (BWA). Supplementary Figure S1: Ray diagram of a water tank split horizontally into two sections. Supplementary Method S1: A detailed methodology for the quantification of plasma lipid profile and hepatic function biomarkers. Supplementary Videos S1–S7: Swimming behavior of zebrafish across different groups.

Author Contributions: Conceptualization, K.-H.C.; methodology, H.-S.K., A.B., J.-E.K., Y.L., S.H.L. and C.-E.Y.; data curation, writing—original draft preparation, K.-H.C.; writing—review and editing, K.-H.C.; supervision, K.-H.C. All authors have read and agreed to the published version of the manuscript.

Funding: This research received no external funding.

Institutional Review Board Statement: The animal study protocol was approved by the Committee of Animal Care and Use of Raydel Research Institute (approval code RRI-23-001, approval date 27 July 2023).

Informed Consent Statement: Not applicable.

Data Availability Statement: The data used to support the findings of this study are available from the corresponding author upon reasonable request.

Conflicts of Interest: The authors declare no conflicts of interest.

References

1. Azman, K.F.; Safdar, A.; Zakaria, R. D-galactose-induced liver aging model: Its underlying mechanisms and potential therapeutic interventions. *Exp. Gerontol.* **2021**, *150*, 111372. [CrossRef] [PubMed]
2. Bo-Htay, C.; Palee, S.; Apaijai, N.; Chattipakorn, S.C.; Chattipakorn, N. Effects of d-galactose-induced ageing on the heart and its potential interventions. *J. Cell. Mol. Med.* **2018**, *22*, 1392–1410. [CrossRef] [PubMed]
3. Sadigh-Eteghad, S.; Majdi, A.; McCann, S.K.; Mahmoudi, J.; Vafaee, M.S.; Macleod, M.R. D-galactose-induced brain ageing model: A systematic review and meta-analysis on cognitive outcomes and oxidative stress indices. *PLoS ONE* **2017**, *12*, e0184122.
4. Umbayev, B.; Askarova, S.; Almabayeva, A.; Saliev, T.; Masoud, A.R.; Bulanin, D. Galactose-Induced Skin Aging: The Role of Oxidative Stress. *Oxid. Med. Cell. Longev.* **2020**, *2020*, 7145656. [CrossRef] [PubMed]
5. Azman, K.F.; Zakaria, R. D-Galactose-induced accelerated aging model: An overview. *Biogerontology* **2019**, *20*, 763–782. [CrossRef]
6. Haider, S.; Liaquat, L.; Shahzad, S.; Sadir, S.; Madiha, S.; Batool, Z.; Tabassum, S.; Saleem, S.; Naqvi, F.; Perveen, T. A high dose of short-term exogenous D-galactose administration in young male rats produces symptoms simulating the natural aging process. *Life Sci.* **2015**, *124*, 110–119. [CrossRef]
7. Ross, K.S.; Smith, C. D-galactose: A model of accelerated ageing sufficiently sensitive to reflect preventative efficacy of an antioxidant treatment. *Biogerontology* **2020**, *21*, 745–761. [CrossRef]
8. Babic Perhoc, A.; Osmanovic Barilar, J.; Knezovic, A.; Farkas, V.; Bagaric, R.; Svarc, A.; Grünblatt, E.; Riederer, P.; Salkovic-Petrisic, M. Cognitive, Behavioral and Metabolic Effects of Oral Galactose Treatment in the Transgenic Tg2576 Mice. *Neuropharmacology* **2019**, *148*, 50–67. [CrossRef]
9. Meydani, M.; Martin, A.; Sastre, J.; Smith, D.; Dallal, G.; Taylor, A.; Blumberg, J. Dose-response characteristics of galactose-induced cataract in the rat. *Ophthalmic Res.* **1994**, *26*, 368–374. [CrossRef]
10. Zhong, L.; Wang, T.; Wang, T.; Chen, H.; Deng, J.; Ye, H.; Li, W.; Ling, S. Characterization of an i.p. D-galactose-induced cataract model in rats. *J. Pharmacol. Toxicol. Methods* **2021**, *107*, 106891. [CrossRef]
11. Du, Z.; Yang, Y.; Hu, Y.; Sun, Y.; Zhang, S.; Peng, W.; Zhong, Y.; Huang, X.; Kong, W. A long-term high-fat diet increases oxidative stress, mitochondrial damage and apoptosis in the inner ear of D-galactose-induced aging rats. *Hear. Res.* **2012**, *287*, 15–24. [CrossRef] [PubMed]
12. Murphy, R.C.; Johnson, K.M. Cholesterol, reactive oxygen species, and the formation of biologically active mediators. *J. Biol. Chem.* **2008**, *283*, 15521–15525. [CrossRef] [PubMed]
13. Li, K.; Deng, Y.; Deng, G.; Chen, P.; Wang, Y.; Wu, H.; Ji, Z.; Yao, Z.; Zhang, X.; Yu, B.; et al. High cholesterol induces apoptosis and autophagy through the ROS-activated AKT/FOXO1 pathway in tendon-derived stem cells. *Stem Cell Res.* **2020**, *11*, 131. [CrossRef] [PubMed]
14. Más, R. D-002: Beeswax alcohols. *Drugs Future* **2001**, *26*, 731–744. [CrossRef]
15. Zamora, Z.; Molina, V.; Mas, R.; Ravelo, Y.; Perez, Y.; Oyarzabal, A. Protective effects of D-002 on experimentally induced gastroesophageal reflux in rats. *World J. Gastroenterol.* **2014**, *20*, 2085–2090. [CrossRef]
16. Carbajal, D.; Molina, V.; Noa, M.; Valdes, S.; Arruzazabala, M.L.; Aguiar, M.; Mas, R. Effects of D-002 on gastric mucus composition in ethanol induced ulcer. *Pharmacol. Res.* **2000**, *42*, 329–332. [CrossRef]
17. Puente, R.; Illnait, J.; Mas, R.; Carbajal, D.; Mendoza, S.; Fernández, J.C.; Mesa, M.; Gámez, R.; Reyes, P. Evaluation of the effect of D-002, a mixture of beeswax alcohols, on osteoarthritis symptoms. *Korean J. Intern. Med.* **2014**, *29*, 191–202. [CrossRef]
18. Korean Food and Drug Administration (KFDA). Available online: https://www.foodsafetykorea.go.kr/portal/healthyfoodlife/searchHomeHFDetail.do?prdlstReportLedgNo=2023021000330706 (accessed on 16 May 2023).
19. Jorat, M.V.; Tabrizi, R.; Mirhosseini, N.; Lankarani, K.B.; Akbari, M.; Heydari, S.T.; Mottaghi, R.; Asemi, Z. The effects of coenzyme Q_{10} supplementation on lipid profiles among patients with coronary artery disease: A systematic review and meta-analysis of randomized controlled trials. *Lipids Health Dis.* **2018**, *17*, 230. [CrossRef]
20. Jorat, M.V.; Tabrizi, R.; Kolahdooz, F.; Akbari, M.; Salami, M.; Heydari, S.T.; Asemi, Z. The effects of coenzyme Q10 supplementation on biomarkers of inflammation and oxidative stress in among coronary artery disease: A systematic review and meta-analysis of randomized controlled trials. *Inflammopharmacology* **2019**, *27*, 233–248. [CrossRef]
21. Cho, K.-H.; Baek, S.-H.; Nam, H.-S.; Bahuguna, A. Enhancement of antioxidant and anti-glycation properties of beeswax alcohol in reconstituted high-density lipoprotein: Safeguarding against carboxymethyllysine toxicity in zebrafish. *Antioxidants* **2023**, *12*, 2116. [CrossRef]

22. Cho, K.-H.; Baek, S.-H.; Nam, H.-S.; Bahuguna, A.; López-González, L.E.; Rodríguez-Cortina, I.; Illnait-Ferrer, J.; Fernández-Travieso, J.C.; Molina-Cuevas, V.; Pérez-Guerra, Y.; et al. Beeswax alcohol prevents low-density lipoprotein oxidation and demonstrates antioxidant activities in zebrafish embryos and human subjects: A clinical study. *Curr. Issues Mol. Biol.* **2024**, *46*, 409–429. [CrossRef] [PubMed]
23. Cho, K.-H.; Bahuguna, A.; Lee, Y.; Lee, S.H.; Kim, J.-E. Twenty-week dietary supplementation with Beeswax alcohol (BWA.; Raydel®) ameliorates high-cholesterol-induced long-term dyslipidemia and organ damage in hyperlipidemic zebrafish in a dose-dependent manner: A comparative analysis between BWA and coenzyme Q_{10}. *Pharmaceuticals* **2024**, *17*, 1434. [CrossRef] [PubMed]
24. Cho, K.-H.; Bahuguna, A.; Kim, J.-E.; Lee, Y.; Lee, S.H. Comparative assessment of Beeswax alcohol and coenzyme Q_{10} (CoQ_{10}) to prevent liver aging, organ damage, and oxidative stress in hyperlipidemic zebrafish exposed to D-galactose: A 12-Week dietary intervention. *Pharmaceuticals* **2024**, *17*, 1250. [CrossRef] [PubMed]
25. Cho, K.-H.; Kim, J.-E.; Bahuguna, A.; Kang, D.-J. Long-term supplementation of ozonated sunflower oil improves dyslipidemia and hepatic inflammation in hyperlipidemic zebrafish: Suppression of oxidative stress and inflammation against carboxymethyllysine toxicity. *Antioxidants* **2023**, *12*, 1240. [CrossRef] [PubMed]
26. Cho, K.-H.; Kim, J.-E.; Lee, M.-S.; Bahuguna, A. Oral Supplementation of Ozonated Sunflower Oil Augments Plasma Antioxidant and Anti-Inflammatory Abilities with Enhancement of High-Density Lipoproteins Functionality in Rats. *Antioxidants* **2024**, *13*, 529. [CrossRef]
27. Fischer, A.H.; Jacobson, K.A.; Rose, J.; Zeller, R. Hematoxylin and eosin staining of tissue and cell sections. *Cold Spring Harb. Protoc.* **2008**, *2008*, prot4986. [CrossRef]
28. Cho, K.-H.; Bahuguna, A.; Kang, D.-J.; Kim, J.-E. Prolonged supplementation of ozonated sunflower oil bestows an antiaging effect, improves blood lipid profile and spinal deformities, and protects vital organs of zebrafish (*Danio rerio*) against age-related degeneration: Two-years consumption study. *Antioxidants* **2024**, *13*, 123. [CrossRef]
29. Al-Ghamdi, T.H.; Atta, I.S. Efficacy of interleukin-6 in the induction of liver cell proliferation after hemi-hepatectomy: Histopathologic and immunohistochemical study. *Int. J. Clin. Exp. Pathol.* **2020**, *13*, 1540–1549.
30. Owusu-Ansah, E.; Yavari, A.; Mandal, S.; Banerjee, U. Distinct mitochondrial retrograde signals control the G1-S cell cycle checkpoint. *Nat. Genet.* **2008**, *40*, 356–361. [CrossRef]
31. Umali, J.; Hawkey-Noble, A.; French, C.R. Loss of foxc1 in zebrafish reduces optic nerve size and cell number in the retinal ganglion cell layer. *Vision Res.* **2019**, *156*, 66–72. [CrossRef]
32. Han, Y.; Zee, S.; Cho, K.-H. Beeswax alcohol and fermented black rice bran synergistically ameliorated hepatic injury and dyslipidemia to exert antioxidant and anti-inflammatory activity in ethanol-supplemented zebrafish. *Biomolecules* **2023**, *13*, 136. [CrossRef] [PubMed]
33. Udomkasemsab, A.; Prangthip, P. High fat diet for induced dyslipidemia and cardiac pathological alterations in Wistar rats compared to Sprague Dawley rats. *Clin. Investig. Arterioscler.* **2019**, *31*, 56–62.
34. Puska, L.G.; Nagy, Z.B.; Giricz, Z.; Onody, A.; Csonka, C.; Kitajka, K.; Hackler, L., Jr.; Zvara, A.; Ferdinandy, P. Cholesterol diet-induced hyperlipidemia influences gene expression pattern of rat hearts: A DNA microarray study. *FEBS Lett.* **2004**, *562*, 99–104. [CrossRef] [PubMed]
35. Ma, X.; Nan, F.; Liang, H.; Shu, P.; Fan, X.; Song, X.; Hou, Y.; Zhang, D. Excessive intake of sugar: An accomplice of inflammation. *Front. Immunol.* **2022**, *13*, 988481. [CrossRef] [PubMed]
36. Haworth, J.C.; Ford, J.D.; Younoszai, M.K. Effect of galactose toxicity on growth of the rat fetus and brain. *Pediatr. Res.* **1969**, *3*, 441–447. [CrossRef] [PubMed]
37. Aragno, M.; Mastrocola, R. Dietary sugars and endogenous formation of advanced glycation end products: Emerging mechanisms of disease. *Nutrients* **2017**, *9*, 385. [CrossRef]
38. Bala, S.; Ganz, M.; Babuta, M.; Zhuang, Y.; Csak, T.; Calenda, C.D.; Szabo, G. Steatosis, inflammasome upregulation, and fibrosis are attenuated in miR-155 deficient mice in a high fat-cholesterol-sugar diet-induced model of NASH. *Lab. Investig.* **2021**, *101*, 1540–1549. [CrossRef]
39. Qu, Z.; Zhang, J.; Yang, H.; Huo, L.; Gao, J.; Chen, H.; Gao, W. Protective effect of tetrahydropalmatine against d-galactose induced memory impairment in rat. *Physiol. Behav.* **2016**, *154*, 114–125. [CrossRef]
40. Budni, J.; Pacheco, R.; da Silva, S.; Garcez, M.L.; Mina, F.; Bellettini-Santos, T.; de Medeiros, J.; Voss, B.C.; Steckert, A.V.; Valvassori, S.d.S.; et al. Oral administration of d-galactose induces cognitive impairments and oxidative damage in rats. *Behav. Brain Res.* **2016**, *302*, 35–43. [CrossRef]
41. Cui, X.; Zuo, P.; Zhang, Q.; Li, X.; Hu, Y.; Long, J.; Packer, L.; Liu, J. Chronic systemic D-galactose exposure induces memory loss, neurodegeneration, and oxidative damage in mice: Protective effects of R-α-lipoic acid. *J. Neurosci. Res.* **2006**, *84*, 647–654. [CrossRef]
42. Lei, M.; Su, Y.; Hua, X.; Ding, J.; Han, Q.; Hu, G.; Xiao, M. Chronic systemic injection of D-galactose impairs the septohippocampal cholinergic system in rats. *Neuroreport* **2008**, *19*, 1611–1615. [CrossRef] [PubMed]
43. Hsieh, H.-M.; WA, W.-M.; Hu, M.-L. Soy isoflavones attenuate oxidative stress and improve parameters related to aging and Alzheimer's disease in C57BL/6J mice treated with D-galactose. *Food Chem. Toxicol.* **2009**, *47*, 625–632. [CrossRef] [PubMed]
44. Pérez, Y.; Oyárzabal, A.; Mas, R.; Molina, V.; Jiménez, S. Protective effect of D-002, a mixture of beeswax alcohols, against indomethacin-induced gastric ulcers and mechanism of action. *J. Nat. Med.* **2013**, *67*, 182–189. [CrossRef] [PubMed]

45. Molina, V.; Ledón, T.; Ravelo, Y.; Zamora, Z.; Mena, L. Effects of D-002 (Beeswax alcohols) on concentrations of prostaglandin E2 in rat gastric mucosa. *J. Pharm. Pharm. Res.* **2017**, *1*, 2.
46. Münch, G.; Westcott, B.; Menini, T.; Gugliucci, A. Advanced glycation endproducts and their pathogenic roles in neurological disorders. *Amino Acids* **2012**, *42*, 1221–1236. [CrossRef] [PubMed]
47. Yaffe, K.; Lindquist, K.; Schwartz, A.V.; Vitartas, C.; Vittinghoff, E.; Satterfield, S.; Simonsick, E.M.; Launer, L.; Rosano, C.; Cauley, J.A.; et al. Advanced glycation end product level, diabetes, and accelerated cognitive aging. *Neurology* **2011**, *77*, 1351–1356. [CrossRef]
48. D'Cunha, N.M.; Sergi, D.; Lane, M.M.; Naumovski, N.; Gamage, E.; Rajendran, A.; Kouvari, M.; Gauci, S.; Dissanayka, T.; Marx, W.; et al. The effects of dietary advanced glycation end-products on neurocognitive and mental disorders. *Nutrients* **2022**, *14*, 2421. [CrossRef]
49. Shwe, T.; Pratchayasakul, W.; Chattipakorn, N.; Chattipakorn, S.C. Role of D-galactose-induced brain aging and its potential used for therapeutic interventions. *Exp. Gerontol.* **2018**, *101*, 13–36. [CrossRef]
50. Khelfaoui, H.; Ibaceta-Gonzalez, C.; Angulo, M.C. Functional myelin in cognition and neurodevelopmental disorders. *Cell Mol. Life Sci.* **2024**, *81*, 1–14. [CrossRef]
51. O'Sullivan, S.A.; Velasco-Estevez, M.; Dev, K.K. Demyelination Induced by Oxidative Stress Is Regulated by Sphingosine 1-Phosphate Receptors. *Glia* **2017**, *65*, 1119–1136. [CrossRef]
52. Jang, G.; Lee, E.M.; Kim, H.J.; Park, Y.; Bang, N.H.; Kang, J.L.; Park, E.M. Visceral adiposity is associated with iron deposition and myelin loss in the brains of aged mice. *Neurochem. Int.* **2024**, *179*, 105833. [CrossRef] [PubMed]
53. Menéndez, R.; Más, R.; Amor, A.M.; Pérez, Y.; González, R.M.; Fernández, J.; Molina, V.; Jiménez, S. Antioxidant effects of D002 on the in vitro susceptibility of whole plasma in healthy volunteers. *Arch. Med. Res.* **2001**, *32*, 436–441. [CrossRef] [PubMed]
54. Lopez, E.; Illnait, J.; Molina, V.; Oyarzabal, A.; Fernandez, L.; Perez, Y.; Mas, R.; Mesa, M.; Fernandez, J.; Mendoza, S. Effects of D-002 (beeswax alcohol) on lipid peroxidation in middle-aged and older subjects. *Lat. Am. J. Pharm.* **2008**, *27*, 695–703.
55. Liu, Z.; Tian, Z.; Zhao, D.; Liang, Y.; Dai, S.; Liu, M.; Hou, S.; Dong, X.; Zhaxinima; Yang, Y. Effects of coenzyme Q_{10} supplementation on lipid profiles in adults: A meta-analysis of randomized controlled trials. *J. Clin. Endocrinol. Metab.* **2023**, *108*, 232–249. [CrossRef] [PubMed]
56. Singh, D.K.; Li, L.; Porter, T.D. Policosanol inhibits cholesterol synthesis in hepatoma cells by activation of AMP-kinase. *J. Pharmacol. Exp. Ther.* **2006**, *318*, 1020–1026. [CrossRef]
57. Lee, J.H.; Jia, Y.; Thach, T.T.; Han, Y.; Kim, B.; Wu, C.; Kim, Y.; Seo, W.D.; Lee, S.J. Hexacosanol reduces plasma and hepatic cholesterol by activation of AMP-activated protein kinase and suppression of sterol regulatory element-binding protein-2 in HepG2 and C57BL/6J mice. *Nutr. Res.* **2017**, *43*, 89–99. [CrossRef]
58. Taylor, J.C.; Rapport, L.; Lockwood, G.B. Octacosanol in human health. *Nutrition* **2003**, *19*, 192–195. [CrossRef]
59. Bai, J.; Yang, T.; Zhou, Y.; Xu, W.; Han, S.; Guo, T.; Zhu, L.; Qin, D.; Luo, Y.; Hu, Z.; et al. Octacosanol modifies obesity, expression profile and inflammation response of hepatic tissues in high-fat diet mice. *Foods* **2022**, *11*, 1606. [CrossRef]
60. Imerb, N.; Thonusin, C.; Pratchayasakul, W.; Arunsak, B.; Nawara, W.; Ongnok, B.; Aeimlapa, R.; Charoenphandhu, N.; Chattipakorn, N.; Chattipakorn, S.C. D-galactose-induced aging aggravates obesity-induced bone dyshomeostasis. *Sci. Rep.* **2022**, *12*, 8580. [CrossRef]
61. Bajaj, S.; Khan, A. Antioxidants and diabetes. *Indian J. Endocrinol. Metab.* **2012**, *16*, S267–S271. [CrossRef]
62. Fernandez-Travieso, J.C.; Rodriguez-Perez, I.; Ruenes-Domech, C.; Fernandez-Dorta, L.; Mendoza-Castano, S. Benefits of the therapy with abexol in patients with non-alcoholic fatty liver disease. *Gastroenterol. Res.* **2020**, *13*, 73. [CrossRef] [PubMed]
63. Amini, M.R.; Kazeminejad, S.; Jalalzadeh, M.; Majd, S.S.; Kavyani, Z.; Askari, G.; Hekmatdoost, A. The effects of policosanol supplementation on blood glucose: A systematic review and dose-response meta-analysis of randomized controlled trials. *Diabetes Res. Clin. Pract.* **2024**, *212*, 111709. [CrossRef] [PubMed]
64. Esteghamati, A.; Jamali, A.; Khalilzadeh, O.; Noshad, S.; Khalili, M.; Zandieh, A.; Morteza, A.; Nakhjavani, M. Metabolic syndrome is linked to a mild elevation in liver aminotransferases in diabetic patients with undetectable non-alcoholic fatty liver disease by ultrasound. *Diabetol. Metab. Syndr.* **2010**, *2*, 65. [CrossRef] [PubMed]
65. Deb, S.; Puthanveetil, P.; Sakharkar, P. A population-based cross-sectional study of the association between liver enzymes and lipid levels. *Int. J. Hepatol.* **2018**, *2018*, 1286170. [CrossRef] [PubMed]
66. Christian, P.; Sacco, J.; Adeli, K. Autophagy: Emerging roles in lipid homeostasis and metabolic control. *Biochim. Biophys. Acta* **2013**, *1831*, 819–824. [CrossRef] [PubMed]
67. Hussain, T.; Tan, B.; Yin, Y.; Blachier, F.; Tossou, M.C.; Rahu, N. Oxidative stress and inflammation: What polyphenols can do for us? *Oxid. Med. Cell. Longev.* **2016**, *2016*, 7432797. [CrossRef]
68. Puig, M.N.; Castaño, S.M.; Ferreiro, R.M.; Clara, M.V.; Hernansez, N.M. Effects of oral administration of D-002 (Beeswax alcohols) on histological and functional outcomes in a rat model of antigen-induced arthritis: Preliminary study. *Int. J. Pharmacol. Phytochem. Ethnomed.* **2016**, *5*, 60–68.
69. Tan, S.-N.; Sim, S.-P.; Khoo, A.S. Potential role of oxidative stress-induced apoptosis in mediating chromosomal rearrangements in nasopharyngeal carcinoma. *Cell Biosci.* **2016**, *6*, 35. [CrossRef]
70. Khosla, S.; Farr, J.N.; Tchkonia, T.; Kirkland, J.L. The role of cellular senescence in ageing and endocrine disease. *Nat. Rev. Endocrinol.* **2020**, *16*, 263–275. [CrossRef]

71. Flensted-Jensen, M.; Oró, D.; Rørbeck, E.A.; Zhang, C.; Madsen, M.R.; Madsen, A.N.; Norlin, J.; Feigh, M.; Larsen, S.; Hansen, H.H. Dietary intervention reverses molecular markers of hepatocellular senescence in the GAN diet-induced obese and biopsy-confirmed mouse model of NASH. *BMC Gastroenterol.* **2024**, *24*, 59. [CrossRef]
72. Yan, M.; Man, S.; Sun, B.; Ma, L.; Guo, L.; Huang, L.; Gao, W. Gut liver brain axis in diseases: The implications for therapeutic interventions. *Signal Transduct. Target. Ther.* **2023**, *8*, 443. [CrossRef] [PubMed]
73. Mandiga, P.; Foris, L.A.; Kassim, G.; Bollu, P.C. *Hepatic Encephalopathy*; StatPearls Publishing: Treasure Island, FL, USA, 2020.
74. How Does the Liver Affect Brain Health? Available online: https://www.alzdiscovery.org/cognitive-vitality/blog/how-does-the-liver-affect-brain-health#ref-4 (accessed on 11 November 2024).

antioxidants

MDPI

Review

Antioxidant Potential of Pollen Polyphenols in Mitigating Environmental Stress in Honeybees (*Apis mellifera*)

Ivana Tlak Gajger [1,*] and Aleksandar Cvetkovikj [2]

[1] Department for Biology and Pathology of Fish and Bees, Faculty of Veterinary Medicine, University of Zagreb, 10000 Zagreb, Croatia

[2] National Veterinary and Food Institute, Faculty of Veterinary Medicine-Skopje, Ss. Cyril and Methodius University in Skopje, 1000 Skopje, North Macedonia; acvetkovic@fvm.ukim.edu.mk

* Correspondence: itlak@vef.unizg.hr; Tel.: +385-912390041

Abstract

Honeybee populations are increasingly threatened by various environmental stressors, including pesticides, pathogens, and climate change. Emerging research highlights the vital role of pollen polyphenols in supporting honeybee health through a network of antioxidants, immune responses, and detoxification mechanisms. This review synthesizes current findings on the chemical diversity, bioactivity, and functional relevance of polyphenolic compounds in honeybee nutrition. Pollen polyphenols, which include flavonoids and phenolic acids, possess remarkably high antioxidant potential, up to 235 times greater than that of nectar. They also significantly increase the expression of antioxidant enzymes, immune system genes, and detoxification pathways such as cytochrome P450s and glutathione-S-transferases. These compounds also demonstrate antimicrobial effects against key pathogens and mitigate the toxic effects of pesticides. The content and composition of polyphenols vary seasonally and geographically, impacting the resilience of honeybee colonies. Field and laboratory studies confirm that polyphenol-rich diets improve survival, gland development, and stress resistance. Advanced analytical techniques, including metabolomics, have expanded our understanding of polyphenol profiles and their effects on honeybee physiology. However, knowledge gaps remain in pharmacokinetics and structure–function relationships. Integrating this evidence into conservation strategies and good beekeeping practices, such as habitat diversification and targeted feed supplementation, is crucial for maintaining honeybee health and ecosystem services in a rapidly changing environment.

Keywords: *Apis mellifera*; reactive oxygen species; honeybee detoxification mechanisms; bee bread composition; honeybee colony resilience; pesticide-induced stress; foraging behavior; thermal stress adaptation; honeybee nutritional requirements

1. Introduction

Honeybees (*Apis mellifera*) face numerous environmental challenges that induce oxidative stress, a physiological condition characterized by an imbalance between the production of reactive oxygen species (ROS) and the antioxidant defense mechanisms. This imbalance represents a significant threat to individual honeybee health and, by extension, colony strength and survival [1,2]. Oxidative stress occurs when normal metabolic processes or external stressors generate excessive free radicals that overwhelm the organism's antioxidant capacity, leading to cellular damage and functional impairment [3–5]. Antioxidants are critical protective compounds that neutralize free radicals, mitigating oxidative damage to cellular components, including proteins, lipids, and nucleic acids. For honeybees, dietary antioxidants derived primarily from pollen and nectar constitute an essential component of the defense mechanism against oxidative stress [6,7]. Among these dietary components, polyphenols as particularly flavonoids, represent a diverse group of plant secondary metabolites with demonstrated antioxidant properties that may significantly enhance honeybee resilience to environmental stressors [8–10]. The importance of polyphenols in honeybee diets extends beyond their antioxidant capabilities, so they also contribute to immune function, nutrient absorption, and overall honeybee colony vitality. Studies indicate that adequate consumption of high-quality pollen can significantly improve honeybee resistance to diseases, enhance immune responses, and mitigate the adverse effects of environmental stressors [11–13]. The global decline in honeybee populations, along with rising environmental pressures from pesticide exposure, inadequate natural food sources, and climate change, necessitates a comprehensive understanding of all factors that support honeybee health, including the role of dietary antioxidants [14]. Despite growing evidence supporting the beneficial effects of pollen-derived antioxidants, a systematic review integrating recent findings on their specific mechanisms of action, bioavailability, and practical applications in beekeeping is still lacking in the scientific literature.

So, this review aims to synthesize current knowledge regarding pollen polyphenols and their role in honeybee antioxidant defense mechanisms against environmental stressors. Specifically, it aims to examine the biochemical basis of oxidative stress in honeybees, characterize key antioxidant compounds in their diet, analyze how climate change affects antioxidant availability, evaluate the influence of dietary antioxidants on immunity and stress resilience, and identify critical knowledge gaps to guide future research. Through this detailed analysis, we aim to enhance understanding of how dietary antioxidants can improve honeybee health and survival in increasingly challenging environmental conditions. The specific objectives of this review are to define oxidative stress in the context of honeybee physiology and biochemistry, identifying its primary endogenous and exogenous sources (including environmental stressors, pathogens, and pesticides) and detailing its consequences for individual honeybee physiology, immune function, and overall colony health; to characterize the key antioxidant compounds in the honeybee diet, focusing on polyphenols (flavonoids and phenolic acids) and essential vitamins (Vitamin C, Vitamin E), detailing their occurrence in pollen and nectar, their metabolic fate, and their mechanisms of action within the honeybee; to analyze the impact of climate change variables (e.g., global warming, altered precipitation, and elevated CO_2) on floral diversity, pollen abundance, and the nutritional quality of pollen, specifically concerning antioxidant composition, and evaluate the potential consequences for honeybee nutrition, immunity, and stress resistance; to elucidate the role of dietary antioxidants in modulating honeybee immunity, including the enhancement of immune responses, influence on resistance to pathogens and parasites, and potential synergistic interactions between different antioxidant compounds; to examine the contribution of dietary antioxidants to honeybee resilience against specific environmental stressors, particularly pesticide-induced oxidative stress and challenges arising from climate change and habitat loss, and their effects on honeybee longevity and colony survival; to identify significant knowledge gaps, conflicting findings, and promising future research directions of honeybee antioxidant biology, including the need for studies on bioavailability and metabolism, and the potential for developing antioxidant-rich dietary supplements for the apiculture sector; and to summarize the key findings of the review and discuss their practical implications for beekeeping practices, honeybee conservation strategies, and future research priorities.

During the writing process, a structured literature search protocol was implemented, which included clearly defined search strategies, specific inclusion and exclusion criteria, and methodologies for data extraction to ensure the relevance and quality of the studies reviewed.

2. Implications of Oxidative Stress in Honeybees

2.1. Biochemical Basis of Oxidative Stress in Honeybees

Oxidative stress represents a fundamental imbalance between pro-oxidant and antioxidant systems, leading to potential cellular damage. At the molecular level, this process involves excessive production of ROS, including superoxide anion (O_2^-), hydrogen peroxide (H_2O_2), and hydroxyl radical (OH), which overwhelm endogenous antioxidant defense mechanisms [3]. This biochemical disruption appears to be particularly problematic for bees due to their high metabolic rate and extensive flight activity [15]. The primary sources of ROS in honeybee physiology include normal cellular respiration, immune responses to pathogens, and exposure to environmental toxins [16]. Mitochondrial respiration generates superoxide anions as natural byproducts of energy metabolism [17]. In healthy bees, these radicals are efficiently neutralized by endogenous antioxidant enzymes, including superoxide dismutase (SOD), catalase (CAT), and glutathione peroxidase (GPx) [18]. However, when ROS production exceeds antioxidant capacity, cellular damage accumulates through lipid peroxidation, protein oxidation, and DNA damage. Recent studies have demonstrated that flight muscle mitochondria in worker bees show particularly high susceptibility to oxidative damage, which may contribute to age-related decline in foraging capacity [15]. In honeybees, as in other aerobic organisms, ROS are continuously generated as byproducts of normal cellular metabolism, particularly through mitochondrial electron transport during respiration [15,19]. As was already mentioned, the honeybee genome encodes several enzymes involved in antioxidant defense, including SOD, CAT, and GPx, which work together to neutralize ROS [20,21]. Additionally, non-enzymatic antioxidants such as glutathione (GSH), vitamin C, and vitamin E provide complementary protection against oxidative damage. When environmental stressors disturb this balance, oxidative stress can lead to lipid peroxidation, protein oxidation, and DNA damage, ultimately compromising cellular function and health [22].

Honeybees encounter numerous sources of oxidative stress throughout their lifecycle. Environmental factors are primary stressors, with pesticide exposure representing a particularly significant threat. The relationship between pesticide exposure and oxidative stress has received considerable attention in the recent literature [23]. Neonicotinoids and organophosphates have been shown to induce oxidative stress in honeybees through mechanisms involving increased ROS production, which in turn modulates the activity of endogenous antioxidant enzymes [24]. Studies have shown that sublethal pesticide exposure can persist in bee tissues for weeks, creating chronic oxidative stress conditions that may compromise long-term health and survival [25]. Environmental toxins, beyond pesticides, also contribute to oxidative burden [26]. Pathogen infections also represent a substantial source of oxidative stress. Viral infections, like deformed wing virus (DWV) and acute bee paralysis virus (ABPV), along with parasitic infestations by *Varroa destructor* and fungal infections by *Nosema ceranae*, trigger oxidative responses in honeybees [27,28]. Alaux et al. (2010) demonstrated that *N. ceranae* infection significantly increases honeybee susceptibility to pesticide toxicity through oxidative stress mechanisms, highlighting the detrimental synergistic effects of multiple stressors [29]. Additionally, exposure to thiamethoxam in food negatively impacts honeybee queen reproductive physiology [30,31]. Nutritional stress from monoculture landscapes and limited floral diversity may contribute to oxidative imbalance by limiting the intake of dietary antioxidants [32,33]. Honeybees perform energetically demanding tasks, such as foraging, and experience elevated metabolic rates that lead to increased ROS production [15]. Climate change factors, including higher temperatures and extreme weather events, further exacerbate oxidative stress through direct physiological effects, the emergence of secondary diseases (in honeybees and wild bees) [34], and indirect impacts on nutritional resource availability [14]. Furthermore, management practices like migratory beekeeping and exposure to radiofrequency electromagnetic fields (RF-EMFs) have been shown to increase markers of oxidative stress [35–37]. Here, we synthesize and critically evaluate the current available scientific literature concerning the role of dietary antioxidants, with a specific focus on pollen-derived polyphenols and essential antioxidant vitamins, in honeybee health, particularly in the context of mitigating oxidative stress induced by environmental factors. So, the scope encompasses the biochemical underpinnings of oxidative stress in honeybees, the identification and function of key dietary antioxidants, the influence of environmental changes, and the intricate links between dietary antioxidants, immunity, and stress tolerance (Table 1).

Table 1. Summary of antioxidant endpoints connected with pollen diets and honeybee health outcomes.

Direct ROS Assay	Enzyme Activity (SOD/CAT/etc.)	Oxidative Damage Markers	Gene/Protein Expression	Link to Polyphenol/ Pollen	Observed Health Outcome	Ref-(s)
no	GST (activity), P450, vitellogenin (expression)	no	yes	yes	not specified	[37]
no	SOD, CAT, GST, GPx	no	no	yes (pollen blends compared)	survival, pesticide clearance	[7]
no	no	no	yes (by pollen type)	yes	none—enzymatic levels only	[21]
TBARS, protein carbonyls	CAT	TBARS, carbonyls	yes (transcriptomics)	yes (fractionated pollen)	survival, overwintering	[38]
not specified	antioxidant capacity	not specified	lysozyme (protein)	no (pollen present; not polyphenol-focused)	survival	[39]
no	not detailed	no	some detox gene upregulation	limited	survival, longevity, digestion, immunity	[40–43]

2.2. Environmental Stressors Affecting Honeybees

Honeybees are affected by various environmental stressors that threaten their health and survival, including pathogens, pesticides, habitat loss, and climate change. They are vulnerable to bacterial pathogens causing diseases like American foulbrood, European foulbrood, spiroplasmosis, and septicemia. These pathogens have long infections and propagate rapidly, causing substantial harm that adversely affects honeybee growth and development [44]. Pesticides in agriculture aim to reduce pest and weed damage, but they can negatively impact honeybees by affecting their behavior, impairing antioxidant abilities, and reducing immunocompetence. Moreover, pesticides can increase the sensitivity of honeybees to viral infections, which compounds the risk of disease [45,46]. Different types of pesticides elicit varied responses from honeybees, with some exhibiting severe negative impacts on their health and behavior [44,47]. Climate change threatens honeybee populations by disrupting weather patterns and the availability of flowering plants. These shifts can cause nutritional stress due to misaligned foraging and food sources. Warmer fall and winter temperatures also extend foraging periods, complicating energy management for honeybee colonies during wintering. Recent reports indicate that managed honeybee colony losses have approached 40%, underscoring the severity of this issue [48]. The degradation of natural habitats further exacerbates the pressures on honeybees. Loss of foraging areas and nesting sites for wild bees limits their access to essential resources needed for survival and reproduction. This environmental change negatively impacts foraging efficiency, immune function, and overall colony health [46,47]. The interaction between these various stressors creates compounded effects on honeybee health. For instance, the presence of pesticides may increase honeybees' vulnerability to diseases caused by pathogens, leading to a cyclical deterioration of their health [45,49]. Understanding the interactions among various stressors is essential for creating strategies to protect honeybee populations in changing environments. Recent decades have seen significant colony losses linked to these multifactorial stresses, which weaken hives and compound health issues. Studies show that colony failure often results from multiple factors, with oxidative stress playing a key mediating role [50].

2.3. Consequences on Physiology, Immunity, and Honeybee Colony Health

Oxidative stress has diverse effects on honeybee physiology, with particularly pronounced impacts on the nervous system. Evidence indicates that oxidative damage to brain tissues impairs cognitive functions essential for foraging, navigation, and learning, thereby compromising individual foraging efficiency and, consequently, honeybee colony nutrition [51]. Chronic oxidative stress has wide-ranging deleterious effects on honeybee physiology. At the cellular level, accumulated oxidative damage leads to inflammation, impaired energy metabolism, and accelerated senescence of tissues [21]. For individual bees, this can translate into shortened lifespan, reduced flight capability, and cognitive impairments (e.g., memory and learning deficits) due to neural damage. Oxidative stress also compromises the bee's immune system; excessive ROS can dysregulate immune signaling or directly damage immune cells and enzymes, resulting in immunosuppression [21,50]. For example, habitat-related stress was shown to induce oxidative stress in bees alongside suppressed immune gene expression [52]. Consequently, oxidatively stressed bees become more susceptible to infections and parasites, creating a vicious cycle of declining health. Neurological function appears particularly vulnerable to oxidative damage, with implications for navigation, communication, and learning behaviors essential for foraging success. Studies have documented decreased performance in learning tasks among bees experiencing chronic oxidative stress, suggesting that cognitive impairment may contribute to reduced foraging efficiency and colony productivity [53]. In winter bees, whose extended lifespan is crucial for colony overwintering success, oxidative stress accelerates senescence processes, potentially reducing survival rates [54]. Immunologically, oxidative stress compromises honeybee defense mechanisms through several pathways. ROS overproduction impairs hemocyte function and antimicrobial peptide production, key components of the innate immune response [55,56]. Di Pasquale et al. (2013) showed that pollen quality affects tolerance to *Nosema* and that survival correlates with the activity of the immune enzyme phenol oxidase [6]. At the colony level, these individual effects result in reduced brood production, decreased honey yields, and ultimately, increased colony mortality [57]. Colonies experiencing oxidative stress exhibit altered division of labor, with premature transitions from hive to foraging duties disrupting the optimal age demographics of honeybee colony members [58]. Additionally, environmental stressors such as electromagnetic fields have been linked to behavioral alterations in colonies, suggesting that stress-induced plasticity may extend to changes in queen-related dynamics [59]. Oxidative stress reduces the resilience and productivity of both individual worker bees and the entire colony (Table 2).

Table 2. A summary of the effects and consequences of oxidative stress in honeybees.

Category	Effect	Mechanism/Description	Ref-(s)
molecular level	↑ ROS	overproduction of O_2^-, H_2O_2, OH· during mitochondrial respiration	[3,15]
	lipid peroxidation	oxidative degradation of membrane lipids → cell damage	[60]
	protein oxidation	oxidative modification of amino acids; impaired enzyme function	[24]
	DNA damage	strand breaks, base modifications may impair gene expression	[61,62]
cellular response	antioxidant enzyme activation	upregulation of SOD, CAT, GPx to detoxify ROS	[17]
	depletion of non-enzymatic antioxidants	GSH, vitamin C, vitamin E are used to neutralize ROS	[24]
neural function	cognitive impairment	ROS damage to neurons → navigation and learning deficits	[51]
	neurodegeneration	structural brain damage in foragers exposed to pesticides	[60]
immune function	suppressed immunity	reduced hemocyte function and antimicrobial peptides	[55]
	↑ susceptibility to pathogens	*Nosema* spp. and viruses proliferate under oxidative stress	[29,57]
	impaired immune gene expression	disrupted expression of immune-related genes (e.g., defensins)	[57]
reproduction	queen fertility decline	exposure to thiamethoxam is linked to oxidative stress in the ovaries	[31]
	queen replacement	behavioral response to declining queen performance	[59]
metabolism and longevity	accelerated aging	especially in winter bees; shorter lifespan due to ROS accumulation	[54]
	↑ metabolic load in foragers	foraging increases ROS via mitochondrial respiration	[15]
colony-level effects	disrupted division of labor	leads to precocious foraging; it weakens the colony's age structure	[58]
	reduced brood production	nutritional and physiological stress impairs brood care	[57]
	lower honey yield	impaired foraging and navigation → reduced nectar collection	[33]
	colony mortality	combined effects of oxidative stress, pathogens, and poor nutrition	[14,57]
environmental triggers	pesticides (e.g., neonicotinoids)	imidacloprid, thiamethoxam increase ROS, suppress antioxidants	[24,60]
	pathogens (DWV, ABPV, *Nosema* spp.)	induce host ROS and weaken antioxidant systems	[27,29]
	nutritional stress	lack of diverse pollen leads to antioxidant deficiency	[33,63]
	high temperature/climate change	enhances ROS production; disrupts bee physiology	[14]
	RF-EMFs (radiofrequency electromagnetic fields)	transport and hive movement increase oxidative load	[36,37]
		induce oxidative biomarkers in bee tissues	[35]

Table 2. *Cont.*

Category	Effect	Mechanism/Description	Ref-(s)
protective role of diet	pollen-derived polyphenols	antioxidant activity reduces ROS, supports detoxification	[64,65]
	vitamin C and E supplementation	enhance resilience against pesticide-induced oxidative stress	[24]
	floral diversity	pollen sources improve antioxidant enzyme expression	[33]

Arrows show the increase, or sequential steps of the reaction/process.

3. Antioxidants in Honeybee Diet—Key Compounds and Their Sources

Bee pollen represents a concentrated source of diverse polyphenolic compounds that vary considerably based on botanical and geographical origin [66–68]. The total phenolic content in bee pollen can range dramatically, with some Brazilian samples reported as containing 1.7–2.2% [69], and other geographical regions reporting lower values, such as 720 mg/100 g (0.72%) [66]. This variation reflects the diverse floral sources from which bees collect pollen and the influence of environmental factors on plant secondary metabolite production [70,71]. The predominant polyphenolic compounds in pollen include flavonoids such as quercetin, kaempferol, isorhamnetin, rutin, and luteolin [66,67,72]. Quercetin levels can vary enormously between samples, ranging from 61.2 to 1221.7 mg/100 g, while (+)—catechin concentrations span from 66.75 to 337.39 mg/100 g [66]. Additional important compounds include phenolic acids like caffeic acid, p-coumaric acid, and chlorogenic acid, as well as specialized metabolites such as hydroxycinnamic acid amide derivatives and spermidine compounds [69,73,74]. The antioxidant activity of pollen correlates strongly with its polyphenolic content, particularly the presence of flavonols with specific structural features [70,72,75]. The planar structure of flavonols caused by hydroxyl groups at position 3 promotes higher radical capture through electron delocalization, while the number of hydroxyl groups directly influences their potency as electron scavengers [70,72]. Research has shown that each type of pure floral pollen has a consistent antioxidant capacity, regardless of its geographic origin. Approximately 50% of this activity is attributed to flavonoids and phenolic acids [76]. Modern extraction techniques such as microwave-assisted extraction can increase polyphenolic yields by 40–60% compared to conventional methods, enhancing the recovery of these bioactive compounds [77]. Pollen polyphenols play a crucial role in the antioxidant capacity, contributing significantly to its potential health benefits and its ability to mitigate environmental stress in honeybees. Both monofloral and polyfloral pollen samples are recognized as rich sources of polyphenols, with substantial antioxidant potential attributed to their complex composition [65,70]. These compounds are essential for mitigating oxidative stress, which is a critical factor in the health and survival of honeybees, particularly under environmental stressors. Pollen and nectar are primary sources of dietary antioxidants for honeybees, with polyphenolic compounds being a dominant class of bioactive constituents [78]. Comprehensive phytochemical analyses reveal substantial diversity in polyphenolic profiles across plant species, with pollen typically containing higher concentrations than nectar [76]. Pollen from families such as Rosaceae, Fabaceae, and Asteraceae has been associated with high antioxidant activity and phenolic content [79,80]. Flavonoids are the most abundant class of polyphenols in honeybee-collected pollen, with quercetin, kaempferol, and rutin consistently found across diverse botanical sources [81]. The predominant group of polyphenols found in bee pollen is flavonoids, which typically constitute about 0.25% to 1.4% of its total composition [21]. Quantitative analyses indicate that the total flavonoid content in honeybee pollen ranges from 2% to

5% dry weight, with significant seasonal and geographical variations reflecting local floral diversity [82]. Notably, the overall polyphenolic content in bee pollen ranges from 3% to 5%, with variations influenced by the origin of the raw material [73]. This variability is important because the antioxidant capacity of pollen samples often correlates positively with their polyphenol concentrations [83,84]. Multifloral pollen typically exhibits higher antioxidant potential compared to monofloral sources, highlighting the nutritional benefits of diverse foraging habitats for honeybee colonies [6]. Anthocyanins, which significantly contribute to the characteristic coloration of many pollen types, exhibit substantial free radical scavenging capacity in other plant products [85] and thus may confer oxidative protection to honeybees consuming diverse pollen diets.

3.1. Role in Honeybee Metabolism

Dietary antioxidants function as essential cofactors in honeybee metabolism, supporting both energy production and cellular protection [86]. The most significant antioxidant compounds in honeybee diets include ascorbic acid (vitamin C), tocopherols (vitamin E), carotenoids, and various polyphenolic compounds [44]. Each class of antioxidants contributes distinct protective mechanisms, and their synergistic interactions appear critical for optimal antioxidant protection [45]. Polyphenolic compounds represent the largest and most diverse group of dietary antioxidants available to honeybees. These compounds are concentrated primarily in pollen, though nectar and propolis also provide significant contributions. The polyphenolic profile of honeybee diets varies substantially based on floral sources, seasonal availability, and geographic location [45]. Following ingestion, pollen polyphenols undergo metabolic processing in the honeybee digestive system, with evidence suggesting selective absorption and biotransformation of specific compounds [8]. Pharmacokinetic studies indicate that p-coumaric acid, a phenolic acid abundant in pollen, is efficiently absorbed and distributed throughout honeybee tissues, including the brain, where it may provide neuroprotective effects against oxidative damage [8,87]. This biochemical support is crucial given that honeybees have a high metabolic rate and correspondingly high baseline ROS production during flight and thermogenesis [88]. Metabolically, dietary polyphenols influence critical detoxification pathways in honeybees. Experimental evidence demonstrates that p-coumaric acid upregulates cytochrome P450 enzyme expression [8], a key component of xenobiotic metabolism, which is discussed further in Section 6.1. Additionally, flavonoids appear to modulate energy metabolism in honeybees, with quercetin supplementation influencing mitochondrial function and ATP production in flight muscles [8]. Emerging evidence suggests that pollen polyphenols can significantly influence gut microbiome composition in honeybees. Previously published articles demonstrated that dietary quercetin and effective microorganisms (EM-1) alter core gut bacterial communities, potentially enhancing colonization resistance against opportunistic pathogens [89,90]. This microbiome-mediated effect represents an additional mechanism through which dietary polyphenols may influence honeybee health.

3.2. Mechanisms of Antioxidative Action

Pollen polyphenols exhibit powerful antioxidant activity through several well-characterized mechanisms that collectively protect against oxidative stress and cellular damage [44,91]. The primary mechanism involves direct free radical scavenging, where polyphenolic compounds neutralize ROS by donating hydrogen atoms or electrons to unstable molecules [92,93]. This electron donation stabilizes free radicals and prevents them from causing further oxidative damage to cellular components. The structural features of pollen polyphenols are crucial for their antioxidant effectiveness. Flavonoids with planar structures caused by hydroxyl groups at position 3 promote higher radical capture through enhanced electron delocalization, while the total number of hydroxyl groups directly influences their potency as electron scavengers [72]. The catechol group on the B ring, the 2,3-double bond, and conjugated groups on the C ring are particularly important structural elements that enhance antioxidant capacity [93,94]. Beyond direct radical scavenging, pollen polyphenols employ additional protective mechanisms, including metal ion chelation, where they bind and inactivate metal catalysts that promote oxidation reactions [44,95]. These compounds can also upregulate endogenous antioxidant enzymes and assist in repairing oxidative damage to biomolecules [72]. Research demonstrates that pollen extracts effectively protect cells against oxidative damage triggered by chemical free radical generators, with this protective effect directly attributed to the phenolic compounds' ability to scavenge reactive species and prevent cellular injury [96]. The synergistic action of multiple antioxidant compounds in pollen, including flavonoids, phenolic acids, carotenoids, and vitamins C and E, creates a comprehensive defense system against oxidative stress [97,98]. This multi-component approach provides robust protection by targeting different types of free radicals and reactive species through complementary mechanisms, making pollen a potent natural antioxidant source [99,100]. Pollen polyphenols enhance oxidative stress resistance in honeybees through their antioxidant effects. They primarily work by scavenging ROS, which can cause cellular damage when excessive [49,83]. The excessive production of ROS can lead to oxidative stress, which is associated with various chronic diseases and accelerates the aging process [39]. In honeybees, factors such as environmental pollution, monoculture diets, and exposure to pesticides can increase ROS levels, thus compromising their immune defense [11,44]. One of the primary mechanisms is direct free radical scavenging, wherein the hydroxyl groups characteristic of flavonoid structures, such as those in quercetin, donate hydrogen atoms to neutralize ROS [101]. Structure–activity relationship studies demonstrate that flavonols with specific hydroxylation patterns, including quercetin, exhibit particularly potent radical scavenging capacity [102], which can be relevant to honeybee oxidative defense. Beyond these direct antioxidant effects, polyphenols also modulate endogenous antioxidant systems in honeybees. Dietary quercetin can mitigate imidacloprid's impacts by lowering the pesticide burden in bees and by engaging detoxification pathways, while imidacloprid alone can alter antioxidant enzymes [103,104]. Similarly, Bryś et al. (2025) observed that pollen supplementation enhanced both SOD activity and glutathione levels in worker bees, establishing a link between dietary polyphenols and endogenous antioxidant capacity [21].

Research demonstrates that pollen consumption significantly enhances honeybee antioxidant enzyme systems, with the magnitude of benefit directly related to the phenolic and flavonoid content of the pollen source [21]. Studies show that bees fed pollen from plants rich in phenolic compounds and flavonoids, such as rapeseed, phacelia, buckwheat, and goldenrod, exhibit increased activities of SOD, CAT, GST, and GPx in both their fat body tissues and hemolymph compared to control groups [21]. The antioxidant capacity of pollen depends on various bioactive compounds, including phenolic acids, flavonoids, and specialized enzymes [74], which work synergistically to modulate antioxidant enzyme expression and activity in bee tissues. Specifically, enzymes like CAT and SOD, along with GSH, play crucial roles in the antioxidant effects observed when bees consume phenolic-rich pollen [74]. Experimental evidence shows that honeybees fed on different pollen types exhibit varying antioxidant responses, with some pollen sources producing higher antioxidant enzyme activity while others increase total antioxidant capacity [40]. The total antioxidant capacity appears closely related to the number and concentration of antioxidant compounds present in the pollen, suggesting that diverse antioxidant profiles provide comprehensive protection [40]. Metal chelation is another significant mechanism through which pollen polyphenols protect honeybees against oxidative damage. Flavonoids containing catechol moieties effectively chelate transition metals, such as iron and copper, preventing these metals from participating in Fenton reactions that produce highly reactive hydroxyl radicals [105]. This metal-chelating property may be particularly relevant in contexts where honeybees encounter environmental pollutants containing heavy metals. Additionally, certain pollen polyphenols directly influence gene expression through interaction with transcription factors, notably Nuclear factor erythroid 2-related factor 2 (Nrf2), which regulates antioxidant response elements in the genome [106]. This epigenetic mechanism may lead to long-term adaptations to oxidative stress by enhancing the activity of endogenous antioxidant pathways. Beyond enzyme enhancement, pollen consumption provides protective effects against oxidative damage in honeybee systems. The high content of flavonoids, particularly flavonols and anthocyanins, promotes healthy cellular function by lowering reactive oxygen species levels and mitigating their adverse effects on bee physiology [107,108]. This protective mechanism is particularly important for maintaining reproductive health and cellular integrity under environmental stress conditions.

3.3. Vitamins with Antioxidant Properties

Honeybees acquire essential vitamins (e.g., vitamin C and vitamin E) with antioxidant properties primarily through pollen consumption, with vitamin content varying substantially across botanical sources. Vitamin E is a vital lipophilic antioxidant, present in both pollen and bee bread [73]. The complex, comprising tocopherols and tocotrienols, provides critical lipid-soluble antioxidant protection in bee body cells [17]. These compounds are particularly important for protecting cell membranes and lipid-rich tissues from peroxidative damage [15]. Vitamin C (ascorbic acid) is another essential antioxidant in the honeybee diet, acting as an electron donor that neutralizes reactive oxygen species and regenerates oxidized vitamin E [109]. The concentration of ascorbic acid declines during storage due to oxidative degradation [76]. B vitamins, including riboflavin (B2), pyridoxine (B6), and pantothenic acid (B5), have antioxidant properties through their roles as enzyme cofactors in redox metabolism. These compounds support metabolic functions essential for oxidative homeostasis in honeybees and are found at nutritionally significant levels in diverse pollen types [110].

3.4. Dietary Sources and Their Bioavailability

The bioavailability of pollen antioxidants to honeybees depends on how effectively they can ingest and assimilate these compounds. Fresh pollen has a tough outer coat (sporopollenin) that can limit digestibility, but adult bees have evolved strategies to access pollen contents, mixing pollen with nectar and enzymes in the mouthparts, packing it into granules, and fermenting it into bee bread in the comb cells. Bee bread, which is fermented pollen, undergoes lactic acid fermentation. It was previously believed that this process significantly enhances nutrient availability. However, recent studies show that the nutritional value and antioxidant effectiveness of pollen do not significantly change during short-term storage. Bees benefit equally from both fresh pollen and bee bread in terms of their physiological effects [21]. The fermentation process primarily preserves pollen and may slightly increase certain vitamin levels (e.g., B vitamins), but polyphenols remain relatively stable and bio-accessible. Key antioxidants in pollen, such as flavonoids, are typically glycosylated (bound to sugars) in plants. When ingested, enzymes from the bee's gut or its microbiota can hydrolyze these glycosides, releasing the active aglycones, which are more easily absorbed. Consequently, honeybee gut microbes may play a significant role in metabolizing dietary polyphenols, affecting their bioavailability and subsequent effects [111]. It is important to note that pollen diversity is significant for antioxidant intake, as different floral pollens contain unique profiles of polyphenols and vitamins [21]. A varied diet ensures a broad spectrum of antioxidants, potentially covering gaps where one plant's pollen is deficient. Monofloral diets, especially from pollen with lower phenolic content (e.g., some wind-pollinated plants like pine), may provide fewer antioxidants than polyfloral diets [21]. Nevertheless, even monofloral pollen rich in polyphenols (e.g., buckwheat or rapeseed) can significantly boost bees' antioxidant enzyme levels. The bioavailability of antioxidant vitamins in honeybees' food depends on several factors, including pollen processing, storage conditions, and interaction with other dietary components. Experimental evidence indicates that bee bread preparation enhances vitamin bioavailability through microbial fermentation processes that partially degrade pollen walls and release encapsulated nutrients [86]. This predigestion process improves extraction efficiency during subsequent consumption by nurse honeybees, potentially maximizing antioxidant vitamin absorption. Seasonal and geographical factors significantly influence the availability of vitamins in honeybee diets. The nutritional composition of honeybee pollen has seasonal variations [112]. Similarly, agricultural intensification and resulting monocultures may limit access to specific vitamin-rich pollen sources, creating nutritional bottlenecks that compromise antioxidant defense [33,78]. Vitamin stability during storage represents another critical factor influencing bioavailability. Ascorbic acid content decreases significantly during pollen storage, with losses of up to 50% reported after three months [76]. Conversely, tocopherols demonstrate greater stability during storage, maintaining antioxidant potential over extended periods. These differential degradation rates suggest potential seasonal variations in vitamin availability for honeybee colonies.

3.5. Role in Honeybee Immune Support and Detoxification

Honeybee immunity is closely tied to nutritional quality, and antioxidants are now recognized as part of bees' immune arsenal. High-quality pollen diets provide not only protein for immune enzyme and peptide synthesis but also antioxidants that shield immune cells from oxidative damage and even directly inhibit pathogens. The antioxidant system itself is considered one of the honeybee's immune mechanisms [21], as it helps maintain immune homeostasis during infection (when ROS are produced as a defense but can also harm host tissues). Polyphenols like p-coumaric acid have been shown to activate the expression of antimicrobial peptide genes in bees, thereby enhancing disease resistance while simultaneously reducing oxidative stress [42]. This dual action improves the overall immune competence of the bee. Additionally, many plant polyphenols have inherent antimicrobial and antiviral properties. In vivo, feeding experiments demonstrate that pollen or supplements high in these compounds reduce pathogen loads: e.g., curcumin (a dietary phenolic from turmeric) significantly lowered *Nosema* spp. spore counts in infected bees and improved their survival [113]. Likewise, grape polyphenols (anthocyanins from pomace) fed to bees enhanced the expression of Relish, a key immune gene in the Toll pathway, and helped bees suppress deformed wing virus replication [114]. In terms of detoxification, antioxidants in the diet mitigate the harmful effects of xenobiotics. They can directly neutralize reactive metabolites of pesticides and prevent oxidative damage to the detoxification organs. More importantly, certain phytochemicals induce the transcription of detoxification enzymes (e.g., glutathione-S-transferases and P450s), which increases the bee's capacity to metabolize and excrete toxins [42]. This has been demonstrated by studies of the same authors showing that bees consuming quercetin or p-coumaric acid along with pesticides had higher survival and lower toxicity symptoms than bees lacking those dietary phytochemicals.

4. Impact of Climate Change on Pollen Antioxidants Availability

4.1. Impact of Global Warming on Floral Diversity and Pollen Abundance

Climate change impacts plant and insect pollinator systems, significantly affecting pollen availability and its nutritional quality. Rising temperatures have shifted flowering phenology in many regions, with earlier bloom times disrupting the synchronization between bee foraging cycles and peak pollen availability [115]. These temporal mismatches may force bees to rely on suboptimal pollen sources or face periods of nutritional scarcity. Geographic range shifts of plant species create additional challenges for honeybee nutrition. As temperature zones migrate northward, traditional pollen sources may disappear from established bee territories while new plant species colonize these areas [116]. The antioxidant profiles of replacement plant species may differ substantially from historical sources, requiring bees to adapt to novel nutritional conditions without evolutionary preparation [14,29,117]. Extreme weather events, increasingly common under climate change scenarios, can dramatically reduce pollen availability during critical periods. Droughts, floods, and unseasonable temperature fluctuations can destroy flowering plants or interrupt their reproductive cycles [118]. These disruptions may create nutritional bottlenecks where bees face simultaneous increases in environmental stress and reductions in antioxidant availability [119]. An observed consequence is nutritional dearth for bees: when key forage plants produce less pollen (or flower at times when bee activity is low), colonies face periods of protein and micronutrient shortage [111]. Beekeepers worldwide have noted colony declines in unusually warm, dry years, often needing to supplement colonies with artificial feed due to lack of natural forage [111]. This reduction in pollen availability inherently means fewer dietary antioxidants for bees, since pollen is their main source of polyphenols and vitamins. Atmospheric carbon dioxide concentrations, a primary driver of climate change, appear to influence pollen production and composition directly [120]. Elevated CO_2 levels generally increase plant biomass production but may dilute the concentration of protein and secondary metabolites in plant tissues [121]. This dilution effect could reduce the antioxidant density of pollen even when total pollen production increases, creating a fundamental challenge for bee nutrition in future climate scenarios.

4.2. Changes in Pollen Composition Affecting Its Antioxidant Levels

Environmental stress factors associated with climate change influence the production of secondary metabolites in plants, directly affecting the antioxidant content of pollen [120,121]. Plants experiencing water stress, temperature extremes, or elevated UV radiation may increase their production of protective polyphenolic compounds [115]. However, these same stress factors can also impair plant metabolism and reduce overall pollen quality [122]. The carbon-to-nitrogen ratio in plant tissues shifts under elevated atmospheric CO_2, potentially altering the balance between structural carbohydrates and bioactive secondary metabolites [123]. This shift may reduce the concentration of nitrogen-containing antioxidant compounds while potentially increasing carbon-based polyphenols [120]. The net effect on total antioxidant capacity depends on the specific compounds affected and their relative contributions to protective activity [45]. For example, phenological shifts have been documented in Central Europe; warmer winters lead to earlier blooming of plants such as hazel [21]. Hazel is a wind-pollinated plant whose pollen, while abundant and often used by bees in late winter, generally has lower polyphenol content than insect-pollinated flowers. If climate change causes bees to rely more on such early wind-pollinated sources (due to mismatches with traditional spring flowers), their intake of antioxidants could diminish compared to a normal spring diet of willow, prunus, and brassica pollen (which are rich in phenolics) [21]. The nutritional quality and safety of pollen critically influence its protective effect on honeybee health. Pollutants such as heavy metals (e.g., lead and cadmium) and pesticide residues can contaminate pollen collected near polluted environments, compromising its antioxidant potential and exerting toxic effects. Heavy metals bioaccumulate in bee tissues, causing oxidative stress and interfering with enzymatic antioxidant defenses. Similarly, contamination with pesticides from agricultural practices presents a dual threat by directly poisoning bees and reducing the functional integrity of their diet [124]. Bees foraging in polluted air or soils demonstrate reduced survival and impaired ability to perform their ecological roles, emphasizing the need for monitoring and mitigating environmental contamination in apiary settings. Strategies for improving pollen quality include selecting foraging areas with minimal pollution, employing contamination screenings, and potentially offering uncontaminated supplemental feed to maintain nutritional adequacy and antioxidant capacity [125]. Assessment of biochemical stress markers such as heat shock proteins and antioxidant enzymes can further inform the impact of environmental contaminants on bee health and guide management interventions [126]. Soil chemistry changes associated with climate change, including altered precipitation patterns and increased weathering rates, influence plant mineral uptake and metabolism. Selenium, zinc, and other trace elements essential for antioxidant enzyme function may become less available to plants, reducing the micronutrient content of pollen [16]. These deficiencies can propagate through the food web to create antioxidant deficiencies in bee colonies [127]. Interactions between climate change and other environmental stressors may create synergistic effects on pollen composition. Air pollution, habitat fragmentation, and agricultural practices interact with climate factors to influence plant health and secondary metabolite production [128]. Understanding these complex interactions is crucial for predicting future changes in pollen antioxidant availability [14].

4.3. Potential Consequences for Honeybee Nutrition and Overall Vitality

Climate-mediated changes in pollen antioxidant availability may compromise honeybee nutritional status through several pathways. Reduced diversity of available pollen types limits the intake of specific antioxidant compounds that may provide complementary or synergistic protective effects [15]. Additionally, honeybees demonstrate selective foraging behavior influenced by pollen nutritional content, but may face constrained choices in degraded floral landscapes resulting from climate change [129]. Nutritional stress resulting from suboptimal antioxidant intake affects immunocompetence in honeybees facing multiple environmental stressors. DeGrandi-Hoffman et al. (2016) showed that honeybee colonies with access to diverse pollen sources exhibited greater immunological resilience to *N. ceranae* infection compared to nutritionally stressed colonies, with effects partially attributed to antioxidant intake [116]. Climate-driven nutritional deterioration may thus exacerbate pathogen pressure through compromised immune function [63]. Furthermore, climate change potentially creates multiple stressor scenarios through simultaneous alterations of temperature regimes, nutritional landscapes, and pathogen dynamics. Experimental evidence indicates that combined thermal and nutritional decline synergistically decreases honeybee survival beyond additive effects [58]. This interaction suggests that climate-mediated nutritional deterioration may be particularly detrimental when occurring concurrently with thermal stress, potentially creating population-level impacts through reduced worker longevity. Adaptation potential exists through both behavioral and physiological mechanisms. Honeybees exhibit remarkable plasticity in foraging behavior, potentially enabling them to exploit novel floral resources as plant communities reorganize under climate change [130]. Additionally, supplemental feeding with antioxidant-rich formulations may partially mitigate climate-induced nutritional deficiencies; however, the economic feasibility of widespread implementation remains uncertain [15,90,131–134].

5. The Role of Dietary Antioxidants in Honeybee Immunity

5.1. Enhancement of Immune Responses Through Dietary Antioxidants

Multiple studies showed that honeybees fed antioxidant-rich diets show elevated immune activity. As mentioned earlier, polyphenols like p-coumaric acid can upregulate genes for antimicrobial peptides. Alaux et al. (2010) found that polyphenol-supplemented diets increased the expression of abaecin and defensin, key antimicrobial peptides in honeybees, suggesting an immunostimulatory effect of dietary phenolics [29]. In a recent experiment, adding grape pomace (high in anthocyanins) to the diet not only reduced virus levels but also significantly increased the expression of Relish, a transcription factor in the NF-κB pathway that controls many immune genes [114]. This indicates that antioxidants can activate immune pathways, priming bees to respond more vigorously to infections. Likewise, bees on pollen diets (as opposed to only sugar) showed higher baseline levels of immune enzymes such as glucose oxidase (which generates antiseptic hydrogen peroxide in honey) and higher hemolymph phenol oxidase activity. Such differences are attributable to the presence of vitamins and polyphenols in natural diets that are absent in pure sugar feeds [39]. Notably, one study observed that bees fed only sucrose had the highest mortality and poor immunocompetence, whereas those consuming pollen or bee bread had lower mortality and better immune indicators despite an increase in certain oxidative stress biomarkers [39]. The same authors interpreted the elevated oxidative enzyme activities in pollen-fed bees not as pathological oxidative stress, but as an adaptive shift in redox balance due to activation of immune and detox pathways. This underscores that a nutrient-rich diet can simultaneously challenge bees (by ramping up metabolism and immune function, hence measurable ROS) but ultimately benefit them by enhancing survival and immunity, as a hormetic effect moderated by dietary antioxidants.

5.2. *Influence on Honeybee Resistance to Diseases*

Dietary antioxidants have been linked to improved resistance against a range of honeybee pathogens. For instance, *N. ceranae* infections are known to cause oxidative stress in bees and are exacerbated by poor nutrition. Polyphenols in pollen appear to help bees tolerate and suppress *Nosema* spp. Feeding experiments show that bees on polyphenol-rich diets have lower spore loads than those on polyphenol-deficient diets [114]. Similar effects were observed after additional feeding of honeybee colonies with protein additives [132,134]. This protective effect appears to be partially mediated by enhanced epithelial integrity in the midgut, the primary site of microsporidia infection, suggesting that antioxidant-mediated preservation of the peritrophic membrane barrier function occurs. Curcumin supplementation is reported to reduce *Nosema* spp. spores load and increase lifespan in infected honeybees [113]. Research shows that p-coumaric acid in the diet decreases *Nosema* spp. infection levels, potentially by upregulating immune genes that fight the parasite [42]. Extracts from native Chilean plants rich in the flavonoids rutin and myricetin significantly decreased *Nosema* spore counts and improved survival of infected bees [114]. Propolis constituents like galangin and pinocembrin (also flavonoids that bees might ingest in small amounts) were similarly associated with lower *Nosema* spp. spores loads and heightened immune response in bees [114]. Viral pathogen dynamics also demonstrate sensitivity to honeybee antioxidant status. For instance, DeGrandi-Hoffman et al. (2010) observed inverse correlations between pollen consumption and DWV titers in honeybee colonies, with effects attributed partially to antioxidant-mediated immune enhancement [116]. Resistance to parasitic mites, particularly *V. destructor*, demonstrates a link between honeybee nutritional status and antioxidant intake. Given the central role of this parasitic mite in global honeybee colony losses, these nutritional interactions warrant further investigation for potential application in integrated pest management strategies [38,135]. Field studies provide ecological validation of laboratory findings regarding antioxidant-mediated resistance to pathogens. For instance, Alaux et al. (2017) demonstrated that honeybee colonies situated in high floral diversity landscapes exhibited lower pathogen prevalence and enhanced survival compared to those in simplified agricultural landscapes, with effects partially attributed to improved nutritional status, including antioxidant intake [136]. These landscape-scale patterns highlight potential conservation strategies focusing on floral resource diversity to enhance honeybee health through dietary mechanisms.

5.3. Synergistic Effects of Different Antioxidants on Immune Function

Honeybees gather antioxidants from various pollens and plants, which may work together to enhance immunity. Different antioxidants affect immune defense in unique ways; for instance, flavonoids can induce detox and immune gene expression, while vitamins neutralize radicals and stabilize immune cell membranes. Research suggests that combinations of phytochemicals can have additive or synergistic benefits. An illustrative case is the interaction of quercetin and p-coumaric acid: bees fed both showed greater protection against pesticide toxicity than those fed either alone [42]. In that scenario, each phytochemical likely induced overlapping but distinct sets of detoxification genes, broadening the protective effect. Similarly, a diverse multifloral pollen diet is often superior to any single pollen type in promoting bee health, precisely because it delivers a wider array of bioactives that can cover each other's gaps. There is also synergism between diet-derived antioxidants and the bees' enzymatic defenses: as one increases, it bolsters the other. For example, vitamin C works synergistically with GSH and CAT, where well-nourished bees with high vitamin C levels tend to possess a more reactive glutathione system, which aids in the rapid detoxification of peroxides [21]. Diverse antioxidant compounds in the honeybee diet show synergistic effects on immune function through complementary mechanisms. Polyphenols and vitamin C exhibit well-documented interactions, with ascorbic acid restoring the antioxidant capacity of oxidized flavonoids, thereby extending their functional longevity [101]. This recycling mechanism may enhance the antioxidant capacity that supports honeybee immune function when consuming diverse pollen mixtures, potentially influencing gut microbiota composition and immunity. Jones et al. (2018) demonstrated that specific polyphenols selectively promoted beneficial microbial taxa in the honeybee gut, potentially enhancing resistance against opportunistic pathogens [117]. This microbiome-mediated mechanism represents an indirect pathway through which dietary antioxidant diversity may influence honeybee immunocompetence beyond direct effects on host physiology. The collective evidence supports nutritional strategies that emphasize pollen diversity over single-compound supplementation for optimal immune benefits. Experimental studies consistently show superior health outcomes in honeybees consuming polyfloral compared to monofloral pollen sources, with effects partially attributed to broader antioxidant profiles [6]. These findings have direct implications for both conservation strategies focusing on floral diversity and the potential development of antioxidant-enriched supplemental feeds for managed honeybee colonies, as part of good beekeeping practices and biosecurity measures.

6. Antioxidants and Honeybee Resilience to Environmental Stressors

Beyond their classical nutritional constituents, pollen polyphenols constitute key antioxidant agents in the honeybee diet. These compounds help mitigate oxidative damage induced by environmental stressors, thereby enhancing bee longevity and colony sustainability. Polyphenol-rich diets have been shown experimentally to improve food consumption and detoxification capacity in worker bees exposed to pesticides, reflecting their role in modulating antioxidant defense systems and promoting survival under toxic insults [43]. Dietary supplementation with phenolic acids and flavonoids has been linked to the upregulation of detoxifying enzymes, including cytochrome P450s and antioxidant enzymes, indicating that these bioactive compounds prime the bees' physiological defenses. Such nutritional strategies can counteract immune suppression and oxidative damage caused by environmental toxins, supporting the use of polyphenol-enriched pollen in apicultural nutrition [137]. However, pollen's composition is influenced by species, climatic and soil factors, thus affecting polyphenol bioavailability, and necessitating further research to optimize formulations for maximal health benefits [138].

6.1. Protection Against Pesticide-Induced Oxidative Stress

Dietary antioxidants provide significant protection against pesticide toxicity in honeybees through multiple mechanisms. Neonicotinoid insecticides represent the most extensively studied pesticide class concerning honeybee oxidative stress [23]. These compounds appear to disrupt mitochondrial electron transport chains, increasing ROS production while simultaneously depleting cellular glutathione levels [139]. Bees with enhanced antioxidant status through dietary supplementation show improved survival rates and reduced behavioral impairments when exposed to sublethal neonicotinoid concentrations [140]. As it was mentioned earlier, pesticides are significant stressors that contribute to declines in honeybee populations, causing neurotoxic and oxidative effects. Pyrethroids, neonicotinoids, and fungicides induce oxidative stress by generating ROS and impairing mitochondrial function, overwhelming endogenous antioxidant defenses. Honeybees deploy cytochrome P450 monooxygenases, glutathione-S-transferases, and other phase I/II detoxification enzymes to metabolize and eliminate such xenobiotics [141]. Notably, polyphenols, such as those found in bergamot polyphenolic fractions, have demonstrated protective effects by reducing mortality in bees exposed to deltamethrin, a pyrethroid pesticide. Such polyphenolic antioxidants mitigate oxidative stress and compensate for energy deficits caused by pesticide-induced metabolic perturbations [141]. Furthermore, nutritional supplementation with polyphenol-rich pollen has been linked to improved detoxification gene expression and survival in pesticide-exposed bees, evidencing a potential nutritional approach to mitigate chemical stress [43]. The combination of understanding detox pathways and integrating antioxidant dietary components forms a critical strategy for protecting the bees from ongoing chemical challenges. Organophosphate and carbamate insecticides create oxidative stress through different mechanisms, primarily by inhibiting acetylcholinesterase and disrupting normal neurotransmission [142,143]. The resulting hyperexcitation increases metabolic demands and ROS production in neural tissues. Antioxidant protection appears particularly important for maintaining cognitive functions essential for navigation and communication under these conditions. Herbicide exposure, while generally less acutely toxic than insecticides, may create chronic oxidative stress through disruption of plant communities and reduction in dietary antioxidant availability. Glyphosate and other widely used herbicides can alter the polyphenolic content of plants, indirectly affecting bee antioxidant nutrition. These indirect effects may be more significant for long-term bee health than direct herbicide toxicity [144]. Fungicide interactions with bee antioxidant systems appear particularly complex, as many fungicides are designed to disrupt cellular respiration in target organisms. While bees possess different cellular machinery than fungi, some fungicides may affect bee mitochondrial function and antioxidant enzyme activity [64]. Synergistic interactions between fungicides and other pesticides may create unpredictable effects on bee oxidative stress levels [145]. Mechanistically, pollen polyphenols enhance detoxification capacity by upregulating metabolic enzymes. P-coumaric acid significantly increases the expression and activity of cytochrome P450 monooxygenases, particularly CYP9Q1, CYP9Q2, and CYP9Q3, which metabolize a wide range of pesticides, including neonicotinoids and organophosphates [8]. Schmehl et al. (2014) showed that honeybee colonies with access to diverse pollen sources experienced reduced mortality following field-realistic pesticide exposure compared to nutritionally restricted colonies [146]. This practical demonstration of antioxidant-mediated resilience highlights potential management strategies emphasizing nutritional interventions to mitigate pesticide impacts in agricultural settings.

6.2. Role in Mitigating Stress from Climate Change and Habitat Loss

With anticipated increases in winter temperature variability due to climate change, antioxidant nutrition could be a crucial factor affecting overwintering success and spring honeybee population strengthening. One indirect evidence is that well-nourished "winter bees" (long-lived bees produced in the autumn) have exceptionally high vitellogenin and antioxidant enzyme levels, which help them survive months of cold [20]; winter bees come from colonies that had good late-season pollen (like goldenrod and aster) rich in antioxidants, supporting the idea that diet contributes to climate resilience. Research demonstrates that pollen polyphenols offer broad protective effects against various environmental stressors that threaten honeybee health and survival. These bioactive compounds exhibit notable antioxidant, anti-inflammatory, anti-mutagenic, anti-allergic, antimicrobial, and antitumor properties, with most of these protective activities linked to their phenolic content and radical scavenging capacity [71,147,148]. The antioxidant potential of pollen extracts has been shown to rival that of pure antioxidant compounds such as quercetin and ascorbate, demonstrating robust capacity for neutralizing harmful free radicals through multiple mechanisms, including iron reduction and radical scavenging [149]. UV radiation represents a significant environmental stressor that can damage bee reproductive systems, but pollen polyphenols provide crucial protection during these conditions. Studies show that flavonoids, particularly flavonols and anthocyanins, promote healthy pollen tube growth by lowering reactive oxygen species levels and mitigating their adverse effects on cellular function [107,108]. Heat stress similarly increases ROS levels in reproductive tissues, but flavonols prevent these reactive species from reaching damaging concentrations, thereby maintaining fertility and reproductive success even under elevated temperatures [108]. The protective effectiveness of pollen polyphenols varies significantly with environmental conditions and collection timing, highlighting the importance of optimal management practices. Environmental factors such as soil conditions and plant physiology impact the antioxidant capacity of bee pollen, while collection during UV-intense periods from early to late summer yields pollen with higher antioxidant activity [68,70]. Proper storage methods are also critical for maintaining protective compounds, with freezing at −20 °C proving most effective for preserving vitamin C, polyphenols, and antioxidant properties compared to drying methods that can cause significant degradation [150]. Habitat fragmentation and resulting nutritional stress interact with antioxidant physiology to influence honeybee resilience. Honeybee colonies in simplified agricultural landscapes show reduced antioxidant enzyme activity and increased oxidative damage markers compared to those in diverse habitats, impacting disease resistance and longevity [136]. These findings highlight the importance of landscape-scale conservation efforts that preserve floral diversity to support honeybee nutritional needs, including antioxidant intake. Beyond direct antioxidant effects, pollen polyphenols provide systemic health benefits that support overall stress resistance in honeybees. These compounds can modulate cardiovascular-like systems by improving endothelial function, reducing inflammatory cytokine synthesis, and enhancing nitric oxide production, which collectively support physiological resilience under environmental pressure [151]. The broad spectrum of biological activities exhibited by bee pollen polyphenols positions them as natural protective agents that can help honeybee colonies cope with the mounting environmental challenges they face in modern ecosystems. To counteract these environmental deficits, nutritional interventions may mitigate climate and habitat-related stressors through antioxidant supplementation. Fleming et al. (2015) demonstrated that supplemental feeding with polyphenol-enriched formulations increased heat tolerance and reduced oxidative damage in honeybees, suggesting practical applications for beekeeping in changing climates [152]. However, economic constraints and limited understanding of appropriate supplementation protocols currently restrict widespread implementation of these approaches.

6.3. Effects on Adult Bees' Longevity and Colony Survival

The ultimate metrics of resilience are how long individual bees live and whether the colony as a whole survives and thrives. Antioxidant intake has a pronounced impact on both. Worker bees on pollen diets live significantly longer than those on protein-poor diets and numerous studies confirm that longevity is linked to diet quality and antioxidant levels [21,39]. For instance, bees fed polyphenol-enriched diets (through natural pollen or spiked supplements) had an extended lifespan even under stress exposure [42]. One mechanism is the preservation of physiological function with age: oxidative damage is a key driver of aging, so antioxidants slow the accumulation of such damage. Additionally, some antioxidants like vitellogenin (which has antioxidant properties and is tied to pollen intake) act as life-extending factors in bees [6,153]. On a colony level, good nutrition with ample antioxidants translates to higher colony population and survival rates. Colonies situated in diverse floral landscapes (e.g., wildflower meadows) typically show higher brood viability, greater adult population heading into winter, and better overwintering success than those in poor landscapes [111]. Colonies with access to pollen high in antioxidants had lower off-season mortality and were more robust against parasites compared to those fed mostly monoculture pollen or syrup. Adding antioxidant supplements to colonies has improved markers of colony health, for example, increased brood rearing and lower *Nosema* spp. infection rates, and higher winter survival percentages [6,39,116,154]. Winter bees, whose extended lifespan is crucial for the success of colony overwintering, demonstrate particular sensitivity to antioxidant status. Colonies fed nutrient-rich autumn pollen showed a winter survival rate above 90%, significantly larger brood area, and in vitro worker longevity extended to about 23 days, contrasting sharply with honeybee colonies fed syrup alone. These results suggest that autumn pollen quality, likely enriched with micronutrients such as vitamin E and other antioxidants, influences the physiological capacity of winter bees and predicts spring colony strength [155]. Colony-level survival studies provide ecological validation of experimental findings regarding antioxidant-mediated resilience. Smart et al. (2016) demonstrated that apiaries situated in landscapes with greater floral diversity experienced reduced annual mortality compared to those in poor agricultural environments, with effects partially attributed to nutritional factors, including antioxidant intake [156]. These landscape-scale patterns highlight potential conservation and management strategies focusing on nutritional resources to enhance honeybee colony survival. The relationship between worker longevity and colony success creates a critical feedback loop for colony survival. When environmental stressors trigger precocious foraging behavior, shortened worker lifespans can prevent colonies from maintaining adequate populations to support larval development, ultimately leading to colony death [157]. Adequate pollen access prevents this cascade by supporting normal worker development, including enhanced hypopharyngeal gland function and improved ovarian development, which directly impacts the colony's ability to rear brood successfully [158]. Table 3 highlights the relationship between pollen-derived polyphenols, environmental stressors, and related antioxidants linked to health outcomes.

Table 3. Summary of polyphenol pollen source links with tested environmental stressors and antioxidants/health endpoints.

Polyphenol Source/ Characterization	Pollen as an Experimental Variable	Stressor(s) Tested	Antioxidant/ Redox Endpoints	Health/Outcome Endpoints	Link(s)	Ref-(s)
+ pollen quantified for p-coumaric acid	+ two distinct pollen blends; measured phenolics	pesticides (azoxystrobin, sulfoxaflor)	GST, P450 gene expression, vitellogenin (expression only); no direct redox damage assays	none directly measured	(−) direct redox damage assays	[37]
+ monofloral pollens (rapeseed, phacelia, etc.); phenolic/flavonoid content noted	+ 6 monofloral pollens vs. sugar-only	nutritional stress (monodiet)	SOD, CAT, GST, GPx (activity) in hemolymph/ fat body	survival, pesticide clearance	+ polyphenol composition-dependent resilience	[7]
+ pollen fractionated (apolar vs. polar)	+ pollen vs. sugar	*V. destructor*, DWV	None beyond enzymatic shifts	none beyond enzymatic levels	(weak) no classic stressor link, but polyphenol-diet effect observed	[21]
not quantified; pollen and bee bread diets	+ rapeseed/ willow pollen and bee bread vs. controls	none applied directly	TBARS, protein carbonyls, CAT activity, lysozyme	survival (lab/field), overwintering, gene expression changes	(weak) survival benefit lost with apolar fraction removal; no redox markers	[38]
+ detailed for 3 pollen types, polyphenol content compared	+ maize, lotus, sunflower pollen	none applied directly	no mechanistic linkage to polyphenols; more diet-induced redox status	digestion, immunity, gut microbiota	(−) polyphenol-focused	[39]
(−) pure compounds (e.g., quercetin, p-coumaric acid) at pollen-level doses	(−) compounds in artificial diets, not pollen	pesticides (imidacloprid, propiconazole, thiacloprid, boscalid)	detox gene upregulation (CYP, GST); no ROS/damage metrics	survival, longevity, food intake, performance	limited: antioxidant effect presumed, no in vivo polyphenol pollen link	[40–43]

Note: + means yes; (−) means no; (weak) means weak connection.

7. Knowledge Gaps and Future Research Directions

Despite progress in understanding dietary antioxidants' role in honeybee health, significant gaps remain. One key issue is the lack of clear dose–response relationships between antioxidant intake and physiological outcomes. While compounds like flavonoids and phenolic acids show protective effects, optimal intake levels and toxicity thresholds are not well defined. Additionally, the absence of standardized methods and metrics hinders conclusive findings across studies. Investigations involving controlled dietary manipulations, graduated concentration ranges, and multiple biomarkers would enable a more mechanistic understanding of how varying antioxidant levels influence immune function, detoxification, longevity, and stress resistance. Such data are essential for formulating evidence-based nutritional guidelines, particularly for beekeepers managing colonies under stress from agricultural intensification or climate variability [6] and veterinarians as crucial stakeholders in animal health under the One Health approach [159]. In addition to intake uncertainties, the molecular mechanisms underlying antioxidant absorption, transport, and utilization in bee physiology remain incompletely characterized [160]. Advanced metabolomics and transcriptomics approaches could provide insights into how different antioxidant compounds are processed and distributed within bee tissues [161]. Few studies have investigated the correlation between different antioxidant doses and physiological endpoints under laboratory-controlled conditions. Current research often focuses on the presence or absence of antioxidants rather than quantifying optimal levels for different physiological states and environmental conditions [6]. Such quantitative clarity would directly inform practical supplementation protocols for managed colonies [86]. The temporal dynamics of antioxidant protection present another area requiring focused research attention [162]. Questions remain regarding how long antioxidant protection persists after consumption, whether there are critical windows when antioxidant supplementation provides maximum benefits, and how antioxidant requirements vary across bee life stages and seasonal cycles [15]. Synergistic and antagonistic interactions among different antioxidant compounds need comprehensive characterization [44]. While some beneficial interactions have been identified, the full complexity of antioxidant interactions in bee nutrition remains unexplored [45]. Understanding these interactions could inform strategies for optimizing antioxidant mixtures rather than focusing on individual compounds. Environmental factors affecting antioxidant bioavailability and efficacy require systematic study [121]. Temperature, humidity, UV exposure, and other environmental variables may influence how effectively bees can utilize dietary antioxidants [120]. These factors become particularly important as climate change alters environmental conditions in bee habitats [115]. The economic implications of antioxidant nutrition for commercial beekeeping operations need quantitative assessment [128]. While enhanced health and productivity appear likely to provide economic benefits, the costs and benefits of different antioxidant intervention strategies require careful analysis to guide commercial adoption [14]. Standardization of methods for measuring antioxidant status in bees and bee products is also urgently needed. Current methods vary substantially among research groups, making it difficult to synthesize findings and establish reference values for different conditions [127,137]. The development of validated biomarker panels, for both field and laboratory use, would significantly enhance the coherence and reproducibility of research findings. Understanding age- and caste-specific nutritional demands could inform strategic feed supplementation during critical phases of colony development, including early brood rearing, drone production, or queen rearing [86].

Another important and unresolved issue is the nature of interactions among antioxidant compounds. While some combinations of polyphenols and flavonoids appear to produce synergistic effects that enhance antioxidant capacity beyond that of individual components, other studies suggest antagonism, where certain compounds interfere with absorption or efficacy when co-administered [29,163]. Such conflicting results underscore the need for systematic evaluation of compound interactions across different doses and stress contexts. This includes assessing additive versus synergistic outcomes and determining whether particular combinations may inadvertently reduce protective benefits or cause redox imbalance. A critical shortcoming of the current literature is its reliance on simplified experimental models that do not capture the multifactorial nature of stress in the real-world environments of honeybee colonies. Most studies isolate a single stressor, such as pesticide exposure or poor nutrition, while in nature, colonies often face multiple simultaneous challenges, including pathogens, chemical contaminants, temperature fluctuations, and forage scarcity. More field studies that incorporate these multifactorial stressors and measure antioxidant dynamics across the season are needed to better understand how nutritional strategies interact with broader ecological pressures [133]. Such research would provide more relevant insights into the resilience of honeybee colonies in commercial and conservation settings. Compounding these knowledge gaps is the limited understanding of the bioavailability and metabolism of ingested antioxidants in honeybees. While certain phenolic compounds, such as p-coumaric acid, are known to be absorbed and utilized by honeybees [8], most dietary antioxidants have not been pharmacokinetically characterized in this species. Critical questions remain about how these compounds are processed post-ingestion, how they are distributed across tissues, whether they are metabolically activated or degraded, and what biological roles their metabolites might serve. The honeybee gut microbiota likely plays an underappreciated role in modulating the efficacy of antioxidants. Several core microbial species possess enzymatic capacities to degrade or transform dietary phytochemicals, potentially altering their biological activity or facilitating absorption [164]. The extent to which these microbial transformations enhance or diminish antioxidant effects in vivo remains unknown. Studies using germ-free bees or controlled microbial reconstitution models could illuminate the symbiotic role of gut microbes in host antioxidant defense, paving the way for microbiome-informed dietary interventions. Methodological inconsistency is a pervasive issue in this field. Current studies rely on a diverse array of assays to evaluate oxidative status and antioxidant function, including enzymatic activity measurements (e.g., SOD and CAT), assessments of oxidative damage (e.g., lipid peroxidation markers), and direct quantification of antioxidant compounds in tissues. This variability hinders cross-study comparisons and slows the meta-analytical synthesis process [165]. The development of standardized biomarker panels and validated assays for field and laboratory use would significantly enhance the coherence and reproducibility of research findings. Technological challenges also hinder the translation of antioxidant research into practical beekeeping solutions. Antioxidant supplements are currently limited by issues such as compound instability, poor palatability, and uncertain bioavailability. Beyond biological and technical issues, socioeconomic and practical barriers must be addressed to facilitate the adoption of antioxidant supplementation in commercial beekeeping. Despite promising laboratory evidence, few studies have evaluated the cost-effectiveness of these interventions at scale. Integration with existing management routines, such as feeding schedules, anti-varroa treatments, and queen replacement protocols, requires logistical compatibility and minimal labor burden. Collaborative research with beekeepers and other stakeholders in apiculture, as well as involvement from the early stages of supplement development, can help ensure that the resulting products meet the needs and constraints of consumers [86]. Ultimately, aligning research priorities with industry perspectives is key to driving meaningful application. In summary, advancing antioxidant-based approaches to honeybee health and welfare will require a more integrated, mechanistic, and translational research agenda. Addressing dose-dependent effects, metabolic pathways, compound interactions, and field-relevant stress dynamics, while simultaneously resolving practical and economic concerns, will position antioxidant nutrition as a viable tool in the broader strategy to enhance pollinator resilience in a changing world (Table 4).

Table 4. Summary of knowledge gaps and future research directions in honeybee antioxidant nutrition.

Research Area	Knowledge Gap/Challenge	Future Direction	Ref-(s)
dose–response relationships	undefined optimal intake, efficacy thresholds, and potential toxicity of antioxidants	controlled dose trials using multiple biomarkers and stressors	[6,134,166]
developmental and caste-specific needs	limited data on stage- and caste-specific antioxidant demands (larvae, queens, drones)	formulation of life-stage-targeted antioxidant diets	[17,167]
compound interactions	unclear whether combinations of antioxidants are synergistic or antagonistic	systematic evaluation of common polyphenol/flavonoid interactions	[15,163]
real-world multifactorial stress models	laboratory studies often isolate a single stressor; field-relevant complexity is missing	seasonal field studies simulating pesticide–pathogen–nutrition–climate combinations	[90,133,168]

Table 4. *Cont.*

Research Area	Knowledge Gap/Challenge	Future Direction	Ref-(s)
bioavailability and metabolism	poor understanding of absorption, transformation, and excretion of dietary antioxidants	use of isotope-labeled compounds and LC-MS/MS profiling	[4,169]
role of microbiota	the gut microbiota interaction with antioxidant metabolism is not well characterized	germ-free bee studies; microbiome-targeted diet design	[164]
standardization of methods	inconsistent biomarkers and oxidative stress assays limit comparability	development of validated standardized testing protocols and field kits	[165]
delivery technologies	antioxidants may degrade, taste bad, or be poorly absorbed	nanoencapsulation, slow-release formulations, pH-stable carriers	[170–172]
socioeconomic and practical adoption	lack of cost-effectiveness data and field-scale validation	participatory research with beekeepers; practical feeding protocols	[86]
precision nutrition and genetics	unknown variability in antioxidant metabolism across colonies or strains	breeding programs selecting enhanced redox resistance traits	[2,17]

When evaluating the use of pollen supplementation for honeybee colonies, it is essential to address several practical aspects. The optimal delivery method (microencapsulation, protein patties, or liquid suspensions) remains unresolved, as each technique offers distinct advantages for nutrient stability and acceptability. It is advisable to limit pollen harvesting to preserve natural food stores and enhance honeybee colony resilience. Additionally, while supplemental feeding during the summer may help address nutritional resource shortages, further research is needed to evaluate its potential effects on the quality of honey and royal jelly.

8. Conclusions

Pollen-derived polyphenols are fundamental to honeybee health, providing essential antioxidant defenses that bolster immunity, detoxification pathways, and overall resilience against a suite of environmental stressors. The availability of these crucial compounds, however, is increasingly threatened by anthropogenic and environmental pressures, including climate change and habitat degradation, which disrupt floral resources. This review synthesizes the evidence suggesting that a diet rich in diverse polyphenols is not merely beneficial but may be critical for colony survival in challenging environments. Therefore, it is crucial to integrate antioxidant-focused nutritional strategies into conservation frameworks and practical beekeeping management. Future efforts must be directed toward translating this biochemical understanding into tangible applications, such as the development of targeted, evidence-based feed supplements and the promotion of floral diversity, to support the long-term health and sustainability of these vital pollinators.

Author Contributions: Conceptualization, A.C. and I.T.G.; project administration, I.T.G.; writing—original draft, A.C. and I.T.G.; writing—review and editing, I.T.G. and A.C. All authors have read and agreed to the published version of the manuscript.

Funding: This research received no external funding.

Data Availability Statement: No new data were created or analyzed in this study. Data sharing is not applicable to this article.

Acknowledgments: The authors would like to acknowledge the COST Action CA22105 entitled "Beekeeping products valorization and biomonitoring for the SAFEty of BEEs and HONEY" (BeSafeBeeHoney).

Conflicts of Interest: The authors declare no conflicts of interest.

References

1. Li-Byarlay, H.; Huang, M.H.; Simone-Finstrom, M.; Strand, M.K.; Tarpy, D.R.; Rueppell, O. Honey bee (*Apis mellifera*) drones survive oxidative stress due to increased tolerance instead of avoidance or repair of oxidative damage. *Exp. Gerontol.* **2016**, *83*, 15–21. [CrossRef]
2. Simone-Finstrom, M.; Li-Byarlay, H.; Huang, M.-H.; Strand, M.K.; Rueppell, O.; Tarpy, D.R. Migratory management and environmental conditions affect lifespan and oxidative stress in honey bees. *Sci. Rep.* **2016**, *6*, 32023. [CrossRef]
3. Halliwell, B.; Gutteridge, J.M.C. *Free Radicals in Biology and Medicine*, 5th ed.; Oxford University Press: Oxford, UK, 2015.
4. Kunat-Budzyńska, M.; Łabuć, E.; Ptaszyńska, A.A. Changes in enzymatic activity and oxidative stress in honeybees kept in the apiary and laboratory conditions during the course of nosemosis. *PLoS ONE* **2025**, *20*, e0317384. [CrossRef] [PubMed]
5. Chouikh, A.; Chenguel, A.; Ali, A.B. Understanding the Role of Free Radicals, Oxidative Stress, and Antioxidants: A Comprehensive Review. *Lett. Appl. NanoBioScience* **2025**, *14*, 66.
6. Di Pasquale, G.; Salignon, M.; Le Conte, Y.; Belzunces, L.P.; Decourtye, A.; Kretzschmar, A.; Suchail, S.; Brunet, J.-L.; Alaux, C. Influence of Pollen Nutrition on Honey Bee Health: Do Pollen Quality and Diversity Matter? *PLoS ONE* **2013**, *8*, e72016. [CrossRef] [PubMed]
7. Barascou, L.; Sene, D.; Barraud, A.; Michez, D.; Lefebvre, V.; Medrzycki, P.; Di Prisco, G.; Strobl, V.; Yañez, O.; Neumann, P.; et al. Pollen nutrition fosters honeybee tolerance to pesticides. *R. Soc. Open Sci.* **2021**, *8*, 210818. [CrossRef]
8. Mao, W.; Schuler, M.A.; Berenbaum, M.R. Honey constituents up-regulate detoxification and immunity genes in the western honey bee *Apis mellifera*. *Proc. Natl. Acad. Sci. USA* **2013**, *110*, 8842–8846. [CrossRef]
9. Tlak Gajger, I.; Pavlović, I.; Bojić, M.; Kosalec, I.; Srečec, S.; Vlainić, T.; Vlainić, J. The Components Responsible for the Antimicrobial Activity of Propolis from Continental and Mediterranean Regions in Croatia. *Czech J. Food Sci.* **2017**, *35*, 376–385. [CrossRef]
10. Tlak Gajger, I.; Dar, S.A. Plant allelochemicals as sources of insecticides. *Insects* **2021**, *12*, 189. [CrossRef]
11. Rodríguez-Pólit, C.; Gonzalez-Pastor, R.; Heredia-Moya, J.; Carrera-Pacheco, S.E.; Castillo-Solis, F.; Vallejo-Imbaquingo, R.; Barba-Ostria, C.; Guamán, L.P. Chemical properties and biological activity of bee pollen. *Molecules* **2023**, *28*, 7768. [CrossRef]
12. Tesfaye, O. Comparative Analysis of Bee Pollen Antioxidant Properties from Western Oromia, Ethiopia. *Austin J. Nutr. Metab.* **2024**, *11*, 1134.
13. Altiner, D.D.; Altunatmaz, S.S.; Sabuncu, M.; Aksu, F.; Şahan, Y. In-vitro bioaccessibility of antioxidant properties of bee pollen in Turkey. *Food Sci. Technol.* **2021**, *41*, 133–141. [CrossRef]
14. Goulson, D.; Nicholls, E.; Botías, C.; Rotheray, E.L. Bee declines driven by combined stress from parasites, pesticides, and lack of flowers. *Science* **2015**, *347*, 1255957. [CrossRef]
15. Williams, J.B.; Roberts, S.P.; Elekonich, M.M. Age and natural metabolically-intensive behavior affect oxidative stress and antioxidant mechanisms. *Exp. Gerontol.* **2008**, *43*, 538–549. [CrossRef]
16. Alburaki, M.; Smith, K.D.; Adamczyk, J.; Karim, S. Interplay between selenium, selenoprotein genes, and oxidative stress in honey bee *Apis mellifera* L. *J. Insect Physiol.* **2019**, *117*, 103891. [CrossRef] [PubMed]
17. Corona, M.; Robinson, G.E. Genes of the antioxidant system of the honey bee: Annotation and phylogeny. *Insect Mol. Biol.* **2006**, *15*, 687–701. [CrossRef]
18. Cervoni, M.S.; Cardoso-Júnior, C.A.M.; Craveiro, G.; Souza, A.D.O.; Alberici, L.C.; Hartfelder, K. Mitochondrial capacity, oxidative damage and hypoxia gene expression are associated with age-related division of labor in honey bee (*Apis mellifera* L.) workers. *J. Exp. Biol.* **2017**, *220*, 4035–4046. [CrossRef]
19. Margotta, J.W.; Roberts, S.P.; Elekonich, M.M. Effects of Flight Activity and Age on Oxidative Damage in the Honey Bee, Apis mellifera. *J. Exp. Biol.* **2018**, *221*, jeb183228. [CrossRef]
20. Strachecka, A.; Staniszewska, P.; Olszewski, K.; Bryś, M.S.; Stec, W.; Bartoń, M. The antioxidant system was unexpectedly strongly suppressed in Apis mellifera worker bees emerged from larvae reared on combs adulterated with paraffin or stearin. *Sci. Rep.* **2025**, *15*, 20363. [CrossRef]

21. Bryś, M.S.; Olszewski, K.; Bartoń, M.; Strachecka, A. Changes in the Activities of Antioxidant Enzymes in the Fat Body and Hemolymph of Apis mellifera L. Due to Pollen Monodiets. *Antioxidants* **2025**, *14*, 69. [CrossRef] [PubMed]
22. Migdał, P.; Murawska, A.; Strachecka, A.; Bieńkowski, P.; Roman, A. Changes in the Honeybee Antioxidant System after 12 h of Exposure to Electromagnetic Field Frequency of 50 Hz and Variable Intensity. *Insects* **2020**, *11*, 713. [CrossRef] [PubMed]
23. Xu, X.; Wang, X.; Yang, Y.; Yang, D.; Zhao, B.; Wang, X.; Jiang, G. Neonicotinoids: Mechanisms of systemic toxicity based on oxidative stress-mitochondrial damage. *Arch. Toxicol.* **2022**, *96*, 1493–1520. [CrossRef]
24. Chakrabarti, P.; Rana, S.; Smith, B.; Sarkar, S.; Basu, P. Pesticide-induced oxidative stress in laboratory and field populations of native honey bees along intensive agricultural landscapes in two Eastern Indian states. *Apidologie* **2015**, *46*, 107–129. [CrossRef]
25. Alburaki, M.; Steckel, S.J.; Williams, M.T.; Skinner, J.A.; Tarpy, D.R.; Meikle, W.G.; Adamczyk, J.J. Agricultural landscape and pesticide effects on honey bee (Hymenoptera: Apidae) biological traits. *J. Econ. Entomol.* **2017**, *110*, 835–847. [CrossRef]
26. Bryden, J.; Gill, R.J.; Mitton, R.A.; Raine, N.E.; Jansen, V.A. Chronic sublethal stress causes bee colony failure. *Ecol. Lett.* **2013**, *16*, 1463–1469. [CrossRef]
27. Dussaubat, C.; Brunet, J.-L.; Higes, M.; Colbourne, J.K.; López, J.; Choi, J.H.; Martin-Hernández, R.; Botías, C.; Cousin, M.; McDonnell, C.; et al. Gut pathology and responses to the microsporidium *Nosema ceranae* in the honey bee (*Apis mellifera*) midgut. *PLoS ONE* **2012**, *7*, e37017. [CrossRef]
28. Dainat, B.; Evans, J.D.; Chen, Y.P.; Gauthier, L.; Neumann, P. Deformed wing virus and *Varroa destructor* reduce the life span of winter honeybees. *Appl. Environ. Microbiol.* **2012**, *78*, 981–987. [CrossRef]
29. Alaux, C.; Ducloz, F.; Crauser, D.; Le Conte, Y. Diet effects on honeybee immunocompetence. *Biol. Lett.* **2010**, *6*, 562–565. [CrossRef] [PubMed]
30. Kozii, I.V.; Barnsley, S.; Silva, M.C.B.d.; Wood, S.C.; Klein, C.D.; Mattos, Í.M.d.; Simko, E. Reproductive fitness of honey bee queens exposed to thiamethoxam during development. *Vet. Pathol.* **2021**, *58*, 1107–1118. [CrossRef] [PubMed]
31. Tlak Gajger, I.; Sakač, M.; Gregorc, A. Impact of Thiamethoxam on Honey Bee Queen (*Apis mellifera*) Reproductive Morphology and Physiology. *Bull. Environ. Contam. Toxicol.* **2017**, *99*, 297–302. [CrossRef]
32. Danner, N.; Keller, A.; Härtel, S.; Steffan-Dewenter, I. Honey bee foraging ecology: Season but not landscape diversity shapes the amount and diversity of collected pollen. *PLoS ONE* **2017**, *12*, e0183716. [CrossRef]
33. Donkersley, P.; Rhodes, G.; Pickup, R.W.; Jones, K.C.; Wilson, K. Nutritional composition of honey bee food stores vary with floral composition. *Oecologia* **2017**, *185*, 749–761. [CrossRef] [PubMed]
34. Piot, N.; Schweiger, O.; Meeus, I.; Yanez, O.; Straub, L.; Villamar-Bouza, L.; De La Rua, P.; Jara, L.; Ruiz, C.; Malmstrom, M.; et al. Honey Bees and Climate Explain Viral Prevalence in Wild Bee Communities on the Continental Scale. *Sci. Rep.* **2022**, *12*, 1904. [CrossRef]
35. Vilić, M.; Žura Žaja, I.; Tkalec, M.; Tucak, P.; Malarić, K.; Popara, N.; Žura, N.; Pašić, S.; Tlak Gajger, I. Oxidative Stress Response of Honey Bee Colonies (*Apis mellifera* L.) during Long-Term Exposure at a Frequency of 900 MHz under Field Conditions. *Insects* **2024**, *15*, 372. [CrossRef]
36. Vilić, M.; Žura Žaja, I.; Tkalec, M.; Štambuk, A.; Šrut, M.; Klobučar, G.; Malarić, K.; Tucak, P.; Pašić, S.; Tlak Gajger, I. Effects of a radio frequency electromagnetic field on honey bee larvae (*Apis mellifera*) differ in relation to the experimental study design. *Vet. Arhiv* **2021**, *91*, 427–435. [CrossRef]
37. Vilic, M.; Tlak Gajger, I.; Tucak, P.; Stambuk, A.; Srut, M.; Klobucar, G.; Malaric, K.; Zura Zaja, I.; Pavelic, A.; Manger, M.; et al. Effects of short-term exposure to mobile phone radiofrequency (900 MHz) on the oxidative response and genotoxicity in honey bee larvae. *J. Apic. Res.* **2017**, *56*, 430–438. [CrossRef]
38. Annoscia, D.; Nazzi, F.; Frizzera, D.; Sablon, L.; Sulotto, F.; Cervo, R.; Rinkevich, F.; Dainat, B. Elucidating the mechanisms underlying the beneficial health effects of dietary pollen on honey bees (*Apis mellifera*) infested by Varroa mite ectoparasites. *Sci. Rep.* **2017**, *7*, 6258. [CrossRef]
39. Yazlovytska, L.S.; Karavan, V.V.; Domaciuk, M.; Panchuk, I.I.; Borsuk, G.; Volkov, R.A. Increased survival of honey bees consuming pollen and beebread is associated with elevated biomarkers of oxidative stress. *Front. Ecol. Evol.* **2023**, *11*, 1098350. [CrossRef]
40. Li, X.-M.; Wu, S.-F.; Wang, Z.-H.; Zhang, Y.; Xu, B.-H. Effects of Three Different Bee Pollen on Digestion, Immunity, Antioxidant Capacity, and Gut Microbes in *Apis mellifera*. *Insects* **2025**, *16*, 505. [CrossRef] [PubMed]
41. Wong, M.J.; Liao, L.-H.; Berenbaum, M.R. Biphasic concentration-dependent interaction between imidacloprid and dietary phytochemicals in honey bees (*Apis mellifera*). *PLoS ONE* **2018**, *13*, e0206625. [CrossRef] [PubMed]
42. Liao, L.-H.; Siegfried, B.D.; Berenbaum, M.R. Increase in longevity and amelioration of pesticide toxicity by natural levels of dietary phytochemicals in the honey bee, *Apis mellifera*. *PLoS ONE* **2020**, *15*, e0243364. [CrossRef]
43. Hýbl, M.; Mráz, P.; Šipoš, J.; Hoštičková, I.; Bohatá, A.; Čurn, V.; Kopec, T. Polyphenols as Food Supplement Improved Food Consumption and Longevity of Honey Bees (*Apis mellifera*) Intoxicated by Pesticide Thiacloprid. *Insects* **2021**, *12*, 572. [CrossRef]

44. Rzepecka-Stojko, A.; Stojko, J.; Kurek-Górecka, A.; Górecki, M.; Kabała-Dzik, A.; Kubina, R.; Stojko, E. Polyphenols from bee pollen: Structure, absorption, metabolism and biological activity. *Molecules* **2015**, *20*, 21732–21749. [CrossRef] [PubMed]
45. Bridi, R.; Echeverría, J.; Larena, A.; Nuñez Pizarro, P.; Atala, E.; De Camargo, A.C.; Montenegro, G. Honeybee pollen from southern Chile: Phenolic profile, antioxidant capacity, bioaccessibility, and inhibition of DNA damage. *Front. Pharmacol.* **2022**, *13*, 775219. [CrossRef]
46. Tlak Gajger, I.; Mutinelli, F. Impact of Environmental Factors and Management Practices on Bee Health. *Insects* **2024**, *15*, 996. [CrossRef]
47. Li, G.; Zhao, H.; Liu, Z.; Wang, H.; Xu, B.; Guo, X. The Wisdom of Honeybee Defenses Against Environmental Stresses. *Front. Microbiol.* **2018**, *9*, 722. [CrossRef]
48. Serra Bonvehí, J.; Soliva Torrentó, M.; Centelles Lorente, E. Evaluation of polyphenolic and flavonoid compounds in honeybee-collected pollen produced in Spain. *J. Agric. Food Chem.* **2001**, *49*, 1848–1853. [CrossRef] [PubMed]
49. El-Seedi, H.R.; Ahmed, H.R.; El-Wahed, A.A.A.; Saeed, A.; Algethami, A.F.; Attia, N.F.; Wang, K. Bee stressors from an immunological perspective and strategies to improve bee health. *Vet. Sci.* **2022**, *9*, 199. [CrossRef]
50. Tahir, F.; Goblirsch, M.; Adamczyk, J.; Karim, S.; Alburaki, M. Honey bee *Apis mellifera* L. responses to oxidative stress induced by pharmacological and pesticidal compounds. *Front. Bee Sci.* **2023**, *1*, 1275862. [CrossRef]
51. Klein, A.M.; Vaissière, B.E.; Cane, J.H.; Steffan-Dewenter, I.; Cunningham, S.A.; Kremen, C.; Tscharntke, T. Why bees are so vulnerable to environmental stressors: Linking stressor-induced neural impairments to colony decline. *Trends Ecol. Evol.* **2016**, *31*, 735–736. [CrossRef]
52. Morimoto, T.; Kojima, Y.; Toki, T.; Komeda, Y.; Yoshiyama, M.; Kimura, K.; Nirasawa, K.; Kadowaki, T. The habitat disruption induces immune-suppression and oxidative stress in honey bees. *Ecol. Evol.* **2011**, *1*, 201–217. [CrossRef]
53. Farooqui, T. Oxidative stress and age-related olfactory memory impairment in the honeybee *Apis mellifera*. *Front. Genet.* **2014**, *5*, 60. [CrossRef] [PubMed]
54. Münch, D.; Amdam, G.V. Chronic inflammation and oxidative stress in honey bees: Winter bees exhibit enhanced oxidative-stress tolerance and negligible senescence. *Genome Biol. Evol.* **2016**, *8*, 495–506. [CrossRef]
55. James, R.R.; Xu, J. Mechanisms by which pesticides affect insect immunity. *J. Invertebr. Pathol.* **2012**, *109*, 175–182. [CrossRef]
56. Chmiel, J.A.; Daisley, B.A.; Pitek, A.P.; Thompson, G.J.; Reid, G. Understanding the effects of sublethal pesticide exposure on honey bees: A role for probiotics as mediators of environmental stress. *Front. Ecol. Evol.* **2020**, *8*, 22. [CrossRef]
57. vanEngelsdorp, D.; Evans, J.D.; Saegerman, C.; Mullin, C.; Haubruge, E.; Nguyen, B.K.; Frazier, M.; Frazier, J.; Cox-Foster, D.; Chen, Y.; et al. Colony collapse disorder: A descriptive study. *PLoS ONE* **2009**, *4*, e6481. [CrossRef] [PubMed]
58. Perry, C.J.; Søvik, E.; Myerscough, M.R.; Barron, A.B. Rapid behavioral maturation accelerates failure of stressed honey bee colonies. *Biol. Open* **2015**, *4*, 27–35. [CrossRef]
59. Tlak Gajger, I.; Vilić, M.; Tucak, P.; Malarić, K. Effect of electromagnetic field on some behaviour modality of honeybee colonies (*Apis mellifera*) in field conditions. *J. Anim. Vet. Adv.* **2019**, *18*, 61–64. [CrossRef]
60. Farjan, M.; Dmitryjuk, M.; Lipiński, Z.; Biernat-Łopieńska, E.; Żółtowska, K. Supplementation of the Honey Bee Diet with Vitamin C: The Effect on the Antioxidative System of *Apis mellifera carnica* Brood at Different Stages. *J. Apic. Res.* **2012**, *51*, 263–270. [CrossRef]
61. Smith, S.M.; Nager, R.G.; Costantini, D. Meta-analysis indicates that oxidative stress is both a constraint on and a cost of growth. *Ecol. Evol.* **2016**, *6*, 2833–2842. [CrossRef]
62. Jansen, E.; Ruskovska, T. Serum Biomarkers of (Anti)Oxidant Status for Epidemiological Studies. *Int. J. Mol. Sci.* **2015**, *16*, 27378–27390. [CrossRef]
63. Majoroš, A.; Tlak Gajger, I.; Smodiš Škerl, M.I. Prehrambeni stres pčelinjih zajednica (*Apis mellifera* L.): Uzroci, učinci i mjere sprječavanja gubitaka. *Vet. Stanica* **2022**, *53*, 461–474. [CrossRef]
64. Liao, L.H.; Wu, W.Y.; Dad, A.; Berenbaum, M.R. Fungicide suppression of flight performance in the honeybee (*Apis mellifera*) and its amelioration by quercetin. *Proc. R. Soc. B* **2019**, *286*, 20192041. [CrossRef]
65. Tlak Gajger, I.; Vlainić, J. Antioxidant Activity of Honey Bee Products. *Antioxidants* **2025**, *14*, 64. [CrossRef]
66. Özcan, M.; Aljuhaimi, F.; Babiker, E.E.; Uslu, N.; Ceylan, D.; Ghafoor, K. Determination of antioxidant activity, phenolic compound, mineral contents and fatty acid compositions of bee pollen grains collected from different locations. *J. Apic. Sci.* **2019**, *63*, 69–79. [CrossRef]
67. Gerçek, Y.C.; Çelik, S.; Bayram, S. Screening of plant pollen sources, polyphenolic compounds, fatty acids and antioxidant/antimicrobial activity from bee pollen. *Molecules* **2021**, *27*, 117. [CrossRef] [PubMed]
68. Araújo, J.; Chambó, É.D.; Costa, M.A.P.; Cavalcante da Silva, S.M.P.; de Carvalho, C.A.L.; Estevinho, L.M. Chemical composition and biological activities of mono- and heterofloral bee pollen of different geographical origins. *Int. J. Mol. Sci.* **2017**, *18*, 921. [CrossRef]

69. Negri, G.; Teixeira, É.; Alves, M.L.T.M.F.; Moreti, A.; Otsuk, I.; Borguini, R. Hydroxycinnamic acid amide derivatives, phenolic compounds and antioxidant activities of extracts of pollen samples from Southeast Brazil. *J. Agric. Food Chem.* **2011**, *59*, 5516–5522. [CrossRef]
70. Boulfous, N.; Belattar, H.; Ambra, R.; Pastore, G.; Ghorab, A. Botanical origin, phytochemical profile, and antioxidant activity of bee pollen from the Mila region, Algeria. *Antioxidants* **2025**, *14*, 291. [CrossRef]
71. Denisow, B.; Denisow-Pietrzyk, M. Biological and therapeutic properties of bee pollen: A review. *J. Sci. Food Agric.* **2016**, *96*, 4303–4309. [CrossRef]
72. Aylanc, V.; Larbi, S.; Calhelha, R.C.; Barros, L.; Rezouga, F.; Rodríguez-Flores, M.; Ferreira, I.C.F.R. Evaluation of antioxidant and anticancer activity of mono- and polyfloral Moroccan bee pollen by characterizing phenolic and volatile compounds. *Molecules* **2023**, *28*, 835. [CrossRef] [PubMed]
73. El Ghouizi, A.; El Menyiy, N.; Falcão, S.I.; Vilas-Boas, M.; Lyoussi, B. Chemical composition, antioxidant activity, and diuretic effect of Moroccan fresh bee pollen in rats. *Vet. World* **2020**, *13*, 1251–1261. [CrossRef] [PubMed]
74. He, Q.; Wang, J.; Li, J.; Yang, W. Polyphenol profile and antioxidant, antityrosinase, and anti-melanogenesis activities of ethanol extract of bee pollen. *Pharmaceuticals* **2024**, *17*, 1634. [CrossRef]
75. Kim, S.; Jo, Y.H.; Liu, Q.; Ahn, J.; Hong, I.; Han, S.; Kim, S. Optimization of extraction condition of bee pollen using response surface methodology: Correlation between anti-melanogenesis, antioxidant activity, and phenolic content. *Molecules* **2015**, *20*, 19764–19774. [CrossRef]
76. Campos, M.G.; Frigerio, C.; Lopes, J.; Bogdanov, S. Bees collect pollen that typically contains significantly higher concentrations of polyphenols than nectar—A diversity driven by floral origin. *J. Apiprod. Apimed. Sci.* **2010**, *2*, 131–144. [CrossRef]
77. Aylanc, V.; Ertosun, S.; Russo-Almeida, P.; Falcão, S.; Vilas-Boas, M. Performance of green and conventional techniques for the optimal extraction of bioactive compounds in bee pollen. *Int. J. Food Sci. Technol.* **2022**, *57*, 3490–3502. [CrossRef]
78. Anjum, S.I.; Ullah, A.; Gohar, F.; Raza, G.; Khan, M.I.; Hameed, M.; Ali, A.; Chen, C.-C.; Tlak Gajger, I. Bee pollen as a food and feed supplement and a therapeutic remedy: Recent trends in nanotechnology. *Front. Nutr.* **2024**, *11*, 1371672. [CrossRef]
79. Mărgăoan, R.; Mărghitaş, L.A.; Dezmirean, D.S.; Dulf, F.V.; Bunea, A.; Socaci, S.; Bobiş, O. Predominant and secondary pollen botanical origins influence the carotenoid and fatty acid profile in fresh honeybee-collected pollen. *J. Agric. Food Chem.* **2014**, *62*, 6306–6316. [CrossRef]
80. Mărgăoan, R.; Özkök, A.; Keskin, Ş.; Mayda, N.; Urcan, A.C.; Cornea-Cipcigan, M. Bee collected pollen as a value-added product rich in bioactive compounds and unsaturated fatty acids: A comparative study from Turkey and Romania. *LWT* **2021**, *149*, 111925. [CrossRef]
81. Mosić, M.; Trifković, J.; Vovk, I.; Gašić, U.; Tešić, Ž.; Šikoparija, B.; Milojković-Opsenica, D. Phenolic composition influences the health-promoting potential of bee-pollen. *Biomolecules* **2019**, *9*, 783. [CrossRef] [PubMed]
82. Feás, X.; Vázquez Tato, M.P.; Estevinho, L.M.; Seijas, J.A.; Iglesias, A. Nectar and bee collected pollen: Phenolic composition and antioxidant properties. *Molecules* **2012**, *17*, 8359–8377. [CrossRef]
83. Nakajima, Y.; Tsuruma, K.; Shimazawa, M.; Mishima, S.; Hara, H. Comparison of bee products based on assays of antioxidant capacities. *BMC Complement. Altern. Med.* **2009**, *9*, 4. [CrossRef] [PubMed]
84. Kostić, A.Ž.; Milinčić, D.D.; Fata, A.; Nedić, N.; Stanojević, S.P.; Tešić, Ž.L.; Pešić, M.B. Polyphenolic profile and antioxidant properties of bee collected sunflower (*Helianthus annuus* L.) pollen. *LWT–Food Sci. Technol.* **2019**, *112*, 108244. [CrossRef]
85. LeBlanc, B.W.; Davis, O.K.; Boue, S.; DeLucca, A.; Deeby, T. Antioxidant activity of Sonoran Desert bee pollen. *Food Chem.* **2009**, *115*, 1299–1305. [CrossRef]
86. Brodschneider, R.; Crailsheim, K. Nutrition and health in honey bees. *Apidologie* **2010**, *41*, 278–294. [CrossRef]
87. Vidkjær, N.H.; Laursen, B.B.; Kryger, P. Phytochemical profiles of honey bees (*Apis mellifera*) and their larvae differ from the composition of their pollen diet. *R. Soc. Open Sci.* **2024**, *11*, 231654. [CrossRef]
88. Moreira, I.R.; Souza, G.D.F.; Astolfi, A.; Lippi, I.C.C.; Scheffer, J.L.; Arruda, R.A.; de Oliveira Orsi, R. The Impact of Pesticides on the Antioxidant System of *Apis Mellifera* Bees-A Systematic Review. *Sociobiology* **2025**, *72*, e10894. [CrossRef]
89. Wu, H.; Ji, C.; Wang, R.; Gao, L.; Luo, W.; Liu, J. Dietary Quercetin Regulates Gut Microbiome Diversity and Abundance in *Apis cerana* (Hymenoptera Apidae). *Insects* **2025**, *16*, 20. [CrossRef] [PubMed]
90. Tlak Gajger, I.; Nejedli, S.; Cvetnić, L. Influence of probiotic feed supplement on *Nosema* spp. infection level and the gut microbiota of adult honeybees (*Apis mellifera* L.). *Microorganisms* **2023**, *11*, 610. [CrossRef] [PubMed]
91. Marinova, M.; Tchorbanov, B. Preparation of antioxidant enzymatic hydrolysates from honeybee-collected pollen using plant enzymes. *Enzyme Res.* **2010**, *2010*, 473298. [CrossRef]
92. Lawag, I.; Yoo, O.; Lim, L.; Hammer, K.; Locher, C. Optimisation of bee pollen extraction to maximise extractable antioxidant constituents. *Antioxidants* **2021**, *10*, 1113. [CrossRef]

93. Gonçalves, A.C.; Aitfella Lahlou, R.; Alves, G.; García-Viguera, C.; Moreno, D.A.; Silva, L.R. Potential activity of Abrantes pollen extract: Biochemical and cellular model studies. *Foods* **2021**, *10*, 2804. [CrossRef]
94. Gonçalves, A.C.; Bento, C.; Nunes, A.R.; Simões, M.; Alves, G.; Silva, L.R. Multitarget protection of *Pterospartum tridentatum* phenolic-rich extracts against a wide range of free radical species, antidiabetic activity and effects on human colon carcinoma (Caco-2) cells. *J. Food Sci.* **2020**, *85*, 4377–4388. [CrossRef]
95. Bava, R.; Castagna, F.; Lupia, C.; Poerio, G.; Liguori, G.; Lombardi, R.; Britti, D. Hive products: Composition, pharmacological properties, and therapeutic applications. *Pharmaceuticals* **2024**, *17*, 646. [CrossRef]
96. Oyarzún, J.E.; Andía, M.; Uribe, S.; Núñez Pizarro, P.; Núñez, G.; Montenegro, G. Honeybee pollen extracts reduce oxidative stress and steatosis in hepatic cells. *Molecules* **2020**, *26*, 6. [CrossRef]
97. Gámbaro, A.; Miraballes, M.; Urruzola, N.; Kniazev, M.; Dauber, C.; Romero, M. Physicochemical composition and bioactive properties of Uruguayan bee pollen from different botanical sources. *Foods* **2025**, *14*, 1689. [CrossRef]
98. Zou, Y.; Hu, J.; Huang, W.; Zhu, L.; Shao, M.; Dordoe, C. The botanical origin and antioxidant, anti-BACE1 and antiproliferative properties of bee pollen from different regions of South Korea. *BMC Complement. Med. Ther.* **2020**, *20*, 236. [CrossRef]
99. Tutun, H.; Kaya, M.; Usluer, M.S.; Kahraman, H. Bee pollen: Its antioxidant activity. *Uludağ Arıcılık Derg.* **2021**, *21*, 119–131. [CrossRef]
100. Oliveira, R.G.; Jain, S.; Freitas, L.; Araújo, E.D. Phenolic compound, nutritional and antioxidant profile of pollen collected by the genus Melipona in North Eastern Brazil. *Braz. J. Food Technol.* **2019**, *22*, e2018079. [CrossRef]
101. Procházková, D.; Boušová, I.; Wilhelmová, N. Antioxidant and prooxidant properties of flavonoids. *Fitoterapia* **2011**, *82*, 513–523. [CrossRef] [PubMed]
102. Ozgen, S.; Kilinc, O.K.; Selamoğlu, Z. Antioxidant activity of quercetin: A mechanistic review. *Turk. J. Agric. Food Sci. Technol.* **2016**, *4*, 1134–1138. [CrossRef]
103. Ardalani, H.; Vidkjær, N.H.; Kryger, P.; Fiehn, O.; Fomsgaard, I.S. Metabolomics unveils the influence of dietary phytochemicals on residual pesticide concentrations in honey bees. *Environ. Int.* **2021**, *152*, 106503. [CrossRef]
104. Ardalani, H.; Vidkjær, N.H.; Laursen, B.B.; Kryger, P.; Fomsgaard, I.S. Dietary quercetin impacts the concentration of pesticides in honey bees. *Chemosphere* **2021**, *262*, 127848. [CrossRef]
105. Perron, N.R.; Brumaghim, J.L. A review of the antioxidant mechanisms of polyphenol compounds related to iron binding. *Cell Biochem. Biophys.* **2009**, *53*, 75–100. [CrossRef] [PubMed]
106. Ma, Q. Role of nrf2 in oxidative stress and toxicity. *Annu. Rev. Pharmacol. Toxicol.* **2013**, *53*, 401–426. [CrossRef]
107. Del Casino, C.; Conti, V.; Licata, S.; Cai, G.; Cantore, A.; Ricci, C.; Del Duca, S. Mitigation of UV-B radiation stress in tobacco pollen by expression of the tardigrade damage suppressor protein (Dsup). *Cells* **2024**, *13*, 840. [CrossRef]
108. Muhlemann, J.K.; Younts, T.L.B.; Muday, G.K. Flavonols control pollen tube growth and integrity by regulating ROS homeostasis during high-temperature stress. *Proc. Natl. Acad. Sci. USA* **2018**, *115*, E11188–E11197. [CrossRef]
109. Parcheta, M.; Świsłocka, R.; Orzechowska, S.; Akimowicz, M.; Choińska, R.; Lewandowski, W. Recent developments in effective antioxidants: The structure and antioxidant properties. *Materials* **2021**, *14*, 1984. [CrossRef]
110. Sultana, N.; Reza, M.E.; Alam, M.N.; Siddiquee, M.N.A.; Islam, M.S.; Rahman, M.A.; Rahman, M.M. Evaluating the efficiency of supplementary feeding as a management strategy for enhancing honeybee (*Apis mellifera* L.) colony growth and productivity. *Front. Bee Sci.* **2024**, *2*, 1386799. [CrossRef]
111. Braglia, C.; Rudelli, C.; Tinti, A.; Porrini, C.; Biró, P.; Mutinelli, F.; Bortolotti, L. Unravelling pollen diet and microbiome influence on honey bee health. *Sci. Rep.* **2025**, *15*, 13474. [CrossRef]
112. Al-Kahtani, S.; Taha, E.K.A. Seasonal variations in nutritional composition of honeybee pollen loads. *J. Kans. Entomol. Soc.* **2021**, *93*, 105–112. [CrossRef]
113. Iorizzo, M.; Letizia, F.; Ganassi, S.; Testa, B.; Petrarca, S.; Albanese, G.; Di Criscio, D.; De Cristofaro, A. Recent Advances in the Biocontrol of Nosemosis in Honey Bees (*Apis mellifera* L.). *J. Fungi* **2022**, *8*, 424. [CrossRef]
114. Pascual, G.; Silva, D.; Vargas, M.; Aranda, M.; Cañumir, J.A.; López, M.D. Dietary Supplement of Grape Wastes Enhances Honeybee Immune System and Reduces Deformed Wing Virus (DWV) Load. *Antioxidants* **2023**, *12*, 54. [CrossRef]
115. Settele, J.; Bishop, J.; Potts, S.G. Climate change impacts on pollination. *Nat. Plants* **2016**, *2*, 16092. [CrossRef]
116. DeGrandi-Hoffman, G.; Corby-Harris, V.; Carroll, M.; Toth, A.L.; Gage, S.; Watkins de Jong, E.; Graham, H.; Chambers, M.; Meador, C.; Obernesser, B. The Importance of Time and Place: Nutrient Composition and Utilization of Seasonal Pollens by European Honey Bees (*Apis mellifera* L.). *Insects* **2021**, *12*, 235. [CrossRef]
117. Jones, L.; Brennan, G.L.; Lowe, A.; Creer, S.; Ford, C.R.; de Vere, N. Shifts in honeybee foraging reveal historical changes in floral resources. *Commun. Biol.* **2021**, *4*, 37. [CrossRef]
118. Chen, M.; Zhang, T.-L.; Hu, C.-G.; Zhang, J.-Z. The Role of Drought and Temperature Stress in the Regulation of Flowering Time in Annuals and Perennials. *Agronomy* **2023**, *13*, 3034. [CrossRef]

119. Walters, J.; Zavalnitskaya, J.; Isaacs, R.; Szendrei, Z. Heat of the moment: Extreme heat poses a risk to bee–plant interactions and crop yields. *Curr. Opin. Insect Sci.* **2022**, *52*, 100927. [CrossRef] [PubMed]
120. Borghi, M.; Perez de Souza, L.; Yoshida, T.; Fernie, A.R. Flowers and climate change: A metabolic perspective. *New Phytol.* **2019**, *224*, 1425–1441. [CrossRef] [PubMed]
121. Descamps, C.; Quinet, M.; Jacquemart, A.L. The effects of drought on plant–pollinator interactions: What to expect? *Environ. Exp. Bot.* **2021**, *182*, 104297. [CrossRef]
122. Rering, C.C.; Franco, J.G.; Yeater, K.M.; Mallinger, R.E. Drought stress alters floral volatiles and reduces floral rewards, pollinator activity, and seed set in a global plant. *Ecosphere* **2020**, *11*, e03254. [CrossRef]
123. Simanonok, M.P.; Otto, C.R.; Smart, M.D. Do the quality and quantity of honey bee-collected pollen vary across an agricultural land-use gradient? *Environ. Entomol.* **2020**, *49*, 189–196. [CrossRef]
124. Yadav, P.; Lata, M. Toxicological effect of environmental pollution on honeybees. *Ecol. Environ. Conserv.* **2023**, *29*, 583–588. [CrossRef]
125. Fatrcov-Ramkov, K.; Nkov, J. Bee Pollen Nutritional and Toxicological Aspects. *Arch. Ecotoxicol.* **2019**, *1*, 41–47. [CrossRef]
126. Nicewicz, Ł.; Bednarek, A.; Kafel, A.; Nakonieczny, M. Set of stress biomarkers as a practical tool in the assessment of multistress effect using honeybees from urban and rural areas as a model organism: A pilot study. *Environ. Sci. Pollut. Res.* **2020**, *28*, 9084–9096. [CrossRef]
127. Castelli, L.; Branchiccela, B.; Garrido, M.; Invernizzi, C.; Porrini, M.; Romero, H.; Antúnez, K. Impact of nutritional stress on honeybee gut microbiota, immunity, and Nosema ceranae infection. *Microb. Ecol.* **2020**, *80*, 908–919. [CrossRef] [PubMed]
128. Potts, S.G.; Biesmeijer, J.C.; Kremen, C.; Neumann, P.; Schweiger, O.; Kunin, W.E. Global pollinator declines: Trends, impacts and drivers. *Trends Ecol. Evol.* **2010**, *25*, 345–353. [CrossRef]
129. Vaudo, A.D.; Tooker, J.F.; Grozinger, C.M.; Patch, H.M. Bee nutrition and floral resource restoration. *Curr. Opin. Insect Sci.* **2015**, *10*, 133–141. [CrossRef] [PubMed]
130. Becher, M.A.; Grimm, V.; Thorbek, P.; Horn, J.; Kennedy, P.J.; Osborne, J.L. BEEHAVE: A systems model of honeybee colony dynamics and foraging to explore multifactorial causes of colony failure. *J. Appl. Ecol.* **2014**, *51*, 470–482. [CrossRef] [PubMed]
131. Tlak Gajger, I.; Ribaric, J.; Matak, M.; Svecnjak, L.; Kozaric, Z.; Nejedli, S.; Smodis Skerl, I.M. Zeolite clinoptilolite as a dietary supplement and remedy for honeybee (*Apis mellifera* L.) colonies. *Vet. Med.* **2015**, *60*, 696–705. [CrossRef]
132. Tlak Gajger, I.; Ribarić, J.; Smodiš Škerl, M.; Vlainić, J.; Sikirić, P. Stable gastric pentadecapeptide BPC 157 in honeybee (*Apis mellifera*) therapy, to control *Nosema ceranae* invasions in apiary conditions. *J. Vet. Pharmacol. Ther.* **2018**, *41*, 614–621. [CrossRef] [PubMed]
133. Tlak Gajger, I.; Vlainić, J.; Šoštarić, P.; Prešern, J.; Bubnič, J.; Smodiš Škerl, M.I. Effects on some therapeutical, biochemical, and immunological parameters of honey bee (*Apis mellifera*) exposed to probiotic treatments, in field and laboratory conditions. *Insects* **2020**, *11*, 638. [CrossRef]
134. Tlak Gajger, I.; Smodiš Škerl, M.I.; Šoštarić, P.; Šuran, J.; Sikirić, P.; Vlainić, J. Physiological and immunological status of adult honeybees (*Apis mellifera*) fed sugar syrup supplemented with pentadecapeptide BPC 157. *Biology* **2021**, *10*, 891. [CrossRef]
135. Traynor, K.S.; Mondet, F.; de Miranda, J.R.; Techer, M.; Kowallik, V.; Oddie, M.A.; McAfee, A. Varroa destructor: A complex parasite, crippling honey bees worldwide. *Trends Parasitol.* **2020**, *36*, 592–606. [CrossRef] [PubMed]
136. Alaux, C.; Allier, F.; Decourtye, A.; Odoux, J.F.; Tamic, T.; Chabirand, M.; Canard, B.; Trivellone, V.; Le Conte, Y. A 'Landscape physiology' approach for assessing bee health highlights the benefits of floral landscape enrichment and semi-natural habitats. *Sci. Rep.* **2017**, *7*, 40568. [CrossRef]
137. Martinello, M.; Mutinelli, F. Antioxidant Activity in Bee Products: A Review. *Antioxidants* **2021**, *10*, 71. [CrossRef]
138. Nascimento, A.M.C.B.; Luz, G.E. Bee pollen properties: Uses and potential pharmacological applications—A review. *J. Anal. Pharm. Res.* **2018**, *7*, 513–515. [CrossRef]
139. Martelli, F.; Zhongyuan, Z.; Wang, J.; Wong, C.O.; Karagas, N.E.; Roessner, U.; Belfield, E.J. Low doses of the neonicotinoid insecticide imidacloprid induce ROS triggering neurological and metabolic impairments in *Drosophila*. *Proc. Natl. Acad. Sci. USA* **2020**, *117*, 25840–25850. [CrossRef]
140. Powner, M.B.; Salt, T.E.; Hogg, C.; Jeffery, G. Improving mitochondrial function protects bumblebees from neonicotinoid pesticides. *PLoS ONE* **2016**, *11*, e0166531. [CrossRef]
141. Bava, R.; Castagna, F.; Ruga, S.; Caminiti, R.; Nucera, S.; Bulotta, R.M.; Naccari, C.; Britti, D.; Mollace, V.; Palma, E. Protective Role of Bergamot Polyphenolic Fraction (BPF) against Deltamethrin Toxicity in Honeybees (*Apis mellifera*). *Animals* **2023**, *13*, 3764. [CrossRef] [PubMed]
142. Fukuto, T.R. Mechanism of action of organophosphorus and carbamate insecticides. Environ. *Health Perspect.* **1990**, *87*, 245–254. [CrossRef]

143. Aroniadou-Anderjaska, V.; Figueiredo, T.H.; de Araujo Furtado, M.; Pidoplichko, V.I.; Braga, M.F.M. Mechanisms of Organophosphate Toxicity and the Role of Acetylcholinesterase Inhibition. *Toxics* **2023**, *11*, 866. [CrossRef]
144. Muñoz, J.P.; Soto-Jiménez, D.; Brito, A.; Quezada-Romegialli, C. Glyphosate-Based Herbicides and Their Potential Impact on the Microbiota of Social Bees. *Toxics* **2025**, *13*, 551. [CrossRef] [PubMed]
145. Pal, E.; Almasri, H.; Paris, L.; Diogon, M.; Pioz, M.; Cousin, M.; Sené, D.; Tchamitchian, S.; Tavares, D.A.; Delbac, F.; et al. Toxicity of the Pesticides Imidacloprid, Difenoconazole and Glyphosate Alone and in Binary and Ternary Mixtures to Winter Honey Bees: Effects on Survival and Antioxidative Defenses. *Toxics* **2022**, *10*, 104. [CrossRef]
146. Schmehl, D.R.; Teal, P.E.; Frazier, J.L.; Grozinger, C.M. Genomic analysis of the interaction between pesticide exposure and nutrition in honey bees (*Apis mellifera*). *J. Insect Physiol.* **2014**, *71*, 177–190. [CrossRef] [PubMed]
147. Chelucci, E.; Chiellini, C.; Cavallero, A.; Gabriele, M. Bio-functional activities of Tuscan bee pollen. *Antioxidants* **2023**, *12*, 115. [CrossRef]
148. Komosińska-Vassev, K.; Olczyk, P.; Kaźmierczak, J.; Mencner, Ł.; Olczyk, K. Bee pollen: Chemical composition and therapeutic application. *Evid. -Based Complement. Altern. Med.* **2015**, *2015*, 297425. [CrossRef]
149. dos Santos, L.N.C.; Malta, S.M.; Franco, R.; Silva, H.C.G.; Silva, M.H.; Rodrigues, T.S.; Silva, T.M. Antioxidant and anti-Alzheimer's potential of Tetragonisca angustula (Jataí) stingless bee pollen. *Sci. Rep.* **2024**, *14*, 308. [CrossRef]
150. Stebuliauskaitė, R.; Liaudanskas, M.; Žvikas, V.; Čeksterytė, V.; Sutkevičienė, N.; Sorkytė, Š. Changes in ascorbic acid, phenolic compound content, and antioxidant activity in vitro in bee pollen depending on storage conditions: Impact of drying and freezing. *Antioxidants* **2025**, *14*, 462. [CrossRef]
151. Rzepecka-Stojko, A.; Stojko, J.; Jasik, K.; Buszman, E. Anti-atherogenic activity of polyphenol-rich extract from bee pollen. *Nutrients* **2017**, *9*, 1369. [CrossRef]
152. Fleming, J.C.; Schmehl, D.R.; Ellis, J.D. Characterizing the impact of commercial pollen substitute diets on the level of *Nosema* spp. in honey bees (*Apis mellifera* L.). *PLoS ONE* **2015**, *10*, e0132014. [CrossRef] [PubMed]
153. Di Pasquale, G.; Alaux, C.; Le Conte, Y.; Odoux, J.F.; Pioz, M.; Vaissière, B.E.; Belzunces, L.P.; Decourtye, A. Variations in the availability of pollen resources affect honey bee health. *PLoS ONE* **2016**, *11*, e0162818. [CrossRef] [PubMed]
154. Jovanovic, N.M.; Glavinic, U.; Ristanic, M.; Vejnovic, B.; Ilic, T.; Stevanovic, J.; Stanimirovic, Z. Effects of Plant-Based Supplement on Oxidative Stress of Honey Bees (*Apis mellifera*) Infected with *Nosema ceranae*. *Animals* **2023**, *13*, 3543. [CrossRef]
155. Topal, E.; Mărgăoan, R.; Bay, V.; Takma, Ç.; Yücel, B.; Oskay, D.; Düz, G.; Acar, S.; Kösoğlu, M. The Effect of Supplementary Feeding with Different Pollens in Autumn on Colony Development under Natural Environment and In Vitro Lifespan of Honey Bees. *Insects* **2022**, *13*, 588. [CrossRef]
156. Smart, M.; Pettis, J.; Rice, N.; Browning, Z.; Spivak, M. Linking Measures of Colony and Individual Honey Bee Health to Survival among Apiaries Exposed to Varying Agricultural Land Use. *PLoS ONE* **2016**, *11*, e0152685. [CrossRef]
157. Morrison, B.; Newburn, L.R.; Fitch, G. Food as Medicine: A Review of Plant Secondary Metabolites from Pollen, Nectar, and Resin with Health Benefits for Bees. *Insects* **2025**, *16*, 414. [CrossRef]
158. Hoover, S.E.; Ovinge, L.P. Pollen collection, honey production, and pollination services: Managing honey bees in an agricultural setting. *J. Econ. Entomol.* **2018**, *111*, 1509–1516. [CrossRef]
159. Tlak Gajger, I.; Mañes, A.M.; Formato, G.; Mortarino, M.; Toporcak, J. Veterinarians and beekeeping: What roles, expectations and future perspectives?-A review paper. *Vet. Arh.* **2021**, *91*, 437–443. [CrossRef]
160. Bankova, V.; Popova, M.; Trusheva, B. The phytochemistry of the honeybee. *Phytochemistry* **2018**, *155*, 1–11. [CrossRef] [PubMed]
161. Kacemi, R.; Campos, M.G. Translational research on bee pollen as a source of nutrients: A scoping review from bench to real world. *Nutrients* **2023**, *15*, 2413. [CrossRef]
162. Aurori, C.M.; Buttstedt, A.; Dezmirean, D.S.; Mărghitaş, L.A.; Moritz, R.F.A.; Erler, S. What is the main driver of ageing in long-lived winter honeybees: Antioxidant enzymes, innate immunity, or vitellogenin? *J. Gerontol. A Biol. Sci. Med. Sci.* **2014**, *69*, 633–639. [CrossRef]
163. Berenbaum, M.R.; Johnson, R.M. Xenobiotic detoxification pathways in honey bees. *Curr. Opin. Insect Sci.* **2015**, *10*, 51–58. [CrossRef]
164. Engel, P.; Martinson, V.G.; Moran, N.A. Functional diversity within the simple gut microbiota of the honey bee. *Proc. Natl. Acad. Sci. USA* **2012**, *109*, 11002–11007. [CrossRef] [PubMed]
165. Korayem, A.M.; Khodairy, M.M.; Abdel-Aal, A.A.; El-Sonbaty, A.A. The protective strategy of antioxidant enzymes against hydrogen peroxide in honey bee, *Apis mellifera* during two different seasons. *J. Biol. Earth Sci.* **2012**, *2*, B93–B109.
166. Roth, A.; Vleurinck, C.; Netschitailo, O.; Bauer, V.; Otte, M.; Kaftanoglu, O.; Page, R.E.; Beye, M. A Genetic Switch for Worker Nutrition-Mediated Traits in Honeybees. *PLoS Biol.* **2019**, *17*, e3000171. [CrossRef]
167. Seehuus, S.C.; Norberg, K.; Gimsa, U.; Krekling, T.; Amdam, G.V. Reproductive Protein Protects Functionally Sterile Honey Bee Workers from Oxidative Stress. *Proc. Natl. Acad. Sci. USA* **2006**, *103*, 962–967. [CrossRef]

168. Tosi, S.; Nieh, J.C.; Sgolastra, F.; Cabbri, R.; Medrzycki, P.; Di Prisco, G.; Porrini, C.; Dainese, M.; Nanetti, A.; Bosca, F.; et al. Long-Term Field-Realistic Exposure to a Next-Generation Pesticide, Flupyradifurone, Impairs Honey Bee Behaviour and Survival. *Commun. Biol.* **2021**, *4*, 805. [CrossRef]
169. Skye, S.M.; Zhu, W.; Romano, K.A.; Guo, C.-J.; Wang, Z.; Jia, X.; Fu, X.; Hicks, K.A.; Wang, Z.; Hazen, S.L. Microbial Transplantation with Human Gut Commensals Containing CutC Is Sufficient to Transmit Enhanced Platelet Reactivity and Thrombosis. *Potential. Circ. Res.* **2018**, *123*, 1164–1176. [CrossRef] [PubMed]
170. Taouzinet, L.; Djaoudene, O.; Fatmi, S.; Bouiche, C.; Amrane-Abider, M.; Bougherra, H.; Rezgui, F.; Madani, K. Trends of Nanoencapsulation Strategy for Natural Compounds in the Food Industry. *Processes* **2023**, *11*, 1459. [CrossRef]
171. Pateiro, M.; Gómez, B.; Munekata, P.E.S.; Barba, F.J.; Putnik, P.; Kovačević, D.B.; Lorenzo, J.M. Nanoencapsulation of Promising Bioactive Compounds to Improve Their Absorption, Stability, Functionality and the Appearance of the Final Food Products. *Molecules* **2021**, *26*, 1547. [CrossRef] [PubMed]
172. Lammari, N.; Louaer, O.; Meniai, A.H.; Elaissari, A. Encapsulation of Essential Oils via Nanoprecipitation Process: Overview, Progress, Challenges and Prospects. *Pharmaceutics* **2020**, *12*, 431. [CrossRef] [PubMed]

antioxidants

MDPI

Review

Antioxidant Capacity and Therapeutic Applications of Honey: Health Benefits, Antimicrobial Activity and Food Processing Roles

Ivana Tlak Gajger [1,*], Showket Ahmad Dar [2], Mohamed Morsi M. Ahmed [3,4], Magda M. Aly [3] and Josipa Vlainić [5]

1 Department for Biology and Pathology of Fish and Bees, Faculty of Veterinary Medicine, University of Zagreb, 10000 Zagreb, Croatia
2 Division of Agricultural Entomology, Sher-e-Kashmir University of Agricultural Sciences and Technology of Kashmir, Srinagar 191111, India; drshowketshameem@gmail.com
3 Department of Biological Sciences, Faculty of Science, King Abdulaziz University, Jeddah 21589, Saudi Arabia; mmmahmed@kau.edu.sa or mmmahmed6@gmail.com (M.M.M.A.); mmmohammad@kau.edu.sa (M.M.A.)
4 Nucleic Acids Research Department, Genetic Engineering and Biotechnology Research Institute (GEBRI), City for Scientific Research and Technological Applications, Alexandria 21934, Egypt
5 Laboratory for Oxidative Stress, Institute Ruđer Bošković, Bijenička Cesta 54, 10000 Zagreb, Croatia; josipa.vlainic@irb.hr
* Correspondence: itlak@vef.unizg.hr; Tel.: +385-912390041

Abstract

Honey is a natural product of honeybees that has been consumed for centuries due to its nutritional value and potential health benefits. Recent scientific research has focused on its antioxidant capacity, which is linked to a variety of bioactive compounds such as phenolic acids, enzymes (e.g., glucose oxidase, catalase), flavonoids, ascorbic acid, carotenoids, amino acids, and proteins. Together, these components work synergistically to neutralize free radicals, regulate antioxidant enzyme activity, and reduce oxidative stress. This review decisively outlines the antioxidant effects of honey and presents compelling clinical and experimental evidence supporting its critical role in preventing diseases associated with oxidative stress. Honey stands out for its extensive health benefits, which include robust protection against cardiovascular issues, notable anticancer and anti-inflammatory effects, enhanced glycemic control in diabetes, immune modulation, neuroprotection, and effective wound healing. As a recognized functional food and dietary supplement, honey is essential for the prevention and adjunct treatment of chronic diseases. However, it faces challenges due to variations in composition linked to climatic conditions, geographical and floral sources, as well as hive management practices. The limited number of large-scale clinical trials further underscores the need for more research. Future studies must focus on elucidating honey's antioxidant mechanisms, standardizing its bioactive compounds, and examining its synergistic effects with other natural antioxidants to fully harness its potential.

Keywords: *Apis mellifera*; reactive oxygen species; honey; diseases; health disorders; antioxidants; phytochemicals

1. Introduction

Honey, a natural product processed by honeybees from the nectar or excretions of plant-sucking insects, has been a staple in human diets and traditional medicine for centuries. Beyond its culinary applications, honey is increasingly recognized for its therapeutic potential, particularly its antimicrobial, anti-inflammatory, and antioxidant properties [1]. In recent decades, scientific attention has turned toward the antioxidant capacity of honey, driven by the rising global occurrence of oxidative stress-related diseases such as cardiovascular disorders, diabetes, cancer, and neurodegenerative conditions [2]. Oxidative stress is an imbalance between reactive oxygen species (ROS) and the body's endogenous antioxidant defenses, leading to cellular damage, inflammation, and disease progression [3]. Antioxidants are crucial in neutralizing ROS and include enzymatic compounds such as glutathione, superoxide dismutase, mycothiol, and bacillithiol, which collectively protect cellular structures and functions from oxidative degradation [4,5]. At the same time, exogenous antioxidants from natural sources, including honey, play a significant complementary function in preventing oxidative damage and supporting overall health [6–8]. Honey contains over 200 bioactive compounds, including sugars, enzymes, amino acids, organic acids, vitamins, minerals, and a rich profile of antioxidants [9,10]. Among these, key antioxidant constituents include phenolic acids, flavonoids, ascorbic acid, catalase, glucose oxidase, and carotenoids, all of which act through various mechanisms such as hydrogen atom donation, radical scavenging, and modulation of redox-sensitive signaling pathways [11–14]. Notably, the antioxidant action of honey arises not only from individual components but also from their synergistic interactions, which enhance their overall bioactivity. These compounds may contribute to the regulation of vital physiological processes: thermogenesis, vasodilation, collagen synthesis, and immune modulation [13,15]. Honey's chemical composition and antioxidant potential are highly variable, influenced by its botanical origin, geographic region, seasonal factors, and beekeeping practices, including processing and temperature control [16]. This complexity poses challenges for standardization and clinical application, although it also provides a broad matrix of therapeutic opportunities (Figure 1). β-carotene and vitamins C and E, which are trace compounds present in honey, further contribute to its health benefits by reducing oxidative stress, improving cardiovascular function, preventing strokes, and exerting anti-carcinogenic effects through the inhibition of tumor necrosis factor [17,18].

Current research has clarified honey's ability to influence signaling pathways related to oxidative stress and inflammation, such as nuclear factor kappa B (NF-κB) and mitogen-activated protein kinases (MAPKs), offering mechanistic insight into its therapeutic actions [19]. Both laboratory-controlled and in vivo studies have supported honey's role in reducing oxidative biomarkers and lipid peroxidation, regulating blood glucose levels, and protecting against cellular apoptosis and DNA damage [20,21]. Furthermore, honey has demonstrated potential in enhancing immune responses by promoting the activity of immune cells and exerting antimicrobial effects [2]. Emerging research also highlights honey's neuroprotective capabilities, indicating its potential role in mitigating age-related cognitive decline and neurodegenerative diseases [22]. Despite these promising findings, gaps remain in fully understanding honey's antioxidant mechanisms and its clinical applicability. There is a critical need for standardized analytical protocols and large-scale, randomized clinical trials to validate honey's therapeutic potential and determine the optimal dosages and formulations [23]. While the historical and ethnopharmacological uses of honey are well-documented [24], its integration into modern healthcare as a scientifically supported antioxidant therapy remains underdeveloped.

In this review, we aim to present a comprehensive analysis of the antioxidant properties of honey, its chemical composition, mechanisms of action, and health benefits. We also discuss the challenges related to standardization and quality assessment, while identifying current research gaps and future directions. By synthesizing recent advances, this work seeks to deepen our understanding of honey as a potent natural antioxidant with significant implications for preventing and managing diseases in humans related to oxidative stress [25–27]. While writing, a structured literature search protocol was followed, which included defined search strategies, inclusion and exclusion criteria, and methods for data extraction to ensure the relevance and quality of the studies reviewed.

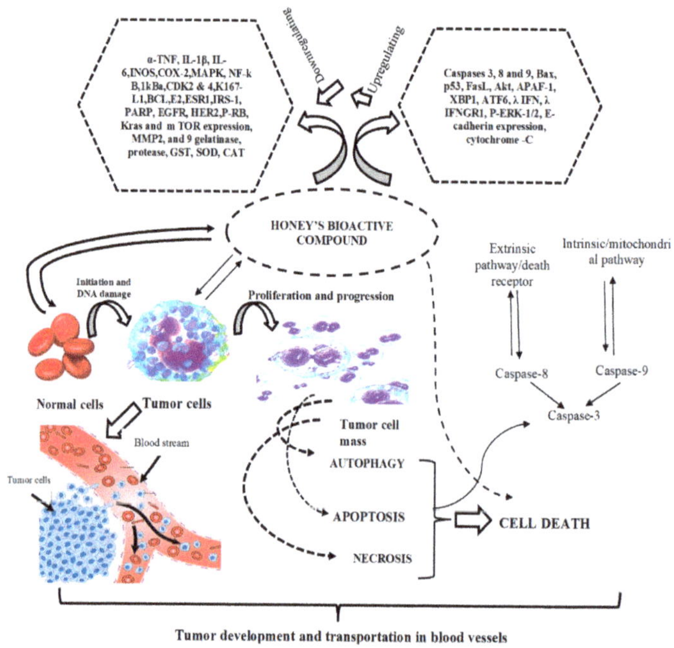

Figure 1. The role of bioactive compounds found in honey serves as a potential antitumor agent, inhibiting the transport of cancer cells and the phases of cell death (adapted from [28]).

2. Chemical Composition and Its Properties

Honey is a complex natural product made up of a variety of bioactive molecules, whose composition is largely influenced by floral origin, environmental conditions, and beekeepers' management practices and processing methods. The most important bioactive constituents are phytochemicals, natural compounds derived from plant sources, which have demonstrated significant health-promoting potential, including antioxidant, antimicrobial, and anti-inflammatory activities [29]. A major class of phytochemicals present in honey is polyphenols, which are well-recognized for their antioxidant capacity [30]. The polyphenolic profile of honey includes a range of flavonoids and phenolic acids such as 4-hydroxybenzoic acid, p-coumaric acid, vanillic acid, caffeic acid, syringic acid, gallic acid, quercetin, kaempferol, myricetin, pinobanksin, pinocembrin, and chrysin. Additional phenolic acids commonly found include hesperetin, ellagic acid, ferulic acid, and chlorogenic acid [31,32]. The concentrations of these compounds are highly variable and are strongly influenced by local forage conditions [33,34]. As such, the polyphenolic and phenolic acid content serve not only as a functional health marker but also as a chemotaxonomic index of botanical origin in monofloral honeys [33].

In addition to polyphenols, volatile organic compounds (VOCs) contribute to both the enchanting therapeutic and sensory qualities of honey. These include terpenes (e.g., linalool, linalool oxide isomers, lilac alcohol and aldehyde isomers), norisoprenoids (e.g., isophorone, vomifoliol), and aromatic alcohols such as benzyl alcohol and 2-phenylethanol. Other notable VOCs include benzaldehyde, furfural, acetic acid, cis-linalool oxide, and 3,5-dimethoxy-4-

hydroxybenzoic acid methyl ester (DMHBME), as well as veratric acid, syringic acid, eudesmic acid, and trimethoxybenzoic acid derivatives [35]. These substances enhance both the antimicrobial and antioxidant properties of honey, as well as its intricate aroma and flavor.

The color and mineral composition of honey are also important indicators of its quality and bioactivity. Darker honeys, such as those from buckwheat, chestnut, wildflower sources, or honeydew, tend to have higher concentrations of antioxidants and minerals. This correlates with a more intense flavor profile and enhanced health benefits [36–39]. Stingless bees, in particular, produce honey with distinct colorations depending on the nectar source, ranging from water white to dark amber hues [40]. From a macronutrient perspective, honey primarily consists of sugars (79.6–80%) and water (17.2–18%). The main sugars are fructose (38.19–40%) and glucose (31.28–35%), with smaller amounts of sucrose (1.3–5%) and maltose (7.3–10%) [41]. These sugars play a role not only in honey's sweetness and preservation but also in modulating its antioxidant potential via Maillard reaction products during processing. Honey's minor constituents, though present in small quantities, play a disproportionately large role in its therapeutic efficacy. These include organic acids (0.57–1.0%), proteins (0.266–0.50%), nitrogenous compounds (0.043%), and amino acids (0.05–0.1%) [36]. Trace elements and minerals (0.04–0.17%) are essential for enzymatic and redox functions [42,43]. In addition, honey contains a variety of micronutrients and bioactive compounds, including carotenoids, flavonoids, vitamins (C, B-complex, K, and E), volatile compounds (e.g., aldehydes, alcohols, esters, ketones, and phenols), sugar alcohols, and colloids, which collectively account for approximately 2.1% of its total biochemical profile [41]. These constituents are essential not only for antioxidant activity but also for wound healing, immunomodulation, and other biological effects. To summarize, honey's chemical composition is rich and diverse, encompassing a wide spectrum of nutritional, sensory, and bioactive compounds (Table 1). This chemical complexity underlies honey's broad pharmacological potential, making it a unique functional food and a promising candidate for further investigation into the prevention and therapy of antioxidant-related diseases.

Table 1. Summary of the main constituents of honey and their functions.

Components Group	Examples	Concentration Range	Function/Relevance
flavonoids	quercetin, kaempferol, myricetin, chrysin, pinobanksin, pinocembrin, hesperetin	trace–varies by floral source	antioxidant, anti-inflammatory, antimicrobial
phenolic acids	gallic acid, caffeic acid, p-coumaric acid, ferulic acid, vanillic acid, syringic acid, chlorogenic acid, ellagic acid, 4-hydroxybenzoic acid	trace–high variability	antioxidant, antimicrobial, and chemotaxonomic markers
volatile organic compounds	linalool, lilac aldehyde/alcohol isomers, benzaldehyde, 2-phenylethanol, acetic acid, furfural, isophorone, vomifoliol, dmhbme	trace (<0.1%)	aroma, flavor, antimicrobial, and antioxidant activity
organic acids	gluconic acid (dominant), acetic acid, formic acid, citric acid	0.57–1.0%	acidity, preservation, antimicrobial, taste
sugars	fructose (38–40%), glucose (31–35%), maltose (7–10%), sucrose (1–5%)	~79.6–80% total sugars	energy source, sweetness, osmotic preservation, Maillard reaction, antioxidants
proteins and enzymes	glucose oxidase, diastase, invertase, catalase, albumins	0.266–0.50%	antimicrobial (via hydrogen peroxide), wound healing, and digestion of sugars

Table 1. *Cont.*

Components Group	Examples	Concentration Range	Function/Relevance
amino acids	proline (dominant), glutamic acid, aspartic acid, phenylalanine	0.05–0.1%	nutritional value, Maillard reaction precursors
vitamins	vitamin C, B-complex (B1, B2, B3, B5, B6, B9), vitamin k, vitamin e	trace–0.5% total	antioxidant, metabolic cofactor, immune support
minerals and trace elements	potassium, calcium, iron, magnesium, zinc, manganese, selenium	0.04–0.17%	enzymatic cofactors, antioxidant defense (e.g., Se, Zn)
other phytochemicals	carotenoids, sugar alcohols, alkaloids, tannins	trace	antioxidant, immune modulation
colloids and nitrogenous compounds	colloidal proteins, peptides, nitrogenous bases	~0.043% nitrogen compounds	structural and biological activity enhancers

Note: Given concentrations are approximate and highly variable depending on botanical source, climatic conditions, and processing management.

3. Antioxidant Capacity of Honey and Its Health Benefits

Oxygen plays a critical role in cellular respiration and energy production; however, its metabolic byproducts, particularly ROS, contribute to oxidative stress, aging, and a wide range of degenerative diseases [44]. ROS and free radicals, as unstable molecules with unpaired electrons, are produced during normal cellular metabolism and exert damaging effects on lipids, proteins, and nucleic acids through complex initiation, propagation, and termination reactions [45,46]. The accumulation of oxidative damage over time accelerates aging and contributes to the "growth" of chronic conditions such as cancer, cardiovascular disorders, and neurodegenerative diseases. Antioxidants counteract oxidative stress by neutralizing free radicals and maintaining redox homeostasis. These compounds may be produced endogenously or obtained from external sources, primarily the diet [47–49]. Among natural dietary sources, honey has received considerable attention due to its rich content of bioactive compounds that exhibit potent antioxidant properties [50]. Honey contains a diverse array of antioxidants, including ascorbic acid, catalase, glucose oxidase, flavonoids, phenolic acids, carotenoids, organic acids, amino acids, proteins, and Maillard reaction products [51]. For example, vitamin C enhances cardiovascular and immune function, while phenolic compounds such as caffeic acid, quercetin, and chrysin exert antioxidant activity through mechanisms including electron donation, hydrogen atom transfer, metal ion chelation, and scavenging of ROS [52].

Polyphenols such as apigenin, kaempferol, pinocembrin, galangin, chrysin, quercetin, and acacetin are widely identified in honey and contribute significantly to its pharmacological effects (e.g., Figure 2), including antioxidant, anti-inflammatory, and chemopreventive actions [53]. The antioxidant properties of honey are also highly dependent on its post-harvest processing and storage conditions. Generally, darker honeys exhibit greater antioxidant activity. Buckwheat honey, for instance, possesses antioxidant potential comparable to 1 mM α-tocopherol in lipid peroxidation assays and exceeds the antioxidant content of various fruits and vegetables [54]. Specific honey varieties such as sage and buckwheat honey exhibit high antioxidant activity, with values reported at 21.3×10^{-5} and 4.32×10^{-3} equivalents, respectively [55].

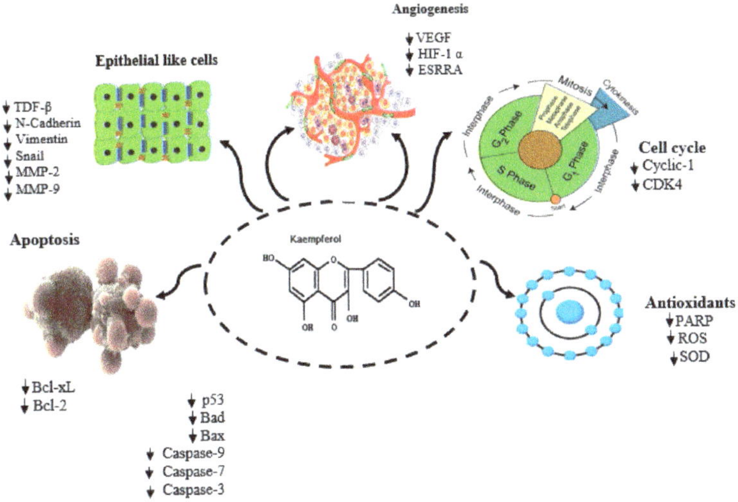

Figure 2. Illustration of the antioxidant mechanism of kaempferol and its role in cancer treatment, along with a schematic representation of the pathways and targets affected by kaempferol identified in vitro cancer experiments (adapted from [53]). Arrows show the sequential steps of the reaction/process.

The phytochemical composition of honey also includes bioactive phenolics like quercetin, kaempferol, pinobanksin, and caffeic acid, which not only enhance antioxidant activity but also serve as chemical indicators that reveal the botanical origin of honey [53]. Additionally, honey contains antimicrobial phytochemicals such as terpenes, benzaldehyde, 2-phenylethanol, and volatile aldehydes that synergize with antioxidant compounds to provide broad-spectrum biological effects [56,57]. Although mentioned minor components represent only about 2.1% of the total honey composition, they include bioactive molecules such as flavonoids, carotenoids, sugar alcohols, and aromatic compounds that are central to honey's antioxidant efficacy. Both in vitro and in vivo studies have substantiated the antioxidant effects of honey. In vitro assays have demonstrated that honey-derived polyphenols possess high oxygen radical absorbance capacity (ORAC) and effectively inhibit lipid peroxidation and oxidative damage to cellular macromolecules [53]. Buckwheat honey has been shown to protect endothelial cell lines (EA.hy926) from oxidative stress induced by cumene hydroperoxide (CuOOH), hydrogen peroxide (H_2O_2), and peroxyl radicals (AAPH), primarily through glutathione (GSH) restoration, inhibition of ROS generation, and prevention of lipid peroxidation [58,59]. These protective effects can be translated into physiological benefits in humans. For instance, individuals consuming 1.5 g/kg of body weight of buckwheat honey exhibited significantly elevated serum antioxidant levels compared to those ingesting corn syrup [60]. Another clinical study reported increases in vitamin C, β-carotene, uric acid, and glutathione reductase by 47%, 3.0%, 12%, and 7%, respectively, following daily honey consumption [25]. The bioavailability of phenolic compounds in honey was confirmed by HPLC-MS analysis [61]. The antioxidant effects of honey are not only the result of individual components but also of synergistic interactions among polyphenols and other bioactive molecules, which enhance their stability, bioefficacy, and reactivity [59]. Even processed honey has demonstrated the capacity to elevate plasma antioxidant levels, suggesting therapeutic relevance for chronic disease prevention and health maintenance [62]. Considering that the average person consumes a few kg of sweeteners annually, replacing refined sugars with honey offers a natural and functional approach to increasing antioxidant intake and reducing oxidative stress. The evidence supports the view that honey, as a complex natural honeybee product, has significant potential as a dietary antioxidant with protective effects against ROS-mediated cellular damage and disease progression.

4. Role of Antioxidants in Preventing Cancer

The antioxidant constituents of honey protect against oxidative damage and contribute significantly to the prevention and potential treatment of cancer (Figure 3). Honey-derived polyphenols have been shown to interfere with key steps in tumorigenesis, including oxidative stress modulation, suppression of angiogenesis, and induction of programmed cell death in malignant cells. Among the most studied bioactive compounds present in honey are flavonoids and phenolic acids such as quercetin, galangin, chrysin, kaempferol, apigenin, acacetin, pinocembrin, artepillin C, and caffeic acid phenethyl ester (CAPE). These molecules display strong radical-scavenging activity and have demonstrated substantial anticancer potential in both in vitro and in vivo models [63,64]. Through standard antioxidant assays including DPPH and FRAP, the capacity of these compounds to counteract oxidative stress has been well characterized, suggesting a strong link between their antioxidant properties and their cytotoxic effects on cancer cells.

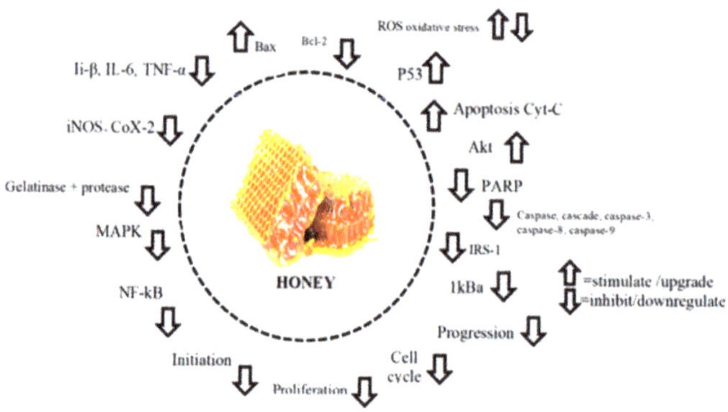

Figure 3. Molecular target modulation as the anticancer effects of honey (adopted from [63]). Legend: Bcl-2 = B cell lymphoma 2; Bcl-xL = B cell lymphoma extra-large; Cyt. C = cytochrome C; MAPK = mitogen-activated protein kinase; NF-κB = nuclear factor kappa B; Akt = altered PI3 kinase; IRS-1 = insulin receptor substrate; IL = interleukin; COX-2 = cyclooxygenase 2; TNF-α = tumor necrosis factor alpha; iNOS = inducible nitric oxide synthase; IκBα = inhibitor of kappa B; PARP = poly ADP-ribose polymerase.

In addition to redox modulation, the antiangiogenic properties of honey polyphenols have gained increasing attention. Using endothelial cell models, Wahid et al. (2024) demonstrated that compounds such as quercetin, caffeine, and phenethyl ester significantly impaired angiogenic activity by reducing endothelial proliferation and tube formation [65]. Artepillin C, galangin, and kaempferol were also found to inhibit neovascularization, albeit with a somewhat reduced potency [66]. Interestingly, polyphenols with only moderate antioxidant scores, such as apigenin and acacetin, were still capable of disrupting endothelial morphogenesis, suggesting that antiangiogenic activity may occur through mechanisms distinct from direct antioxidant action. These findings have prompted proposals to explore honey polyphenols as potential leads for the development of antiangiogenic drugs. Alarjani et al. [67] and Mahani et al. [68] emphasized the therapeutic relevance of these natural compounds in oncology. At the same time, Zulkifli et al. further highlighted their pharmaceutical applicability, particularly in the context of multi-targeted chemoprevention strategies and sleep [69]. A synergistic role of honey-derived compounds with other bioactive substances has also been suggested. For instance, Qu et al. demonstrated that co-administration of caffeine with honey antioxidants significantly suppressed colon tumorigenesis in rats, despite caffeine's previously reported oncogenic potential when isolated [70]. This finding underscores the importance of matrix effects and compound interactions within the honey phytochemical profile. Several honey constituents act directly on tumor cells by activating intrinsic apoptotic pathways (Figure 4). CAPE, for example, has been shown to exert antimitogenic, cytostatic, and immunomodulatory effects, further supporting its utility in anticancer applications [71].

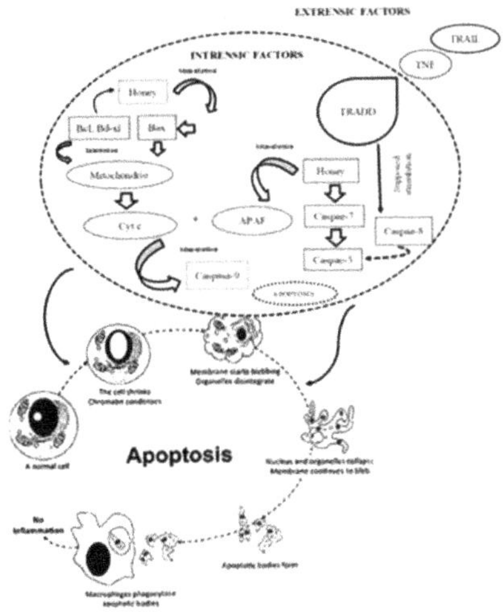

Figure 4. Effect of honey on apoptotic pathways (adopted from [6,63]). Legend: Bcl-2 = B cell lymphoma 2; Bcl-xL = B cell lymphoma extra-large; Cyt. C = cytochrome C; APAF-1 = apoptotic protease-activating factor 1; TNF = tumor necrosis factor; TRAIL = TNF-related apoptosis-inducing ligand; TRADD = TNFR-associated death domain protein.

Chrysin induces apoptosis through proteasome inhibition and activation of the p38-MAPK signaling cascade, leading to upregulation of p21^Waf1/Cip1 in glioma cells [72,73], while also modulating the Akt pathway and enhancing caspase-3 activation in leukemia models [74]. Galangin has demonstrated selective cytotoxicity in HL-60 leukemia cells, where it induces DNA fragmentation without disrupting membrane integrity, indicative of programmed cell death [64]. Quercetin, one of the most extensively studied honey polyphenols, has shown the ability to inhibit various kinases involved in cancer cell signaling, including membrane tyrosine kinases and protein kinase C. At lower concentrations, quercetin can paradoxically promote proliferation in A549 lung carcinoma cells, while higher doses induce apoptosis and cell cycle arrest [75]. Furthermore, Carrillo-Martinez et al. (2024) noted that quercetin and structurally related flavonoids enhance Fas-ligand and p53 expression, which are crucial regulators of apoptosis in cancer cells such as A549, HepG2, and H460 [76]. Kaempferol has been reported to trigger mitochondrial apoptosis by elevating ATP production and enhancing the expression of apoptosis-inducing factor (AIF) and caspase-3 in H460 cells [77], and by inducing cell cycle arrest in leukemia cell lines [78]. Pinocembrin disrupts mitochondrial membrane potential in HCT-116 colon cancer cells, promoting cytochrome c release and downstream caspase activation [79]. Similarly, apigenin has demonstrated cytotoxic effects in a range of tumor types, including hepatic, breast, cervical, and neuroblastoma cells, through mechanisms involving DNA damage, mitochondrial dysfunction, and caspase-mediated apoptosis. Taken together, these findings support the concept that honey contains a complex matrix of polyphenolic compounds capable of exerting multifaceted anticancer effects (e.g., Figure 5). By acting at the intersection of oxidative stress modulation, angiogenesis inhibition, and apoptotic signaling, these compounds provide a compelling rationale for further exploration in both preventive and therapeutic oncology settings.

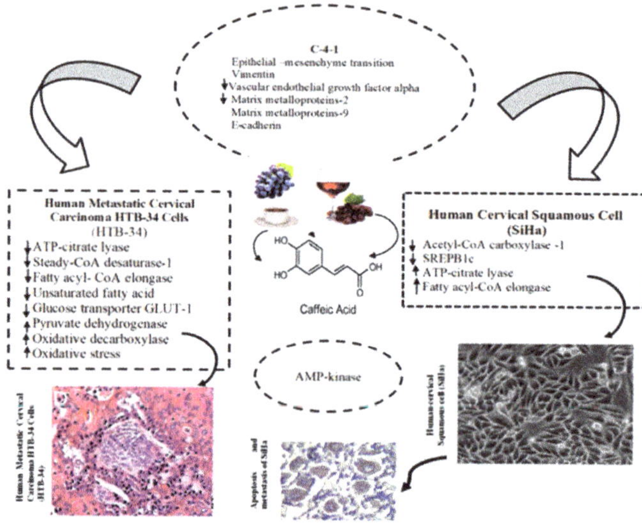

Figure 5. Illustration of the effects of caffeic acids upregulating the expression of enzymes involved in the production of fatty acids, followed by AMPK expression in cervical cancer cells (adapted from [80]). Arrows show the sequential steps of the reaction/process.

5. Antioxidants and Cardiovascular Diseases

The role of antioxidants in the prevention and management of cardiovascular diseases (CVDs) has been extensively documented, particularly regarding natural sources such as honey. Honey possesses a rich array of enzymatic and non-enzymatic antioxidants, including polyphenols, flavonoids, ascorbic acid (vitamin C), and reduced glutathione, which collectively contribute to its significant cardioprotective properties [81]. These antioxidants play a crucial role in neutralizing ROS, which are implicated in the pathogenesis of various cardiovascular conditions like myocardial infarction, ischemia-reperfusion injury, atherosclerosis, and hypertension [82]. Experimental studies using animal models have provided strong evidence that honey consumption offers cardiovascular benefits. In rats anesthetized with urethane, pre-treatment with honey (5 g/kg) one hour before administration of epinephrine (100 μg/kg) significantly mitigated hyperadrenergic-induced cardiovascular dysfunction. Improvements were observed in electrocardiographic parameters, vasomotor function, and inotropic and chronotropic responses, particularly under conditions simulating anaphylaxis, cardiac arrest, asthma, and hemorrhage [83,84]. These effects are attributed to the high total antioxidant capacity (TAC) of honey, particularly due to its elevated levels of SOD and reduced glutathione, despite lower catalase activity [85].

The cardioprotective potential of honey extends beyond its radical-scavenging activity. Polyphenolic compounds in honey, including quercetin, caffeic acid phenethyl ester (CAPE), kaempferol, galangin, chrysin, and pinocembrin, have been shown to modulate multiple redox-sensitive signaling pathways. These include enhancing nitric oxide (NO) bioavailability, suppressing superoxide generation, and preserving endothelial integrity [86,87]. For instance, kaempferol and galangin have stabilized mitochondrial function, reduced cardiomyocyte apoptosis, and maintained cardiac tissue structure following ischemic insults [88]. Additionally, clinical studies indicate that moderate antioxidant supplementation significantly enhances cardiovascular health by reducing oxidative stress biomarkers such as malondialdehyde (MDA) and conjugated dienes (CD), and increasing overall plasma antioxidant capacity [89]. In one controlled study involving myocardial infarction survivors and healthy middle-aged men, daily supplementation with 100 mg of vitamin E, 100 mg of vitamin C, 6 mg of β-carotene, and 50 μg of selenium led to measurable improvements in antioxidant status and lipid peroxidation markers [90,91]. These outcomes support the inclusion of antioxidant-rich foods such as honey as adjunctive therapy in cardiovascular care, particularly in populations at elevated risk. Furthermore, honey can influence lipid metabolism. For example, studies in animal models of hypertension and hyperlipidemia showed that honey administration reduced total cholesterol and LDL levels while increasing HDL concentrations. These effects are mediated by upregulation of endogenous antioxidant enzymes, including SOD, CAT, and GPx, and by reducing lipid peroxidation [92–94].

From a vascular health perspective, honey enhances endothelial function through NO-mediated vasodilation. This is facilitated by the presence of minerals such as magnesium, sodium, and chlorine, and is potentiated by the antioxidant activity of vitamin C [95]. Flavonoids, such as apigenin and pinocembrin, further contribute to this effect by inhibiting platelet aggregation through the modulation of thromboxane A2 and the enhancement of prostacyclin synthesis, thereby reducing the risk for thrombosis [96,97]. In a prospective cohort study involving 34,492 postmenopausal women, higher flavonoid intake, including from honey, was associated with a significantly reduced risk of coronary artery disease (CHD) [98]. Adjusted risk assessments indicated a downward trend in CHD-related deaths across increasing quintiles of flavonoid consumption, with broccoli and other flavonoid-rich foods contributing substantially to this effect. The benefits of honey are also evident in surgical and pediatric cardiovascular contexts. Elevated levels of endothelin-1, a potent vasoconstrictor linked to hypertension and heart failure, are attenuated by honey polyphenols such as quercetin and kaempferol. This has particular relevance in postoperative care following cardiopulmonary bypass, where antioxidant therapies, such as Salvia miltiorrhiza injections rich in phenolic compounds, have demonstrated effectiveness in reducing endothelin-1 levels [99]. In summary, honey's diverse antioxidant components confer multifaceted cardiovascular benefits. These include reducing oxidative stress, enhancing endothelial and mitochondrial function, improving lipid profiles, regulating inflammatory responses, and modulating key signaling pathways involved in vascular homeostasis and cardiac protection. Given its natural origin and broad therapeutic range, honey emerges as a promising adjunct in both the prevention and management of cardiovascular diseases.

5.1. *Quercetin*

Quercetin, a widely studied flavonoid found in apples, onions, green tea, and honey, exerts cardiovascular protective effects by inhibiting vascular smooth muscle cell (VSMC) hypertrophy and reducing oxidative stress and inflammation. Angiotensin II (Ang-II)-induced VSMC hypertrophy involves activation of MAPK pathways such as JNK, ERK1/2, and p38. Also, quercetin selectively inhibits Ang-II-induced JNK activation, attenuating hypertrophic signaling and apoptosis in VSMCs without affecting ERK1/2 and p38 [100–102]. Also, quercetin inhibits tyrosine phosphorylation of adaptor proteins such as Shc and blocks PI3-K/Akt pathway activation, which are key drivers of VSMC hypertrophy [103]. The quercetin analog LY294002 also inhibits Akt phosphorylation, corroborating this pathway's significance [104]. In vivo studies have demonstrated that quercetin lowers blood pressure and improves endothelial function in spontaneously hypertensive rats (SHR) by activating eNOS and suppressing NADPH oxidase [105]. Clinical trials reveal quercetin's antihypertensive effects in stage-1 hypertension, though not in prehypertension. The mode of quercetin administration critically affects its bioavailability and efficacy, with oral gavage providing superior cardiovascular benefits compared to dietary supplementation [106,107]. Quercetin's pleiotropic effects, including antioxidant, anti-inflammatory, and anti-proliferative actions, underpin its potential as a cardioprotective agent [108].

5.2. Acacetin

Acacetin, a natural flavone, acts as a selective inhibitor of atrial-specific potassium currents, particularly K+ current (IKur), making it a promising agent for atrial fibrillation treatments [109]. Unlike many antiarrhythmics that prolong the QT interval and risk ventricular arrhythmias, acacetin prolongs atrial action potential duration and refractory period without significant ventricular effects [110,111]. In anesthetized dogs, intraduodenal acacetin effectively suppressed atrial fibrillation without QT prolongation or adverse ventricular effects, unlike non-selective agents such as sotalol [112]. This atrial selectivity stems from acacetin's targeted blockade of atrial ion channels and calcium-activated SK channels [109,113].

5.3. Caffeic Acid Phenethyl Ester (CAPE)

Caffeic acid phenethyl ester (CAPE), a bioactive component of honeybee propolis, exerts potent antioxidant and anti-inflammatory effects relevant to cardiovascular health. One of CAPE's primary molecular actions is the inhibition of NF-κB signaling, a central pathway in oxidative stress and vascular inflammation [114]. Additionally, CAPE suppresses both NF-κB and JNK signaling cascades, which have been shown to improve insulin sensitivity and reduce inflammatory responses in diabetic animal models [115]. These findings highlight CAPE's therapeutic potential in modulating intracellular inflammatory pathways associated with cardiometabolic disorders.

5.4. Kaempferol

Kaempferol, a flavonol abundant in various plants and honey, exhibits antioxidant, anti-inflammatory, and anti-platelet activities, contributing to its cardiovascular protective effects [108]. Early studies demonstrated kaempferol-induced vasorelaxation in isolated arteries, mediated via both endothelium-dependent nitric oxide and endothelium-independent mechanisms. More recent investigations reveal that kaempferol protects cardiac myocytes against ischemia-reperfusion (I/R) injury by modulating endoplasmic reticulum (ER) stress responses and apoptosis. It decreases the expression of ER stress markers such as GRP78 and CHOP and regulates the Bcl-2/Bax ratio, reducing cardiomyocyte apoptosis after I/R insult. This modulation of ER stress and apoptotic pathways represents a novel mechanism for kaempferol's cardio protection role [116].

5.5. Galangin

Galangin (3,5,7-trihydroxyflavone), a naturally occurring flavonol found in honey and the rhizome of *Alpinia officinarum*, exhibits potent antioxidant and anti-inflammatory properties relevant to cardiovascular health. Recent studies demonstrate that galangin attenuates endothelial dysfunction by activating the heme oxygenase-1 (HO-1) signaling pathway, reducing oxidative stress, and suppressing proinflammatory mediators such as VCAM-1 and TNF-α in vascular endothelial cells [117]. In animal models of metabolic syndrome, galangin improves aortic function by restoring endothelial nitric oxide synthase (eNOS) activity, enhancing nitric oxide (NO) availability, and inhibiting the angiotensin II/AT1R/TGF-β signaling axis [118]. These findings highlight galangin's therapeutic potential in preventing vascular damage associated with hypertension and cardiometabolic disorders.

6. Antioxidants and Neural Diseases

Flavonoids, a diverse class of polyphenolic compounds found in honey, various fruits, vegetables, and plant-based products, have been widely investigated for their pharmacological properties, including anti-inflammatory, anticancer, and antioxidant activities. However, their mechanisms in neuroprotection, particularly in the context of neurodegenerative disorders, are still under active investigation [119]. Neurodegenerative diseases such as Alzheimer's and Parkinson's are now increasingly understood to involve chronic neuroinflammation, largely mediated by the activation of microglia, the resident immune cells of the central nervous system (CNS). Microglial activation leads to the release of pro-inflammatory cytokines, NO, and ROS, contributing to neuronal injury and disease progression [120]. Flavonoids, including apigenin, have been shown to modulate this inflammatory microenvironment. In murine BV-2 microglia cells, apigenin suppresses LPS-induced activation by inhibiting nuclear factor kappa B (NF-κB) signaling and promoting nuclear translocation of Nrf2, a master regulator of antioxidant responses [121]. This dual action results in decreased expression of inducible nitric oxide synthase (iNOS) and cyclooxygenase-2 (COX-2), ultimately reducing NO and prostaglandin E2 (PGE2) levels. Furthermore, apigenin exerts neuroprotective effects in models of cerebral ischemia-reperfusion (I/R) injury, such as middle cerebral artery occlusion (MCAO). In both in vitro and in vivo models, it mitigates oxidative stress, reduces infarct volume, and limits neuronal apoptosis. This protection is mediated via attenuation of endoplasmic reticulum (ER) stress, modulation of Bcl-2 family proteins, and suppression of JNK and p38 MAPK phosphorylation, while sparing extracellular signal-regulated kinase (ERK) pathways [122]. Beyond its anti-inflammatory and antioxidant effects, apigenin also modulates central neurotransmission, contributing to its sedative, antidepressant, and anxiolytic properties. Behavioral studies have shown that apigenin reduces anxiety-like behavior and decreases locomotor activity in rodents, effects likely mediated by its interaction with GABA_A receptors [123]. Patch-clamp experiments in HEK293 cells expressing α1β2γ2 GABA_A receptor subunits have revealed that apigenin inhibits GABA-evoked chloride currents, albeit with less potency compared to classical benzodiazepines. This supports a modulatory, rather than agonistic, role at these receptor sites. Apigenin also affects excitatory neurotransmission. In cultured cortical neurons, it inhibits NMDA receptor-mediated currents while sparing AMPA receptor activity, suggesting a selective antagonism of the NMDA receptor complex. This is particularly relevant since excessive glutamate signaling and NMDA receptor activation contribute to excitotoxicity, a common mechanism in stroke and neurodegeneration. Apigenin's inhibition of NR2 subunits of NMDA receptors offers a novel target for its neuroprotective profile. In animal models of aging and neuroinflammation, chronic apigenin administration reduces hippocampal microglial activation and improves synaptic integrity without causing overt sedation or cognitive impairment, making it a promising candidate for neurodegenerative disease prevention and adjunct therapy [123].

Pinocembrin

Pinocembrin, a flavonoid primarily found in honey and propolis, is recognized for its broad pharmacological properties, including antioxidant, anti-inflammatory, and neuroprotective activities. Among the polyphenolic compounds in honeybee products, pinocembrin is one of the most abundant and biologically active, contributing significantly to the radical-scavenging capacity of honey and its neuroprotective properties [124]. Recent studies have demonstrated that pinocembrin crosses the blood–brain barrier and offers neuroprotection in models of cerebral ischemia/reperfusion (I/R) injury. In a rat middle cerebral artery occlusion (MCAO) model, administration of pinocembrin at the onset of reperfusion significantly reduced infarct volume, improved neurological deficit scores, and increased survival rates [125,126]. These effects are associated with its ability to reduce ROS, suppress neuronal (nNOS) and inducible nitric oxide synthase (iNOS) expression, and increase intracellular glutathione levels, thereby enhancing redox homeostasis [126].

Pinocembrin also modulates apoptotic pathways by downregulating caspase-3 activity, limiting PARP cleavage, and reducing neurite retraction and lactate dehydrogenase (LDH) release, ultimately supporting neuronal survival [126]. Additionally, in global cerebral ischemia and chronic cerebral hypoperfusion models, pinocembrin has been shown to preserve blood–brain barrier (BBB) integrity, reduce cerebral edema, and restore cerebral blood flow [127]. Mitochondrial protection is a key mechanism underlying these effects. Pinocembrin enhances mitochondrial respiratory chain complex I and III activities, stabilizes the mitochondrial membrane potential, reduces mitochondrial swelling, and limits mitochondrial ROS production [128]. Restoration of cytochrome oxidase expression further contributes to mitochondrial integrity and neuronal energy homeostasis. These molecular and functional effects translate into improved cognitive outcomes, as evidenced by enhanced performance in behavioral assays such as the Morris water maze [128]. Collectively, these findings underscore the potential of pinocembrin as a promising therapeutic candidate for neurovascular protection, particularly in the context of stroke and vascular cognitive impairment.

7. Antimicrobial Activities of Honey

Phenolic compounds extracted from both dark and light honey types possess notable antioxidant and antimicrobial activities. Their antioxidant capacity is typically evaluated by assays like DPPH free radical scavenging and ferric reducing antioxidant power (FRAP), reflecting their ability to donate electrons and neutralize reactive species [129]. Bioactive phenolics also exhibit robust in vitro antibacterial effects. Studies on honeys, including *Apis laboriosa* subspecies, have demonstrated significant inhibition of Gram-negative (*Escherichia coli*, *Salmonella enterica* serovar *Typhimurium*) and Gram-positive (*Staphylococcus aureus*, *Bacillus subtilis*) bacteria. The potency correlates strongly with total phenolic and flavonoid content [130]. The antimicrobial action of honey is multifactorial, arising from both peroxide and non-peroxide mechanisms. Hydrogen peroxide, generated enzymatically by glucose oxidase, is a primary antibacterial agent. However, catalase-insensitive phenolic constituents, such as caffeic and ferulic acids, also play a crucial role, particularly in dark honeys [1,131]. Further research confirms that isolated phenolic acids and flavonoids, major contributors found in honey, inhibit bacterial growth through membrane disruption, enzyme inactivation, and synergistic antioxidant activity [132]. These findings support the view that phenolic-rich honeys deliver enhanced antimicrobial efficacy, highlighting their therapeutic and food safety potential.

8. Wound Healing

Honey has increasingly attracted attention in clinical and regenerative medicine due to its broad-spectrum antimicrobial activity and wound healing potential [133,134]. Traditionally, honey has been used to treat various wounds, including burns, ulcers, diabetic wounds, and surgical injuries, due to its ability to inhibit microbial growth and facilitate tissue repair [133,135]. Bioactive compounds in honey, such as flavonoids, phenolic acids, organic acids, enzymes, and vitamins, work synergistically to enhance antioxidant defenses, reduce inflammation, and promote tissue regeneration [136,137]. Electron paramagnetic resonance

(EPR) spectroscopy studies have elucidated the generation and scavenging of free radicals in different honey types for wound healing. For instance, commercial heat-processed honey typically lacks antibacterial activity, whereas pasture honeys exhibit antibacterial effects despite minimal peroxide production. Manuka honey is well-known for its strong antibacterial properties independent of peroxide generation [138,139]. The capacity of honey to neutralize ROS correlates with its antioxidant potential, with manuka honey rapidly scavenging free radicals within minutes [139,140]. Peroxide-producing honeys generate free radicals upon dilution, contributing to antimicrobial activity, while some honey types also help resolve chronic inflammation characteristics of non-healing wounds by modulating ROS production and quenching, thereby reducing oxidative stress and supporting tissue recovery [141,142]. Figure 6 presents summarized mechanisms of antimicrobial and wound healing effects of honey.

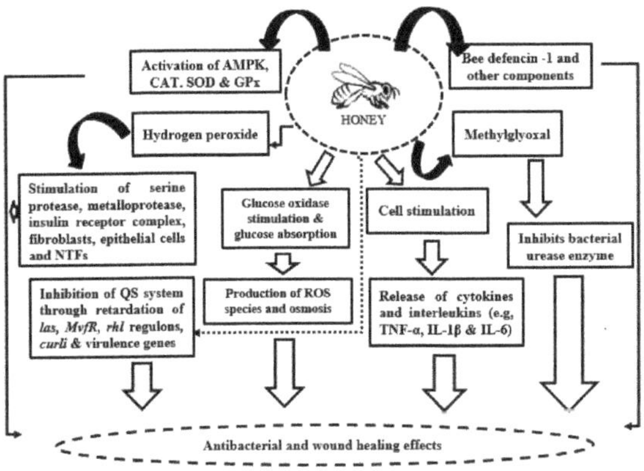

Figure 6. Mechanisms of antimicrobial and wound healing effects of honey (modified from [143]). Legend: AMPK = 5′adenosine monophosphate-activated protein kinase; QS = quorum sensing; SOD = superoxide dismutase; GPx = glutathione peroxidase; NTFs = nuclear transcription factors; TNF-α = tumor necrosis factor alpha; IL = interleukin.

9. Food Processing

Honey from various floral sources has gained considerable attention as a functional ingredient in the food industry due to its natural antioxidant properties, sweetening capacity, and potential to replace synthetic additives. Recent studies have evaluated its application in salad dressings, meat preservation, and food formulations, where oxidative stability and sensory quality are critical [144,145]. Rasmussen et al. [144] conducted a comprehensive evaluation using clover and blueberry honeys, selected based on their high oxygen radical absorbance capacity (ORAC) values and phenolic compound profiles identified through high-performance liquid chromatography (HPLC). These honeys were incorporated into a French salad dressing system and subjected to storage under accelerated oxidative conditions at 37 °C for six weeks, and at ambient (23 °C) and refrigerated (4 °C) conditions for one year. The study found that salad dressings containing honey exhibited oxidation resistance comparable to that provided by EDTA, a synthetic chelator widely used in commercial formulations to inhibit lipid oxidation. Further supporting the antioxidative role of honey, McKibben and Engeseth [145] evaluated its effectiveness in mitigating lipid peroxidation in ground turkey. Using thiobarbituric acid reactive substances (TBARS) as a marker of oxidation, they showed that acacia honey at 5% (*w/w*) reduced TBARS levels by approximately 34%, while buckwheat honey at the same concentration reduced it by nearly 70% after three days of storage at 4 °C. These reductions were significantly greater than those observed with tocopherol and butylated hydroxytoluene (BHT), both commonly used antioxidants in meat products. Such findings underline honey's dual role as a natural preservative and healthier sweetener alternative, supporting its potential to replace high fructose corn syrup (HFCS) and synthetic antioxidants in food processing. Beyond meat and dressings, the antioxidant potential of different honeys has been quantified using assays such as the antioxidant capacity of water-soluble (ACW) and lipid-soluble (ACL) substances, showing wide applicability in lipid-rich food and functional formulations [146,147]. In a broader context, replacing synthetic additives with honey aligns with the consumer demand for "clean label" products and supports food industry efforts to reduce exposure to chemical preservatives while maintaining oxidative stability and sensory quality. Moreover, since honey contains a complex mixture of flavonoids, phenolic acids, enzymes, and organic acids, its antioxidant function may also confer additional health benefits beyond shelf-life extension, such as protection against oxidized lipids and related inflammatory conditions [148].

A summary of the wide range of therapeutic properties of honey is presented in Table 2.

Table 2. Summary of therapeutic properties of honey.

Type of Study	Honey Type/Compound	Therapeutic Area	Key Findings/Effects	Mechanism/Activity	References
clinical trial	multifloral honey	wound healing	enhanced healing of diabetic ulcers	antimicrobial, anti-inflammatory	[25]
clinical and in vitro	buckwheat honey	antioxidant	increased serum antioxidant capacity; protected endothelial cells	ROS inhibition, GSH restoration, lipid peroxidation prevention	[58–60]
in vitro	quercetin, caffeine, phenethyl ester	anticancer (antiangiogenic)	inhibited tube formation, endothelial proliferation	suppression of angiogenesis	[65]
in vitro	quercetin, flavonoids	anticancer	enhanced p53 and fas-ligand expression in tumor cells	induced apoptosis	[76]
animal model	honey + caffeine	anticancer	reduced colon tumorigenesis	synergistic inhibition of tumor growth	[70]
in vitro	chrysin	anticancer (glioma, leukemia)	triggered apoptosis, p21 activation, caspase-3 activation	p38-MAPK, proteasome inhibition	[72–74]
	galangin	anticancer	induced selective DNA fragmentation in HL-60 cells	apoptotic signaling	[64]
review	honey polyphenols	cancer, sleep regulation	emphasized the multifunctionality of polyphenols	multi-targeted pathways	[69]
	various honeys	antimicrobial	effective against bacteria and fungi	enzymatic activity, H_2O_2, low pH, phenolics	[53]
clinical study	raw honey	gastrointestinal health	relieved gastritis symptoms	antibacterial, mucosal protection	[56,57]
animal study	Iranian honey	anti-inflammatory/neuroprotective	reduced neuroinflammation markers	cytokine suppression, antioxidant enzymes	[53]

Table 2. *Cont.*

Type of Study	Honey Type/Compound	Therapeutic Area	Key Findings/Effects	Mechanism/Activity	References
animal model	tualang honey	antidiabetic cardiovascular	lowered blood glucose, improved lipid profile	antioxidant, insulin sensitivity	[47–49]
in vitro	gelam honey	antioxidant	strong free radical scavenging	high flavonoid content	[53]
	manuka honey	antibacterial	inhibited *S. aureus*, *E. coli* growth	MGO content, H_2O_2 release	[51,52]
	various honey types	antioxidant/food processing	High ORAC values; effective in lipid stabilization	polyphenol content	[53,54]
clinical study	honey vs. corn syrup	cardiovascular antioxidant support	improved levels of vitamin C, β-carotene, uric acid, GSH	serum antioxidant response	[25,60]
biochemical	button sage and buckwheat	food/antioxidant quality	high antioxidant equivalents (21.3×10^{-5}; 4.32×10^{-3})	phenolic marker analysis	[54,55]
analytical	various honeys	bioavailability	phenolics detectable in plasma after consumption	confirmed absorption	[61]
clinical	processed honeys	food/health maintenance	still elevated plasma antioxidant levels	synergistic polyphenol interaction	[62]
reviews	honey phytochemicals	oncology	emphasized chemopreventive relevance	multi-pathway targeting	[67,68]
in vitro/in vivo	CAPE, chrysin, pinocembrin	anti-inflammatory immunomodulatory	reduced pro-inflammatory cytokines, enhanced immune response	NF-κB inhibition, MAPK modulation, caspase activation	[71–74]

10. Anti-Inflammatory and Immunological Properties of Honey

Honey has demonstrated significant anti-inflammatory and immunomodulatory effects, primarily attributed to its rich content of flavonoids, phenolic acids, and other bioactive compounds. These molecules modulate immune responses through a variety of mechanisms, including the downregulation of pro-inflammatory cytokines, suppression of oxidative stress pathways, and modulation of intracellular signaling cascades such as NF-κB and MAPKs. In vitro and in vivo studies have consistently shown that honey inhibits the production of key pro-inflammatory mediators such as tumor necrosis factor-alpha (TNF-α), interleukin-1 beta (IL-1β), and interleukin-6 (IL-6), which play central roles in the initiation and propagation of inflammation. For instance, it is demonstrated that manuka honey significantly reduced levels of inflammatory cytokines and oxidative stress markers in a rat model of inflammation, suggesting its potential as a natural anti-inflammatory agent [149]. Caffeic acid phenethyl ester (CAPE), a bioactive component found in some honey types, has been particularly well studied for its immunomodulatory effects. CAPE suppresses the nuclear translocation of NF-κB, thereby downregulating inflammatory gene expression and inhibiting the release of cytokines such as IL-2 and IFN-γ from T-cells [150,151]. Similarly, chrysin, another honey-derived flavonoid, was found to inhibit lipopolysaccharide (LPS)-induced production of nitric oxide (NO) and cytokines in macrophages by interfering with NF-κB and Akt signaling pathways [152]. Moreover, honey has shown promise in enhancing immune responses during infection and tissue damage. In animal models, honey treatment resulted in increased proliferation of lymphocytes and enhanced phagocytic activity of immune cells [153]. The immunostimulatory effects may also be partially attributed to trace amounts of honeybee-derived peptides and enzymes such as defensin-1 and glucose oxidase, which contribute to antimicrobial defense while modulating host immunity [132]. The ability of honey to balance immune function, suppressing excessive inflammation while supporting innate immune defense, makes it particularly attractive in the context of chronic inflammatory diseases and wound healing. Table 3 summarizes the antioxidant activities of different honey types.

Table 3. Comparative antioxidant activity of different types of honey and methods used for evaluation.

Honey Type	Country/Floral Source	Antioxidant Method(s)	Total Antioxidant Activity/Results	References
manuka honey	New Zealand/*Leptospermum scoparium*	DPPH, FRAP, ORAC	high antioxidant activity; strong radical scavenging in DPPH assay	[149]
tualang honey	Malaysia/wild rainforest	DPPH, FRAP	comparable to vitamin C in scavenging activity; high phenolic content	[150]
buckwheat honey	USA, Europe/*Fagopyrum* spp.	DPPH, ORAC	one of the highest antioxidant capacities; similar to 1 mM α-tocopherol	[151,152]
gelam honey	Malaysia/*Melaleuca cajuputi*	FRAP, ABTS	high reducing power and flavonoid concentration	[153]
heather honey	Europe/*Calluna vulgaris*	DPPH, FRAP	high total phenolic content; strong metal chelating activity	[154]
acacia honey	Europe/*Robinia pseudoacacia*	DPPH, FRAP	low to moderate antioxidant activity due to lower polyphenol levels	[155]
sidr honey	Yemen/*Ziziphus spina-christi*	DPPH, FRAP	moderate activity; antioxidant potential varies by region	[156]
eucalyptus honey	Mediterranean/*Eucalyptus* spp.	DPPH, FRAP, TAC	rich in phenolic acids; moderate to high activity	[157]
multifloral honey	various regions	DPPH, FRAP	highly variable depending on floral origin and processing	[158]

11. Conclusions

Honey's antioxidant capacity arises from its rich and diverse bioactive compounds. Key antioxidant compounds not only exhibit strong free radical scavenging abilities but also modulate various cellular and molecular pathways relevant to health and disease. The neuroprotective properties of flavonoids like apigenin and pinocembrin underscore honey's potential in mitigating neurodegenerative disorders by attenuating oxidative stress, neuroinflammation, and apoptosis. Moreover, honey's antimicrobial activities are broad-spectrum, and these properties render honey an effective natural agent against pathogens involved in wound infections and food spoilage. Advanced applications, including tissue-engineered scaffolds incorporating honey, are expanding their clinical utility in regenerative medicine. While the therapeutic potential of honey is well-supported by numerous studies, several important limitations must be considered when interpreting these findings. A major challenge lies in the compositional variability of honey, which depends heavily on botanical origin, geographical region, and pre-/post-harvest processing practices. This variability affects the concentration and profile of bioactive compounds, leading to significant inter-study heterogeneity and limiting reproducibility across both experimental and clinical studies. Moreover, batch-to-batch consistency is seldom reported or controlled for in research, further complicating efforts to compare findings or formulate standardized therapeutic products. Although honey has a lower glycemic index than refined sugars, it raises concerns about its glycemic load, particularly in diabetic or insulin-resistant consumers. Additionally, dosing inconsistencies are apparent across the literature, with studies employing a wide range of dosages, from small dietary quantities to high pharmacological concentrations, without standardized guidelines or reference ranges. In summary, the multifaceted antioxidant, neuroprotective, antimicrobial, wound healing, immunological, anti-inflammatory, and food preservative properties of honey establish it as a valuable natural product with significant potential in both health-related and industrial applications. Mentioned limitations underscore the need for continued research into its bioactive compounds and mechanisms of action, which will further elucidate honey's role in disease prevention and therapy, as well as in sustainable food preservation strategies.

Author Contributions: Conceptualization, S.A.D. and I.T.G.; project administration, I.T.G.; writing—original draft, S.A.D. and I.T.G.; writing—review and editing, I.T.G., J.V., M.M.M.A. and M.M.A. All authors have read and agreed to the published version of the manuscript.

Funding: This research received no external funding.

Institutional Review Board Statement: Not applicable.

Conflicts of Interest: The authors declare no conflicts of interest.

References

1. Alvarez-Suarez, J.M.; Tulipani, S.; Diaz, D.; Estevez, Y.; Romandini, S.; Giampieri, F.; Damiani, E.; Astolfi, P.; Bompadre, S.; Battino, M. Antioxidant and antimicrobial capacity of several monofloral Cuban honeys and their correlation with color, polyphenol content and other chemical compounds. *Food Chem. Toxicol.* **2010**, *48*, 2490–2499. [CrossRef]
2. Alvarez-Suarez, J.M.; Giampieri, F.; Battino, M. Honey as a Source of Dietary Antioxidants: Structures, Bioavailability and Evidence of Protective Effects Against Human Chronic Diseases. *Curr. Med. Chem.* **2013**, *20*, 621–638. [CrossRef] [PubMed]
3. Pisoschi, A.M.; Pop, A. The Role of Antioxidants in the Chemistry of Oxidative Stress: A Review. *Eur. J. Med. Chem.* **2015**, *97*, 55–74. [CrossRef] [PubMed]
4. Halliwell, B. Understanding mechanisms of antioxidant action in health and disease. *Nat. Rev. Mol. Cell Biol.* **2024**, *25*, 13–33. [CrossRef] [PubMed]
5. Kumar, D.; Hazra, K.; Prasad, P.V.V.; Bulleddu, R. Honey: An important nutrient and adjuvant for maintenance of health and management of diseases. *J. Ethn. Food* **2024**, *11*, 19. [CrossRef]
6. Al-Madhagi, H.; Masoud, A. Limitations and Challenges of Antioxidant Therapy. *Phytother. Res.* **2024**, *38*, 5549–5566. [CrossRef]
7. Tumilaar, S.G.; Hardianto, A.; Dohi, H.; Kurnia, D. A Comprehensive Review of Free Radicals, Oxidative Stress, and Antioxidants: Overview, Clinical Applications, Global Perspectives, Future Directions, and Mechanisms of Antioxidant Activity of Flavonoid Compounds. *J. Chem.* **2024**, *2024*, 1–21. [CrossRef]
8. Koo, T.H.; Zakaria, A.D.; Mustafa, M.Z. Honey clinical applications in complementary medicine: A critical review. *J. Pharm. Pharmacogn. Res.* **2024**, *12*, 1040–1055. [CrossRef]
9. Tafere, D.A. Chemical composition and uses of honey: A Review. *J. Food Sci. Nutr. Res.* **2021**, *4*, 194–201. [CrossRef]
10. Dar, S.A.; Farook, U.B.; Rasool, K.; Ahad, S. Chapter one—Honey: Classification, composition, safety, quality issues and health benefits. In *Advanced Techniques of Honey Analysis—Characterization, Authentication, and Adulteration*, 1st ed.; Nayik, G.A., Uddin, J., Nanda, V., Eds.; Academic Press: Cambridge, MA, USA, 2024; pp. 1–37.
11. Tlak Gajger, I.; Vlainić, J. Antioxidant Activity of Honey Bee Products. *Antioxidants* **2025**, *14*, 64. [CrossRef]
12. Yangoua, H.; Dibacto, R.E.K.; Tchuente, B.R.T.; Nyobe, E.C.; Nguedjo, M.W.; Dimitri, T.K.A.; Kamini, M.F.G. Physicochemical properties and antioxidant potential of honey from Cameroon agroecological zones. *Heliyon* **2024**, *10*, e40232. [CrossRef] [PubMed]
13. Mahesar, S.A.; Sidhu, A.R.; Naz, S.; Kandhro, A.A. Phenolic and Mineral Contents in Honey and Their Associated Health Benefits. In *Honey in Food Science and Physiology*, 1st ed.; Kumar, R., Hajam, Y.A., Bala Dhull, S., Giri, A., Eds.; Springer: Singapore, 2024; pp. 155–179.

14. Osés, S.M.; Rodríguez, C.; Valencia, O.; Fernández-Muiño, M.A.; Sancho, M.T. Relationships among Hydrogen Peroxide Concentration, Catalase, Glucose Oxidase, and Antimicrobial Activities of Honeys. *Foods* **2024**, *13*, 1344. [CrossRef] [PubMed]
15. Quinn, A.; El Chazli, Y.; Escrig, S.; Daraspe, J.; Neuschwander, N.; McNally, A.; Genoud, C.; Meibom, A.; Engel, P. Host-derived organic acids enable gut colonization of the honey bee symbiont *Snodgrassella alvi*. *Nat. Microbiol.* **2024**, *9*, 477–489. [CrossRef] [PubMed]
16. Zammit Young, G.-W.; Blundell, R. A Review on the Phytochemical Composition and Health Applications of Honey. *Heliyon* **2023**, *9*, e12507. [CrossRef]
17. Shakoori, Z.; Salaseh, E.; Mehrabian, A.R.; Tehrani, D.M.; Dardashti, N.F.; Salmanpour, F. The amount of antioxidants in honey has a strong relationship with the plants selected by honey bees. *Sci. Rep.* **2024**, *14*, 351. [CrossRef] [PubMed]
18. Jiang, Y.; Ye, J.; Hu, Y.; Zhang, J.; Li, W.; Zhou, X.; Yu, M.; Yu, Y.; Yang, J.; Yang, W.; et al. Extraction and Synthesis of Typical Carotenoids: Lycopene, β-Carotene, and Astaxanthin. *Molecules* **2024**, *25*, 4549. [CrossRef] [PubMed]
19. Ahmed, S.; Sulaiman, S.A.; Baig, A.A.; Ibrahim, M.; Liaqat, S.; Fatima, S.; Jabeen, S.; Shamim, N.; Othman, N.H. Honey as a Potential Natural Antioxidant Medicine: An Insight into Its Molecular Mechanisms of Action. *Oxid. Med. Cell. Longev.* **2018**, *2018*, 8367846. [CrossRef]
20. Yunus, M. Antioxidant Properties of Honey: Mechanisms and Clinical Applications. *Free Radic. Antioxid.* **2024**, *13*, 50–53. [CrossRef]
21. Saad, B. Immunomodulatory and Anti-Inflammatory Properties of Honey and Bee Products. *Immuno* **2025**, *5*, 19. [CrossRef]
22. Fadzil, M.A.M.; Mustar, S.; Rashed, A.A. The Potential Use of Honey as a Neuroprotective Agent for the Management of Neurodegenerative Diseases. *Nutrients* **2023**, *15*, 1558. [CrossRef]
23. Martinello, M.; Mutinelli, F. Antioxidant Activity in Bee Products: A Review. *Antioxidants* **2021**, *10*, 71. [CrossRef] [PubMed]
24. Kumar, R.; Kumar, S.; Kanwar, S.S. Honey: Introduction, History, Composition, and Its Uses. In *Biomedical Perspectives of Herbal Honey*, 1st ed.; Springer: Singapore, 2024; pp. 1–17.
25. Balkrishna, A.; Saini, A.; Yadav, P.; Yadav, P.; Kumar, B.; Arya, V. Cardioprotective, Anti Atherogenic and Anti-Cancerous Potential of Honey. In *Honey in Food Science and Physiology*, 1st ed.; Kumar, R., Hajam, Y.A., Bala Dhull, S., Giri, A., Eds.; Springer: Singapore, 2024; pp. 255–265.
26. Liu, B.; Nguyen, P.L.; Li, H.Y.X.; Wang, H.; Nguyen, T.G.B.; Sahoo, P.K.; Sur, M.; Reddy, J.; Sillman, S.; Kachman, S.D.; et al. Honey vesicle-like nanoparticles protect aged liver from non-alcoholic steatohepatitis. *Acta Pharm. Sin. B* **2024**, *4*, 3661–3679. [CrossRef] [PubMed]
27. Bonsignore, G.; Martinotti, S.; Ranzato, E. Honey Bioactive Molecules: There Is a World Beyond the Sugars. *BioTech* **2024**, *13*, 47. [CrossRef] [PubMed]
28. Spoială, A.; Ilie, C.-I.; Ficai, D.; Ficai, A.; Andronescu, E. Synergic Effect of Honey with Other Natural Agents in Developing Efficient Wound Dressings. *Antioxidants* **2023**, *12*, 34. [CrossRef] [PubMed]
29. Safitri, E.; Purnobasuki, H.; Al-Anshori, A.A.; Lestari, T.D.; Utama, S.; Yimer, N. Honey-derived phytochemicals: Implications for stem cell activation and health benefits. *J. Funct. Foods* **2024**, *114*, 106058. [CrossRef]
30. Ndungu, N.N.; Kegode, T.M.; Kurgat, J.K.; Baleba, S.B.S.; Cheseto, X.; Turner, S.; Taboue, G.C.T.; Kasina, J.M.; Subramanian, S.; Nganso, B.T. Bio-functional properties and phytochemical composition of selected *Apis mellifera* honey from Africa. *Heliyon* **2024**, *10*, e30839. [CrossRef]
31. Wang, S.; Qiu, Y.; Zhu, F. An updated review of functional ingredients of Manuka honey and their value-added innovations. *Food Chem.* **2024**, *440*, 138060. [CrossRef]
32. Kumar, S.; Verma, M.; Hajam, Y.A.; Kumar, R. Honey infused with herbs: A boon to cure pathological diseases. *Heliyon* **2024**, *10*, e23302. [CrossRef]
33. Ntakoulas, D.D.; Tsagkaris, A.S.; Raptis, S.; Pasias, I.N.; Raptopoulou, K.G.; Kharoshka, A.; Schulzova, V.; Proestos, C. Study of authenticity, quality characteristics and bioactivity in honey samples from different botanical origins and countries. *J. Food Comp. Anal.* **2024**, *136*, 106716. [CrossRef]
34. Ameer, K.; Murtaza, N.A.; Jiang, G.; Zhao, C.C.; Siddique, F.; Kausar, T.; Mueen-ud-Din, G.; Mahmood, S. Chapter six-Fluorescence and ultraviolet–visible spectroscopy in the honey analysis. In *Advanced Techniques of Honey Analysis-Characterization, Authentication, and Adulteration*, 1st ed.; Nayik, G.A., Uddin, J., Nanda, V., Eds.; Academic Press: Cambridge, MA, USA, 2024; pp. 153–191.
35. Nyarko, K.; Mensah, S.; Greenlief, C.M. Examining the Use of Polyphenols and Sugars for Authenticating Honey on the U.S. Market: A Comprehensive Review. *Molecules* **2024**, *29*, 4940. [CrossRef]
36. Dżugan, M.; Tomczyk, M.; Sowa, P.; Grabek-Lejko, D. Antioxidant activity as biomarker of honey variety. *Molecules* **2018**, *23*, 2069. [CrossRef]
37. Filipe, M.S.; Kowalczyk, T.; Kukula-Koch, W.; Wieczfinska, J.; Bangay, G.; Diaz-Lanza, A.M.; Cardoso, R.V.C.; Mandim, F.; Falcão, S.I.; Vilas-Boas, M.; et al. Evaluating the quality, physicochemical properties, and biological activities of Centauri® honey from Turkey. *Food Biosci.* **2024**, *62*, 105028. [CrossRef]

38. Halagarda, M.; Groth, S.; Popek, S.; Rohn, S.; Pedan, V. Antioxidant activity and phenolic profile of selected organic and conventional honeys from Poland. *Antioxidants* **2020**, *9*, 44. [CrossRef] [PubMed]
39. Al-Farsi, M.; Al-Amri, A.; Al-Hadhrami, A.; Al-Belushi, S. Color, flavonoids, phenolics and antioxidants of Omani honey. *Heliyon* **2018**, *4*, e00874. [CrossRef] [PubMed]
40. Rao, P.V.; Krishnan, K.T.; Salleh, N.; Gan, S.H. Biological and therapeutic effects of honey produced by honey bees and stingless bees: A comparative review. *Rev. Bras. Farmacogn.* **2016**, *26*, 657–664. [CrossRef]
41. Mračević, S.Đ.; Krstić, M.; Lolić, A.; Ražić, S. Comparative study of the chemical composition and biological potential of honey from different regions of Serbia. *Microchem. J.* **2020**, *152*, 104420. [CrossRef]
42. Alves, A.; Ramos, A.; Gonçalves, M.M.; Bernardo, M.; Mendes, B. Antioxidant activity, quality parameters and mineral content of Portuguese monofloral honeys. *J. Food Comp. Anal.* **2013**, *30*, 130–138. [CrossRef]
43. Tlak Gajger, I.; Pavliček, D.; Oreščanin, V.; Varenina, I.; Sedak, M.; Bilandžić, N. Mineral Concentrations in Different Types of Honey Originating from Three Regions of Continental Croatia. *Foods* **2024**, *13*, 2754. [CrossRef]
44. Wang, X.; Liu, H.; Xue, Y.; Cui, L.; Chen, L.; Ho, K.; Huang, Y. Formation of environmentally persistent free radicals and their risks for human health: A review. *Environ. Chem. Lett.* **2024**, *22*, 1327–1343. [CrossRef]
45. Hassan, H.A.; Ahmed, H.S.; Hassan, D.F. Free radicals and oxidative stress: Mechanisms and therapeutic targets. *Hum. Antibodies* **2024**, *32*, 151–167. [CrossRef]
46. Yang, H.; Leng, J.; Liu, N.; Huang, L. Free radicals and antioxidants in diseases associated with immune dysfunction, inflammatory process, and aberrant metabolism. *Front. Endocrinol.* **2024**, *15*, 1363854. [CrossRef]
47. Bose, D.; Famurewa, A.C.; Akash, A.; Othman, E.M. The Therapeutic Mechanisms of Honey in Mitigating Toxicity from Anticancer Chemotherapy Toxicity: A Review. *J. Xenobiot.* **2024**, *14*, 1109–1129. [CrossRef] [PubMed]
48. Tan, B.L.; Konsue, N.; Bennett, L.L.; El-Kenawy, A.E.-M. Preventive potential of antioxidants in age-related diseases. *Front. Pharmacol.* **2024**, *15*, 1515004. [CrossRef] [PubMed]
49. Kardas, M.; Staśkiewicz-Bartecka, W.; Sołtys, K.; Dul, L.; Sapała, A.M.; Kiciak, A.; Bielaszka, A.; Kardas, J. The quality of selected raw and pasteurized honeys based on their sensory profiles and consumer preferences. *Front. Nutr.* **2024**, *16*, 1330307. [CrossRef] [PubMed]
50. Dossena, S.; Marino, A. Oxidative Stress and Antioxidants in Aging. *Antioxidants* **2024**, *13*, 1288. [CrossRef] [PubMed]
51. Yan, S.; Zhang, M.; Yuan, Y.; Mu, G.; Xu, H.; Zhao, T.; Wang, Y.; Xue, X. Chaste honey in long term-storage: Occurrence and accumulation of Maillard reaction products, and safety assessment. *Food Chem.* **2023**, *30*, 136457. [CrossRef]
52. Ahmad, A.; Younas, K. Microbial Production of Apigenin and Chrysin. In *Microbial Production of Food Bioactive Compounds*; Jafari, S.M., Harzevili, F.D., Eds.; Springer: Berlin, Germany, 2024; pp. 1–29.
53. Ayoub, W.S.R.; Zahoor, I.; Dar, A.H.; Farooq, S.; Mir, T.A.; Ganaie, T.A.; Srivastava, S.; Pandey, V.K.; Altaf, A. Exploiting the polyphenolic potential of honey in the prevention of chronic diseases. *Food Chem. Advan.* **2023**, *3*, 100373. [CrossRef]
54. Vani, S.N.; Soni, D. Nutritional Profile and Health Benefits of Buckwheat: A Miracle Pseudo Cereal. CARAS. 4, Centre for Advanced Research in Agricultural Sciences. *Res. J. Agricul. Sci.* **2024**, *15*, 1204–1209.
55. Sawarkar, A.B. Color intensity and antioxidant properties of honey produced from North Maharashtra, India. *J. Food Sci. Technol.* **2024**, *1*, 1–9. [CrossRef]
56. Mahmoud, M.A.A.; Kılıç-Büyükkurt, O.; Aboul Fotouh, M.A.; Selli, S. Aroma active compounds of honey: Analysis with GC-MS, GC-O, and molecular sensory techniques. *J. Food Comp. Anal.* **2024**, *134*, 106545. [CrossRef]
57. Hajj, E.R.; Estephan, N. Advances in infrared spectroscopy and chemometrics for honey analysis: A comprehensive review. *Cri. Rev. Food Sci. Nutr.* **2024**, *12*, 1–14. [CrossRef]
58. Zhou, Z.Y.; Wu, L.; Liu, Y.F. IRE1α: From the function to the potential therapeutic target in atherosclerosis. *Mol. Cell Biochem.* **2024**, *479*, 1079–1092. [CrossRef] [PubMed]
59. Zhou, S.; Cai, H.; He, X.; Tang, Z.; Lu, S. Enzyme-mimetic antioxidant nanomaterials for ROS scavenging: Design, classification, and biological applications. *Coord. Chem. Rev.* **2024**, *500*, 215536. [CrossRef]
60. Bani, C.; Peñas, E.; Baron, G.; Martínez-Villaluenga, C.; Mercogliano, F.; Aldini, G.; Piazza, S.; Lorenzo, C.D.; Restani, P. Characterization of the phenolic profile and in vitro antioxidant potential of different varieties of common buckwheat (*Fagopyrum esculentum* Moench) and Tartary buckwheat (*Fagopyrum tataricum* (L.) Gaertn.). *LWT* **2025**, *215*, 117261. [CrossRef]
61. Alcoléa, M.; Santana Junior, M.B.; Oliveira, K.A.M.; Tussolini, L.; Leite, M.A.G.; Honorio-França, A.C.; França, E.L., Pertuzatti, P.B. Bioactive compounds of honey from different regions of Brazil: The effect of simulated gastrointestinal digestion on antioxidant and antimicrobial properties. *Food Funct.* **2024**, *15*, 1310–1322. [CrossRef]
62. Bok, V.V.; Šola, I.; Rusak, G.; Budisavljević, A.; Nguyen, R.; Ludwig-Müller, J.; Maleš, Ž. Phenolic content and antioxidant activity of Croatian and German honey. *Acta Pharm.* **2024**, *74*, 673–692. [CrossRef] [PubMed]

63. Ahmed, S.; Othman, N.H. Honey as a Potential Natural Anticancer Agent: A Review of Its Mechanisms. *Evid.-Based Complement. Altern. Med.* **2013**, *2013*, 829070. [CrossRef]
64. Martinotti, S.; Bonsignore, G.; Ranzato, E. Understanding the Anticancer Properties of Honey. *Int. J. Mol. Sci.* **2024**, *25*, 11724. [CrossRef]
65. Wahid, M.; Nazeer, M.; Qadir, A.; Azmi, M.B. Investigating the Protein-Based Therapeutic Relationship between Honey Protein SHP-60 and Bevacizumab on Angiogenesis: Exploring the Synergistic Effect through In Vitro and In Silico Analysis. *ACS Omega* **2024**, *9*, 17143–17153. [CrossRef]
66. Garcıa, L.M.; Caicedo-Garzon, V.; Riveros, A.J. Oral administration of phytochemicals protects honey bees against cognitive and motor impairments induced by the insecticide fipronil. *PLoS ONE* **2024**, *19*, e0300899. [CrossRef]
67. Alarjani, W.M.A.; Mohammed, M.E.A. Antioxidant activities of Saudi honey samples related to their content of short peptides. *Sci. Rep.* **2024**, *14*, 24318. [CrossRef] [PubMed]
68. Mahani, P.R.F.; Ghibran, H.M.; Herlina, A.F.; Nurhasanah, S.; Nurjanah, N.; Elfirta, R.R.; Pribadi, A.; Rahadian Amalia, R.L.; Samudra, I.M. A report on the physicochemical and antioxidant properties of three Indonesian forest honeys produced by *Apis dorsata*. *Food Chem.* **2025**, *25*, 102156. [CrossRef]
69. Zulkifli, M.F.; Masimen, M.A.A.; Ridzuan, P.M.; Wan Ismail, W.I. Exploring honey's potential as a functional food for natural sleep aid. *Food Funct.* **2024**, *15*, 9678–9689. [CrossRef] [PubMed]
70. Qu, Y.; Cheng, Y.; Chen, F. The relationship between caffeine consumption and colon cancer prevalence in a nationally representative population. *Front. Nutr.* **2024**, *11*, 1375252. [CrossRef] [PubMed]
71. Balhara, N.; Arup Giri, R.; Naryal, A.; Kumar, R. Golden Nectar: Unveiling Honey's Healing Touch as Nature's Immunity Booster. In *Honey in Food Science and Physiology*; Kumar, R., Hajam, Y.A., Bala Dhull, S., Giri, A., Eds.; Springer: Singapore, 2024; pp. 89–119.
72. Cutty, S.; Hughes, F.; Ortega-Prieto, P.; Desai, S.; Thomas, P.; Fets, L.; Secrier, M.; Barr, A. Pro-survival roles for p21(Cip1/Waf1) in non-small cell lung cancer. *Br. J. Cancer* **2024**, *132*, 421–437. [CrossRef]
73. Elmitwalli, O.; Darwish, R.; Al-Jabery, L.; Algahiny, A.; Roy, S.; Butler, A.E.; Hasan, A.S. The Emerging Role of p21 in Diabetes and Related Metabolic Disorders. *Int. J. Mol. Sci.* **2024**, *25*, 13209. [CrossRef]
74. Hassan, D.; Menges, C.W.; Testa, J.R.; Bellacosa, A. AKT kinases as therapeutic targets. *J. Exp. Clin. Cancer Res.* **2024**, *43*, 313. [CrossRef] [PubMed]
75. Nguyen, T.T.T.; Tran, E.; Nguyen, T.H.; Do, P.T.; Huynh, T.H.; Huynh, H. Role of Activated MEK–ERK Pathway in Quercetin-Induced Growth Inhibition and Apoptosis in A549 Lung Cancer Cells. *Carcinogenesis* **2004**, *25*, 647–659. [CrossRef] [PubMed]
76. Carrillo-Martinez, E.J.; Flores-Hernández, F.Y.; Salazar-Montes, A.M.; Nario-Chaidez, H.F.; Hernández-Ortega, L.D. Quercetin, a Flavonoid with Great Pharmacological Capacity. *Molecules* **2024**, *29*, 1000. [CrossRef] [PubMed]
77. Fang, J.; Zhou, G.; Zhao, H.; Xie, D.; Zhang, J.; Kües, U.; Xiao, Y.; Fang, Z.; Liu, J. An apoptosis-inducing factor controls programmed cell death and laccase expression during fungal interactions. *Appl. Microbiol. Biotechnol.* **2024**, *108*, 135. [CrossRef]
78. Mustafa, M.; Ahmad, R.; Tantry, I.Q.; Ahmad, W.; Siddiqui, S.; Alam, M.; Abbas, K.; Moinuddin; Hassan, M.I.; Habib, S.; et al. Apoptosis: A Comprehensive Overview of Signaling Pathways, Morphological Changes, and Physiological Significance and Therapeutic Implications. *Cells* **2024**, *13*, 1838. [CrossRef]
79. Naponelli, V.; Rocchetti, M.T.; Mangieri, D. Apigenin: Molecular Mechanisms and Therapeutic Potential against Cancer Spreading. *Int. J. Mol. Sci.* **2024**, *25*, 5569. [CrossRef] [PubMed]
80. Alam, M.; Ashraf, G.M.; Sheikh, K.; Khan, A.; Ali, S.; Ansari, M.M.; Adnan, M.; Pasupuleti, V.R.; Hassan, M.I. Potential Therapeutic Implications of Caffeic Acid in Cancer Signaling: Past, Present, and Future. *Front. Pharmacol.* **2022**, *13*, 845871. [CrossRef] [PubMed]
81. Kocot, J.; Kiełczykowska, M.; Luchowska-Kocot, D.; Kurzepa, J.; Musik, I. Antioxidant Potential of Propolis, Bee Pollen, and Royal Jelly: Possible Medical Application. *Oxid. Med. Cell. Longev.* **2018**, *2018*, 7074209. [CrossRef] [PubMed]
82. Plasay, M.; Muslimin, L. The Effect of Administration of Honey on Physical Activity in Malondialdehyde, Glutathione, and Superoxide Dismutase Blood Levels in Male Rat. *J. Med. Chem. Sci.* **2024**, *7*, 590–597.
83. Dalal, R.; Grujic, D. Epinephrine. In *StatPearls*; StatPearls: Treasure Island, FL, USA, 2024. Available online: https://www.ncbi.nlm.nih.gov/books/NBK482160/ (accessed on 7 July 2025).
84. Rakha, M.K.; Nabil, Z.I.; Hussein, A.A. Cardioactive and Vasoactive Effects of Natural Wild Honey Against Cardiac Malperformance Induced by Hyperadrenergic Activity. *Glob. J. Pharmacol.* **2008**, *2*, 20–25. [CrossRef]
85. Chu, J.; Hua, L.; Liu, X.; Xiong, H.; Jiang, F.; Zhou, W.; Wang, L.; Xue, G. Superoxide dismutase alterations in COVID-19: Implications for disease severity and mortality prediction in the context of omicron variant infection. *Front. Immunol.* **2024**, *15*, 1362102. [CrossRef]

86. Zhang, S.; Li, S.; Xie, S.; Cui, L.; Gao, Y.; Wang, Y. The Role of Ca^{2+}/PI3K/Akt/eNOS/NO Pathway in Astragaloside IV–Induced Inhibition of Endothelial Inflammation Triggered by Angiotensin II. *Mediat. Inflamm.* **2024**, *2024*, 3193950. [CrossRef]
87. Zhang, F.; Yan, Y.; Zhang, L.; Li, D.; Li, L.; Lian, W.; Xia, C.; He, J.; Xu, J.; Zhang, W. Pharmacological activities and therapeutic potential of galangin, a promising natural flavone, in age-related diseases. *Phytomedicine* **2023**, *120*, 155061. [CrossRef]
88. Kamisah, Y.; Jalil, J.; Yunos, N.M.; Zainalabidin, S. Cardioprotective Properties of Kaempferol: A Review. *Plants* **2023**, *12*, 2096. [CrossRef]
89. Abdelbagi, O.; Taha, M.; Al-Kushi, A.G.; Alobaidy, M.A.; Baokbah, T.A.S.; Sembawa, H.A.; Azher, Z.A.; Obaid, R.; Babateen, O.; Bokhari, B.T.; et al. Protective Effects of Galangin Against Cyclophosphamide-Induced Cardiotoxicity via Suppressing NF-κB and Improving Mitochondrial Biogenesis in Rats. *Medicina* **2025**, *61*, 335. [CrossRef]
90. Hercberg, S.; Galan, P.; Preziosi, P.; Bertrais, S.; Mennen, L.; Malvy, D.; Roussel, A.-M.; Favier, A.; Briançon, S. The SU.VI.MAX Study: A Randomized, Placebo-Controlled Trial of the Health Effects of Antioxidant Vitamins and Minerals. *Arch. Intern. Med.* **2004**, *164*, 2335–2342. [CrossRef] [PubMed]
91. Nagyová, A.; Krajčovičová Kudláčková, M.; Horská, A.; Smolková, B.; Blažíček, P.; Rašlová, K.; Collins, A.; Dušinská, M. Lipid peroxidation in men after dietary supplementation with a mixture of antioxidant nutrients. *Free Radic. Res.* **2004**, *38*, 597–602.
92. El-Shafey, A.; Seliem, M.; El-Zawahry, S.; Shahen, E.; Mahmoud, D. Effect of honey bee on some antioxidant enzymes and lipid profile in hypercholesterolemic male albino rats. *Zagazig Univ. Med. J.* **2015**, *21*, 1–11. [CrossRef]
93. Erejuwa, O.O.; Sulaiman, S.A.; Ab Wahab, M.S.; Sirajudeen, K.N.S.; Md Salleh, M.S.; Gurtu, S. Differential Responses to Blood Pressure and Oxidative Stress in Streptozotocin-Induced Diabetic WKY and Spontaneously Hypertensive Rats: Effects of Antioxidant (Honey) Treatment. *Int. J. Mol. Sci.* **2011**, *12*, 1888–1907. [CrossRef] [PubMed]
94. Erejuwa, O.O.; Sulaiman, S.A.; Ab Wahab, M.S. Honey: A Novel Antioxidant. *Molecules* **2012**, *17*, 4400–4423. [CrossRef]
95. Erejuwa, O.O.; Sulaiman, S.A.; Ab Wahab, M.S.; Sirajudeen, K.N.S.; Salleh, M.S.M.; Gurtu, S. Honey improves antioxidant defense system in kidneys of streptozotocin-induced diabetic rats. *Int. J. Mol. Sci.* **2011**, *12*, 861–870. [CrossRef]
96. Guerrero, J.A.; Navarro-Nuñez, L.; Lozano, M.L.; Martínez, C.; Vicente, V.; Gibbins, J.M.; Rivera, J. Flavonoids inhibit the platelet TxA_2 signalling pathway and antagonize TxA_2 receptors (TP) in platelets and smooth muscle cells. *Br. J. Clin. Pharmacol.* **2007**, *64*, 133–144. [CrossRef]
97. Zhu, H.; Xia, Q.; Jiang, H.; Wang, F.; Yu, L.; Zhang, L. Pinocembrin shows antithrombotic activity in thrombosis-induced experimental rats. *Pharmacogn. Mag.* **2022**, *18*, 746–751. [CrossRef]
98. Mink, P.J.; Scrafford, C.G.; Barraj, L.M.; Harnack, L.; Hong, C.-P.; Nettleton, J.A.; Jacobs, D.R., Jr. Flavonoid intake and cardiovascular disease mortality: A prospective study in postmenopausal women. *Am. J. Clin. Nutr.* **2007**, *85*, 895–909. [CrossRef]
99. Rangel-López, A.; González-Cabello, H.; Paniagua-Medina, M.E.; López-Romero, R.; Arriaga-Pizano, L.A.; Lozano-Ramírez, M.; Pérez-Barragán, J.J.; Márquez-González, H.; López-Sánchez, D.M.; Mata-Rocha, M.; et al. Levels of Plasma Endothelin-1, Circulating Endothelial Cells, Endothelial Progenitor Cells, and Cytokines after Cardiopulmonary Bypass in Children with Congenital Heart Disease: Role of Endothelin-1 Regulation. *Int. J. Mol. Sci.* **2024**, *25*, 8895. [CrossRef]
100. Nagata, S.; Yoshizumi, M.; Tsuchiya, K.; Kirima, K.; Kyaw, M.; Tamaki, T. Quercetin inhibits Shc- and phosphatidylinositol 3-kinase-mediated c-Jun N-terminal kinase activation by angiotensin II in cultured rat aortic smooth muscle cells. *Mol. Pharmacol.* **2001**, *60*, 656–665.
101. Wang, Y.; Jo, E.K.; González-Zamora, J.; Ohta, K.; Griendling, K.K.; Liao, J.K. Differential activation of mitogen-activated protein kinases in smooth muscle cells by angiotensin II. Arterioscler. *Thromb. Vasc. Biol.* **2000**, *20*, 940–946. [CrossRef]
102. Wei, Y.; Qiu, X.; Qiu, Y.; Qu, L.; Chu, X.; Luo, L. Quercetin reduces oxidative stress and inhibits activation of c-Jun N-terminal kinase/activator protein-1 signaling in experimental mouse model of abdominal aortic aneurysm. *Mol. Med. Rep.* **2014**, *10*, 2353–2360. [CrossRef]
103. Datta, S.R.; Dudek, H.; Tao, X.; Masters, S.; Fu, H.; Gotoh, Y.; Greenberg, M.E. Akt phosphorylation of BAD couples survival signals to the cell-intrinsic death machinery. *Cell* **1997**, *91*, 231–241. [CrossRef] [PubMed]
104. Vlahos, C.J.; Matter, W.F.; Hui, K.Y.; Brown, R.F. A specific inhibitor of phosphatidylinositol 3-kinase, 2-(4-morpholinyl)-8-phenyl-4H-1-benzopyran-4-one (LY294002). *J. Biol. Chem.* **1994**, *269*, 5241–5248. [CrossRef] [PubMed]
105. Duarte, J.; Pérez-Palencia, R.; Vargas, F.; Ocete, M.A.; Pérez-Vizcaíno, F.; Zarzuelo, A.; Tamargo, J. Antihypertensive effects of the flavonoid quercetin in spontaneously hypertensive rats. *Br. J. Pharmacol.* **2001**, *133*, 117–124. [CrossRef]
106. Egert, S.; Bosy-Westphal, A.; Seiberl, J.; Wagner, A.E.; Frank, J.; Schrezenmeir, J.; Rimbach, G.; Müller, M.J. Quercetin reduces systolic blood pressure and plasma oxidised LDL concentrations in overweight subjects with high cardiovascular risk: A double-blind, placebo-controlled, crossover study. *Br. J. Nutr.* **2009**, *102*, 1065–1074. [CrossRef]
107. Graefe, E.U.; Wittig, J.; Mueller, S.; Riethling, A.K.; Uehleke, B.; Drewelow, B.; Pforte, H.; Jacobasch, G.; Derendorf, H.; Veit, M. Pharmacokinetics and bioavailability of quercetin glycosides in humans. *J. Clin. Pharmacol.* **2001**, *41*, 492–499. [CrossRef]

108. Middleton, E.; Kandaswami, C.; Theoharides, T.C. The effects of plant flavonoids on mammalian cells: Implications for inflammation, heart disease, and cancer. *Pharmacol. Rev.* **2000**, *52*, 673–751. [CrossRef]
109. Wang, S.-Y.; Wang, Y.-J.; Dong, M.-Q.; Li, G.-R. Acacetin Is a Promising Drug Candidate for Cardiovascular Diseases. *Am. J. Chin. Med.* **2024**, *52*, 1661–1692. [CrossRef] [PubMed]
110. Kodama, I.; Kamiya, K.; Toyama, J. Cellular Electropharmacology of Antiarrhythmic Drugs. *Jpn. J. Pharmacol.* **1997**, *74*, 195–223. [CrossRef]
111. Hohnloser, S.H.; Singh, B.N. Proarrhythmia with Class III antiarrhythmic drugs: Definition, electrophysiologic mechanisms, incidence, predisposing factors, and clinical implications. *J. Cardiovasc. Electrophysiol.* **1995**, *6 Pt 2*, 920–936.
112. Li, G.-R.; Wang, H.-B.; Qin, G.-W.; Jin, M.-W.; Tang, Q.; Sun, H.-Y.; Lau, C.P. Acacetin, a Natural Flavone, Selectively Inhibits Human Atrial Repolarization Potassium Currents and Prevents Atrial Fibrillation in Dogs. *Circulation* **2008**, *117*, 2449–2457. [CrossRef] [PubMed]
113. Zhou, Z.; Li, T.; Qin, H.; Wang, X.; He, S.; Fan, Z.; Ye, Q.; Du, Y. Acacetin as a Natural Cardiovascular Therapeutic: Mechanisms and Preclinical Evidence. *Front. Pharmacol.* **2025**, *16*, 1493981. [CrossRef]
114. Aggarwal, B.B.; Shishodia, S. Suppression of Nuclear Factor-κB Activation as a Therapeutic Strategy for Inflammatory Disorders. *Biochim. Biophys. Acta* **2004**, *1697*, 69–84. [CrossRef]
115. Nie, J.; Chang, Y.; Li, Y.; Zhou, Y.; Qin, J.; Sun, Z.; Wang, Z.; Li, X.; Yu, D.; Zhang, Y.; et al. Caffeic Acid Phenethyl Ester (Propolis Extract) Ameliorates Insulin Resistance by Inhibiting JNK and NF-κB Inflammatory Pathways in Diabetic Mice and HepG2 Cell Models. *J. Agric. Food Chem.* **2017**, *65*, 9041–9053. [CrossRef]
116. Xu, Y.C.; Leung, S.W.S.; Leung, G.P.H.; Man, R.Y.K. *Kaempferol enhances endothelium-dependent relaxation in porcine coronary artery through BK* channels. *Br. J. Pharmacol.* **2015**, *172*, 3003–3014. [CrossRef]
117. Luo, R.; Tian, L.; Yang, Y. Galangin inhibits oxidized low-density lipoprotein-induced angiogenic activity in human aortic endothelial cells by downregulating lncRNA H19. *Nan Fang Yi Ke Da Xue Xue Bao* **2024**, *44*, 52–59. [CrossRef]
118. Pakdeechote, P.; Poasakate, A.; Prasatthong, P.; Wongmekiat, O. Galangin Mitigates Aortic Dysfunction and Hypertrophy in Rats with Metabolic Syndrome. *Heliyon* **2023**, *9*, e16500. [CrossRef]
119. Santoso, M.I.; Paradiso, G.; Russo, G.L. Role of Flavonoids in Brain Health: Antioxidant, Anti-inflammatory, and Neuroprotective Effects. *Int. J. Mol. Sci.* **2024**, *25*, 5041. [CrossRef]
120. Block, M.L.; Zecca, L.; Hong, J.-S. Microglia-mediated neurotoxicity: Uncovering the molecular mechanisms. *Nat. Rev. Neurosci.* **2007**, *8*, 57–69. [CrossRef]
121. Chen, P.; Huo, X.; Liu, W.; Li, K.; Sun, Z.; Tian, J. Apigenin exhibits anti-inflammatory effects in LPS-stimulated BV-2 microglia via activation of GSK3β/Nrf2 and suppression of NF-κB. *Immunopharmacol. Immunotoxicol.* **2020**, *42*, 9–16. [CrossRef] [PubMed]
122. Ling, C.; Lei, C.; Zou, M.; Cai, X.; Xiang, Y.; Xie, Y.; Li, X.; Huang, D.; Wang, Y. Neuroprotective effect of apigenin against cerebral ischemia/reperfusion injury in vitro and in vivo. *J. Int. Med. Res.* **2020**, *48*, 300060520945859. [CrossRef] [PubMed]
123. Chesworth, J.; Gamage, T.; Ullah, F.; Sonego, M.; Millington, P.; Fernández, I.; Liang, H.; Karl, T.; Münch, G.; Niedermayer, G.; et al. Chronic apigenin supplementation reduces microglia activation but does not reverse memory decline in GFAP-IL6 mice. *Front. Neurosci.* **2021**, *15*, 699329. [CrossRef]
124. Spiegel, M. Theoretical Insights into the Oxidative Stress-Relieving Properties of Pinocembrin—An Isolated Flavonoid from Honey and Propolis. *J. Phys. Chem. B* **2023**, *127*, 8769–8779. [CrossRef] [PubMed]
125. Wang, P.; Pan, L.; Liu, Q.; Huang, Y.; Tang, Y.; Lin, B.; Liao, Y.; Luo, H.; Meng, X. Pinocembrin Activation of DPP9 Inhibits NLRP1 Inflammasome Activation to Alleviate Cerebral Ischemia/Reperfusion-Induced Lung and Intestinal Injury. *Immunol. Res.* **2024**, *73*, 13. [CrossRef]
126. Liu, R.; Gao, M.; Yang, Z.; Du, G. Pinocembrin Protects Rat Brain Against Oxidation and Apoptosis Induced by Ischemia–Reperfusion Both In Vivo and In Vitro. *Brain Res.* **2008**, *1216*, 136–146. [CrossRef]
127. Meng, F.; Liu, R.; Gao, M.; Wang, Y.; Yu, X.; Xuan, Z.; Sun, J.; Yang, F.; Wu, C.; Du, G. Pinocembrin Attenuates Blood–Brain Barrier Injury Induced by Global Cerebral Ischemia–Reperfusion in Rats. *Brain Res.* **2011**, *1391*, 93–101. [CrossRef]
128. Ye, Y.; Wang, H.; Lin, Y.; Song, L.; Xie, W. Pinocembrin Ameliorates Post-Infarct Heart Failure through Activation of Nrf2/HO-1 Signaling Pathway. *Mol. Med.* **2021**, *27*, 63. [CrossRef]
129. Santos, S.; Maia, M.; Barros, A.; Gouvinhas, I. Assessment of Phenolic Content, Antioxidant and Anti-Aging Activities of Honey from Pittosporum undulatum Vent. Naturalized in the Azores Archipelago (Portugal). *Appl. Sci.* **2023**, *13*, 1788. [CrossRef]
130. Tan, W.; Tian, Y.; Zhang, Q.; Miao, X.; Kuang, H.; Yang, W. Antioxidant and Antibacterial Activity of *Apis laboriosa* Honey against Salmonella enterica Serovar Typhimurium. *Front. Nutr.* **2023**, *10*, 1181492. [CrossRef] [PubMed]
131. Weston, R.J. The Contribution of Catalase and Other Natural Products to the Antibacterial Activity of Honey: A Review of the Evidence. *Food Chem.* **2000**, *71*, 495–498. [CrossRef]
132. Kwakman, P.H.S.; Zaat, S.A.J. Antibacterial Components of Honey. *IUBMB Life* **2012**, *64*, 48–55. [CrossRef] [PubMed]

133. Bezerra, A.; Fonseca, H.; Rodrigues, F.; Delerue-Matos, C.; Gouvinhas, I.; Garcia, J. Honey Therapy in Diabetic Foot Ulcers: A Promising Strategy for Effective Wound Healing. *Appl. Sci.* **2023**, *13*, 12820. [CrossRef]
134. Molan, P.C. The role of honey in the management of wounds. *J. Wound Care* **1999**, *8*, 415–418. [CrossRef]
135. White, J.W.; Subers, M.H.; Schepartz, A.I. The identification of inhibine, the antibacterial factor in honey, as hydrogen peroxide and its origin in a glucose-oxidase system. *Biochim. Biophys. Acta* **1963**, *73*, 57–70. [CrossRef]
136. Mandal, M.D.; Mandal, S. Honey: Its medicinal property and antibacterial activity. *Asian Pac. J. Trop. Biomed.* **2011**, *1*, 154–160. [CrossRef]
137. Cianciosi, D.; Forbes-Hernández, T.Y.; Afrin, S.; Gasparrini, M.; Reboredo-Rodriguez, P.; Manna, P.P.; Zhang, J.; Bravo Lamas, L.; Martínez Flórez, S.; Agudo Toyos, P.; et al. Phenolic Compounds in Honey and Their Associated Health Benefits: A Review. *Molecules* **2018**, *23*, 2322. [CrossRef]
138. Jenkins, R.; Burton, N.; Cooper, R. Manuka honey inhibits cell division in methicillin-resistant *Staphylococcus aureus*. *J. Antimicrob. Chemother.* **2011**, *66*, 2536–2542. [CrossRef]
139. Kwakman, P.H.S.; Te Velde, A.A.; de Boer, L.; Speijer, D.; Vandenbroucke-Grauls, C.M.J.E.; Zaat, S.A.J. How honey kills bacteria. *FASEB J.* **2011**, *25*, 2576–2582. [CrossRef]
140. Tonks, A.J.; Cooper, R.A.; Jones, K.P.; Blair, S.; Parton, J.; Tonks, A. Honey stimulates inflammatory cytokine production from monocytes. *Cytokine* **2003**, *21*, 242–247. [CrossRef] [PubMed]
141. Efem, S.E. Clinical observations on the wound healing properties of honey. *Br. J. Surg.* **1988**, *75*, 679–681. [CrossRef] [PubMed]
142. Majtan, J. Honey: An immunomodulator in wound healing. *Wound Repair Regen.* **2023**, *31*, 14–26. [CrossRef] [PubMed]
143. Tashkandi, H. Honey in Wound Healing: An Updated Review. *Open Life Sci.* **2021**, *16*, 1091–1100. [CrossRef] [PubMed]
144. Rasmussen, C.N.; Wang, X.-H.; Leung, S.; Andrae-Nightingale, L.M.; Schmidt, S.J.; Engeseth, N.J. Selection and Use of Honey as an Antioxidant in a French Salad Dressing System. *J. Agric. Food Chem.* **2008**, *56*, 8650–8657. [CrossRef] [PubMed]
145. McKibben, D.; Engeseth, N.J. Honey as a Protective Agent against Lipid Oxidation in Ground Turkey. *J. Agric. Food Chem.* **2002**, *50*, 592–595. [CrossRef] [PubMed]
146. Wesołowska, M.; Dżugan, M. The Use of the Photochem Device in Evaluation of Antioxidant Activity of Polish Honey. *Foods* **2017**, *10*, 1568–1574. [CrossRef]
147. Brudzynski, K.; Miotto, D. The Relationship between the Content of Polyphenols in Honey and Its Antioxidant Activity. *Pol. J. Food Nutr. Sci.* **2015**, *65*, 71–80. [CrossRef]
148. Decker, E.A.; Park, Y. Healthier Meat Products as Functional Foods. *Meat Sci.* **2010**, *86*, 49–55. [CrossRef]
149. Weston, R.J.; Mitchell, K.R.; Allen, K.L. Antibacterial phenolic components of New Zealand manuka honey. *Food Chem.* **1999**, *64*, 295–301. [CrossRef]
150. Khalil, M.I.; Sulaiman, S.A.; Boukraa, L. Antioxidant properties of honey and its role in preventing health disorders. *Open Nutraceuticals J.* **2010**, *3*, 6–16. [CrossRef]
151. Gheldof, N.; Wang, X.H.; Engeseth, N.J. Identification and quantification of antioxidant components of honeys from various floral sources. *J. Agric. Food Chem.* **2002**, *50*, 5870–5877. [CrossRef] [PubMed]
152. Schramm, D.D.; Karim, M.; Schrader, H.R.; Holt, R.R.; Cardetti, M.; Keen, C.L. Honey with high levels of antioxidants can provide protection to healthy human subjects. *J. Agric. Food Chem.* **2003**, *51*, 1732–1735. [CrossRef] [PubMed]
153. Khalil, M.I.; Mahaneem, M.; Jamalullail, S.M.S.; Alam, N.; Sulaiman, S.A. Evaluation of radical scavenging activity and colour intensity of nine Malaysian honeys of different origin. *J. ApiProd. ApiMed. Sci.* **2011**, *3*, 4–11. [CrossRef]
154. Ferreira, I.C.F.R.; Aires, E.; Barreira, J.C.M.; Estevinho, L.M. Antioxidant activity of Portuguese honey samples: Different contributions of the entire honey and phenolic extract. *Food Chem.* **2009**, *114*, 1438–1443. [CrossRef]
155. Meda, A.; Lamien, C.E.; Romito, M.; Millogo, J.; Nacoulma, O.G. Determination of the total phenolic, flavonoid and proline contents in Burkina Fasan honey, as well as their radical scavenging activity. *Food Chem.* **2005**, *91*, 571–577. [CrossRef]
156. Al-Waili, N.S.; Salom, K.; Butler, G.; Al Ghamdi, A.A. Honey and microbial infections: A review supporting the use of honey for microbial control. *J. Med. Food* **2011**, *14*, 1079–1096. [CrossRef]
157. Bertoncelj, J.; Doberšek, U.; Jamnik, M.; Golob, T. Evaluation of the phenolic content, antioxidant activity and colour of Slovenian honey. *Food Chem.* **2007**, *105*, 822–828. [CrossRef]
158. Alvarez-Suarez, J.M.; Tulipani, S.; Romandini, S.; Bertoli, E.; Battino, M. Contribution of honey in nutrition and human health: A review. *Mediterr. J. Nutr. Metab.* **2010**, *3*, 15–23. [CrossRef]

antioxidants

MDPI

Systematic Review

The Physicochemical Parameters, Phenolic Content, and Antioxidant Activity of Honey from Stingless Bees and *Apis mellifera*: A Systematic Review and Meta-Analysis

Ana Karen Zaldivar-Ortega [1], Antonio de Jesús Cenobio-Galindo [1], Nuria Morfin [2], Gabriel Aguirre-Álvarez [1], Rafael G. Campos-Montiel [1,†], Nuria Esturau-Escofet [3], Angel Garduño-García [4,*] and Juan Carlos Angeles-Hernandez [5,*]

[1] Instituto de Ciencias Agropecuarias, Universidad Autónoma del Estado de Hidalgo, Avenida Universidad Km. 1 s/n Exhacienda Aquetzalpa, Tulancingo 43600, Mexico; ana_saldivar@uaeh.edu.mx (A.K.Z.-O.); antonio_cenobio@uaeh.edu.mx (A.d.J.C.-G.); aguirre@uaeh.edu.mx (G.A.-Á.); rcampos@uaeh.edu.mx (R.G.C.-M.)

[2] Michael Smith Laboratories, Department of Biochemistry & Molecular Biology, The University of British Columbia, Vancouver, BC V6T1Z4, Canada; nuriamorfin@ttp-bchpa.ca

[3] Instituto de Química, Universidad Nacional Autónoma de México, Ciudad de México 04510, Mexico; nesturau@iquimica.unam.mx

[4] Departamento de Ingeniería Mecánica Agrícola, Universidad Autónoma Chapingo, Carretera México-Texcoco, Km 38.5, Texcoco 56230, Mexico

[5] Departamento de Medicina y Zootecnia de Rumiantes, Facultad de Medicina Veterinaria y Zootecnia, Universidad Nacional Autónoma de México, Ciudad de México 04510, Mexico

* Correspondence: agardunog@chapingo.mx (A.G.-G.); juanangeles@fmvz.unam.mx (J.C.A.-H.); Tel.:+52-772-106-74-38 (J.C.A.-H.)

† This author has passed away.

Abstract: The most common bee species used for honey production is *Apis mellifera* (*A. mellifera*), followed by stingless bees. This study included scientific articles using the PRISMA approach. A random effect model was implemented and the effect size (ES) was calculated and reported as the standardized mean difference (SMD) and raw mean difference (RMD). The mean phenolic content in *A. mellifera* honey was 61.21 ± 28.3 mg GAE/100 g and stingless bee honey +33.69 mg GAE/100 g; $p = 0.01$. The antioxidant activity, discovered by the Ferric Reducing Antioxidant Power (FRAP) method, showed a mean of 97.34 ± 7.84 µmol Fe(II)/100 g in *A. mellifera* and stingless bee honey +63.39 µmol Fe(II)/100 g; $p = 0.009$. The physicochemical properties showed significant differences in moisture (*A. mellifera* honey 19.54 ± 3.65%; stingless bee honey +8.02%; $p = 0.0001$), hydroxymethylfurfural (HMF) (*A. mellifera* honey 20.14 ± 16.27 mg/kg; stingless bee honey −11.25 mg/kg; $p = 0.001$), and free acidity (*A. mellifera* honey 31.32 ± 16.67 meq/kg; stingless bee honey +34.76 meq/kg; $p = 0.01$). The variability in the trials was explained by the heterogeneity, and a meta-regression analysis incorporated four covariates: (1) stingless bee species; (2) floral source; (3) country, and (4) latitude. This study highlights the importance of conducting further studies on stingless bee honey.

Keywords: honey; *Apis mellifera*; stingless bees; bioactive compounds; antioxidant activity; physicochemical composition

1. Introduction

The most common bee species used for honey production and other bee products (i.e., pollen, beeswax, propolis, and royal jelly) is the species *Apis mellifera* L., followed by stingless bees [1]. Both bees are classified in the family *Apidae* and the subfamily *Apinae*, but the most popular European honey bee is classified in the tribe *Apini*, genus *Apis*, species *mellifera*, while stingless bees are classified within the tribe *Meliponini* [2,3], which has different genera including *Melipona*, *Scaptotrigona*, and *Trigona* [4]. There are more than

500 species of stingless bees found in tropical and subtropical regions, with most of these species found in Latin America, Africa, Asia, and Australia [5,6].

The bee genera and species have been shown to have a significant impact on both the content of phenolic content and physicochemical parameters of honey [7], but this has not been extensively studied. The composition of honey produced by *Apis mellifera* has been studied and its quality parameters have been established. However, this is not the case for honey produced by stingless bees [5,8]. In fact, the presence of honey from stingless bees on the world market is limited compared to honey from *Apis mellifera*, partly as a result of the lack of legislation on the product and information on its nutritional value [2]. Honey standards have been modified to align with the botanical origin, but not according to different species of bees. Thus, it is important that these factors are considered and that quality standards are established for stingless bee honey [9].

The commercial market for *Apis mellifera* honey and stingless bee honey is made up of different types of consumers. For example, between 80 and 90% of the total honey harvested from stingless bees is sold by producers to local buyers and middlemen in the southeast region of Mexico [10], whereas 90% of the honey from *Apis mellifera* produced in the same region is exported to European countries [11]. Additionally, stingless bee honey is commercialized in traditional free markets at significantly higher prices compared to *Apis mellifera* honey [12]. For instance, in 2016, stingless bee honey prices were three times higher compared to conventional honey from *Apis mellifera* [10], which might seem advantageous for meliponiculturists (beekeepers specializing in the management of stingless bees). The honey from stingless bees is widely used in various industries, including food, pharmaceuticals, and cosmetics [13]. On the other hand, stingless bees also play an important ecological role in ecosystem conservation [6,14]. They are the main pollinators of wild and cultivated tropical plants, such as orchids, macadamia nuts, mangoes, and coffee [15,16]. Culturally, stingless bees are important because of the hive products obtained from them, such as honey, pollen, and cerumen, which have been used for medicinal purposes by the indigenous communities in many parts of the world due to their nutritional and therapeutic properties attributed to the presence of phenolic compounds [17–19].

Nutritionally, honey is mainly composed of carbohydrates and water, but other important components include proteins, organic acids, enzymes, minerals, amino acids, vitamins, and phenolic compounds [20,21]. Regarding phenolic content, flavonoids and phenolic acids are the main chemical compounds studied in honey [22]. These compounds confer nutraceutical properties to stingless bee honey [12]. Bees transfer phenolic compounds from nectar, pollen or propolis to honey [23]. Phenolic compounds are secondary metabolites synthesized by plants to counteract biotic (e.g., pathogenic or competing species) and abiotic (e.g., UV radiation, desiccation, temperature) stressors and help attract bees as pollinators [24]. Additionally, phenolic compounds can influence the organoleptic properties (i.e., color, taste, or flavor) of honey [21], and they have been used as chemical indicators to determine different botanical and geographical origins of honey [25,26]. Several studies have shown higher phenolic content in honey produced by stingless bees (e.g., *Scaptotrigona* spp. and *Melipona* spp.) compared to *Apis mellifera* honey [1,18,27]. Phenolic compounds have been associated with beneficial antimicrobial, anti-inflammatory, and antioxidant activities [4,19,20,28].

The antioxidant activity of honey refers to the potential to reduce oxidative reactions in food systems and human health [29]. Abu et al. [30] reported that stingless bee honey has higher antioxidant activity compared to *Apis mellifera* honey, and the phenolic compounds contribute to the antioxidant capacity. The phenolic profiles of honey depend on the floral sources visited by the bee and consequently the antioxidant capacity [29]. The antioxidant activity in honey has been demonstrated using different tests, for instance, 1,1-diphenyl-2-picrylhydrazyl (DPPH) scavenging assay and Ferric Reducing Antioxidant Power (FRAP) assay [31]. The phenolic compounds, antioxidant activity, and physicochemical composition of honey are influenced by many factors, such as botanical origin, climate, environmental conditions, geographical area, latitude, season, and bee species [8,18,32,33]. The physicochemical parameters of stingless honey differ from those of *Apis mellifera* honey, such as color, higher moisture content, and higher acidity [2,5]. In contrast, the pH, hydroxymethylfurfural, and diastase activity are typically lower in stingless bee honey compared to *Apis mellifera* honey [8,9,14]. Additionally, the species of bee could have an influence on the variety and concentration of bioactive compounds and, consequently, on the antioxidant activity [12]. The aim of this systematic review was to analyze and compare the phenolic content, antioxidant activity, and physicochemical properties of the honey of *Apis mellifera* and stingless bees.

2. Materials and Methods

2.1. Search Strategy

In this study, we conducted an exhaustive and structured search of scientific articles using the predetermined protocol in accordance with the Preferred Reporting Items for Systematic Reviews and Meta-analysis (PRISMA) [34] (Figure 1), and the protocol was registered using the Open Science Framework platform (https://osf.io/s9zt8). The focus of the search was on studies comparing the phenolic compound content, antioxidant activity, and physicochemical composition of honey from *Apis mellifera* versus honey produced by stingless bees. To collect the data, we obtained scientific publications from various databases, including Google Scholar, Primo-UAEH, PubMed, and DOAJ. The article search was performed by three researchers to avoid reviewer bias. A search strategy was developed using the following terms: "honey", "*Apis mellifera*", "stingless bees", "phenols", "flavonoids", "antioxidant activity", and "physicochemical composition" in combination with the Boolean operators ('and' or 'or'). Additionally, some articles were identified from the reference lists of the articles found in the initial search and included in the comprehensive database.

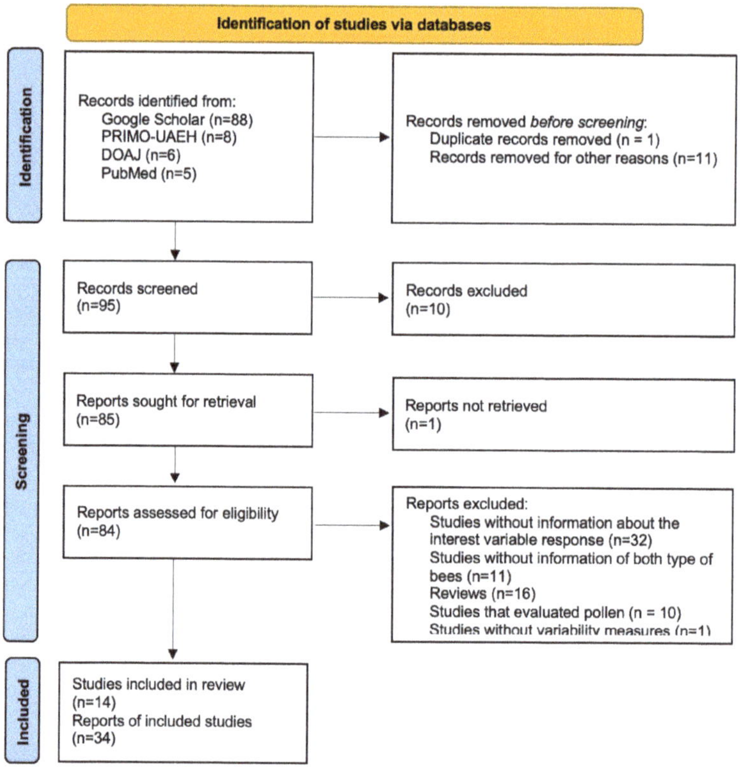

Figure 1. Preferred Reporting Items for Systematic Reviews and Meta-analysis (PRISMA) study flow diagram of the systematic review from the initial search and screening to the final selection of publications to be included in the meta-analysis.

2.2. Inclusion Criteria

The parameters of the inclusion criteria were the following:

(a) Studies published in an internationally peer-reviewed scientific journal. Therefore, unpublished manuscripts, conference proceedings, and dissertations were excluded.
(b) Studies that contained information about the physicochemical composition, phenolic content, and antioxidant activity of stingless bees and *Apis mellifera* honey.
(c) Studies that reported the sample size for each group.
(d) Studies that provided least squares means and a measure of data variability such as the coefficient of variation, standard deviation, or standard error.

2.3. Database

A total of 14 studies were selected to be included in the meta-analysis procedure, which comprised 34 trials (Figure 1). One researcher extracted the data from the 14 studies that met the inclusion criteria into an Excel spreadsheet (version 2016, Microsoft Corp., Redmond, WA, USA), and two researchers verified the information to identify discrepancies. An Excel spreadsheet was elaborated for each of the following response variables: phenols (mg GAE/100 g), flavonoids (mg QE/100 g), 2,2 diphenyl 1 picrylhydrazyl (DPPH; µmol TE/100 g), Ferric Reducing Antioxidant Power (FRAP; µmol Fe(II)/100 g), moisture (%), hydroxymethylfurfural (HMF; mg/kg), free acidity (meq/kg), color (mm/Pfund), and protein content (g/kg). To explore the sources of between-study variability, the following explanatory variables were selected a priori: the genera of stingless bees (*Heterotrigona, Melipona, Hypotrigona, Plebeia, Ptilotrigona, Nannotrigona, Partamona, Scaptotrigona,* and *Trigona*), floral source (monofloral and polyfloral), country (Malaysia, Cuba, Nigeria, Brazil, Peru, and Portugal), and latitude.

2.4. Statistical Analysis

The first step of the analytical review was to perform a descriptive analysis using the 'psych' package version 2.1.9 [35] in the R statistical computing environment (version 04.2+764, R Core Team, 2024).

2.4.1. Meta-Analysis Procedure

A random effect model was implemented, which assumed that the observed difference among studies was a combination of chance and genuine variation in the intervention effects. The effect size (ES) was calculated and reported as the standardized mean difference (SMD) and raw mean difference (RMD). An SMD allows for the summarization of studies with biologically comparable traits but on different scales. The SMD was calculated according to the method proposed by Hedges [36] and the pooling of analyzed studies was carried out using the inverse variance weighting. The RMD was calculated for response variables reporting outcomes on the same scales, which permits the interpretation of the summary effect under the original measures' units [37]. All analyses were performed in the R environment for statistical computing (version 04.2+764, R Core Team, 2024) using the package 'meta' version 4.13-0 [38].

2.4.2. Assess of Heterogeneity

The current study analyzed honey samples from experiments implemented under different regions, environmental conditions, laboratory methodologies, and experimental designs; hence, the heterogeneity was calculated. Heterogeneity was assessed through the estimation of between-study random effects variance (t^2) and the percentage of variability explained by heterogeneity rather than simple variance (I^2 index), which represents the proportion of between-study variability and is calculated as follows (Equation (1)) [39].

$$I^2 = \frac{Q-(k-1)}{Q} \times 100 \quad (1)$$

where Q is the χ^2 heterogeneity statistic and k is the number of trials. I^2 values of 25%, 50%, and 75% represent small, moderate, and high levels of heterogeneity, respectively.

2.4.3. Meta-Regression

The first step of the meta-regression analysis was to incorporate all explanatory variables in a full model; secondly, multi-predictor models were manually reduced by the backward selection of variables until all predictors were significant ($p < 0.05$). Mixed-effects regression models (meta-regression analysis) were constructed in the "metafor" package. Models were compared and selected by means of information-theoretic criteria using Akaike´s Information Criterion and Bayesian Information Criterion; also, the goodness of fit was evaluated by the coefficient of correlation analog (R^2) for meta-regression which was calculated as follows (Equation (2)):

$$R^2 = \frac{\tau^2{}_{REM-}\tau^2{}_{MEM}}{\tau^2{}_{REM}} \tag{2}$$

where $\boldsymbol{\tau}^2{}_{\boldsymbol{REM}-}$ is the estimated total heterogeneity based on the random effects model and $\boldsymbol{\tau}^2{}_{\boldsymbol{MEM}}$ is the total heterogeneity of the mixed-effects regression model. To evaluate evidence for publication bias, a Begg´s tests were performed [40]. The sensitivity analysis was carried out using a leave-one-out meta-analysis.

3. Results

An analytical review and meta-analysis were carried out to assess the effect of the type of bee (*Apis mellifera* vs. stingless bees) on the phenolic content (phenols and flavonoids), antioxidant activity (DPPH and FRAP assays), and physicochemical composition (moisture, HMF, free acidity, color, and protein) of honey. The main causes of article exclusion were studies without results for the response variables of interest (n = 34), articles without comparison of both types of bees (n = 11), review articles (n = 16), studies that evaluated pollen (n = 10), and articles without measures of variability (n = 1). After applying the inclusion and exclusion criteria, a total of 14 articles published between 2000 and 2022 were included in the meta-analysis (Figure 1). After a sensitivity analysis using the leave-one-out meta-analysis, the pooled result was made stable. No single study had a disproportionate effect on the summary estimate.

3.1. Results of Meta-Analysis

3.1.1. Phenols, Flavonoids, and Antioxidant Activity

The RMD, SMD, heterogeneity, and bias of the response variables are presented in Table 1. The honey produced by stingless bees showed a higher phenolic content compared to *Apis mellifera* honey with an RMD of +33.69 mg GAE/100 g (p = 0.01; CI 95% +7.11, +60.27 mg GAE/100 g) and a high heterogeneity (I^2 = 90.4%). The flavonoid content in honey from the different genera of stingless bees was higher (Figure 2) compared to honey from *Apis mellifera* by +3.59 mg QE/100 g (p = 0.05; CI 95% −0.01, +7.19 mg QE/100 g), with a high heterogeneity (I^2 = 89.3%) (Table 1).

Figure 3 shows the mean difference in flavonoid content between the two types of honey of both monofloral and polyfloral origin. The honey of monofloral origin showed a mean difference of 13.22 (95% CI: [13.15, 13.29]), suggesting a higher flavonoid content in this type of honey and a combined effect of 6.07 (95% CI: +0.48, +11.67). Our results depict a significant difference in the antioxidant activity based on the FRAP values among types of bees, with a high antioxidant potential of stingless bee honey (+63.39 µmol Fe(II)/100 g; p = 0.009) (Table 1).

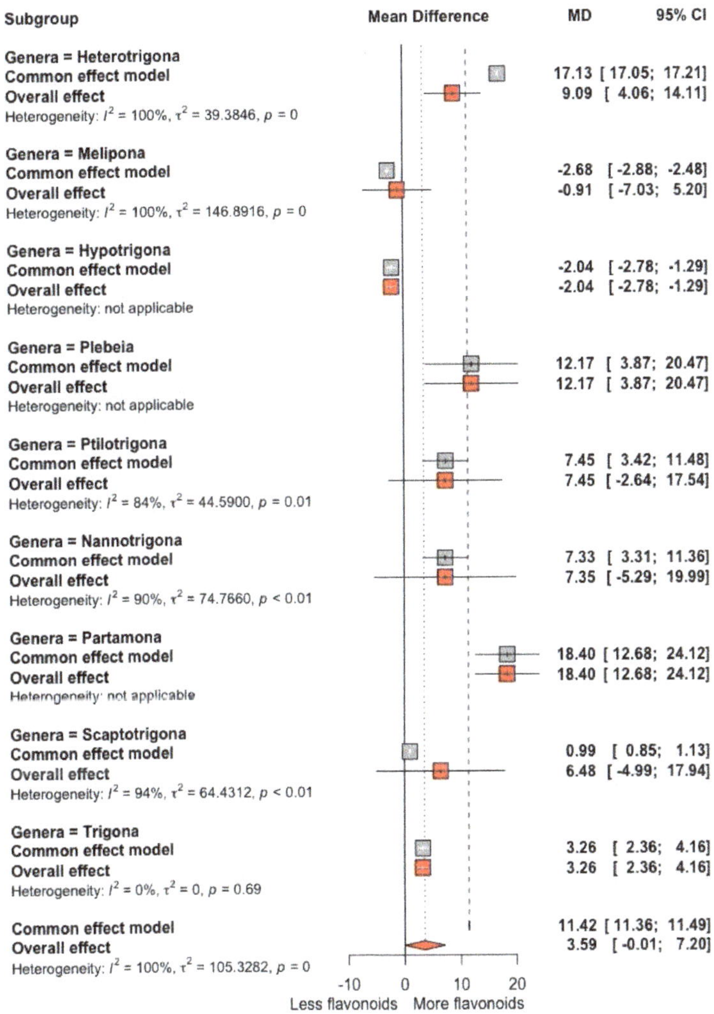

Figure 2. Forest plot of flavonoid content (mg QE/100 g) for trials that included different genera of from stingless bees. The point size reflects the relative weighting of the study to the overall effect size estimate, where a larger point size represents a greater weight in the combined effect size estimated, including the confidence intervals. The diamond represents the overall effect as the raw mean difference (RMD).

The antioxidant activity based on the FRAP value shows an SMD with a significant difference among stingless bee genera (Figure 4). The genus *Heterotrigona* showed an SMD of 3.55 (95% CI: +1.93, +5.17), suggesting a significant increase in antioxidant activity. The overall effect was 7.38 (95% CI: +2.51, +12.25), also indicating a large increase in antioxidant activity (I^2 = 67%.). The genus *Hypotrigona* showed a high effect with an SMD of 11.53 (95% CI: +9.12, +19.94), suggesting an increase in antioxidant activity in this bee species. The heterogeneity of the studies was high, showing that the results varied between studies.

Table 1. Standardized mean difference (SMD) and raw mean difference (RMD) of random effect model with 95% CI and a heterogeneity test of phenolic compounds, antioxidant activity, and physicochemical parameters outcomes of *Apis mellifera* and stingless bee honey.

Item	Trials (*n*)	*Apis mellifera* Mean (SD) [a]	Effect Size RMD (95% CI) [b]	*p*-Value	SMD (95% CI) [c]	*p*-Value	Heterogeneity I^2 (%)	*p*-Value	Bias *p*-Value [d]
Phenols (mg GAE/100 g)	34	61.21 (28.30)	+33.69 (+7.11, +60.27)	0.01	+2.45 (−4.78, +9.70)	0.50	90.4	0.0001	0.21
Flavonoids (mg QE/100 g)	33	9.94 (8.7)	+3.59 (−0.01, +7.19)	0.05	+2.56 (−0.70, +5.83)	0.12	89.3	0.001	0.06
DPPH (µmol TE/100 g)	30	78.10 (20.21)	+37.57 (−13.60, 88.76)	0.15	+2.0 (−3.41, +7.41)	0.46	90.4	0.001	0.19
FRAP (µmol Fe(II)/100 g)	23	97.34 (7.84)	+63.39 (+15.52, +111.26)	0.009	+4.51 (+1.60, +7.42)	0.002	89.7	0.0001	0.08
Moisture (%)	31	19.54 (3.65)	+ 8.02 (+5.93, +10.11)	0.0001	+10.09 (+6.52, +13.66)	0.0001	92.6	0.0001	0.0006
HMF (mg/kg)	17	20.14 (16.27)	−11.25 (−15.88, −6.62)	0.001	−3.96 (−6.07, −1.85)	0.0002	93.5	0.0001	0.008
Free acidity (meq/kg)	19	31.32 (16.67)	+34.76 (+7.17, +62.34)	0.01	+3.17 (−0.45, +6.81)	0.08	93.6	0.0001	0.52
Color (mm/Pfund)	18	64.48 (11.93)	+10.24 (−10.22, 30.71)	0.32	+6.31 (−12.23, −0.39)	0.036	95.1	0.0001	0.101
Protein (g/kg)	22	3.49 (4.08)	−0.31 (−1.18, −0.55)	0.48	−0.83 (−3.98, +2.30)	0.60	92.6	0.0001	0.77

[a] Mean and standard deviation of *Apis mellifera* honey. [b] RMD is the raw mean difference that estimates the global effect expressed in the original unit measures. [c] SMD is the standardized mean difference estimated of the random model. [d] *p*-value of Begg's test; *p*-values > 0.05; I^2 is measure of heterogeneity of random model.

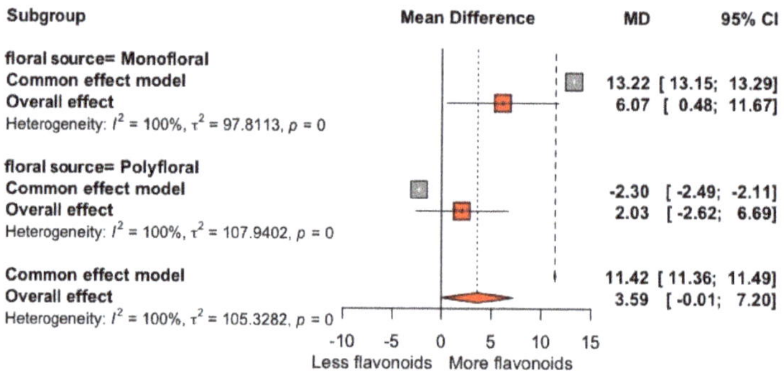

Figure 3. Forest plot of flavonoid content (mg QE/100 g) for floral source trials divided into monofloral and polyfloral origins. The point size reflects the relative weighting of the study to the overall effect size estimate, where a larger point size represents a greater weight in the combined effect size estimated, including the confidence intervals. The diamond represents the overall effect as the raw mean difference (RMD).

The antioxidant activity was analyzed according to the floral origin of the honey, divided into monofloral and polyfloral (Figure 5). The results indicated an SMD of 2.19 (95% CI: +1.46, +2.92), with a higher overall effect of 6.62 (95% CI: +2.34, +10.91), showing a statistically significant increase in antioxidant activity in polyfloral honey (I^2 > 90%).

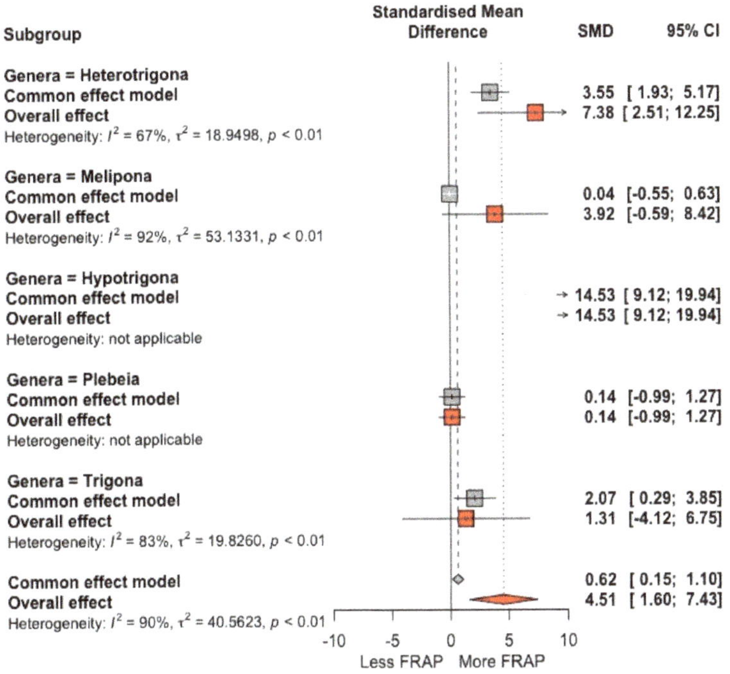

Figure 4. Forest plot antioxidant activity with FRAP (μmol Fe (II)/100 g) for trials that included different genera of stingless bees. The point size reflects the relative weighting of the study to the overall effect size estimate, where a larger point size represents a greater weight and the combined effect size estimated, including the confidence intervals. The diamond represents the overall effect as the raw mean difference (RMD).

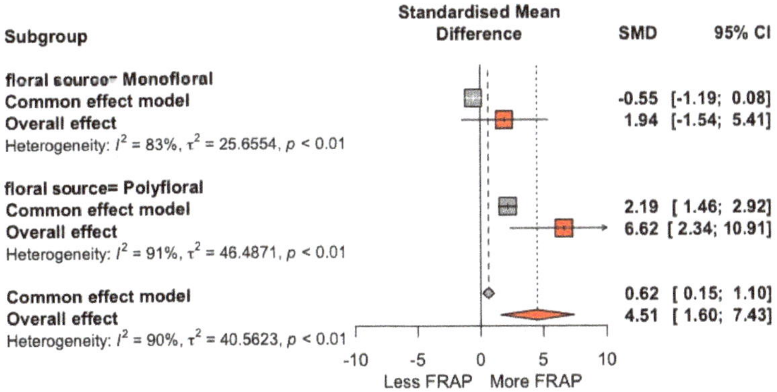

Figure 5. Forest plot of antioxidant activity with FRAP (μmol Fe (II)/100 g) floral source trials, divided into monofloral and polyfloral origin. The point size reflects the relative weighting of the study to the overall effect size estimate, where a larger point size represents a greater weight in the combined effect size estimated, including the confidence intervals. The diamond represents the overall effect as the raw mean difference (RMD).

3.1.2. Physicochemical Composition of Honey

The meta-analysis showed significant differences in the physicochemical parameters (Table 1). The moisture (%) was higher in stingless bee honey by +8.02% (p = 0.0001; CI 95%: +5.93, +10.11%), with the highest SMD (+10.09; p = 0.0001) compared to *Apis mellifera* honey.

Additionally, stingless bee honey showed a higher free acidity compared to *Apis mellifera* honey by +34.76 meq/kg ($p = 0.01$; CI 95%: +7.17; +62.34). The color intensity was higher in stingless bee honey compared to *Apis mellifera* (+10.24; CI 95%: +10.22, +30.71), with evidence of significant heterogeneity ($I^2 = 95.1\%$). Conversely, stingless bee honey showed a lower content of HMF (−11.25 meq/100 g; $p = 0.001$; CI 95%: −15.88, −6.62) compared to *Apis mellifera* honey.

3.2. Meta-Regression Analysis

A meta-regression analysis was performed to explain the significant heterogeneity among the studies ($I^2 > 90\%$) for the variables analyzed, the physicochemical parameters, phenolic content, flavonoids, and antioxidant activity. Four covariates were identified a priori to explain the variability: (1) stingless bee species; (2) floral source for honey production; (3) country, and (4) latitude. The slopes of the meta-regression and the statistical significance of the covariates, for heterogeneity levels of $I^2 > 25\%$, are shown in Table 2. Various bee genera, the floral source, and country influenced the phenolic content. The results of the meta-regression showed that the genera *Melipona* could produce honey with lower phenolic content (−26.01), ($p < 0.05$). Regarding the country, Cuba ($\beta = +69.58$; $p < 0.001$) and Peru ($\beta = +82.04$; $p < 0.001$) were associated with a significant increase in the phenolic content of honey. Latitude also had an effect, as seen in Nigeria ($\beta = -1.19$; $p < 0.01$), which showed a decrease in the phenolic content.

Table 2. Meta-regression slopes and significance of covariables to outcomes with a moderate or higher heterogeneity ($I^2 > 25\%$).

	Covariates (β)				R^2	I^2
Item	**Genera**	**Floral Source**	**Country**	**Latitude**		
Phenols	*Melipona* −26.01 * *Nannotrigona* −29.43 * *Partamona* −27.41 *	Polyfloral −45.60 **	Cuba +69.58 *** Nigeria +63.61 *** Peru +82.04 ***	−1.19 **	87.85	59.41
Flavonoids			Cuba +18.07 *** Nigeria +19.87 *** Peru +15.31 ***	−0.46 **	9.34	97.76
FRAP	*Hypotrigona* +12.11 * *Melipona* −8.59 ** *Plebeia* −6.98 * *Trigona* −14.42 ***	Polyfloral +13.60 ***	Cuba −10.03 *** Nigeria −17.47 ***		83.78	74.66
Moisture	*Trigona* −14.44 *	Polyfloral +14.65 *	Cuba −15.02 ** Nigeria −16.87 ** Peru −9.89 *		53.86	91.55
HMF	*Hypotrigona* +26.55 *** *Melipona* +19.43 *** *Plebeia* +19.82 ***	Polyfloral −3.91 ***	Cuba +2.35 *		100.0	0.0
Free acidity	*Hypotrigona* +27.47 *** *Melipona* −7.76 *** *Scaptotrigona* −5.09 * *Trigona* −45.95 **	Polyfloral −2.53 *	Brazil +6.14 *** Cuba +2.95 * Nigeria −10.85 ***	−0.31 *	96.58	70.06
Color	*Hypotrigona* +24.48 ** *Melipona* +23.64 ** *Plebeia* +28.13 ***		Brazil −69.59 * Nigeria −30.80 *	−2.39 *	99.93	11.69

R^2 = amount of heterogeneity accounted for the mixed model. Significance codes: *** < 0.001; ** < 0.01; * < 0.05.

The flavonoid content was significantly higher in stingless bee honey from Cuba ($\beta = +18.07$; $p < 0.001$), Nigeria ($\beta = +19.87$; $p < 0.001$), and Peru ($\beta = +15.31$; $p < 0.001$) compared to honey from Malaysia, Brazil, and Portugal. In this sense, the country significantly impacted the phenolic content of honey.

Our results showed a positive relationship between the *Hypotrigona* genera and antioxidant activity ($\beta = +12.11$; $p < 0.05$). The polyfloral source showed a positive association with the high antioxidant activity of honey ($\beta = +13.60$; $p < 0.001$). The honey from the polyfloral source had a higher moisture percentage ($\beta = +14.65$; $p < 0.05$) compared to the honey from the monofloral source. Additionally, the moisture was significantly lower in honey from Cuba ($\beta = -15.02$; $p < 0.01$), Nigeria ($\beta = -16.87$; $p < 0.01$), and Peru ($\beta = -9.89$; $p < 0.05$) compared to the rest of the studied locations. Additionally, the hydroxymethylfurfural content in stingless bee honey was lower for honey with a polyfloral botanical origin compared to honey from monofloral sources ($\beta = -3.91$; $p < 0.001$). Furthermore, the free acidity was lower in honey from the species *Melipona* ($\beta = -7.76$; $p < 0.001$), *Scaptotrigona* ($\beta = -5.09$; $p < 0.05$), and *Trigona* ($\beta = -45.95$; $p < 0.01$) compared to *Heterotrigona*, *Hypotrigona*, *Plebeia*, *Ptilotrigona*, *Nannotrigona*, and *Partamona* bees. The honey that showed the highest free acidity was produced in Brazil ($\beta = +6.14$; $p < 0.001$) and Cuba ($\beta = +2.95$; $p < 0.05$), and the honey from Nigeria had the lowest values of free acidity ($\beta = -10.85$; $p < 0.001$) compared to the rest of the countries. The color intensity was significantly lower in honey from Brazil ($\beta = -69.59$; $p < 0.05$) and Nigeria ($\beta = -30.80$; $p < 0.05$). Lastly, our results revealed that the phenolic content ($\beta = -1.19$; $p < 0.01$), flavonoids ($\beta = -0.46$; $p < 0.01$), free acidity ($\beta = -0.31$; $p < 0.05$), and the color intensity ($\beta = -2.39$ $p < 0.05$) of honey were negatively correlated with latitude.

4. Discussion

The current analytical review evaluated the phenolic content (phenols and flavonoids), antioxidant activity (DPPH and FRAP assays), and physicochemical composition of honey (moisture, HMF, free acidity, color, and protein) between *Apis mellifera* and nine genera of stingless bees. The species of honey bee can influence the composition of physicochemical and phenolic content. However, the magnitude of the differences in the honey composition depends mainly on environmental factors (latitude), the floral source (polyfloral and monofloral), and the genera of the producing bee [41]. This investigation highlights the importance of conducting more studies on stingless bee honey.

4.1. Phenolic Content and Antioxidant Activity

4.1.1. Phenolic Content

The present study reports differences in the phenolic content between *Apis mellifera* and stingless bee honey. We report on the phenolic and flavonoid content as these are considered to be the primary phenolic compounds present in honey [2,42]. These compounds are particularly important in honey bee products because of their association with anti-inflammatory, antimicrobial, and antioxidant activities, as demonstrated in in vitro assays [43–45].

The meta-regression analyses depicted that phenolic and flavonoid content could vary according to the country, floral source, latitude, and the bee type (i.e., *Apis mellifera* and stingless bees). In this sense, the bee type influences honey production due to the different morphological and physiological characteristics among bees that could influence the chemical characteristics of honey. For example, stingless bees are smaller than *Apis mellifera* bees, which allows them to pollinate small-sized flowers and gives them access to different pollen and nectar sources [6,46]. *Apis mellifera* has foraging preferences for specific hues and saturations. In contrast, stingless bees show little preference for color hue or saturation [47]. Foraging choices are influenced by foragers returning to the nest, conveying olfactory and gustatory information about the food source used to nestmates [48]. In addition, stingless bees tend to make shorter flights and prefer low flowering plants for nectar collection [49]. Thus, foraging preferences could be related to the species included in this analysis. Furthermore, the storage conditions of honey in the hive could influence its composition; for example, stingless bees build pots with cerumen to store honey [50,51].

Cerumen is made from the secretions of the bees' abdominal wax glands, mixed with salivary secretions and plant resins collected by foragers [52]. Similarly, honey bees use beeswax produced by their abdominal glands to build the comb used to store honey [53]. In both cases, the nectar collected by foragers is mixed with the bees' salivary secretions at the time that trophallaxis occurs, a process that is also useful for reducing the water content of nectar [52]. During the handling of nectar to produce honey, the product is exposed to the pollen collected by foragers; indeed, pollen grains are used to determine the floral origin of honey (melissopalynology) [54]. Additionally, stingless bees and honey bees collect resins to produce propolis, which in both cases is used to seal and sanitize the hive [52,55]. Therefore, the phytochemical compounds found in propolis and pollen could influence the chemical composition of honey. In agreement with Wakhungu et al. [56], *Hypotrigona* bees provide good pollination services within a radius of up to 250 m and can collect pollen from diverse floral sources, which could increase the variability of phenolic compounds in honey collected by this bee. Alvarez-Suarez et al. [18] carried out a study comparing 16 honeys from two different species of bees (*Apis mellifera* and *Melipona beecheii*); the honey samples belonged to municipalities in Cuba, selected based on their melliferous and pollination potential as well as their geographical proximity.

Based on the floral maps developed by the National Centre for Apicultural Research of Cuba, which ensure the similarity of plants that serve as pollen and nectar sources for bees, the results showed a significantly higher content of phenolic and flavonoid compounds in honey from *Melipona bececheii* (stingless bee) compared to *Apis mellifera*. Actually, phenolic compounds have been used as chemical indicators to determine different botanical and geographical origins of honey [25,26] and can influence the organoleptic properties (color, taste, or flavor) of honey [23]. Some studies have reported high levels of phenolic compounds in stingless bee honey, like quercetin, kaempferol, p-coumaric acid, hesperidin, ferulic acid, ellagic acid, trans-cinnamic, rutin, catechin, chrysin, and hesperidin [12,29]. Therefore, it is important to mention that the variation in the phenolic and flavonoid content in honey also depends on several factors such as the botanical source of nectar, season, storage conditions, geographical area, environmental conditions, beekeeping management, foraging area, and bee species [2,49,57]. However, further studies on the preferences of honey bees and stingless bees for floral sources and resins, as well as the influence of other hive products such as propolis and pollen on the chemical composition of honey, should be further investigated.

4.1.2. Antioxidant Activity

This systematic review compares the antioxidant activity of *Apis mellifera* and stingless honey using the DPPH and FRAP techniques. The results indicate a higher antioxidant activity in stingless honey compared to *Apis mellifera* honey, as evidenced by the FRAP values. Several studies have linked the phenolic acids and flavonoids in honey to its antioxidant activity [21,28,44]. This activity is strongly influenced by the molecular structure of the phenolic compounds [58], especially the number and arrangement of the hydroxyl groups and the position of the phenolic rings, which are crucial for their antioxidant activity [59]. The antioxidant activity consists of the mechanism by which the antioxidants scavenge or reduce the formation of free radicals (reactive oxygen species—ROS) in the cell [42,58]. The antioxidants protect cells from oxidative damage caused by free radicals and contribute to the prevention of certain diseases [59], including cardiovascular disease, aging, heart disease, cancer, and inflammation [58]. The antioxidant activity in honey can be estimated using different techniques, such as the DPPH and FRAP assays. DPPH is used in a radical scavenging assay; based on electron transfer, the antioxidant donates an electron to the DPPH radical, neutralizing and converting it to the reduced form, while the antioxidant is oxidized [60]. The DPPH radical produces an intense violet solution. When a DPPH solution is mixed with a test compound that can donate a hydrogen atom, this changes the color of the compound to colorless or light yellow as the free radicals are scavenged [61]. The FRAP assay measures the ability of the antioxidant to reduce ferric ions (Fe^{3+}) to ferrous ions (Fe^{2+}) [21]. An antioxidant assay assesses the ability of these phenolic and flavonoid compounds to donate an electron from a hydroxyl group to an unpaired electron of free radicals and then the antioxidant activity occurs [43,62]. The reaction of radical neutralization depends on many factors, among them the hydrogen ion concentration in the measuring system [63].

The differences in antioxidant activity found in the current systematic review and meta-analysis between *Apis mellifera* and stingless honey may be related to the presence of components such as vitamin C, vitamin E, and carotenoids, which may contribute to the higher antioxidant activity of stingless honey [8,64]. Together, these components can increase antioxidant activity and stimulate biomolecules (e.g., proteins, carbohydrates, nucleic acids, and lipids) in cells to produce reactive oxygen species, resulting in a high antioxidant response [12]. Thus, antioxidant activity results from the combined activity of a number of compounds, not just phenolic and flavonoid compounds [44]. Accordingly, Shamsudin et al. [21] found that the phenolic content, flavonoids, and antioxidant activity were higher in stingless honey than in *Apis mellifera* honey. In addition, our results showed that the *Hypotrigona* bee genera produced honey with higher antioxidant activity compared to the other stingless bee species ($p < 0.05$) (Table 2).

4.2. Physicochemical Composition

Our results revealed significant differences in the physicochemical parameters between stingless honey and *Apis mellifera* honey, highlighting the high moisture percentage and free acidity in contrast to the lower HMF content and diastase activity in stingless honey. In this sense, the physicochemical properties such as the color, free acidity, moisture, and HMF of honey are some of the parameters used for the authentication, the measurement of quality, and the detection of any adulteration in the product [65]. The physicochemical quality criteria of honey from the honey bee *Apis mellifera* are well specified, set by the Codex Alimentarius Commission (2001) [66], but these could not be applied for stingless bee honey. The moisture, free acidity, diastase activity, and HMF content have been shown to vary between *Apis mellifera* honey and stingless bee honey [8]. The physicochemical difference found in the current study could be related to several factors such as botanical origin, geographical area, climate, harvesting method, bee genera, and species [46]. An implication of the current study is the possibility to contribute to the determination of quality parameters that are important to know, to set quality standards for honey from stingless bees, and to prevent its adulteration [67].

4.2.1. Moisture

Based on the available studies, it was found that the stingless honey had high moisture levels (>20%) to be compared with *Apis mellifera* honey; the relevance of this finding lies in the fact that moisture is a quality parameter in the set standards of honey. In addition, moisture affects viscosity, flavor, and crystallization, and is considered a useful indicator for improving storage and preservation [1,67]. The high moisture content in stingless honey is not surprising, as it has been widely reported for this type of honey [68]. This higher moisture content of stingless honey could be related to the fact that stingless bees collect nectar from the flowers of low-growing plants, which tend to produce water-rich nectar [69,70]. According to Jimenez et al. [8], the moisture content of honey from stingless bees varies according to the geographical area and the predominant vegetation. In tropical areas, polyfloral vegetation is influenced by high annual rainfall and high humidity, which may increase the water content of honey, whereas in a rainforest, honey is characterized by low water content [1,46]. Stingless bees store honey in pots made of cerumen, whereas *Apis mellifera* store honey in a hexagonal comb constructed with beeswax, which is a complex material with more than 300 identified substances [65]. During honey storage, stingless bees and honey bees use trophallaxis to reduce the water content of the collected nectar to produce honey, but the moisture content of stingless bee honey is usually higher (~31%) [71].

The high moisture content influences the characteristics of honey, as it allows microorganisms to proliferate during storage, especially yeasts, which could cause honey fermentation [7,46]. In contrast, the honey produced by *Apis mellifera* has a lower water content (<20%), which prevents the fermentation process from occurring [53]. The fermentation process could be considered as contributing to the risk of spoilage in *Apis mellifera* honey, but in stingless bee honey, it does not seem to affect the quality of the product [22]. The Codex Alimentarius establishes 20% as the maximum limit of moisture in *Apis mellifera* honey [72], so these actual standards cannot be applied to stingless bee honey due to the natural water content in each product; therefore, it is necessary to establish specific standards for stingless bee honey.

4.2.2. Free Acidity

Our results showed that free acidity was higher in stingless honey compared to *Apis mellifera* honey. Free acidity is considered one of the most important parameters for assessing the freshness of honey [67]. Free acidity refers to the free natural acids present in honey, which contribute to honey flavor, stability against microorganisms, and antibacterial and antioxidant activities [73]. The high acidity could be beneficial for prolonging the shelf life of honey, as there would be no favorable conditions for the development of undesirable microorganisms [46], such as *Escherichia coli* and *Streptococcus aureus* [74,75]. Furthermore, the differences in free acidity found between honeys could be related to the floral source, the harvesting season, the storage process, and the geographical area [76,77]. The maximum limit of free acidity according to Codex Alimentarius [72] is 50 meq/kg for *Apis mellifera* honey. Stingless honey has higher levels and therefore does not comply with these regulations [57]. As mentioned above, stingless honey is not included in international honey standards. Given the potential for stingless honey to be a part of regional and international markets, it is imperative to establish parameters for this product to ensure its safe consumption, quality, and marketability.

4.2.3. Hydroxymethylfurfural (HMF)

Our analysis showed lower concentrations of HMF in honey from stingless bees compared to *Apis mellifera* honey. HMF is an important parameter for freshness, overheating, and honey quality [33,78]. The differences found in HMF content can be explained by several factors such as heat, prolonged storage time, content of simple sugars, content of acids, minerals, floral sources, and environmental conditions [8,29]. Hydroxymethylfurfural is a six-carbon heterocyclic organic compound containing aldehyde and alcohol (hydroxymethyl) functional groups, formed by the degradation of sugars [70]. In relation to our results, HMF can be formed by the Maillard reaction; thus, due to the high water activity (Aw) and acidity in stingless honey, the Maillard reaction could be slow and HMF formation could be inhibited, which partially explains the reduced HMF content in stingless honey [79]. Finally, in addition to the importance of HMF content in assessing honey quality, HMF has been found to be toxic to rats and mice; potential health risks to humans associated with high concentrations cannot be excluded [80]. In this sense, the lower levels that are found in stingless honey are an advantage compared to *Apis mellifera* honey.

4.2.4. Color

The high content of phenolic acids and flavonoids may explain the higher color intensity found in stingless honey compared to *Apis mellifera* honey. The importance of honey color lies in the fact that it indicates the presence of pigments, such as flavonoids and carotenoids [17]; also, the color is one of the main characteristics that consumers take into account when buying honey [81]. The color of honey varies from light yellow to dark brown; light-colored honey is usually low in mineral and phenolic compounds, while dark-colored honey is usually high in mineral and phenolic compounds [57], and the differences in honey color are associated with botanical origin, phenolic content, biochemical reactions during honey maturation, and exposure to high temperatures or light [2].

4.3. Limitations and Opportunity Areas

This study has certain limitations, mainly the reduced number of available studies on stingless honey and *Apis mellifera* honey that specifically address the phenolic content, antioxidant activity, and physicochemical differences between these two types of honey. Fortunately, the available data were sufficient to estimate the global effect and explore the sources of heterogeneity in the current systematic review. In addition, there is a lack of uniformity in the variables analyzed, as not all studies included the parameters relevant to this research. For this reason, some chemical characteristics could not be analyzed using a meta-analytic approach. For these variables, only descriptive statistics are presented in the supplementary materials. Finally, although the diversity of stingless bee species is considerable, the lack of studies focusing on honey produced by individual species represents a significant research gap and an area for future studies.

5. Conclusions

Several studies have reported a higher concentration of bioactive compounds and antioxidant activity in stingless bee honey. However, they are very heterogeneous, so it is necessary to carry out more studies on the content and profile of phenolic acids and flavonoids in stingless bee honey. The phenolic compounds are related to their antioxidant activity in vitro, and in vivo studies concerning the nutraceutical properties of honey are still scarce. It is also vital to note that the floral source, geographical area, climate, latitude, and bee species influence variances in the bioactive component content, antioxidant activity, and physicochemical composition of honey. The honey produced by stingless bees has a high level of free acidity and moisture, which does not comply with the standards laid down by the legislation and applies exclusively to honey produced by *Apis mellifera*. Therefore, it is necessary to have an exclusive quality standard for honey from stingless bees to enable better production and commercialization that will guarantee the authenticity of the products for consumers.

Supplementary Materials: The following supporting information can be downloaded at https://www.mdpi.com/article/10.3390/antiox13121539/s1.

Author Contributions: A.K.Z.-O.: data curation, resources, investigation, writing—original draft. A.d.J.C.-G.: review and editing, supervision, validation. N.M.: investigation, review and editing, validation, supervision. G.A.-Á. review, validation, supervision. R.G.C.-M.: conceptualization, supervision, validation, visualization. N.E.-E.: review and editing, validation, supervision. A.G.-G.: review and editing, validation, supervision. J.C.A.-H.: conceptualization, methodology, software, formal analysis, writing, review and editing, project administration. All authors have read and agreed to the published version of the manuscript.

Funding: This research received no external funding.

Informed Consent Statement: Not applicable.

Data Availability Statement: The relevant data are available from J.C.A.-H. upon reasonable request.

Conflicts of Interest: The authors declare no conflicts of interest.

References

1. Rozman, A.S.; Hashim, N.; Maringgal, B.; Abdan, K. A Comprehensive Review of Stingless Bee Products: Phytochemical Composition and Beneficial Properties of Honey, Propolis, and Pollen. *Appl. Sci.* **2022**, *12*, 6370. [CrossRef]
2. Kek, S.P.; Chin, N.L.; Yusof, Y.A.; Tan, S.W., Chua, L.S. Total phenolic contents and colour intensity of Malaysian honeys from the *Apis* spp. and *Trigona* spp. bees. *Agric. Agric. Sci. Proc.* **2014**, *2*, 150–155. [CrossRef]
3. Michener, C.D. *The Bees of the World*, 2nd ed.; The Johns Hopkins University Press: Baltimore, MD, USA, 2007; pp. 667–830.
4. Ali, H.; Abu Bakar, M.F.; Majid, M.; Muhammad, N.; Lim, S.Y. In Vitro Anti-Diabetic Activity of Stingless Bee Honey from Different Botanical Origins. *Food Res.* **2020**, *4*, 1421–1426. [CrossRef]
5. Chuttong, B.; Chanbang, Y.; Sringarm, K.; Burgett, M. Physicochemical profiles of stingless bee (Apidae: Meliponini) honey from South east Asia (Thailand). *Food Chem.* **2016**, *192*, 149–155. [CrossRef]

6. Abd Jalil, M.A.; Kasmuri, A.R.; Hadi, H. Stingless bee honey, the natural wound healer: A review. *Skin Pharmacol. Physiol.* **2017**, *30*, 66–75. [CrossRef] [PubMed]
7. Nordin, A.; Sainik, N.Q.A.V.; Chowdhury, S.R.; Saim, A.B.; Idrus, R.B.H. Physicochemical properties of stingless bee honey from around the globe: A comprehensive review. *J. Food Compos. Anal.* **2018**, *73*, 91–102. [CrossRef]
8. Jimenez, M.; Beristain, C.I.; Azuara, E.; Mendoza, M.R.; Pascual, L.A. Physicochemical and antioxidant properties of honey from *Scaptotrigona mexicana* bee. *J. Apic. Res.* **2016**, *55*, 151–160. [CrossRef]
9. Ramón-Sierra, J.; Ruíz-Ruíz, J.; Ortiz-Vázquez, E. Electrophoresis characterisation of protein as a method to establish the entomological origin of stingless bee honeys. *Food Chem.* **2015**, *183*, 43–48. [CrossRef]
10. Pat Fernández, L.A.; Anguebes Franceschi, F.; Pat Fernández, J.M.; Hernández Bahena, P.; Ramos Reyes, R. Condición y perspectivas de la meliponicultura en comunidades mayas de la reserva de la biósfera Los Petenes, Campeche, México. *Estud. Cult. Maya* **2018**, *52*, 227–254. [CrossRef]
11. Soto-Muciño, L.E.; Chiatchoua, C.; Castañeda-González, Y. National and international panorama of honey production in Mexico. *ECORFAN J.* **2015**, *1*, 15–33.
12. Pimentel, T.C.; Rosset, M.; de Sousa, J.M.B.; de Oliveira, L.I.G.; Mafaldo, I.M.; Pintado, M.M.E.; Magnani, M. Stingless bee honey: An overview of health benefits and main market challenges. *J. Food Biochem.* **2022**, *46*, e13883. [CrossRef] [PubMed]
13. Julika, W.N.; Ajit, A.; Naila, A.; Sulaiman, A.Z. The effect of storage condition on physicochemical properties of some stingless bee honey collected in malaysia local market. *Mater Today Proc.* **2022**, *57*, 1396–1402. [CrossRef]
14. Chen, Y.H.; Chuah, W.C.; Chye, F.Y. Effect of drying on physicochemical and functional properties of stingless bee honey. *J. Food Process. Preserv.* **2021**, *45*, e15328. [CrossRef]
15. Heard, T.A. The role of stingless bees in crop pollination. *Annu. Rev. Entomol.* **1999**, *44*, 183–206. [CrossRef] [PubMed]
16. Marconi, M.; Luna, J.O.; Giove, C.D.V. Physicochemical and microbiological quality of honeys produced by stingless bees *Scaptotrigona polysticta*, *Melipona illota* and *Tetragonisca angustula* (Apidae: Meliponini) in San Martín, Peru. *Peruv. J. Agron.* **2020**, *4*, 55–60. [CrossRef]
17. Otmani, A.; Amessis-Ouchemoukh, N.; Birinci, C.; Yahiaoui, S.; Kolaylı, S.; Rodríguez-Flores, M.; Escuredo, O.; Seijo, M.; Ouchemoukh, S. Phenolic compounds and antioxidant and antibacterial activities of Algerian honeys. *Food Biosci.* **2021**, *42*, 101070. [CrossRef]
18. Alvarez-Suarez, J.M.; Giampieri, F.; Brenciani, A.; Mazzoni, L.; Gasparrini, M.; González-Paramás, A.M.; Battino, M. *Apis mellifera* vs *Melipona beecheii* Cuban polifloral honeys: A comparison based on their physicochemical parameters, chemical composition and biological properties. *LWT* **2018**, *87*, 272–279. [CrossRef]
19. Biluca, F.C.; da Silva, B.; Caon, T.; Mohr, E.T.B.; Vieira, G.N.; Gonzaga, L.V.; Costa, A.C.O. Investigation of phenolic compounds, antioxidant and anti-inflammatory activities in stingless bee honey (*Meliponinae*). *Food Res. Int.* **2020**, *129*, 108756. [CrossRef]
20. Cruz, L.C.; Batista, J.E.S.; Zemolin, A.P.P.; Nunes, M.E.M.; Lippert, D.B.; Royes, L.F.F.; Franco, J.L. A study on the quality and identity of Brazilian Pampa biome honey: Evidences for its beneficial effects against oxidative stress and hyperglycemia. *Int. J. Food Sci.* **2014**, *2014*, 470214. [CrossRef]
21. Shamsudin, S.; Selamat, J.; Abdul Shomad, M.; Ab Aziz, M.F.; Haque Akanda, M.J. Antioxidant properties and characterization of *Heterotrigona itama* honey from various botanical origins according to their polyphenol compounds. *J. Food Qual.* **2022**, *2022*, 2893401. [CrossRef]
22. Dos Santos, A.C.; Biluca, F.C.; Braghini, F.; Gonzaga, L.V.; Costa, A.C.O.; Fett, R. Phenolic composition and biological activities of stingless bee honey: An overview based on its aglycone and glycoside compounds. *Food Res. Int.* **2021**, *147*, 110553. [CrossRef]
23. Makawi, S.Z.A.; Gadkariem, E.A.; Ayoub, S.M.H. Determination of antioxidant flavonoids in Sudanese honey samples by solid phase extraction and high performance liquid chromatography. *J. Chem.* **2009**, *6*, S429–S437. [CrossRef]
24. Di Marco, G.; Gismondi, A.; Panzanella, L.; Canuti, L.; Impei, S.; Leonardi, D.; Canini, A. Botanical influence on phenolic profile and antioxidant level of Italian honeys. *J. Food Sci. Technol.* **2018**, *55*, 4042–4050. [CrossRef] [PubMed]
25. Cavazza, A.; Corradini, C.; Musci, M.; Salvadeo, P. High-performance liquid chromatographic phenolic compound fingerprint for authenticity assessment of honey. *J. Sci. Food Agric.* **2013**, *93*, 1169–1175. [CrossRef] [PubMed]
26. Das, A.; Datta, S.; Mukherjee, S.; Bose, S.; Ghosh, S.; Dhar, P. Evaluation of antioxidative, antibacterial and probiotic growth stimulatory activities of *Sesamum indicum* honey containing phenolic compounds and lignans. *LWT-Food Sci. Technol.* **2015**, *61*, 244–250. [CrossRef]
27. Morroni, G.; Alvarez-Suarez, J.M.; Brenciani, A.; Simoni, S.; Fioriti, S.; Pugnaloni, A.; Giovanetti, E. Comparison of the antimicrobial activities of four honeys from three countries (New Zealand, Cuba, and Kenya). *Front. Microbiol.* **2018**, *9*, 1378. [CrossRef]
28. Ruiz-Ruiz, J.C.; Matus-Basto, A.J.; Acereto-Escoffié, P.; Segura-Campos, M.R. Antioxidant and anti-inflammatory activities of phenolic compounds isolated from *Melipona beecheii* honey. *Food Agric. Immunol.* **2017**, *28*, 1424–1437. [CrossRef]
29. Sousa, J.M.; De Souza, E.L.; Marques, G.; Meireles, B.; de Magalhães Cordeiro, Â.T.; Gullón, B.; Magnani, M. Polyphenolic profile and antioxidant and antibacterial activities of monofloral honeys produced by *Meliponini* in the Brazilian semiarid region. *Food Res. Int.* **2016**, *84*, 61–68. [CrossRef]
30. Abu, B.M.; Sanusi, S.B.; Abu, B.F.; Cong, O.J.; Mian, Z. Physicochemical and antioxidant potential of raw unprocessed honey from Malaysian stingless bee. *Pak. J. Nutr.* **2017**, *16*, 888–894. [CrossRef]
31. Erejuwa, O.O.; Sulaiman, S.A.; Ab Wahab, M.S. Honey: A novel antioxidant. *Molecules* **2012**, *17*, 4400–4423. [CrossRef]

32. Maringgal, B.; Hashim, N.; Tawakkal, I.S.M.A.; Mohamed, M.T.M.; Shukor, N.I.A. Phytochemical Compositions and Antioxidant Activities of Malaysian Stingless Bee Honey. *Pertanika J. Sci. Technol.* **2019**, *27*, 15–28. [CrossRef]
33. Ya'akob, H.; Norhisham, N.F.; Mohamed, M.; Sadek, N.; Endrini, S. Evaluation of Physicochemical Properties of *Trigona* sp. Stingless Bee Honey from Various Districts of Johor. *J. Kejuruteraan.* **2019**, *2*, 59–67. [CrossRef]
34. Page, M.J.; McKenzie, J.E.; Bossuyt, P.M.; Boutron, I.; Hoffmann, T.C.; Mulrow, C.D.; Moher, D. The PRISMA 2020 statement: An updated guideline for reporting systematic reviews. *Int. J. Surg.* **2021**, *88*, 105906. [CrossRef]
35. Revelle, W. *Psych: Procedures for Personality and Psychological Research, Version 2.1.9*; Northwestern University: Evanston, IL, USA, 2021; Available online: https://CRAN.R-project.org/package=psych (accessed on 4 October 2021).
36. Hedges, L.V. Distribution theory for Glass's estimator of effect size and related estimators. *J. Educ. Behav. Stat.* **1981**, *6*, 107–128. [CrossRef]
37. Appuhamy, J.R.N.; Strathe, A.B.; Jayasundara, S.; Wagner-Riddle, C.; Dijkstra, J.; France, J.; Kebreab, E. Anti-methanogenic effects of monensin in dairy and beef cattle: A meta-analysis. *J. Dairy Sci.* **2013**, *96*, 5161–5173. [CrossRef] [PubMed]
38. Balduzzi, S.; Rücker, G.; Schwarzer, G. How to perform a meta-analysis with R: A practical tutorial. *BMJ Ment Health* **2019**, *22*, 153–160. [CrossRef]
39. Higgins, J.P.; Thompson, S.G.; Deeks, J.J.; Altman, D.G. Measuring inconsistency in meta-analyses. *BMJ* **2003**, *327*, 557–560. [CrossRef] [PubMed]
40. Begg, C.B.; Mazumdar, M. Operating characteristics of a rank correlation test for publication bias. *Biometrics* **1994**, *50*, 1088–1101. [CrossRef]
41. de Almeida-Muradian, L.B.; Stramm, K.M.; Horita, A.; Barth, O.M.; da Silva de Freitas, A.; Estevinho, L.M. Comparative study of the physicochemical and palynological characteristics of honey from *Melipona subnitida* and *Apis mellifera*. *Int. J. Food Sci. Technol.* **2013**, *48*, 1698–1706. [CrossRef]
42. da Silva, P.M.; Gauche, C.; Gonzaga, L.V.; Costa, A.C.O.; Fett, R. Honey: Chemical composition, stability and authenticity. *Food Chem.* **2016**, *196*, 309–323. [CrossRef]
43. Ge, L.; Li, S.P.; Lisak, G. Advanced sensing technologies of phenolic compounds for pharmaceutical and biomedical analysis. *J. Pharm. Biomed. Anal.* **2020**, *179*, 112913. [CrossRef] [PubMed]
44. Tuksitha, L.; Chen, Y.L.S.; Chen, Y.L.; Wong, K.Y.; Peng, C.C. Antioxidant and antibacterial capacity of stingless bee honey from Borneo (Sarawak). *J. Asia-Pac. Entomol.* **2018**, *21*, 563–570. [CrossRef]
45. Mahmood, R.; Asif, J.A.; Shahidan, W.N.S. Stingless-bee (*Trigona itama*) honey adversely impacts the growth of oral squamous cell carcinoma cell lines (HSC-2). *Eur. J. Integr. Med.* **2020**, *37*, 101162. [CrossRef]
46. Mokaya, H.O.; Nkoba, K.; Ndunda, R.M.; Vereecken, N.J. Characterization of honeys produced by sympatric species of Afrotropical stingless bees (Hymenoptera, Meliponini). *Food Chem.* **2022**, *366*, 130597. [CrossRef]
47. Koethe, S.; Fischbach, V.; Banysch, S.; Reinartz, L.; Hrncir, M.; Lunau, K. A comparative study of food source selection in stingless bees and honeybees: Scent marks, location, or color. *Front. Plant Sci.* **2020**, *11*, 516. [CrossRef]
48. Farina, W.M.; Grüter, C.; Díaz, P.C. Social learning of floral odours inside the honeybee hive. *Proc. R. Soc. B Biol. Sci.* **2005**, *272*, 1923–1928. [CrossRef]
49. Moreira, F.I.N.; de Medeiros, L.L.; de Carvalho, L.M.; Olegario, L.S.; de Sousa Galvão, M.; da Franca, S.A.M.; Madruga, M.S. Quality of Brazilian stingless bee honeys: *Cephalotrigona capitata*/*mombucão* and *Melipona scutellaris* Latrelle/uruçu. *Food Chem.* **2023**, *404*, 134306. [CrossRef] [PubMed]
50. Wille, A. Biology of the stingless bees. *Annu. Rev. Entomol.* **1983**, *28*, 41–64. [CrossRef]
51. Khongkwanmueang, A.; Nuyu, A.; Straub, L.; Maitip, J. Physicochemical profiles, antioxidant and antibacterial capacity of honey from stingless bee *Tetragonula laeviceps* species complex. *E3S Web Conf.* **2020**, *141*, 03007. [CrossRef]
52. Layek, U.; Das, N.; Kumar De, S.; Karmakar, P. The botanical origin of cerumen and propolis of Indian stingless bees (*Tetragonula iridipennis* Smith): Pollen spectrum does not accurately indicate latex and resin sources. *Apidologie* **2023**, *54*, 18. [CrossRef]
53. Winston, M.L. *The Biology of the Honey Bee*; Harvard University Press: Cambridge, MA, USA, 1991.
54. Herrero, B.; María Valencia-Barrera, R.; San Martín, R.; Pando, V. Characterization of honeys by melissopalynology and statistical analysis. *Can. J. Plant Sci.* **2002**, *82*, 75–82. [CrossRef]
55. Borba, R.S.; Klyczek, K.K.; Mogen, K.L.; Spivak, M. Seasonal benefits of a natural propolis envelope to honey bee immunity and colony health. *J. Exp. Biol.* **2015**, *218*, 3689–3699. [CrossRef] [PubMed]
56. Wakhungu, D.A.; Namikoye, E.S.; Lattorff, H.M.G. Foraging range of an African stingless bee, *Hypotrigona gribodoi* (Apidae: Meliponini). *Afr. J. Ecol.* **2022**, *60*, 1094–1098. [CrossRef]
57. Biluca, F.C.; Braghini, F.; Gonzaga, L.V.; Costa, A.C.O.; Fett, R. Physicochemical profiles, minerals and bioactive compounds of stingless bee honey (*Meliponinae*). *J. Food Compos. Anal.* **2016**, *50*, 61–69. [CrossRef]
58. Zehiroglu, C.; Ozturk Sarikaya, S.B. The importance of antioxidants and place in today's scientific and technological studies. *J. Food Sci. Technol.* **2019**, *56*, 4757–4774. [CrossRef]
59. Martinello, M.; Mutinelli, F. Antioxidant Activity in Bee Products: A Review. *Antioxidants* **2021**, *10*, 71. [CrossRef]
60. Jan, N.U.; Kifayatullah, M.; Amin, F.; Rahim, H.; Abbas, S.; Mohani, S.N.U.H.; Raza, M. Antioxidant and Cytotoxic activity of Steroidal Alkaloids Isolated from *Sarcococca saligna* against DPPH and HeLa Cell Lines. *Indian J. Pharm. Educ. Res.* **2022**, *56*, 489–496. [CrossRef]

61. Kedare, S.B.; Singh, R.P. Genesis and development of DPPH method of antioxidant assay. *J. Food Sci. Technol.* **2011**, *48*, 412–422. [CrossRef]
62. Chua, L.S.; Rahaman, N.L.; Adnan, N.A.; Tan, T.T.E. Antioxidant activity of three honey samples in relation with their biochemical components. *J. Anal. Methods Chem.* **2013**, *2013*, 45673. [CrossRef]
63. Dawidowicz, A.L.; Olszowy, M. Mechanism change in estimating of antioxidant activity of phenolic compounds. *Talanta* **2012**, *97*, 312–317. [CrossRef]
64. Escuredo, O.; Seijo, M.; Salvador, J.; González-Martín, M. Near infrared spectroscopy for prediction of antioxidant compounds in the honey. *Food Chem.* **2013**, *1401*, 3409–3414. [CrossRef] [PubMed]
65. Kek, S.P.; Chin, N.L.; Yusof, Y.A.; Tan, S.W.; Chua, L.S. Classification of entomological origin of honey based on its physicochemical and antioxidant properties. *Int. J. Food Prop.* **2017**, *20*, S2723–S2738. [CrossRef]
66. Wong, P.; Ling, H.S.; Chung, K.C.; Yau, T.M.S.; Gindi, S.R.A. Chemical Analysis on the Honey of *Heterotrigona itama* and *Tetrigona binghami* from Sarawak, Malaysia. *Sains Malaysiana* **2019**, *48*, 1635–1642. [CrossRef]
67. Silva, I.P.; Caldas, M.J.M.; Machado, C.S.; Nascimento, A.S.D.; Lordêlo, M.S.; Barbara, M.F.S.; Carvalho, C.A.L.D. Antioxidants activity and physicochemical properties of honey from social bees of the Brazilian semiarid region. *J. Apic. Res.* **2021**, *60*, 797–806. [CrossRef]
68. Lira, A.F.; de Mello Sousa, J.P.L.; Lorenzon, M.C.A.; Vianna, C.A.F.J.; Castro, R.N. Estudo comparativo do mel de *Apis mellifera* com méis de meliponíneos de diferentes regiões. *Acta Vet. Bras.* **2014**, *8*, 169–178. [CrossRef]
69. Bertsch, A. Nectar production of *Epilobium angustifolium* L. at different air humidities; nectar sugar in individual flowers and the optimal foraging theory. *Oecologia* **1983**, *59*, 40–48. [CrossRef]
70. Ávila, S.; Beux, M.R.; Ribani, R.H.; Zambiazi, R.C. Stingless bee honey: Quality parameters, bioactive compounds, health-promotion properties and modification detection strategies. *Trends Food Sci. Technol.* **2018**, *81*, 37–50. [CrossRef]
71. Roubik, D.W. Nest and colony characteristics of stingless bees from Panama (Hymenoptera: Apidae). *J. Kans. Entomol. Soc.* **1983**, *56*, 327–355. Available online: http://www.jstor.org/stable/25084419 (accessed on 1 August 2024).
72. Codex Alimentarius Commission Revised. In *Codex Standard for Honey*; CODEX STAN 12–1981; Codex Alimentarius Commission: Rome, Italy; FAO/OMS: Rome, Italy, 2001; p. 7.
73. Cavia, M.M.; Fernández-Muino, M.A.; Alonso-Torre, S.R.; Huidobro, J.F.; Sancho, M.T. Evolution of acidity of honeys from continental climates: Influence of induced granulation. *Food Chem.* **2007**, *100*, 1728–1733. [CrossRef]
74. Chan-Rodríguez, D.; Ramon-Sierra, J.; Lope-Ayora, J.; Sauri-Duch, E.; Cuevas-Glory, L.; Ortiz-Vazquez, E. Antibacterial properties of honey produced by Melipona beecheii and *Apis mellifera* against foodborn microorganisms. *Food Sci Biotechnol.* **2012**, *21*, 905–909. [CrossRef]
75. Ng, W.J.; Sit, N.W.; Ooi, P.A.C.; Ee, K.Y.; Lim, T.M. The antibacterial potential of honeydew honey produced by stingless bee (*Heterotrigona itama*) against antibiotic resistant bacteria. *Antibiotics* **2020**, *9*, 871. [CrossRef] [PubMed]
76. Alves, A.; Ramos, A.; Goncalves, M.M.; Bernardo, M.; Mendes, B. Antioxidant activity, quality parameters and mineral content of Portuguese monofloral honeys. *J. Food Compos. Anal.* **2013**, *30*, 130–138. [CrossRef]
77. Habib, H.M.; Al Meqbali, F.T.; Kamal, H.; Souka, U.D.; Ibrahim, W.H. Physicochemical and biochemical properties of honeys from arid regions. *Food Chem.* **2014**, *153*, 35–43. [CrossRef] [PubMed]
78. Godocikova, J.; Bugarova, V.; Kast, C.; Majtan, V.; Majtan, J. Antibacterial potential of Swiss honeys and characterisation of their bee-derived bioactive compounds. *J. Sci. Food Agric.* **2020**, *100*, 335–342. [CrossRef] [PubMed]
79. Biluca, F.C.; Della Betta, F.; de Oliveira, G.P.; Pereira, L.M.; Gonzaga, L.V.; Costa, A.C.O.; Fett, R. 5-HMF and carbohydrates content in stingless bee honey by CE before and after thermal treatment. *Food Chem.* **2014**, *159*, 244–249. [CrossRef]
80. Zirbes, L.; Nguyen, B.K.; de Graaf, D.C.; De Meulenaer, B.; Reybroeck, W.; Haubruge, E.; Saegerman, C. Hydroxymethylfurfural: A possible emergent cause of honey bee mortality. *J. Agric. Food Chem.* **2013**, *61*, 11865–11870. [CrossRef]
81. Ramly, N.S.; Sujanto, I.S.R.; Tang, J.Y.H.; Abd Ghani, A.; Alias, N.; Bakar, M.F.A.; Ngah, N. Correlation Between the Color Lightness and Sweetness of Stingless Bee Honey with Its Minerals Content. *J. Agrobiotechnol.* **2021**, *12*, 88–96. [CrossRef]

MDPI AG
Grosspeteranlage 5
4052 Basel
Switzerland
Tel.: +41 61 683 77 34

Antioxidants Editorial Office
E-mail: antioxidants@mdpi.com
www.mdpi.com/journal/antioxidants

www.ingramcontent.com/pod-product-compliance
Lightning Source LLC
Chambersburg PA
CBHW061817210926
51850CB00006B/733
9783725871483